Two Manuals Combined into One Book

Static Line Parachuting

And

Marine Scouting and Patrolling

Department of Defense

Pentagon
Publishing

***FM 3-21.220**(FM 57-220)/
MCWP 3-15.7/AFMAN11-420/
NAVSEA SS400-AF-MMO-010

FIELD MANUAL
No. 3-21.220(57-220)/
MCWP 3-1.5.7/AFMAN11-420/
NAVSEA SS400-AF-MMO-010

HEADQUARTERS
DEPARTMENT OF THE ARMY
Washington, DC, 23 September 2003

STATIC LINE PARACHUTING TECHNIQUES AND TACTICS

CONTENTS

DISTRIBUTION RESTRICTION: Approved for public release; distribution is unlimited.

*This publication supersedes FM 57-220/MCWP 3-1.5.7/TO 14D1-2-1-121, 19 August 1996.

Part Five. DROP ZONES

CHAPTER 20. PROCEDURES ON THE DROP ZONE

PREFACE

This manual contains basic and advanced training and techniques for static line parachuting. It is designed to standardize procedures for initial qualification and training of personnel in their duties and responsibilities in airborne operations. The jumpmaster, assistant jumpmaster, safeties, DACO, DZSTL, and DZSO occupy key positions in airborne operations. This manual contains the initial training and qualifications of the personnel designated to occupy these critical positions.

SOF unit personnel must meet the requirements for static line parachuting contained in this manual as well as provide special training and instruction for nonstandard equipment, aircraft, and personnel procedures. These procedures are documented in FM 31-19, TC 31-24, TC 31-25, and USASOC Reg 350-2.

Individual service components that deviate from this manual will use approved procedures, techniques, equipment, and equipment-attaching methods specified by their respective service. All deviations must be approved in writing by the using unit commanders.

The proponent of this publication is HQ U.S. Army Infantry School. Submit changes for improving this publication to doctrine@benning.army.mil or on DA Form 2028 (Recommended Changes to Publications and Blank Forms) directly to Commandant, U.S. Army Infantry School, ATTN: ATSH-TPP-A, Fort Benning, Georgia 31905-5593.

Unless this publication states otherwise, masculine nouns and pronouns do not refer exclusively to men.

NOTE: The terms *jumper* and *parachutist* are used interchangeably in this manual.

PART ONE

Basic Airborne Techniques and Training

CHAPTER 1

AIRBORNE TRAINING

The purpose of airborne training is to qualify personnel in the use of the parachute as a means of combat deployment. This training also develops leadership, self-confidence, and aggressive spirit through tough mental and physical conditioning.

1-1. STANDARDS

Airborne training initiates and sustains a high standard of proficiency through repetition and time-proven techniques. Valid results are obtained when the following training standards are employed:

- Strict discipline.
- High standards of proficiency on each training apparatus and during each phase of training.
- A vigorous physical conditioning program to ensure parachutists are capable of jumping with a minimum risk of injury.
- A strong sense of esprit de corps and camaraderie among parachutists.
- Emphasis on developing mental alertness, instantaneous execution of commands, self-confidence, and confidence in the equipment.

1-2. PHASES

The three-week airborne course is divided into two training phases. Weeks 1 and 2 form the ground and tower training phase, and Week 3 is the jump training phase.

a. **Ground and Tower Phase**. Each of the five basic jump techniques pertains to a particular area of military parachuting and provides a sequence for dividing the ground phase into six instructional segments.

(1) ***Actions Inside the Aircraft.*** To ensure that the maximum number of parachutists can safely exit an aircraft, a means of controlling their actions inside the aircraft just before exiting is necessary. The jumpmaster maintains control by issuing jump commands. Each command calls for specific action on the part of each parachutist.

(2) ***Body Control Until Opening Shock.*** Due to aircraft speed and air turbulence around the rear of the aircraft, the parachutist must exit properly and maintain the correct body position after exiting. This action reduces spinning and tumbling in the air and allows for proper parachute deployment.

(3) ***Parachute Control During Descent.*** Parachute control is essential to avoid other parachutists in the air and to avoid hitting obstacles on the ground.

(4) ***Parachute Landing Fall Execution.*** The PLF is a landing technique that enables the parachutist to distribute the landing shock over his entire body to reduce the impact and the possibility of injury.

(5) ***Parachute Control on Landing.*** The parachutist releases one canopy release assembly after landing. Winds on the drop zone may cause a parachutist to be injured from being dragged along the ground.

(6) ***Physical Training.*** Prior to reporting for airborne training, volunteers must achieve APFT standards for the 17- to 21-year-old level (Table 1-1 and DA Pam 351-4). Physical training is included in each day of ground training. Students who cannot progress in daily physical training are referred to a board that decides either to recycle them or to return them to their unit. Daily exercises are designed to condition the muscle groups that play a significant part in jumping (Table 1-2).

EVENT	REPETITIONS		TIME LIMIT
	MALE	FEMALE	
PUSH-UPS	42	19	2 MINUTES
SIT-UPS	53	53	2 MINUTES
TWO-MILE RUN		MALE	15.54 MINUTES
		FEMALE	18.54 MINUTES

Table 1-1. APFT standards for the 17- to 21-year-old level.

WARM-UP EXERCISES	SETS	REPETITIONS
CHIN-UPS (MALE)	1	10
CHIN-UPS (FEMALE)	1	10
ROTATION EXERCISES	**SETS**	**INTERVAL**
NECK ROTATION	1	10 seconds
ARM/SHOULDER ROTATION	1	10 seconds
HIP ROTATION	1	10 seconds
KNEE/ANKLE ROTATION	1	10 seconds
STRETCHING EXERCISES	**SETS**	**INTERVAL**
ABDOMINAL STRETCH	1	15 seconds
OVERHEAD ARM PULL	1	15 seconds
UPPER BACK STRETCH	1	15 seconds
CALF STRETCH	1	15 seconds
HAMSTRING STRETCH	1	15 seconds
HIP/BACK STRETCH	1	15 seconds
GROIN STRETCH (SEATED)	1	15 seconds

Table 1-2. Daily physical training exercises.

NOTE: The following calisthenic exercises are conducted with a 15-second break between each exercise.		
CALISTHENIC EXERCISES CATEGORY A	**SETS**	**REPETITIONS**
SKI JUMPER	1	15
FOUR-COUNT PUSH-UP	2	15
SIT-UP	2	15
MOUNTAIN CLIMBER	2	15
KNEE BENDER	2	15
CALISTHENIC EXERCISES CATEGORY B	**SETS**	**REPETITIONS**
SKI JUMPER	1	15
EIGHT-COUNT PUSH-UP	2	15
SUPINE BICYCLE	2	15
SQUAT BENDER	2	15
FLUTTER KICK	2	15

RUN DISTANCES AND TIMES PER MILE			
WEEK	**DISTANCE**	**TIME PER MILE**	**TOTAL TIME**
GROUND	3.2 MILES	Min. 8:45 - Max. 9:15	Min. 28:00 - Max. 29:36
TOWER	4.0 MILES	Min. 8:45 - Max. 9:15	Min. 35:00 - Max. 37:00

8:45-MIN MILE		**9:00-MIN MILE**		**9:15-MIN MILE**	
.2 MILE	1 MIN 45 SEC	.2 MILE	1 MIN 50 SEC	.2 MILE	1 MIN 51 SEC
.4 MILE	3 MIN 30 SEC	.4 MILE	3 MIN 35 SEC	.4 MILE	3 MIN 42 SEC
.6 MILE	5 MIN 15 SEC	.6 MILE	5 MIN 15 SEC	.6 MILE	5 MIN 33 SEC
.8 MILE	7 MIN 00 SEC	.8 MILE	7 MIN 15 SEC	.8 MILE	7 MIN 24 SEC
1 MILE	8 MIN 45 SEC	1 MILE	9 MIN 00 SEC	1 MILE	9 MIN 15 SEC

Table 1-2. Daily physical training exercises (continued).

During ground week, students must complete three 3.2-mile runs at a 8:45- to 9:15-minute pace. The runs are completed in formation and the student must not fall more than three steps behind the original formation. During tower week, students must complete two 4-mile runs at the same standard (Table 1-2) and one 5-mile off-track run.

b. **Jump Phase**. Students who meet training proficiency in the basic jump techniques and physical fitness requirements during ground and tower week training are advanced to the jump training phase. During jump phase training, the student makes five qualifying jumps from aircraft at an altitude of 1,250 feet AGL (Table 1-3, page 1-4).

JUMP NUMBER	EQUIPMENT	TYPE EXIT
1	Ballistic helmet	ADEPT Option 1 Individual Exit
2	Ballistic helmet, combat equipment (HSPR, ALICE pack, and M1950 weapons case)	ADEPT Option 2 Both Doors
3	Ballistic helmet	Mass Exit Both Doors
4	Ballistic helmet	Mass Exit Both Doors
5 (Night)	Ballistic helmet, combat equipment (HSPR, ALICE pack, and M1950 weapons case)	Mass Exit Both Doors

Table 1-3. Typical jump week schedule.

1-3. PREJUMP ORIENTATIONS

Students are thoroughly briefed before performing their qualification jumps. The topics include—

- A review of the five points of performance, collisions and entanglements, towed parachutist, malfunctions, activation of the reserve, and emergency landings.
- Maintenance of the T-10-series or MC1-series parachute to include shakeout and storage after landing.
- Donning the parachute "by the numbers" on the first jump. Additional instructors are available for close supervision and JMPI.
- Aircraft orientation to include enplaning and jump procedures.
- Drop zone and approximate point of impact.

CHAPTER 2

PERSONAL EQUIPMENT

This chapter discusses the types of parachutes, donning the parachutes, and the headgear used in airborne training.

Section I. PARACHUTES

The T-10-series and MC1-series parachutes are used during static line airborne operations. The T-10-series is a nonsteerable canopy and the MC1-series is a steerable canopy. The main parachute consists of five major components—the harness assembly, the riser assembly, the deployment bag, the pack tray, and the canopy assembly. A reserve—either the modified improved reserve parachute system (MIRPS) or the T-10 troop chest reserve parachute—is used in conjunction with the main parachute.

2-1. HARNESS ASSEMBLY

The T-10-series harness assembly (Figure 2-1, page 2-2) is used with the T-10-series and MC1-series main canopies.

a. **Components**. The harness assembly is made of a flexible framework of Type XIII nylon webbing. The components attached to it are as follows:

(1) Female fitting, canopy release assembly with a safety clip, cable loop, and latch. Rated capacity of the assembly is 5,000 pounds.

(2) Canopy release assembly pads permanently attached behind the canopy release assembly.

(3) Main lift web constructed of two plies of Type XIII nylon with a rated capacity of 6,500 pounds.

(4) Chest strap with an ejector snap with activating lever, ball detent, and opening gate with a rated capacity of 2,500 pounds.

- An ejector snap pad is attached behind the ejector snap.
- The quick-fit V-ring has a rated capacity of 2,500 pounds.
- A webbing retainer is used for stowing excess webbing.

(5) D-rings used for attaching the MIRPS or T-10 reserve and additional combat equipment, with a rated capacity of 5,000 pounds.

(6) Triangle links sewn into the main lift web of the harness (about 5 1/2 inches below the D-rings) for attaching equipment and lowering lines.

(7) Leg straps with an ejector snap with activating lever, ball detent, and opening gate with a rated capacity of 2,500 pounds. The parts of the leg straps (ejector snap pads, quick-fit V-rings, and webbing retainers) are identical to the chest strap.

(8) Saddle (Type XIII nylon) with two attached leg straps.

(9) Diagonal back straps with six sizing channels (S, 1, 2, 3, 4, and L).

(10) Back strap adjusters with attached free-running ends of the horizontal back strap (rolled and sewn).

b. **Adjustment Points**. The harness assembly has five points of adjustment: the chest strap, two leg straps, and two free-running ends of the horizontal back strap.

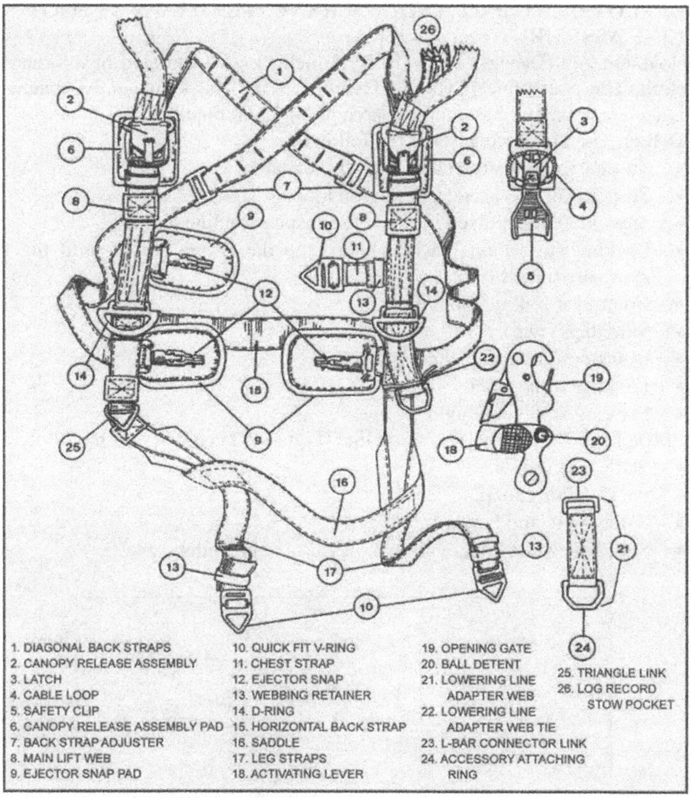

Figure 2-1. T-10-series harness assembly and nomenclature.

2-2. RISER ASSEMBLY

The riser assembly on the T-10-series consists of the following:

- Two riser assemblies, with a finished length of 30 inches (Type XIII nylon) and a tensile strength of 6,500 pounds. A male fitting canopy release assembly is permanently attached to the center of the webbing. When attached to the canopy, the riser assemblies provide four individual risers.
- Log record stow pocket. Opening the log record book by non-parachute rigger personnel is prohibited.
- Connector link loops.
- L-bar connector links.

The riser assembly on the MC1-series is identical to the T-10-series, but it also has a guide ring retainer strap, a guide ring, and upper and lower control line channels.

2-3. DEPLOYMENT BAG AND PERMANENTLY SEWN 15-FOOT STATIC LINE ASSEMBLY

The deployment bag (D-bag) (18 by 12 by 5 inches) is constructed of 8.8-ounce cotton sateen cloth. The static line (Type VIII yellow nylon) is permanently attached to the D-bag, is 15 feet long, and has a tensile strength of 3,600 pounds.

a. **D-Bag**. The D-bag consists of the following:

- Suspension line protective flap with data block.
- Suspension line protective flap tie loop.
- Stow loop panel (used to retain the suspension lines).
- Locking stow loops (two, which keep the D-bag closed until the first two stows are pulled free).
- Connector link tie loops (four).
- Side flaps (two).
- Break cord attaching strap pocket.
- Locking stow panel.
- Locking stow loop hood.

b. **Static Line Assembly**. The static line (Figure 2-2) consists of the following:

- Static line sleeve.
- Pack opening loop.
- Safety wire and lanyard.
- Static line snap hook with locking button and sliding sleeve.

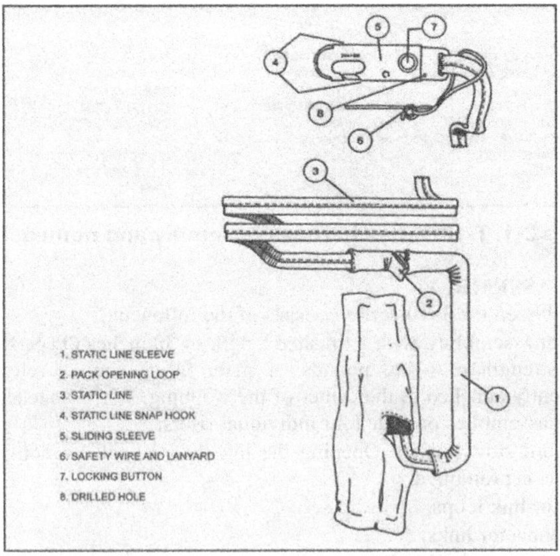

Figure 2-2. Static line and nomenclature.

2-4. DEPLOYMENT BAG WITH UNIVERSAL STATIC LINE ASSEMBLY

The deployment bag (D-bag) (Figure 2-3) consists of the following:

- Suspension line protective flap with data block.
- Suspension line protective flap tie loop.
- Stow loop panel (used to retain the suspension lines).
- Two locking stow loop, which keep the D-bag closed until the first stows are pulled free.
- Four connector link tie loops.
- Two side flaps.
- Break cord attaching strap pocket.
- Locking stow panel.
- Locking stow loop hood.

a. **Universal Static Line.** A 15-foot universal static line (USL) is secured to the D-bag by a girth hitch. The opposite end of the USL has a snap hook, which is also secured by a girth hitch. To configure the USL to the 20-foot length, the snap hook is removed and the extension is secured to the USL by a girth hitch. The snap hook is then secured to the opposite end of the extension. The length of the 15-foot static line after being secured to the deployment bag and snap hook is 14 feet 7 inches. The 20-foot configuration is 20 feet 1 inch long. There are four separate components to the USL, which must be purchased separately: deployment bag, static line, static line extension (Figure 2-4), and static line snap hook (Figure 2-5, page 2-6).

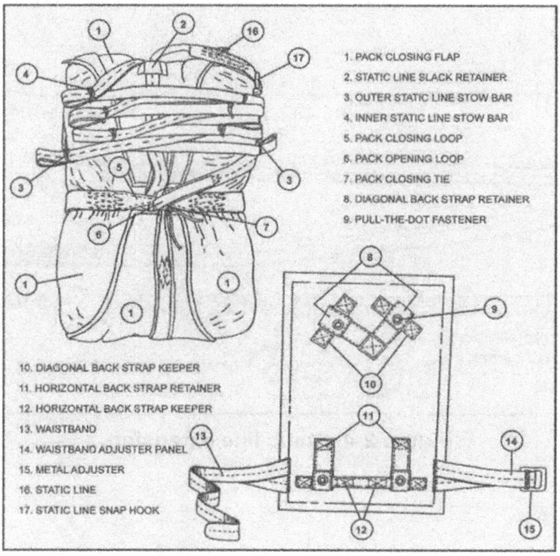

Figure 2-3. Deployment bag components with universal static line.

b. **Deployment Bag.** The deployment bag is currently used on the T10- and MC1-series personnel parachutes. The deployment bag comes without the static line and snap hook.

c. **Static Line.** The static line is 14 feet 10 inches long (+ - 1 inch). The static line webbing material is 3/4-inch wide, nylon, tube edge, class 1A (nylon 6.6), class R (resin treated), with "Dooley" finish. It has a 4,000-pound breaking strength and 30 percent minimum elongation. It is yellow in color with one green color marker yarn located at the center of the warp.

d. **Static Line Extension.** The static line extension is 70 1/2 inches long (+ - 1/2 inch) (Figure 2-4). It is made from the same webbing as the static line.

e. **Snap Hook.** The snap hook is 5 1/2 inches long, approximately 1/4-inch thick, and weights 0.67 pounds (Figure 2-5, page 2-6). It is made from alloy steel with Dacroment 320, grade A plus L finish for abrasion/humidity/salt/fog protection. The finish is environmentally compliant and a standard in the automotive industry. The snap hook gate and rivet pin are made from the same alloy steel, whereas the pullback spring is made from steel spring wire. The snap hook is rated to withstand a 1,750-pound load and has an ultimate failure rate of no less than 8,000 pounds.

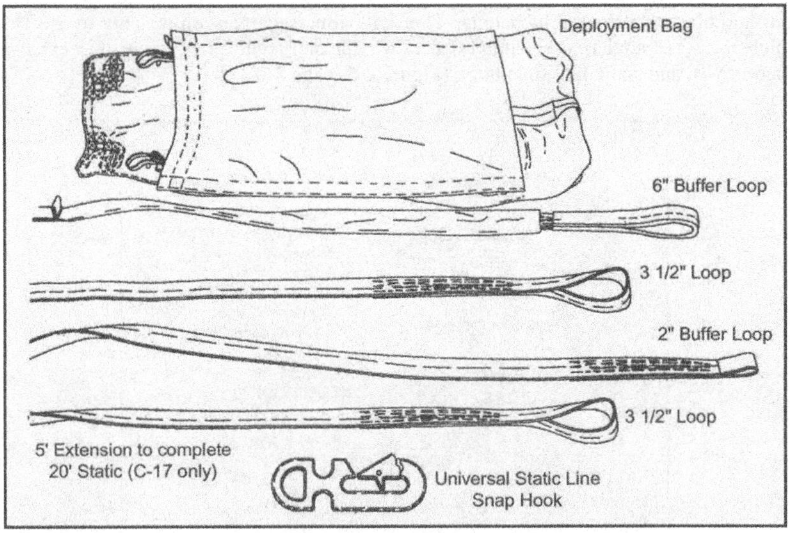

Figure 2-4. Static line extension.

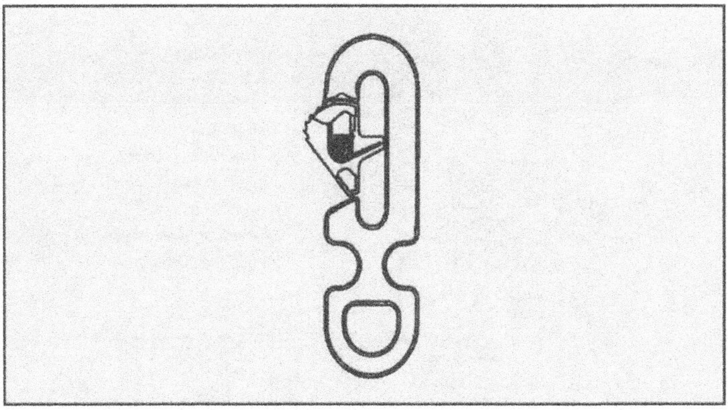

Figure 2-5. Static line snap hook.

2-5. PACK TRAY

The pack tray (20 by 14 by 5 inches) (Figure 2-6) is constructed of 7.25-ounce nylon duck material and consists of the following:

- Pack closing flaps (four): right and left side flaps, and upper and lower end flaps.
- Pack closing loops (four): right and left side pack closing loops, and upper and lower end pack closing loops.
- Static line stow bar.
- Waistband adjuster panel.
- Metal adjuster.
- Waistband (43 inches long).
- Pack closing tie (one turn 1/4-inch cotton webbing) tied in a surgeon's knot and a locking knot; the knot is between the 3 and 6 o'clock positions.
- Diagonal back-strap retainer.
- Diagonal back-strap keeper.
- Horizontal back-strap retainer.
- Horizontal back-strap keeper.

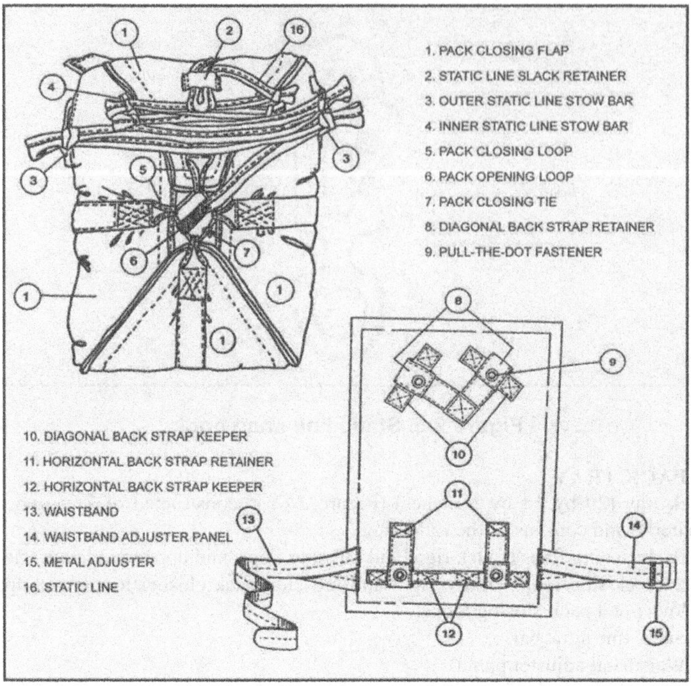

Figure 2-6. T-10-series parachute pack tray and nomenclature.

2-6. CANOPY ASSEMBLY

The T-10- and MC1-series parachutes are static line deployed. The parachute deployment sequence is the same for both types of canopies.

- The parachutist falls to the end of the static line. His body weight breaks the pack closing tie, and the deployment bag is pulled from the pack tray.
- Two connector link ties break, and the suspension lines are pulled from the deployment bag.
- Two locking stows disengage, and the canopy is pulled from the deployment bag to its full length.
- The break cord tie securing the apex of the canopy to the static line ending loop breaks, and the parachute begins to inflate, retarding the parachutist's rate of descent.

a. **General Characteristics**. Following are characteristics of both T-10-series and MC1-series canopies:

(1) ***Shape and Weight.*** Shape is parabolic; weight is between 28 and 31 pounds.

(2) ***Rates of Descent.*** Depending on the jumper's total weight and relative air density, the average rates of descent for the different canopies are as follows: MC1-series, 14 to 22 feet per second and T-10-series, 19 to 23 feet per second.

(3) ***Diameter.*** Nominal diameter is 35 feet (measured 3 feet up from the skirt) and 24.5 feet at the skirt.

(4) ***Anti-Inversion Nets.*** The anti-inversion net is sewn 18 inches down on each suspension line and is made of 3 3/4-inch square mesh, knotless, braided nylon.

(5) ***Shelf and Service Life.*** Combined shelf life and service life is 16.5 years; service life is 12 years, and shelf life is 4.5 years.

(6) ***Repacking.*** Both canopies are repacked every 120 days.

(7) ***Use.*** Both canopies are suitable for airdropping personnel from as high as 10,000 feet MSL.

b. **MC1-1B Parachute Design Characteristics.** The MC1-1B has an estimated 8.8-second turn rate.

(1) The bridle loop is 3 inches in diameter; it is made of Type VIII nylon with a tensile strength of 3,600 pounds.

(2) The 15 apex vent lines are 19 inches long and made of Type II nylon cord with a tensile strength of 375 pounds. The apex vent lines with centering lines keep the bridle loop in place and the canopy even during deployment.

(3) The apex vent cap has a 3-inch diameter opening, which reduces oscillation and assists in a positive opening of the canopy.

(4) The upper lateral band is 1-inch tubular nylon with a tensile strength of 4,000 pounds. It is sewn in as reinforcement, since the first 64 inches of the canopy is a high-pressure area.

(5) Each canopy has 30 gores with five sections in each gore. Each gore is numbered 1 through 30, and each section is sewn diagonally to prevent rips and holes from spreading throughout the canopy.

(6) The 30 radial seams are 9/16 inch wide and 17 feet 2 7/32 inches long, measured from the upper lateral band to the lower lateral band.

(7) The T-U shaped configuration has 11 gores removed (25 to 5) from the rear (100.4 square feet of canopy), which enables the canopy to turn 360 degrees in 8.8 seconds and gives a forward thrust of 8 knots (9.5 mph), or 14 feet per second.

(8) Two 28-foot-long control lines are attached to a control bridle that in turn is attached to radial seams 5 and 6. They are 6 feet long and attached to seams 25 and 26. They run down and out to the front of the rear set of risers and through the control line channel and control line guide ring. They are attached to a toggle that is a 5/8-inch diameter hardwood dowel.

(9) The lower lateral band is a 1-inch nylon tape with a tensile strength of 525 pounds.

(10) The 15 pocket bands ensure positive opening of the canopy.

(11) The 30 V-tabs are 9/16 inch wide and are sewn over the suspension lines to the lower lateral band for reinforcement.

(12) The 30 suspension lines are Type II nylon with a tensile strength of 375 pounds. They are 25 feet 6 inches long when measured from the lower lateral band to the L-bar connector link.

c. **MC1-1C Parachute Design Characteristics**. The MC1-1C canopy has the same basic design as the MC1-1B with the following exceptions:

(1) It has an estimated 7.7-second turn rate.

(2) It is made of nonporous (NOPO) material (0-3 CFM).

(3) The vent cap is removed.

(4) The suspension lines are shortened to 22 feet.

(5) The H-TC modification is a 60-square-foot opening.

d. **T-10-Series Parachute Design Characteristics**. The T-10-series parachute is designed with the following characteristics:

(1) The bridle loop is 3 inches in diameter and made of Type VIII cotton or nylon with a tensile strength of 3,600 pounds.

(2) The 15 apex vent lines are 19 inches long and made of Type II nylon cord with a tensile strength of 375 pounds.

(3) The two apex centering loops are 9 inches long and made of Type II nylon cord with a tensile strength of 375 pounds.

(4) The apex vent is 20 inches in diameter and expands to 22 inches in diameter when the canopy is inflated.

(5) The upper lateral band is 1-inch tubular nylon with a tensile strength of 4,000 pounds.

(6) The 30 gores have five sections each.

(7) The 30 radial seams are 9/16 inch wide and 17 feet 2 7/32 inches long.

(8) The lower lateral band is made of 1-inch nylon tape with a tensile strength of 525 pounds.

(9) The 15 pocket bands have been lengthened to 7 1/2 inches to provide a more positive opening and a 4.37 foot (overall) increase in the canopy to reduce descent to about 15 feet per second.

(10) The 30 V-tabs are 9/16 inch wide.

(11) The 30 suspension lines are 25 feet 6 inches long.

2-7. MODIFIED IMPROVED RESERVE PARACHUTE SYSTEM

The MIRPS (used with the T-10- and MC1-series) is an emergency-type parachute designed to be activated by the parachutist if the main parachute malfunctions. It is chest-mounted, manually operated, and loaf-shaped when packed. It weighs between 13 and 13.5 pounds, has a combined shelf life and service life not to exceed 16.5 years, and is 24 feet in diameter. It is repacked every 365 days. The MIRPS consists of six major components: pilot parachute, bridle line with weight and apex weight, ejector spring, canopy assembly, pack assembly, and rip cord assembly.

a. **Pilot Parachute Assembly**. The pilot parachute assists in deployment of the parachute canopy by serving as an air anchor. It is a five-foot, flat, circular parachute with bridle line and is constructed from zero-porosity nylon parachute cloth and marquisette netting. The netting is reinforced with six radial tapes that form the bridle attachment loop. A centerline is attached to the parachute cloth to speed pilot chute inflation and also forms part of the bridle attachment loop. The pilot chute does not have suspension lines. It resembles a large ball.

b. **Bridle Line Assembly**. The bridle line assembly is 13 feet long and is constructed from two-inch-wide polyester webbing with a four-inch loop at each end. One end of the bridle line is fitted with an apex sock that aids in pressurizing the reserve main canopy during low-speed deployments. Adjacent to the apex sock, the bridle line is fitted with two curved metal pins that are used to secure the canopy staging flaps located in the pack assembly. The other end of the bridle line is fitted with a five-ounce deployment weight. The deployment weight provides the necessary mass to cause positive launch of the pilot chute once the ejector spring reaches full extension.

c. **Deployment Assistance Device**. The ejector spring is a 30-inch-long helical spring encased in marquisette netting and fitted with an end cap at each end. On one end cap, two grommet tabs are attached. These are used only during packing to keep the spring compressed. Before the final closing of the pack, the spring compression aid is removed, and the grommet tabs are no longer used.

d. **Canopy Assembly**. The canopy assembly is a 24-foot, flat, circular parachute constructed of 1.1-ounce olive drab ripstop nylon parachute cloth. Depending on the jumper's total weight, its rate of descent varies from 15 to 22 feet per second. It is described as follows:

(1) An apex vent, 20 inches in diameter.

(2) Twelve suspension lines, 57 feet 6 inches long (measured from connector snap to connector snap), made of Type III nylon cord with a tensile strength of 550 pounds. On this parachute, the suspension lines serve three purposes:

- From the connector snaps that double as connector links, the lines are 20-foot suspension lines.
- Where the suspension lines go through the radial seams of the canopy, they become canopy lines.
- Across the apex vent, the canopy lines become apex vent lines until they again go into the upper radial seam; there they again become canopy lines.

(3) An upper lateral band of 1-inch tubular nylon with a tensile strength of 4,000 pounds.

(4) A lower lateral band of 1-inch nylon tape with a tensile strength of 525 pounds.

(5) Twenty-four gores with four sections in each gore.

(6) A 9/16-inch nylon V-tab reinforcement sewn into the lower lateral band of the parachute and wrapped around each suspension line.

(7) Twenty-four pocket bands.

e. **Pack Tray**. The MIRPS and the T-10 troop chest reserve use different pack trays that are almost identical on the outside (Figure 2-7 and Figure 2-8, page 2-12). Thus, an identifying yellow binding tape on the rip cord protector flap distinguishes the MIRPS. The major parts of the pack tray used for the MIRPS are the main panel and two end panels. The main panel forms a top and bottom flap, and the end panels form a right end flap and a left end flap. A rectangular metal pack frame is enclosed in a pocket formed in the bottom. The pack has two holes with lift-the-dot strap fasteners for attaching the MIRPS to the primary parachute harness using the connector snaps. A suspension-line free-bag deployment pouch is located on the inside of the container on the pack bottom. Canopy staging flaps are attached to the inside of the top and bottom flaps; these provide canopy retention.

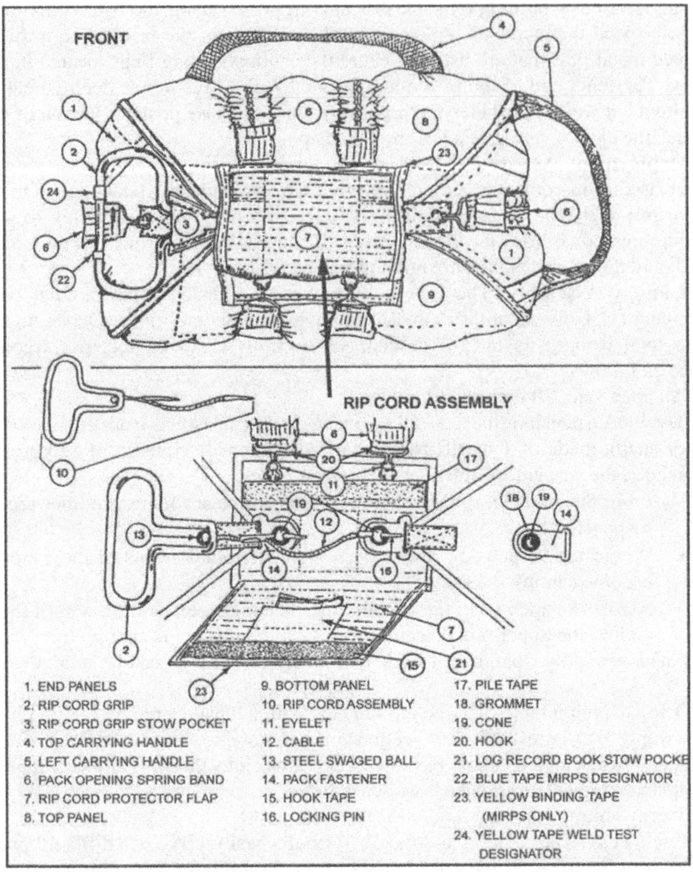

Figure 2-7. MIRPS/T-10 reserve parachute pack tray and nomenclature (front).

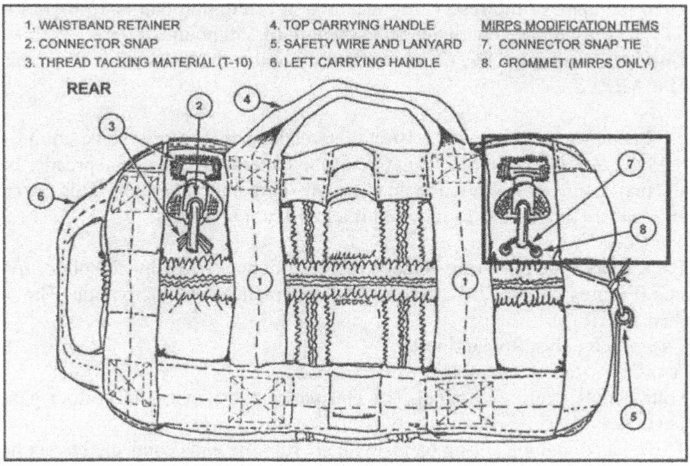

Figure 2-8. MIRPS/T-10 reserve parachute pack tray and nomenclature (back).

f. **Rip Cord Assembly**. The rip cord assembly is used to activate the reserve parachute and requires about 27 pounds of pull pressure. It is composed of a cloverleaf-shaped, stainless steel rip cord grip attached to a 7.25-inch-long flexible steel cable by means of a steel swaged ball. The cable has two permanently fastened locking pins (1 1/4 inches long) that are inserted into holes in the pack tray bottom panel cones; they keep the reserve pack tray closed until activation.

2-8. T-10 TROOP CHEST RESERVE PARACHUTE

The T-10 reserve parachute (used with T-10- and MC1-series) is chest-mounted, manually operated, and loaf-shaped when packed. It weighs about 12 pounds, has a combined shelf life and service life not to exceed 16.5 years, and is 24 feet in diameter. It is repacked every 365 days. It is an emergency-type parachute designed to be activated by the parachutist if the main parachute malfunctions. The T-10 reserve parachute consists of four major components: spring-activated pilot parachute assembly, canopy assembly, pack tray assembly, and rip cord assembly. The rip cord grip must be marked with a blue and yellow piece of tape. Blue signifies that it is for the MIRPS and yellow shows that it has been weld tested.

a. **Pilot Parachute Assembly**. The pilot parachute acts as an air anchor to assist the canopy in deploying quickly. It is spring-activated, made of 1.1-ounce ripstop nylon parachute cloth, and is 40 inches in diameter. It is described as follows:

(1) Octagon-shaped, with reinforcements sewn to the inside of the canopy.

(2) Four pockets sewn on the inside of the canopy to stow the spring framework.

(3) Eight suspension lines of Type I nylon cord. Each pair of suspension lines is formed by one continuous line that runs through the canopy and is stitched together at the lowest point to form the connector loop. The connector loop is used to attach the pilot

parachute to the apex of the reserve by means of a bridle line that is 15 inches long and made of Type III nylon cord with a tensile strength of 550 pounds.

b. **Canopy Assembly**. The canopy assembly in the T-10 reserve is the same canopy used for the MIRPS.

NOTE: Inside the pack tray is a 10-inch spreader bar. If one reserve connector snap becomes disconnected from one of the harness D-rings, the spreader bar keeps that connector snap attached to the parachute harness. This arrangement enables the reserve to function if a malfunction occurs.

c. **Pack Tray**. The pack tray is made of 12.29-ounce nylon or 7.25-ounce nylon on a rigid frame (Figures 2-7 and 2-8). It contains the parachute until activation. The pack tray is described as follows:

(1) Carrying handles (top and left).

(2) A safety wire and lanyard.

(3) Four panels (right end panel, left end panel, top panel, and bottom panel) with attached eyelets.

(4) Three pack opening spring bands with six running ends connected to six hooks.

(5) A rip cord grip stow pocket sewn to the right-end panel.

(6) A rip cord protector flap sewn to the bottom panel, with log record stow pocket on the inside of the rip cord protector flap.

(7) Pack fasteners sewn to the left- and right-end panels.

(8) Grommets and cones that keep the pack closed by inserting the locking pins of the rip cord assembly through the cones. (Grommets are on the top panel and cones are on the bottom panel.)

(9) Two connector snaps to attach the reserve to the D-rings.

(10) Two nylon waistband retainers through which the waistband is routed. The waistband retainers consist of a single piece of material sewn into the pack tray, which forms the reserve top carrying handle.

d. **Rip Cord Assembly**. The rip cord assembly for the T-10 reserve is the same as for the MIRPS with one exception. The length of the cable is .25 inch shorter (7 inches rather than 7.25).

2-9. BA-18 BACK AUTOMATIC PARACHUTE

The USAF emergency parachute BA-18, back automatic parachute, is used by non-jumping jumpmasters and safety personnel onboard USAF aircraft. This USAF parachute has a repack cycle of 180 days and must pass a routine inspection at the home unit every 30 days. Once the parachute is activated, it must be destroyed. The BA-18 is equipped with either an FLB-Model 7000 or FLB-Model 11000 automatic release system (ARS). The owning USAF unit's life support department sets the time delay release (from 1 to 13 seconds) and then attaches a small white tag on the back of the BA-18 to indicate what delay time has been set (3 seconds, 4 seconds, 5 seconds, and so on). To employ the BA-18, the jumper either pulls the red arming cable knob to activate the time delay feature or overrides the ARS by pulling the T-shaped blast handle. Personnel scheduled to wear the BA-18 must inspect it before each use. If they note any discrepancies during the inspection, they must not use the parachute.

a. **BA-18 Inspection Criteria**. The inspection procedures are simple, and their sequence is not important. However, the parachute must be inspected before each donning.

(1) Check the canopy release assemblies in the same manner as on the T-10-series or MC1-series main parachutes.

(2) Ensure that no straps (chest, leg, and horizontal back strap) are cut or frayed.

(3) Ensure all three ejector snaps are serviceable.

(4) Between the parachute harness assembly and the pack tray, a pad is held with four pull-the-dot fasteners (one at each corner). Open the pull-the-dot fasteners, expose the long zipper that is behind the pad, and unzip the zipper. Ensure that the locking pins are not bent and are routed through the white nylon cord loops, and that the cable is free of rust or other corrosion. Re-secure the zipper and pad once the inspection has been completed.

(5) Inspect the risers and ensure they are not cut or frayed and are secured to the canopy release assemblies.

(6) Make an overall inspection on the outside of the pack tray to ensure that the pack tray is free of rips, tears, oil, grease, dirt, or water and that no canopy or suspension lines protrude from the pack tray.

(7) Open the slide fastener on the back of the pack tray. Ensure that the spring has both hooks (one at each end) routed through the white nylon cord loops; then secure the slide fastener.

(8) Ensure the rip cord is properly stowed and free of rust or other corrosion.

(9) Ensure the red time delay arming cable knob is properly stowed.

(10) If the parachute is equipped with a personnel lowering device, inspect it to ensure that the associated hardware is secure inside the stow pocket. Inspect the lowering line from where it exits the stow pocket to the point it disappears into the pack tray.

b. **Planned and Unplanned Exits**.

(1) ***Above 14,000 Feet.***

(a) Once the last jumper has exited the aircraft, pull the red automatic release arming knob to arm the ARS, and exit the aircraft.

(b) When you reach an altitude of 14,000 feet, the canopy will open. However, at anytime during descent, you can activate the rip cord and override the ARS.

(2) ***Below 14,000 Feet.***

(a) If the parachute is equipped with an ARS set for a delay of 4 seconds or more, pull the automatic release arming knob and exit the aircraft.

(b) If the parachute is equipped with an ARS set for less than 4 seconds, exit the aircraft, assume the correct body position, and pull the automatic release arming knob. To assume the correct body position, place your chin on your chest, your elbows into your sides, your feet and knees together, and visually locate and place your hand(s) on the rip cord. Allow sufficient time (one second, minimum) to clear the aircraft, then pull the rip cord.

NOTE: When bailing out of an aircraft at extremely low levels, the most important factor is to pull the rip cord as soon as you clear the aircraft, regardless of your body position.

c. **Proper Method to Pull the Rip Cord.**

(1) The automatic release may be overridden by pulling the rip cord. Grasp the "T" handle with your right hand and guide or assist it with your left hand. Pull the rip cord down toward your feet and away from your body hard and fast. Pull it to arm's length and then bring your arms back close to your body immediately. Ensure the rip cord clears the housing.

(2) After exiting the aircraft, complete the normal remaining four points of performance—check canopy and gain canopy control, keep a sharp lookout during your entire descent, prepare to land, and land (see Chapter 3). To avoid obstacles and other jumpers in the air, or maneuver the canopy, pull a vigorous two-riser slip in the desired direction of travel. To execute a two-riser slip with this canopy, reach up high onto the risers to the elbow-locked position, grasp a set of risers in the direction of the desired movement, and pull them down to your chest.

d. **Personnel Lowering Device (PLD).** If you become hung in wires, trees, or rough terrain and the parachute is equipped with a PLD, use it to climb down to safety.

(1) Visually check to see that you are securely hung in the obstacle.

(2) Take the hardware out of the stow pocket. Grasp the braking device and snap it into the top portion of the V-ring of the chest strap with the hook side facing your midsection.

(3) Pass the snap hook at the end of the tape through the "V" of both risers.

(4) Attach the snap hook to the O-ring (which should be in front of your face).

(5) Grasp the tape with your left hand and disconnect the right canopy release assembly with the right hand. Exchange hands on the tape, with your left hand disconnecting the left canopy release assembly.

(6) Lower yourself to the ground by feeding the tape through the braking device.

(7) To stop the descent, pull the tape up vertically with your right hand.

2-10. CARE OF THE PARACHUTE BEFORE JUMPING

Troop parachute assemblies and reserves may be issued in kit bags to aid handling and to prevent damage or unintentional opening. Until removed for fitting by parachutists, parachutes and reserves should remain in the kit bags and protected from moisture during storage to prevent mildew. (Kit bags are not waterproof and do not provide adequate protection from wet weather or damp ground.) Parachutes must be stored in weatherproof areas such as adequate storage buildings, trucks, tents, or transport aircraft.

2-11. CARE OF THE PARACHUTE AFTER JUMPING

The parachute is recovered and properly cared for to minimize damage. Upon landing, parachutists activate the canopy release assembly (while lying on their backs to observe other landing parachutists), take off the harness, and place the parachute in the kit bag using either the tactical recovery method or by executing a series of figure-eight folds with the arms (Figure 2-9, page 2-16). The specific actions are as follows:

a. Remove all air items. Place harness in the aviator kit bag with the smooth side up, leaving the waistband out. Place released riser underneath the harness.

b. Move to the apex of the canopy, grasp the bridle loop, elongate the parachute into the wind to straighten the canopy and suspension lines, and remove all foreign objects and debris from suspension lines and canopy.

c. Fold the canopy and suspension lines into a series of figure eights, using both arms. Do **not** twist the canopy unnecessarily because friction can cause the nylon to fuse.

d. Lay the canopy on the top of the harness and, before closing the bag fasteners, ensure the bridle loop is on top of the canopy and the waistband is routed through the bridle loop. Close bag fasteners; **do not zip** the bag because the canopy may become entangled in the zipper and damaged. Attach a reserve connector snap to each kit bag handle. Carry the equipment so the reserve parachute is to the (parachutist's) front and the kit bag to the rear. (Reverse the carry when jumping with combat equipment.)

NOTE: If it is necessary to activate both canopy release assemblies upon landing, then fold the canopy in figure eights by itself and place it in the kit bag on top of the harness.

Figure 2-9. Stowing the parachute.

2-12. SHAKEOUT PROCEDURES

The parachute is suspended from a rope passed over a pulley suspended from a ceiling (or from pole's) high enough to allow the canopy to clear the surface area.

a. **Main Canopy**. A two-man shakeout team is recommended. Number 1 holds the bridle loop, while number 2 fastens the rope to the loop. Number 2 pulls the rope until the skirt is about 1 foot above number 1's head. They accomplish the rest of the shakeout by taking the following steps:

STEP 1: The team leaves the bulk of the suspension lines and the parachute harness in the kit bag. Number 2 grasps the rope attached to the suspended canopy while number 1 shakes the parachute.

STEP 2: Number 1 grasps two adjacent suspension lines at the lower lateral band, one in each hand, and vigorously shakes the gore, making certain no grass, twigs, insects, or other foreign matter are left on the fabric or tangled in the anti-inversion net.

STEP 3: Number 1 then transfers both suspension lines to his left hand, grasps the suspension lines of the next gore with his right hand and continues as in step 2, working counterclockwise, until each gore has been shaken and all suspension lines are in his left hand. He must pay particular attention to the anti-inversion net to ensure no foreign material remains. Debris left in the net can result in a total malfunction.

STEP 4: Number 2 begins to slowly pull the canopy up, elongating the suspension lines. Number 1 shakes the suspension lines and dusts them by hand and then turns the kit bag inside out and cleans it thoroughly to ensure no debris is in the bag.

STEP 5: Number 1 puts the harness in the bag.

STEP 6: Number 2 then slowly lowers the parachute while number 1 coils the suspension lines on top of the harness and places the canopy inside the bag.

b. **Reserve Parachute**. The shakeout procedure for the reserve parachute (if used) is the same as that for the main canopy. Do the shakeout as soon after jumping as practicable.

c. **Wet Parachute Procedures**. Parachutes used in wet weather or exposed to moisture will be hung to dry within 24 hours of the jump. Once dry, shakeout procedures will then occur.

Section II. DONNING THE PARACHUTES

Using the buddy system to properly don and adjust the troop parachute harness provides an additional safety check, prevents delays during JM inspection, and provides minimum discomfort to the parachutist while aboard the aircraft or when receiving the opening shock of the parachute. The buddy system method provides the best combination of speed and accuracy for parachutists to adjust and check each other's parachutes.

2-13. TROOP PARACHUTE HARNESS

Each parachutist first checks the parachute assembly for visible defects.

STEP 1: The parachutist lays the assembly out with the pack tray face down. Then he—

(1) Activates the waistband quick-release and pulls up each of the activating levers on the ejector snaps, releasing the leg straps and the chest strap.

(2) Checks for appropriate size and, if necessary—

- Lifts the pull-the-dot fasteners on the diagonal back-strap retainer and frees the diagonal back straps from the sizing channel (using fingers only; do not use tools).
- Sizes the parachute to one of the six sizing channels (S, 1, 2, 3, 4, and L).
- Rechannels the diagonal back strap retainer and fastens the pull-the-dot fastener.

- Lets out about half of the slack in the horizontal back strap, leg straps, and chest strap; straightens the leg straps and chest strap; and folds the kit bag, leaving the outermost carrying handle extended.

STEP 2: The parachutist (number 1) bends slightly forward at the waist to don the parachute. A second parachutist (number 2) holds the parachute assembly by the main lift web under the canopy release assemblies and places it on the back of number 1.

STEP 3: Number 1 remains bent forward at the waist; number 2 pushes the pack tray high on number 1's back and pulls the saddle well down over the buttocks. As the adjustment is being made, number 1 fastens the chest strap and ensures that the activating lever is closed over the ball detent.

STEP 4: Number 2 calls out "Left leg strap," grasps the leg strap by the quick-fit V-ring with one hand, and with his other hand starts from the saddle (with thumb and index finger) and feels the length of the leg strap, removing any twists and turns, and hands the left leg strap to the jumper. Number 1 inserts the left leg strap through (over the bottom and under the top) the kit bag carrying handle and snaps the quick-fit V-ring into the left ejector snap. The right leg strap is passed over the other end of the kit bag (securing it in place), and the quick-fit V-ring is snapped into the right ejector snap. The parachutist ensures that both the left and right activating levers are closed over the ball detents.

STEP 5: Number 1 stands erect and checks to ensure the canopy release assemblies are in the pockets of the shoulders.

STEP 6: Number 2 locates the free-running ends of the horizontal back strap and tightens the harness until number 1 indicates it fits snugly and comfortably. The horizontal back strap is the main point of adjustment for the harness. After final adjustment, number 1 should be able to stand fully erect without straining (Figure 2-10).

Figure 2-10. Troop parachute harness fitted.

STEP 7: Number 1 and number 2 then change positions and repeat steps 1 through 6. When both parachutists have donned their parachute harnesses, they face each other and make a visual inspection. They correct any discrepancies before securing the reserve parachute.

All excess webbing is stowed in webbing retainers.

2-14. MIRPS/T-10 RESERVE PARACHUTE

The parachutist attaches the reserve parachute by cradling the parachute in his left arm with the connector snaps up and the rip cord grip in the palm of the left hand.

STEP 1: Using the right hand, start at the pack tray and run out the waistband between the thumb and index finger to remove any twists or turns.

STEP 2: Thread the waistband through the two reserve waistband retainers and fasten the right connector snap to the right D-ring. Insert the safety wire in the right connector snap and then bend the wire down to safety it. Then connect the left connector snap to the left D-ring.

STEP 3: Parachutists help each other in securing the waistband and forming the quick-release. They ensure that all slack is pulled out of the waistband, and the slack in the quick-release loop is about the width of two to three fingers.

NOTE: During the initial periods of airborne training, students receive thorough training in the nomenclature, fitting, and wearing of the parachute assemblies. Demonstration, followed by student participation, is the key to the success of this instruction. Instructors constantly check to ensure students know the proper nomenclature as well as the proper methods of wearing and fitting the parachutes.

Section III. PROTECTIVE HEADGEAR

The ballistic helmet or the advanced combat helmet (ACH) is used during airborne operations. The helmets must be fitted properly to ensure they stay on the parachutist's head during deployment of the parachute and during the parachute landing fall.

2-15. BALLISTIC HELMET DESCRIPTION

The ballistic helmet is a laminated, one-piece helmet. It has a low profile, a close fit, and a low center of gravity. The helmet is available in four sizes: extra small, small, medium, and large. It weighs from 2 pounds 8 ounces (extra small) to 3 pounds 4 ounces (large).

a. **Parachutist Helmet Modifications**. The ballistic helmet is modified for airborne operations by adding a parachutist retention strap and a foam impact pad (Figure 2-11, page 2-20). These items provide maximum safety and stability.

NOTE: Be sure the hook-pile tape on the ends of the parachutist retention strap faces the rear of the helmet.

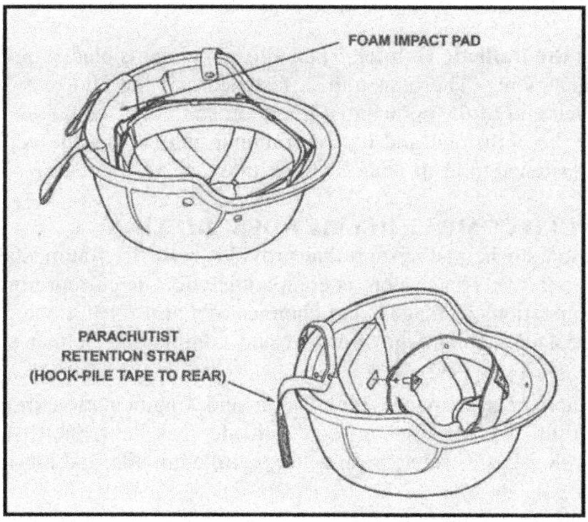

Figure 2-11. Ballistic helmet with retention strap and foam impact pad.

b. **Camouflage Cover**. A camouflage cover (Figure 2-12) is secured to the helmet by placing the cover over the helmet and threading both sides of the unfastened chin strap through the corresponding slits in the cover. The two rear attachment tabs of the cover are tucked between the cover and the helmet. The two ends of the parachutist retention strap are threaded through the corresponding slits in the rear of the cover. The four remaining attachment tabs are fastened around the nylon webbing of the suspension band with drawstring and adjustable tab.

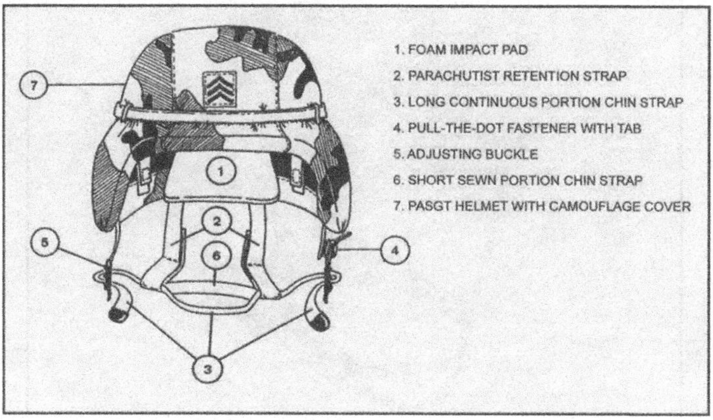

Figure 2-12. Ballistic helmet and nomenclature.

c. **Donning the Ballistic Helmet.** The ballistic helmet is placed on the head to make any comfort adjustments. The chinstrap is fastened and the chin cup is adjusted to fit snugly. Then, each end of the parachutist retention strap is placed alongside the face and wrapped around the chinstrap, and its free-running end (below the adjusting buckle) is pulled tight and fastened to itself with the hook-pile tape attachment.

2-16. ADVANCED COMBAT HELMET DESCRIPTION

The ACH is a modular helmet system that provides ballistic, fragmentation, and impact protection (Figure 2-13). This system is compatible with the current night vision devices (NVGs), communications packages, and chemical defense (NBC) equipment. The ACH is intended to replace standard government and commercial helmet systems currently in use.

- The helmet system provides 9-mm and fragmentation protection within a spectrum of environments (cold [-40 degrees Fahrenheit]; hot [140 degrees Fahrenheit]; salt water; fresh water; petroleum, oils, and lubricants [POL]; and so on).
- The ACH allows maximum sensory awareness for the operator. which includes an unobstructed field of view and increased ambient hearing capabilities.
- The helmet's retention/suspension system provides unsurpassed balance, stability, and comfort. This unique system provides for proper size, fit, and ventilation.
- The pad suspension system provides superior impact protection throughout all operational requirements.

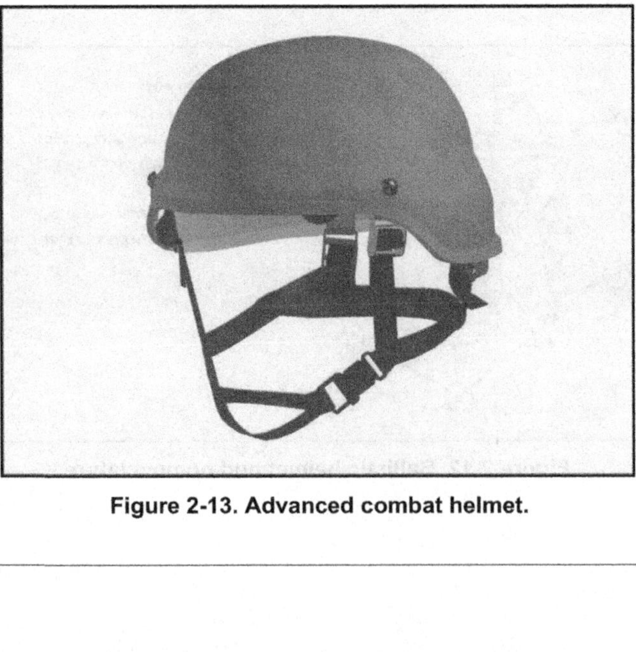

Figure 2-13. Advanced combat helmet.

a. **Helmet Assembly.** The ACH has a unique, fully adjustable pad suspension system. All seven pads are worn with the helmet (Figure 2-14).

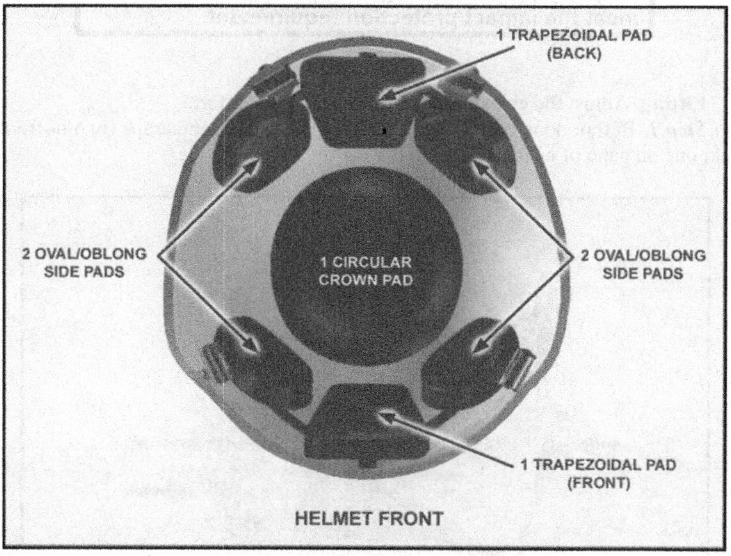

Figure 2-14. ACH pads.

(1) The placement of the oblong side pads must cover the four bolt ends in the four internal buckles in the helmet shell.

(2) The crown pad may be replaced by two oblong/oval pads. The direction of the side oblong/oval pads may be changed to maximize comfort: they may be routed vertically from bolt to crown, which maximizes airflow for better temperature regulation; or, they may be routed horizontally to make a seal around the user's head, which is better for cold weather environments.

NOTES:
1. If hot spots or discomfort are experienced, try rearranging the pad system to accommodate a more comfortable fit. If discomfort persists, try resizing the shell.
2. See the sizing and fitting troubleshooting guidelines (paragraph c, page 2-26) for problems with fit, such as tightness or looseness, or if the helmet profile is too high or too low.
3. When trying on the helmet for the first time in a cold environment, wear the helmet for a few minutes to allow the pads to warm up and conform to the shape of your head.

WARNING
Helmet must be worn with crown pad or two properly placed oblong/oval pads (size 6 or 8) to meet the impact protection requirement.

b. **Fitting.** Adjust the chinstrap to optimize fit and comfort.

(1) ***Step 1.*** Before donning the helmet, loosen all adjustment straps (two in front, two in back, one on nape of neck) (Figure 2-15).

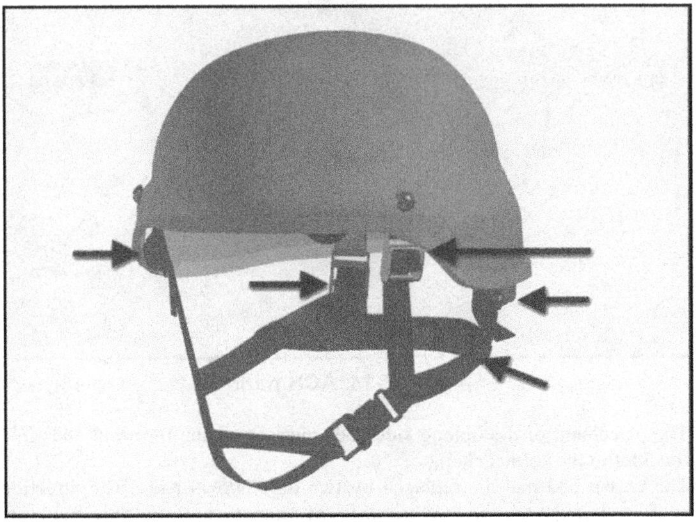

Figure 2-15. ACH adjustable straps.

(2) ***Step 2.*** Position the helmet on the head and hold in place with one hand on top of the helmet for initial adjustment (Figure 2-16, page 2-24).

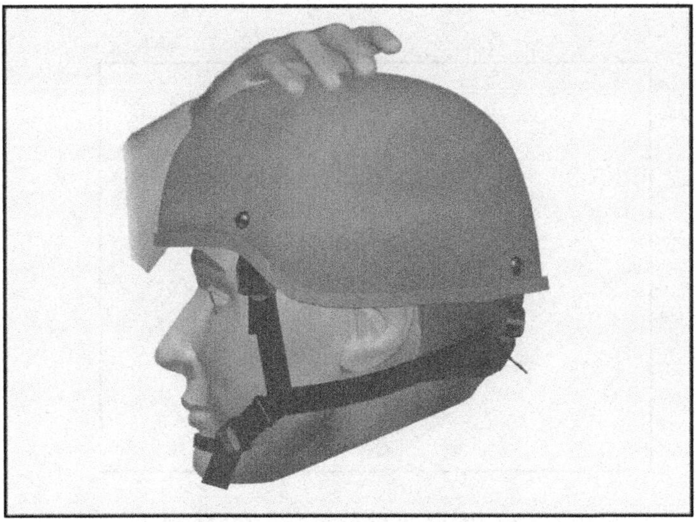

Figure 2-16. ACH held in place for initial adjustment.

(3) ***Step 3.*** Partially tighten the two back adjustment straps (one side at a time) (Figure 2-17).

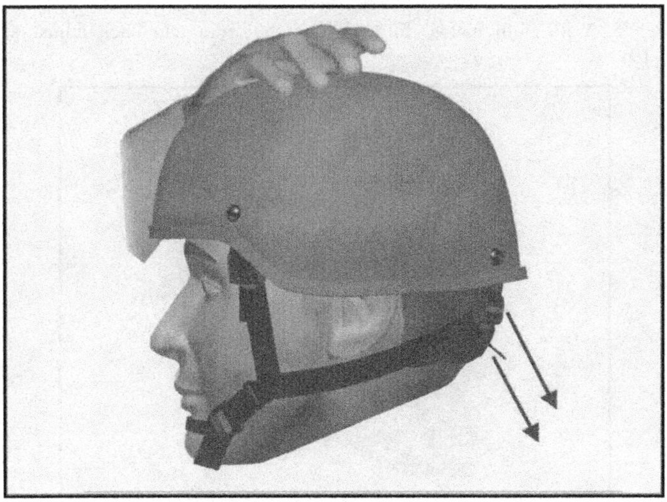

Figure 2-17. Partially tighten two back straps.

(4) ***Step 4.*** Partially tighten the two front adjustment straps (one side at a time) (Figure 2-18).

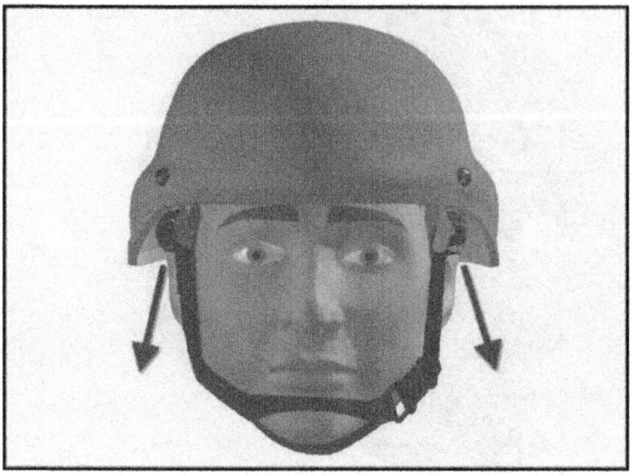

Figure 2-18. Partially tighten two front straps.

NOTE: If any one strap is pulled too tightly during steps 3 and 4, the helmet may become uncomfortable and tilted on the head.

(5) ***Step 5.*** With both hands, fully tighten the front and back adjustment straps (Figure 2-19).

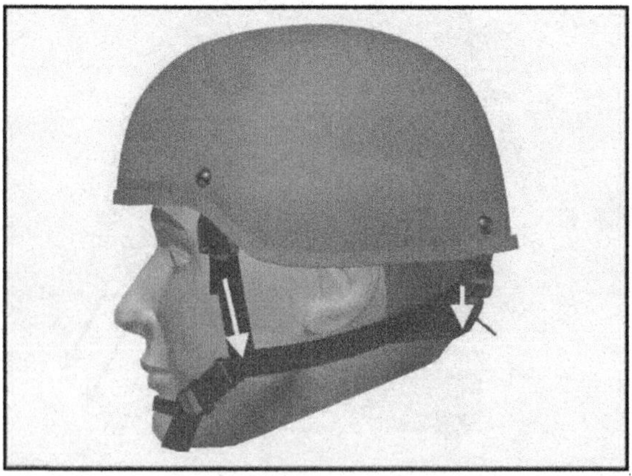

Figure 2-19. Fully tighten front and back straps.

(6) ***Step 6.*** Position the nape pad up and down according to personal comfort (Figure 2-20).

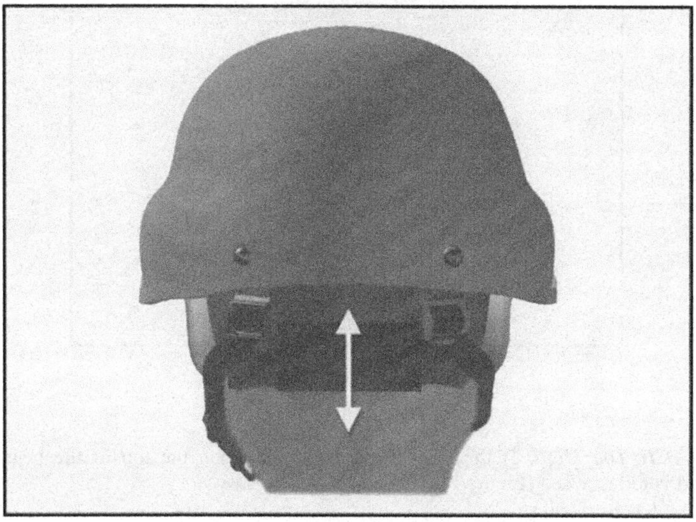

Figure 2-20. Position nape pad.

c. **Sizing and Fitting Troubleshooting Guidelines.** Use the following guidelines to solve the problems discussed herein.

(1) ***ACH Too Tight.*** If the ACH feels too tight—

- Try arranging the side pads in a vertical configuration.
- Try removing the back pad and rearranging the side pads.
- Try to create space in the area that feels tight.
- Try the next smallest pad size if rearranging the pads does not alleviate the tightness.

(2) ***ACH Too Loose.*** If the ACH feels too loose (Figure 2-21)—

- Try a larger sized pad set or increase the current number of pads in use.
- If wearing a size 8 pad set and a size large shell, downsize to a size medium shell. If the helmet is still too loose, increase the number of pads.

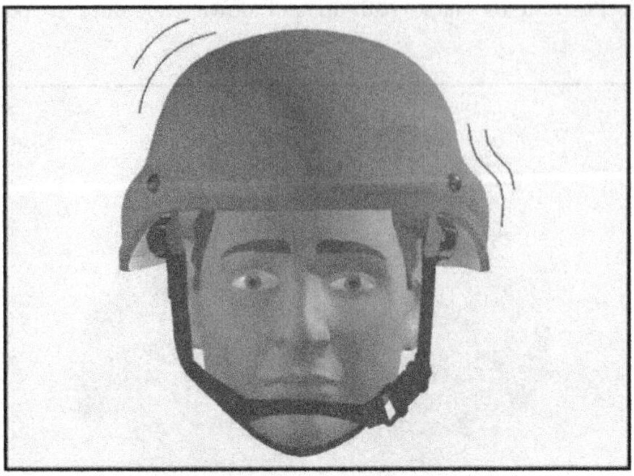

Figure 2-21. ACH too loose.

(3) ***ACH Too High.*** If the crown pad does not touch the top of the head or the forehead is too exposed (Figure 2-22)—
Try a smaller sized pad set.
Try a larger shell size if smaller pads do not work.

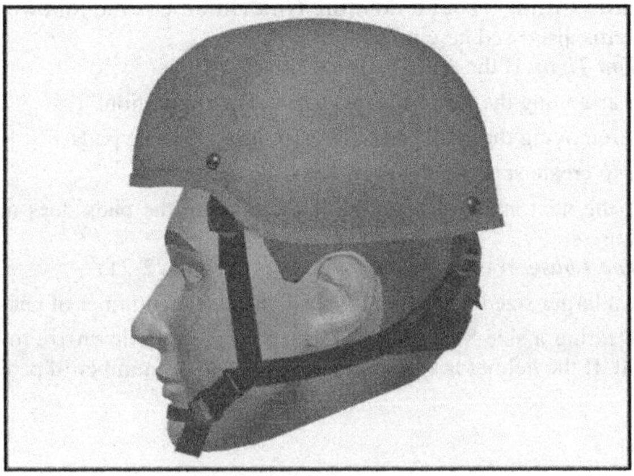

Figure 2-22. ACH sits too high.

(4) ***ACH Too Low.*** If the ACH sits too low on the brow, is not compatible with eyewear, or other compatibility issues exist (Figure 2-23)—

- Try larger suspension pads.
- Try a smaller ballistic helmet shell if the larger pads do not solve the problem.

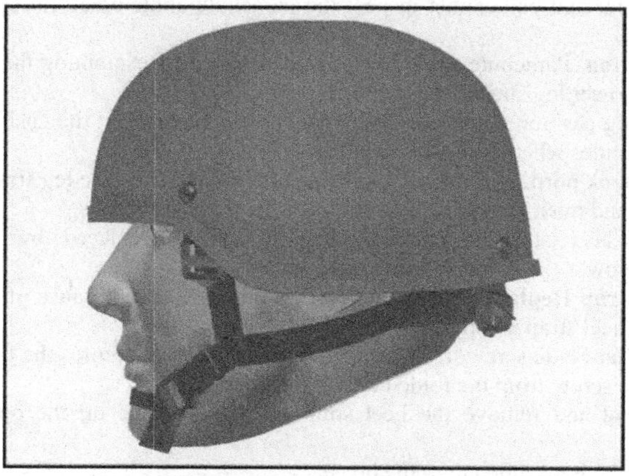

Figure 2-23. ACH sits too low.

Section IV. PARACHUTIST ANKLE BRACE

The parachutist ankle brace (PAB) stabilizes the jumper's ankle during PLFs and reduces the potential for ankle injury. The PAB, which is worn outside the combat boot, consists of sidewalls that extend vertically to encircle the ankle and the lower leg. The side and back portions are connected by a U-shaped support that fits over the boot heel. The support has a hook-pile strap system that attaches the brace to the leg and boot. The PAB is lined with a three-section aircell to cushion the lower leg.

NOTE: The PAB is not a required item of equipment.

2-17. OBTAINING THE PARACHUTIST ANKLE BRACE

The unit orders the PAB or the heel strap replacement kit directly from AIRCAST, P.O. Box 709, Summit, NJ 07902-0709, commercial 1-800-526-8785. All sizes are based on men's shoe sizes. The size codes for ordering are small, less than men's size 8, 02G (NSN 8465-01-416-6217); medium, men's sizes 8 through 11, 02H (NSN 8465-01-417-4002); or large, larger than men's size 11, 02I (NSN 8465-01-416-6218). European sizes are small, less than 42; medium, 42 through 45; and large, larger than 45. The cost is about $60 per pair.

NOTE: The PAB must be worn with standard-heeled boots. When worn with boots with flat or rippled soles, the PAB tends to slip and does not provide the proper support for the jumper's ankles.

2-18. INSPECTING THE PARACHUTIST ANKLE BRACE

Both jumpmaster and parachutist inspect the parachute ankle brace for serviceability as follows:

a. **Inspection**. Parachute ankle braces are inspected by examining the PAB cushion, the upper and lower leg straps, and the heel strap.

(1) If the leg cushion is missing, the PAB is unserviceable. If the cushion is torn, the jumpmaster decides whether the PAB is serviceable.

(2) If the hook portion of the hook-pile tape is missing from the leg straps, the PAB is unserviceable and must be replaced.

(3) If the heel strap is unserviceable, it can be replaced by following the instructions below.

b. **Heel Strap Replacement**. If the PAB is unserviceable because of a worn or torn heel strap, the heel strap is replaced as follows:

(1) Use a flat-blade screwdriver to remove the screw that secures the heel strap to the brace. Save the screw from the folded tab on the heel strap.

(2) Unthread and remove the heel strap from the buckle on the opposite side of the brace.

(3) Fasten the hook-pile tab to the brace.

NOTE: Ensure the new heel strap size corresponds to the brace size (small, medium, or large).

(4) Flip the brace over so the heel strap buckle is facing up. Thread the heel strap through the buckle top slot and pull through the slack. Loop the strap through the bottom slot and pull until snug.

(5) Flip the brace back over, remove the hook-pile tab to expose the screw hole, fold the strap end over, line up the screw holes, insert the screw through the strap, and refasten the screw to the brace.

2-19. DONNING THE PARACHUTIST ANKLE BRACE

To reduce the potential for jumper ankle injury, the parachutist ankle brace must be adjusted to fit snugly around the foot, ankle, and lower leg. The PAB is adjusted to the jumper's foot as follows:

a. Loosen the two leg straps and the heel strap, step into the PAB with the heel strap under the boot's instep (the area in front of the boot heel), and tighten the heel strap and then the two leg straps. The bottom of the shell should align with the top of the sole.

b. Fasten the leg straps using only the outer portion of the leg strap buckle. If both inner and outer portions of the buckle are used, the PAB could be difficult to remove in a tactical situation.

NOTE: The PAB is designed so the jumper can run short distances while wearing the PAB if necessary due to the tactical situation.

2-20. DOFFING THE PARACHUTIST ANKLE BRACE

The parachutist ankle brace is removed by loosening the leg straps and pulling the PAB off the leg and foot. The PAB is an air item and must be retained by the jumper for use during subsequent jumps.

CHAPTER 3

FIVE POINTS OF PERFORMANCE

The five points of performance are specific actions the parachutist performs between the time of exit from the aircraft and the recovery after landing. These points of performance are individual actions and are essential on every parachute jump. Failure to perform any one point correctly could result in a jump injury. They are stressed during jumper training, and each point is taught using one or more of the training apparatuses.

3-1. FIRST POINT OF PERFORMANCE: PROPER EXIT, CHECK BODY POSITION AND COUNT

A proper exit, body position, and count are essential to lessen the possibility of a parachute malfunction/bodily injury during the deployment and inflation of the parachute. The duration of the 4000-count corresponds to the approximate time it takes the main parachute to fully deploy when used by a jumper exiting an aircraft flying 130 knots per hour. The following must be trained reflex actions as the parachutist exits the aircraft:

a. The parachutist starts the 4000-count at ONE THOUSAND and snaps his feet and legs together, locking his knees and pointing his boot toes toward the ground. He lowers his head and places his chin firmly against his chest.

b. At the same time, he rotates his elbows firmly into his sides (with the palms of his hands on the ends of the reserve parachute, fingers spread, and right hand over the rip cord grip), and he bends his body forward at the waist to look over the reserve and to see his boot toes while he continues to count, TWO THOUSAND, THREE THOUSAND, FOUR THOUSAND, at normal cadence. (He keeps his eyes open to react to situations around him.)

WARNING

If no opening shock is felt by the parachutist at the end of the 4000-count, he must activate the reserve parachute as for a total malfunction.

3-2. SECOND POINT OF PERFORMANCE: CHECK CANOPY AND GAIN CANOPY CONTROL

When he finishes the 4000-count, the parachutist feels the parachute open, checks the canopy for malfunction/damage, and controls the parachute.

a. **T-10-Series**. He grasps the risers (thumbs up), spreads the risers apart, and throws his head back to inspect the entire canopy.

b. **MC1-Series**. He throws his head back to inspect the entire canopy and at the same time grasps the control line toggles, with his elbows well back, for immediate canopy control.

c. **Twists**. The main parachute may have twisted suspension lines, risers, or both. This condition may be caused by a single action or a combination of actions. The most common causes include the following:

- The deployment bag spinning before the canopy deploys.
- The canopy spinning when it comes out of the deployment bag and before it inflates.
- The parachutist tumbling or spinning (caused by improper exit and body position) during his descent.

If the suspension lines are twisted and the parachutist cannot raise his head enough to check the canopy properly, he compares his rate of descent with that of nearby parachutists.

(1) ***Rate of descent same as others around him.*** If his descent is the same as other jumpers around him, the parachutist untwists his suspension lines by reaching behind his neck, grasping each pair of risers (thumbs down, knuckles to the rear), and exerting an outward pull on each pair. He kicks his legs in a bicycle motion, continues to pull outward on the risers, and kicks until the twists are out of the suspension lines. When the twists are out of the lines, he checks the canopy and gains canopy control.

(2) ***Partial malfunction and rate of descent too fast.*** If the parachutist's main canopy has a partial malfunction and his descent is too fast (when compared to nearby parachutists), he activates the reserve parachute. (If using the T-10 reserve parachute, he uses the down-and-away method of activation.) See Chapter 6 for more information on activation of the reserve.

(3) ***No comparison can be made.*** When other parachutists are not close enough to compare rates of descent, he activates the reserve parachute (down-and-away method if using the T-10).

3-3. THIRD POINT OF PERFORMANCE: KEEP A SHARP LOOKOUT DURING THE ENTIRE DESCENT

The ability to hit a specific landing spot and to avoid other parachutists during descent is essential to successful airborne operations.

> **WARNING**
>
> **During descent, the parachutist must watch to avoid collisions and entanglements with other parachutists and to avoid obstacles on the DZ. Jumpers stay 25 feet away from other jumpers.**

a. **T-10-Series Parachute**. The degree of maneuverability with a T-10-series parachute is limited compared to the MC1-series. The jumper maneuvers the T-10-series parachute using slips.

(1) ***Use of Slips.*** The parachutist performs slips to avoid other parachutists, to avoid obstacles on the ground, and to prepare to land.

(2) ***Types of Slips.*** The two types of slips are the **two-riser** and the **one-riser**. For an effective slip of either type, the parachutist must ensure his hands are not placed through or behind the riser(s).

(3) ***Execution.*** When slipping, the parachutist looks in the direction that the slip is being made, makes a sharp initial pull to effectively spill air from the canopy, and releases the riser(s) slowly to prevent spinning or oscillations.

(a) Two-riser slip. A two-riser slip is made by reaching up to the elbow-locked position, grasping a pair of risers in the direction of the desired movement, and pulling them down to your chest.

(b) One-riser slip. A one-riser slip is made by pulling down three full arm lengths of the one riser nearest the desired direction of movement with a hand-over-hand motion.

b. **MC1-Series Parachute**. Depending on the wind conditions and his skill, the parachutist can steer his parachute to a selected point of impact on the DZ to avoid other parachutists in the air, to avoid obstacles on the ground, or to use a preferred PLF. To control MC1-series movement, the parachutist must know the principles by which the canopy operates and the factors that govern its control. The movement of the canopy is controlled by the action of the wind, the position of the canopy orifice (opening) relative to wind direction, and the way in which the control lines are manipulated.

WARNING

Due to the canopy's steerability, parachutists should stay at least 50 feet apart in the air to prevent collisions.

(1) ***Opening Away from Wind.*** When the orifice is located directly opposite the wind, the thrust of the orifice will be acting against the wind. This reduces the effect of wind velocity on the canopy and will retard the lateral movement of the canopy in the direction of the wind. This technique is called **holding**.

(2) ***Opening with the Wind.*** When the orifice is located directly with the wind, the thrust of the orifice combines with the thrust of the wind to speed the movement of the canopy in the direction of the wind. This technique is called **running**.

(3) ***Opening at an Angle.*** When the orifice is at an angle to wind direction, then the force of the wind from one direction and the thrust of the orifice at the angle moves the canopy in a direction near to a right angle to the direction of orifice thrust. The direction of movement varies with wind velocity and the angle at which the orifice is pointed. This technique is called **crabbing**.

WARNING

Before attempting any maneuvers, the parachutist must check around him to prevent collisions with other parachutists.

(4) ***Canopy Manipulation.*** Properly executed MC1-series maneuvers require correct canopy manipulation to combine the force of the wind and the thrust of the canopy orifice to move the parachute in a given direction. To maneuver the parachute to a certain point on the ground or to avoid ground obstacles, the parachutist may have to turn and hold into the wind, run with the wind, or crab to the left or right while running or holding.

(a) *Turning*. This is accomplished by pulling down on one control line toggle. The farther down the toggle is pulled, the faster the turn. Pulling the right toggle causes a right turn. Pulling the left toggle causes a left turn. Pulling both at the same time reduces forward speed but increases the rate of descent. This is called **braking**. To deliberately lose altitude quickly, the parachutist pulls down on both toggles. This maneuver should be stopped before he is less than 250 feet above the ground.

(b) *Holding into the Wind.* Holding into the wind is done by rotating the MC1-series canopy until the orifice is on the downwind side. Thereafter, the parachutist manipulates the control line toggles to retain this position.

(c) *Running with the Wind.* Running with the wind is accomplished when the parachutist rotates the canopy until the orifice is on the upwind side. Thereafter, control line toggles are manipulated as needed to retain the position.

WARNING

Running with the wind just prior to landing can cause injury and must be avoided below 100 feet above the ground.

(d) *Crabbing*. Maneuvering while holding into, or running with or at an angle to the wind, is performed by rotating the canopy to the left or right. As the canopy begins to move in the desired direction, the parachutist manipulates the control line toggles to maintain this direction.

(e) *Maneuvering with a Broken Control Line.* If a right or left control line is broken, the canopy can still be maneuvered, though more slowly. The parachutist reaches high on the right or left rear riser, on the same side as the broken control line, and pulls down.

(f) *Maneuvering with an Inversion.* If the canopy has inverted while opening, the parachutist reverses the maneuvering technique. To turn left, the parachutist pulls down on the right control line; to turn right, he pulls down on the left control line.

(g) *Maneuvering with a Tangled Control Line.* If the control line becomes tangled in a suspension line, the parachutist uses the same procedure as with a broken control line.

c. **Collisions and Entanglements**. A collision is the physical impact or contact, however slight, of one parachutist or parachutist's equipment with that of another parachutist. An entanglement is the entwining or attachment of a parachutist or parachutist's equipment with that of another parachutist during descent, whether or not the entanglement lasts until the parachutists contact the ground.

(1) ***Collisions.*** Parachutists must be alert in the air and warn each other of impending collisions. If a collision cannot be avoided by slipping or turning, the parachutist attempts to bounce off the other parachutist's suspension lines or canopy by spreading his arms and legs just before making contact.

(2) ***Entanglements.*** If a parachutist becomes entangled with one or more suspension lines of another parachute, the parachutist does one of the following, depending on the type of parachute being used.

(a) *T-10-Series.* Release entanglements as follows:

- The upper parachutist firmly grasps a portion of the lower parachute and moves hand under hand down the suspension lines of the lower parachute until each parachutist can grasp and hold the main lift web of the other's parachute, being careful not to grip the canopy release assemblies.
- If neither parachutist has a fully inflated canopy, both parachutists will push away from each other and activate their reserves using the pull-drop method.
- When the balls of the feet strike the ground, both parachutists release their grips and make either right, left, or rear PLFs away from each other. No front PLFs will be made.
- With the T-10-series, both jumpers can ride one good canopy to the ground. If both canopies collapse, both jumpers must activate their reserves using the pull-drop method.

(b) *MC1-Series.* Both jumpers remain where they are and activate their reserves for a partial malfunction. When using the MIRPS, jumpers must have a clear path to their front for the spring-assisted deployment device.

d. **Stealing Air**. A descending parachute causes an area of partial air compression immediately below the canopy and an area of partial vacuum and descending turbulent air above the canopy. **This turbulent air extends about 50 feet above the canopy**.

(1) A parachute falling into an area of partial vacuum (from a parachute below) does not capture enough air to stay fully inflated. The top parachute may partially collapse and drop below the other parachutist's canopy until the force of unaffected air reinflates it. Then this canopy, being lower, "steals" the air from the canopy above; this causes the canopy above to partially collapse and the jumper to drop past the lower canopy. This "leap-frogging" action continues unless corrective action is taken by the parachutist. Depending on the type of parachutes involved, the parachutist does one of the following:

(a) *T-10-Series.* He slips vigorously to maintain a lateral distance of at least 25 feet between the parachutes.

(b) *MC1-Series.* He turns in the opposite direction to provide at least a 50-foot distance between the parachutes. (When facing another parachutist, both parachutists execute a right turn.)

(2) When 250 feet or less above the ground, parachutists must exercise care to avoid stealing air from another parachute, because a deflated canopy will not be high enough above the ground to reinflate completely. If this situation occurs, the parachutist immediately prepares to land and to execute a PLF.

3-4. FOURTH POINT OF PERFORMANCE: PREPARE TO LAND

A proper landing attitude is necessary to lessen the risk of injury to the parachutist when he hits the ground (Figure 3-1, page 3-6). The preliminary movements of the parachutist vary, depending on the type of parachute used. However, lowering his individual equipment is the same with either parachute. He lowers the equipment on a lowering line when he is between 200 and 100 feet above the ground.

a. **T-10-Series**. When he is about 100 feet above the ground, the parachutist checks the direction of drift and pulls a two-riser slip into the wind. He holds the risers firmly against his chest and presses his elbows against his body. He keeps his head erect with his eyes on the horizon. He keeps his legs slightly bent and knees unlocked, and he keeps his feet and knees together with the balls of his feet pointed slightly toward the ground. He maintains moderate muscular tension in the legs, which absorb a significant portion of the landing impact, and he avoids becoming stiff or tense.

b. **MC1-Series**. When he is about 100 feet above the ground, the parachutist turns and holds into the wind. When nearing the ground, he holds the control line toggles at eye level. On impact, he holds the toggles, rotates his arms into his body, and executes a PLF.

c. **Obstacles**. The parachutist slips or turns to avoid obstacles. If obstacles (trees, water, or high tension wires) cannot be avoided, the parachutist takes the following precautions.

Figure 3-1. Landing attitude.

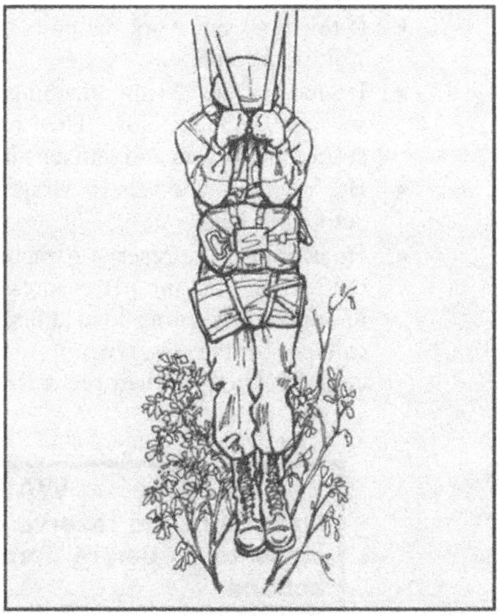

Figure 3-2. Tree landing attitude.

(1) ***Tree Landing.*** Initial precautions taken depend on the type parachute worn.

(a) *T-10-Series.* The parachutist continues to execute a diagonal slip to avoid the trees. Once he sees he cannot avoid them, he immediately assumes a normal prepare-to-land attitude. Just before he makes contact with the trees, he brings his hands up in front of his head and elbows in front of his chest while he continues to grasp the risers (Figure 3-2). He keeps his equipment attached. If he has lowered his equipment before realizing a tree landing is imminent, he looks below prior to jettisoning his equipment and then jettisons the equipment. He continues to watch the ground and trees. He keeps his feet and knees together and prepares to make a good PLF should he pass through the trees.

(b) *MC1-Series.* The parachutist keeps his feet and knees together and toes pointed downward. He continues to control the canopy to avoid the trees if possible. He keeps his equipment attached and wears his ballistic helmet. If he has lowered his equipment before realizing a tree landing is imminent, he looks below prior to jettisoning his equipment and then jettisons the equipment. He continues to watch the ground and trees. Just before impact, he assumes a normal prepare-to-land attitude, but he rotates his arms inward with his elbows high. Upon impact with a tree, the jumper places his hands in front of his face. He must be prepared to execute a parachute landing fall. If he gets hung up in a tree, he takes the following action:

- He reaches up high on both sets of risers and tugs on them three or four times to determine if he is securely hung. He prepares to do a good PLF in case he drops.
- He tries to reach the tree trunk or a large limb to allow him to climb down to the ground.
- If this does not work, he pulls the saddle down and over his buttocks and sits well in the saddle.
- He locates the D-ring attaching straps on his combat equipment and looks to see if it is clear below. Then he pulls down and out on the D-ring attaching straps and lowers and jettisons his combat equipment.
- He releases the chest strap by pulling outward on the ejector snap activating lever.
- He activates the reserve parachute by pulling the rip cord grip. (When using the MIRPS, the jumper ensures his left hand covers the ripcord protector flap to control the spring-loaded deployment assistance device.) He helps feed the canopy of the reserve out to ensure that all of the suspension lines are completely out of their retainers.

WARNING

Make sure the reserve reaches the ground, or is close to it, before continuing with the following actions.

- He activates the quick release in the waistband and frees it from the metal adjuster.
- He unfastens the left connector snap of the reserve from the left D-ring and pushes the reserve behind his right arm.
- He seats himself well into the saddle.
- He wraps his legs around the suspension lines of the reserve parachute and then carefully gets out of the harness.
- With one hand, he grasps the main lift web and holds it firmly.
- With the other hand, he grasps the activating lever of either the left or right leg strap and pulls outward, releasing the leg strap. He releases the other leg strap in the same manner.

- He climbs down the suspension lines and canopy, staying to the outside of the canopy.

(2) ***Water Landing.*** As soon as the parachutist realizes he is going to land in water, he does the following:

- He tries to slip, or steer, away from the water.
- He looks below to be sure the area is clear. Then he jettisons his helmet.
- He releases all equipment tie-downs.
- He looks below to be sure the area is clear. Then, he lowers any attached equipment.
- He activates the waistband quick release.
- He unhooks the left connector snap of the reserve parachute from the D-ring and rotates the reserve parachute to his right side.

(a) *Without a Life Preserver.* When wearing the troop parachute harness and a water landing **without** a life preserver is imminent, the parachutist does the following (Figure 3-3):

- He pulls the saddle well under his buttocks.
- He releases the chest strap by pulling on the activating lever of the ejector snap.
- He makes all possible attempts to remove the pistol belt and all equipment attached to his body that may hinder movement in the water.
- He releases the leg straps ejector snaps when his feet touch the water.
- He prepares to make a PLF in case the water is shallow (2 feet or less in depth).

Figure 3-3. Landing without a life preserver.

Figure 3-4. Landing with a B-7 life preserver.

(b) *With a Life Preserver.* When wearing a troop parachute harness and jumping **with** a B-7 or service approved life preserver (Figure 3-4), the parachutist does the following:

- He activates the B-7 after checking the canopy. If the B-7 fails to inflate, the parachutist inflates the B-7 manually by blowing air into the inflation valve hose. The parachutist activates one canopy release assembly after entering the water and signals "All okay" to the recovery boat.
- He pulls the safety clip out and away from his body (exposing the cable loops) and activates the canopy release assembly using one of the two methods used in the recovery from the drag as his feet touch the water.
- He does not remove the harness and equipment, since the B-7 will support up to 500 pounds.

NOTE: For more information on life preservers, see Chapter 12, Section II.

WARNING
When wearing the B-5, the parachutist does not inflate the B-5 until the parachute harness is removed. If restricted by the harness, the inflation force may crush his ribs.

(3) ***High Tension Wire Landing.*** The parachutist does the following if unable to avoid high tension lines when landing (Figure 3-5):

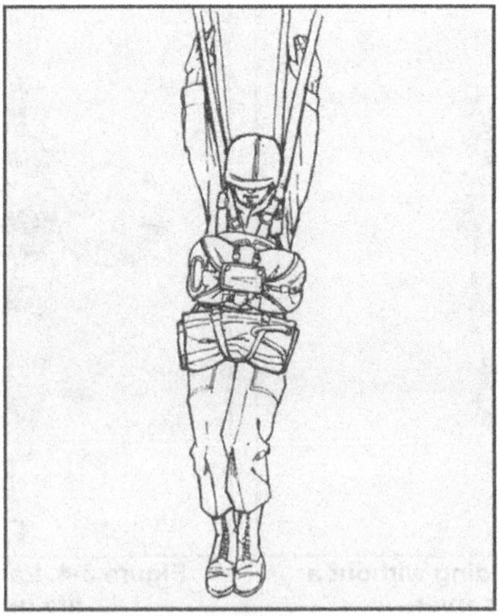

Figure 3-5. Wire landing attitude.

- He tries to slip away from the wires.
- He keeps his feet and knees together and toes pointed downward.
- He looks below and checks for fellow jumpers.
- He jettisons his combat equipment.
- He holds his hands high, inside the front set of risers with palms out and thumbs behind the risers, elbows back, with the fingers extended and joined.
- He keeps his chin on his chest, his body straight, with an exaggerated bend to his knees.
- He prepares to make a normal parachute landing fall.

If he contacts the wires, he begins a rocking motion with his body by pushing forward on the front risers and kicking back with his legs; this may keep him from getting entangled in the wires. He prepares to execute a PLF should he pass through the wires.

NOTE: If the jumper becomes entangled in the wires, he makes no attempt to climb down, but waits to be rescued by a recovery team.

3-5. FIFTH POINT OF PERFORMANCE: LAND

Most jump injuries occur because of improper PLF techniques. To lessen the possibility of injuries, the parachutist is trained to absorb the impact of landing by executing a proper PLF. To do this, the following five fleshy portions of the body must contact the ground in sequence: balls of feet, calf, thigh, buttock, and pull-up muscle(s). The three basic PLFs are **side** (right or left), **front** (right or left), and **rear** (right or left). The type of fall to be made is dictated by the direction of the wind drift. Before the landing attitude is assumed, the parachutist judges the direction of drift by looking at the ground. Then he prepares to make the appropriate PLF.

a. **Side PLF**. As the balls of his feet strike the ground, the parachutist begins several actions at the same time. As the fall continues, he does the following to complete a left-side PLF. (The right-side PLF is similar to the left-side PLF, except the points of contact on the right side of the body are used.)

(1) He lowers his chin firmly to his chest and tenses his neck. He brings his hands up in front of his head and elbows in front of his chest, continuing to grasp the risers (T-10-series) or the toggles (MC1-series). Then he bends and twists his torso sharply to the right. This movement forces the body into an arc. The twisting motion of the hips pushes both knees to the left as the fall continues, and it exposes the second through the fifth points of contact (calf, thigh, buttock, side).

(2) As the PLF is completed in the direction of drift, the parachutist maintains tension in his neck to prevent his head from striking the ground. The momentum caused by drift brings his feet around to the right and into the line of drift. After completing the PLF, he activates the canopy release assembly to keep from being dragged.

b. **Front PLF**. The two types of front falls are **right-front fall** and **left-front fall**. The right-front fall is used if the direction of (wind) drift is slightly to the right. The left-front fall is used if the direction of drift is slightly to the left. If the direction of drift is directly to the front, the parachutist selects either PLF. For a left-front PLF, he rotates from the waist down 45 degrees to his right, exposing his second and third points of contact to the line of drift. Upon contact, he continues to rotate his body to the right, exposing the second, third, fourth, and fifth points of contact. (When executing the right-front PLF, he rotates to the left.)

c. **Rear PLF**. The two variations of the rear PLF are **right-rear PLF** and **left-rear PLF**.

(1) The parachutist determines what PLF to make by checking the direction of drift. If the drift is directly to the rear, he selects the appropriate PLF.

(2) For a left-rear PLF, he rotates from the waist down 45 degrees to his left, exposing the second and third points of contact to the line of drift. Upon contact, he continues to rotate his body and bend his upper torso away in the opposite direction, exposing the second, third, fourth, and fifth points of contact. When executing the right-rear PLF, he rotates to the right.

CHAPTER 4

TRAINING APPARATUSES

The apparatuses described in this chapter are used in basic airborne training. They allow the student to demonstrate proficiency in the tasks necessary to complete the course.

Section I. PARACHUTE LANDING FALL DEVICES

The three types of PLF training devices are the **2-foot high platform**, the **lateral drift apparatus**, and the **swing landing trainer**. The 2-foot high platform may be portable or permanently fixed. A soft landing area of pea gravel or like material is used with all of the training apparatuses. These devices are used to teach front, side (right and left), and rear PLFs. They are high enough to simulate the shock the student will feel when he contacts the ground during parachute jumps. Initial instruction for each PLF is given without using the apparatuses. Once students are familiar with the techniques, they progress to the 2-foot high platform, the lateral drift apparatus, and the swing landing trainer. The swing landing trainer provides a means for gaining forward momentum and simulating the lateral movement experienced during a parachute landing.

4-1. INSTRUCTOR CRITIQUES

Table 4-1, on page 4-3, lists common PLF errors, their causes, and ways to correct them. Instructor PLF critiques should be brief and clear, and should emphasize the following points to students (Figure 4-1, page 4-2).

a. When the balls of your feet make contact with the ground, several actions occur at the same time: Place your chin on your chest, tense your neck muscles, bring your elbows high in front of your face, and expose the second (calf) and third (thigh) points of contact by shifting and bending your knees, maintaining pressure with your opposite knee.

b. Rotate your upper body (from the waist up) around toward the opposite direction of drift. Your body should be contorted in an arc, and the four remaining points of contact should be exposed at this time.

c. Lay the points of contact down on the ground in sequence and complete the fall by bringing your feet up and around your opposite shoulder, completing the fall on your back.

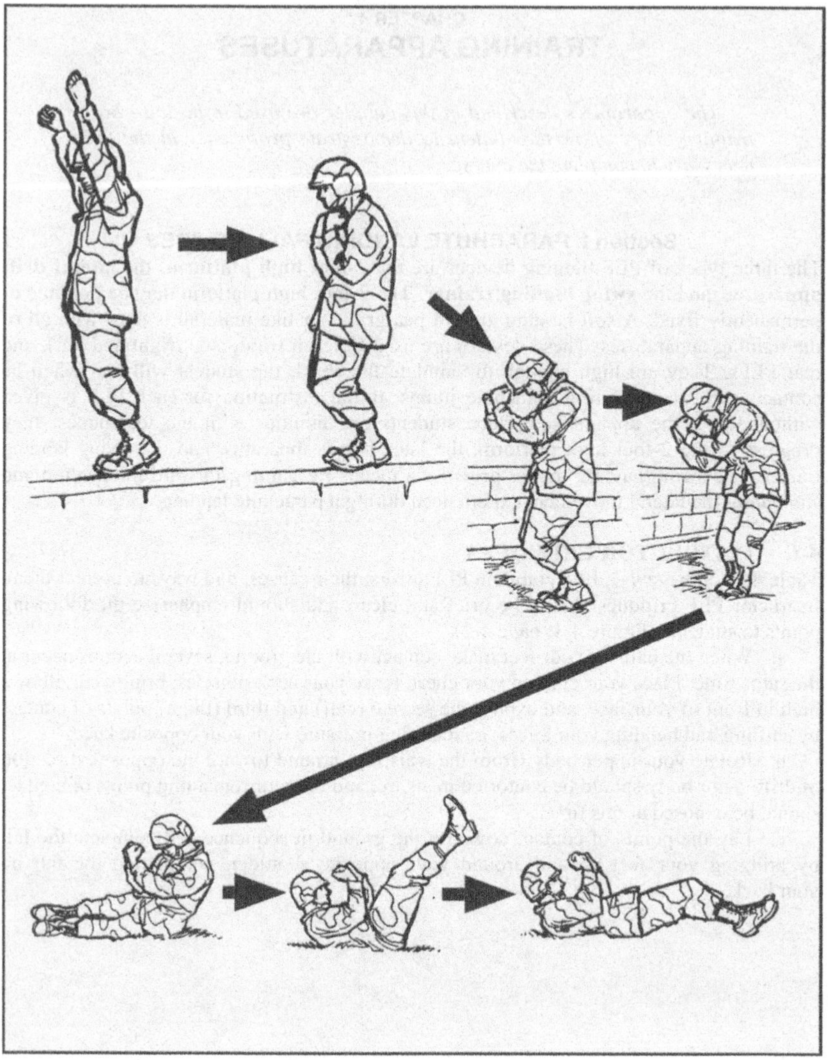

Figure 4-1. PLF sequence.

ERROR	CAUSE	CORRECTION
Feet apart	Anticipation of landing.	Keep moderate muscle tension in the legs. Keep feet and knees together.
Drawing the legs up under the buttocks	Anticipation of landing. Relaxing the knees. Exaggeration of the bend in the knees. Pulling feet up upon landing.	Keep moderate muscle tension in legs. Maintain a proper prepare-to-land attitude, naturally exposing the balls of the feet to the ground.
Missing the second (calf) and third (thigh) points of contact	Feet and knees apart. Straightening the legs. Failure to shift the knees.	Bend the knees slightly. Keep feet and knees together. Shift and bend the knees throughout the fall.
Knees into the ground	Hesitation upon landing. Pushing knees forward toward ground. Excessive bend in knees. Turning the feet toward the direction of drift.	Do not hesitate upon landing. Shift knees to the side. Keep moderate muscle tension in legs. Maintain the proper prepare-to-land attitude.
Elbows hit the ground	Leaning forward toward ground. Failure to rotate upper body away from ground. Breaking fall with elbows.	Rotate upper body away from the ground with elbows up in front of face.
Head strikes the ground	Relaxing the neck. Arching the back during the prepare-to-land attitude. Rotating the upper body into the ground. Missing the points of contact.	Keep chin on chest. Tense neck muscle throughout the PLF. Assume proper prepare-to-land attitude. Rotate upper body away from the ground.

Table 4-1. Common PLF errors.

4-2. TWO-FOOT HIGH PLATFORM

Each platform is divided into dismount points. One instructor controls each point.

a. On the command READY, the student rocks up on the balls of the feet with feet and knees together and knees slightly bent, naturally exposing the balls of the feet to the ground. Arms are skyward, simulating the grasping and spreading of the risers for the T-10-series parachute, or, for the MC1-series, palms are facing forward at eye level with elbows out to the side and back.

b. On the command SLIP (TURN for MC1-series), the student assumes a landing attitude by simulating the grasping of a set of risers opposite the direction of drift and pulling down into his chest with his head and eyes on the horizon, back straight, elbows tight into the side, and feet and knees together. If executing front or rear parachute landing falls, the student rotates his lower body (from the waist down) 45 degrees to expose the second and third points of contact.

c. On the command LAND, the student jumps straight off the platform, executes the PLF, and makes a quick recovery.

d. The instructor critiques each PLF immediately, emphasizing the significance of the five points of contact.

4-3. LATERAL DRIFT APPARATUS

A platoon is distributed among several apparatuses. Each apparatus requires a parachutist, a ropeman, and a safety officer. The safety officer is positioned on the top step of the platform to catch the trolley when it is returned by the ropeman, who is located to the left side at the base of the platform (Figure 4-2). The parachutist mounts the platform, grasps the bar with his palms facing toward his face, and assumes a good landing attitude. On the command CLEAR THE PLATFORM, the parachutist maintains a grasp on the trolley with both hands, picks up his feet, and drifts off the platform. On the command LAND, the parachutist releases the bar and executes the PLF.

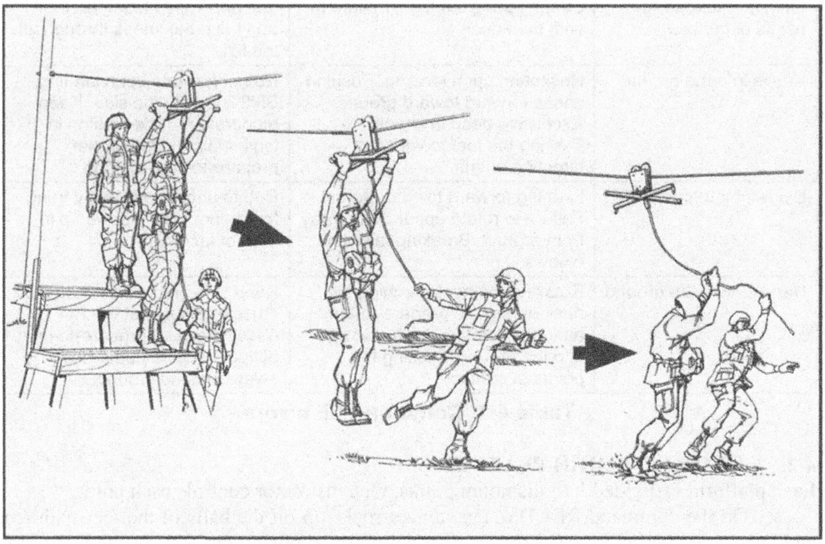

Figure 4-2. Lateral drift apparatus.

4-4. SWING LANDING TRAINER

The swing landing trainer (SLT) apparatus is suspended above a 12-foot-high platform from which students, wearing a parachute harness, descend to practice PLFs (Figure 4-3). The apparatus provides a downward motion and oscillation similar to that experienced during a parachute jump. The suspension is placed so that the student swings when stepping off the platform. Using a control line, the instructor controls the rate of descent. Students receive practical work in two fundamental training objectives: assuming the correct landing attitude (T-10-series and MC1-series) and executing front, side, and rear PLFs.

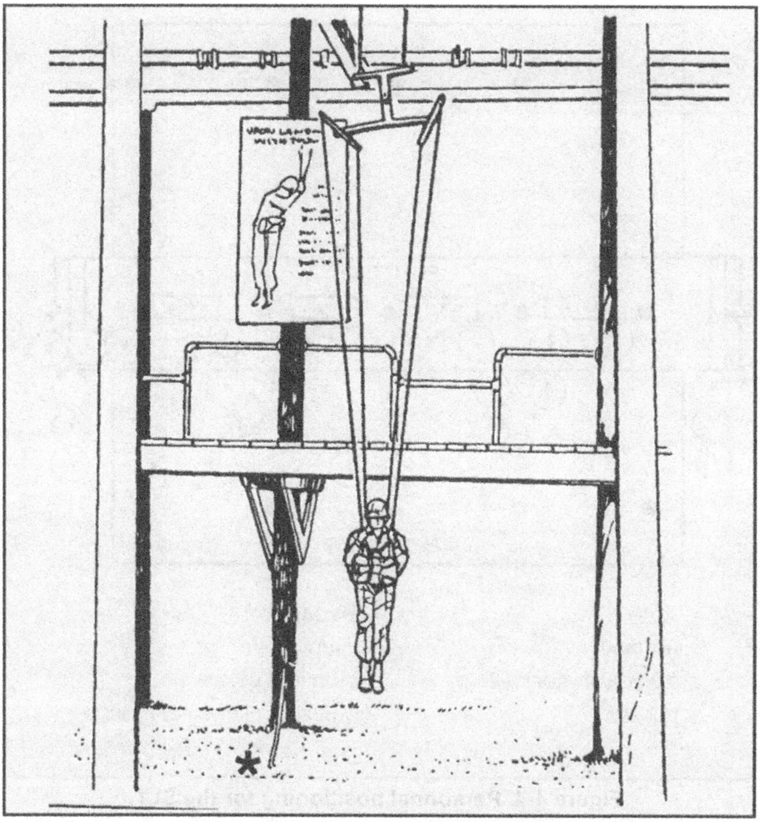

Figure 4-3. Swing landing trainer.

a. **Personnel and Equipment Requirements.** Personnel required to train on the SLT apparatus include one instructor for every two dismount points in use and four detail personnel for each unit (Figure 4-4, page 4-6). Enough harnesses to accommodate the students are also required.

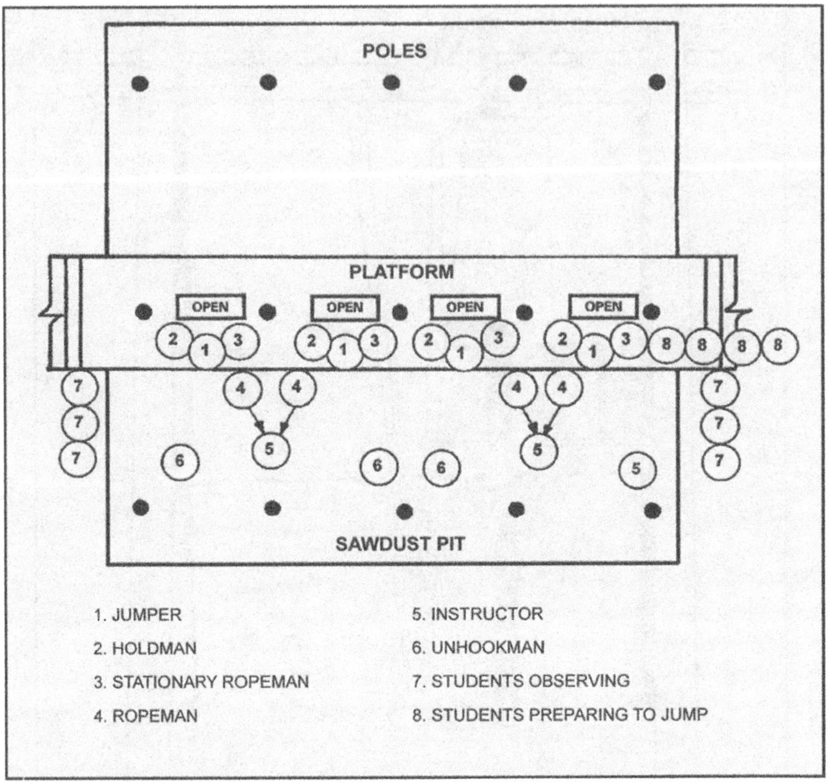

Figure 4-4. Personnel positioning for the SLT.

b. **Training.** The swing landing trainer provides practical exercise in the various PLFs and may be used to practice the last two points of performance. Students must—

- Secure a harness and reserve parachute, and put them on using the buddy system.

NOTE: The instructor inspects students before they mount the steps.

- Move to and mount the stairs at an assigned point. One student covers each open point. If a point is not open, the students wait on the stairs. Students stand on alternate steps of the stairs while waiting.
- Jump from two even or odd points in a section. They do not change sections or switch sides of the apparatus unless directed to by an instructor.

Detail personnel hook up students, who await commands from instructors.

c. **Platform Personnel Duties.** The following detail personnel are needed for training with the SLT.

(1) ***Stationary Ropeman***. The stationary ropeman stands by the stationary rope and—

- Pulls up the stationary rope with risers hooked to it.
- Unhooks the risers from the rope.
- Passes one set of risers to the holdman.
- Hooks one set of risers to one of the jumper's D-rings and butterfly snaps.
- Grasps the horizontal back strap of the parachutist to prevent him from clearing the platform before being told to do so. The stationary ropeman grasps the platform handhold with his free hand.
- Releases the parachutist on the command, CLEAR THE PLATFORM.

(2) ***Holdman.*** The holdman on the platform stands beside the rail padding, allowing space for the parachutist between himself and the stationary ropeman. He—

- Receives one set of risers from the stationary ropeman and removes any twists.
- Hooks the risers to the jumper's D-rings and butterfly snaps.
- Grasps the horizontal back strap of the parachutist to prevent the parachutist from clearing the platform before being told to do so. The holdman grasps the platform handhold with his free hand.
- Releases the parachutist on the command, CLEAR THE PLATFORM.

(3) ***Unhookman.*** The unhookman stands in the pit at parade rest near the outside pole at each point. The unhookman—

- Commands the ropeman, TAKE UP THE SLACK, ROPEMAN.
- Unhooks the parachutist's risers after each PLF.
- Hooks the risers to the stationary rope and returns to his assigned position.

(4) ***Ropeman.*** The ropeman stands beneath the platform and—

- On the command TAKE UP THE SLACK, ROPEMAN, grasps the rope in both hands and moves toward the instructor while taking up the slack.
- Gives the rope to the instructor and sounds off, "Rope, Sergeant"; he then returns to his assigned position.

d. **Instructor Duties.** When ready to drop a student, the instructor takes the rope from the ropeman and tells the parachutist the direction of drift and type of parachute (T-10-series or MC1-series) before commanding CLEAR THE PLATFORM. Then the following occurs:

- The platform detail personnel release the parachutist.
- The parachutist executes a half chin-up on the risers, clears the platform, and assumes the correct landing attitude.
- The instructor lowers the student to the ground.
- At the completion of the PLF, the parachutist activates one canopy release assembly, makes a quick recovery, and reports to the instructor for a critique and grade.
- The parachutist sounds off "Clear" or "Not clear" at the completion of the critique and moves directly out of the pit.
- The detail personnel perform their duties in sequence to prepare another student for the exercise.

4-5. SAFETY CONSIDERATIONS

The following precautions are taken to ensure the student's safety.

- The landing area must be constructed of at least 12 inches of pea gravel or like material.
- The pea gravel must be loosened by raking before each period of instruction.
- The ropes on the apparatus must be checked daily for wear.
- The spreader bars and risers must be checked for wear.
- The harnesses and canopy release assemblies must be checked for completeness and serviceability.
- The student must not be dropped from more than 3 feet.
- The student must not be dropped on the initial oscillation or when unprepared for the PLF.
- The student must be dropped at a point in oscillation that aids in executing the desired PLF.
- Any student with a prior head injury will have the letter H on his helmet to allow close instructor monitoring.

Section II. MOCK DOOR

The mock door is a replica of the cargo/troop compartment of a troop carrier aircraft. This apparatus includes openings about the size of the aircraft door and anchor line cables for each door (Figure 4-5). For training purposes, the instruction is divided into a basic phase and an advanced phase. The basic phase teaches basic jump techniques and familiarizes students with equipment, aircraft terms, and safety procedures. The advanced phase provides instruction in the sequence of jump commands and the mass exit technique.

Figure 4-5. Mock door apparatus.

4-6. BASIC PHASE

Students are taught the terms and nomenclature of military parachuting and the use of aircraft equipment. Teaching objectives include the shuffle position, static line control, the STAND BY position, the exit and body positions, and the 4000-count.

a. **Personnel and Equipment Requirements**. One instructor is required for each mock door in use. One section of static line (about 4 feet long and not attached to the parachute assembly) is required for each student. Loudspeakers will be needed if the class or facilities are large.

b. **Training**. The initial instruction in the mock door apparatus includes a lecture and demonstration followed by practical exercises.

(1) Each student is given the commands STAND BY and GO. The student is critiqued and corrected by the instructor on the movement to the door, his exit, his body position, and the 4000-count.

(2) Repetition is the key to this training; however, its value is decreased if allowed to become boring. Correct and automatic reactions by each student is the goal.

(3) The "Hit It" exercise is given as a test of mental alertness and for practical work in assuming the proper body position. On the command HIT IT, the student—

- Snaps into the proper body position and at the same time commences the 4000-count.
- Remains in the body position until commanded to RECOVER, or until he is told to CHECK CANOPY and GAIN CANOPY CONTROL. The student simulates checking the canopy. Then he is told to RECOVER or that there is a MALFUNCTION. In the latter case, the student returns to a modified body position and simulates activation of the reserve. The command RECOVER is given.
- Receives instruction on the right and left jump doors of the mockup.

c. **Terms**. The following terms are important in mock door training and are explained during the initial phase of instruction.

(1) ***Left and Right Door.*** When the parachutist is facing the pilot's compartment, the door on his right is the right door; the door to his left is the left door.

(2) ***Anchor Line Cable.*** A cable is normally extended along the long axis of the cargo/troop compartment and secured at both ends. The cable is designed to accommodate the static line snap hook and to initiate parachute deployment.

(3) ***Stick of Parachutists.*** This is a group of parachutists exiting from the same door (or from one side of a ramp) during one pass over the DZ.

(4) ***Drop Zone.*** This is a designated area where personnel or equipment are delivered by means of parachute or free drop. The GUC designates the DZ location.

(5) ***Shuffle Position.*** This is a method of moving toward the jump door, used to avoid losing balance or tripping (Figure 4-6, page 4-10).

(a) The guide hand is extended down and out to maintain balance; the other hand grasps the static line. When jumping the left door, the parachutist has the static line over the left shoulder; when jumping the right door, he has the static line over the right shoulder.

(b) Facing the rear of the aircraft, the parachutist keeps both feet directly beneath the body and staggered with the outboard foot forward; this is the shuffle foot. The inboard foot is the trail foot.

(c) The parachutist moves by stepping forward 6 to 8 inches with his shuffle foot and then his trail foot. Both feet are staggered in the same heel-and-toe position.

Figure 4-6. Shuffle position.

(6) ***Bight.*** The parachutist forms a bight of about 6 inches in the static line by making one fold and grasping the loop at eye level about 6 inches to the front (Figure 4-7). The remainder of the static line is routed over the shoulder. The free hand is used to steady the parachutist while moving toward the door.

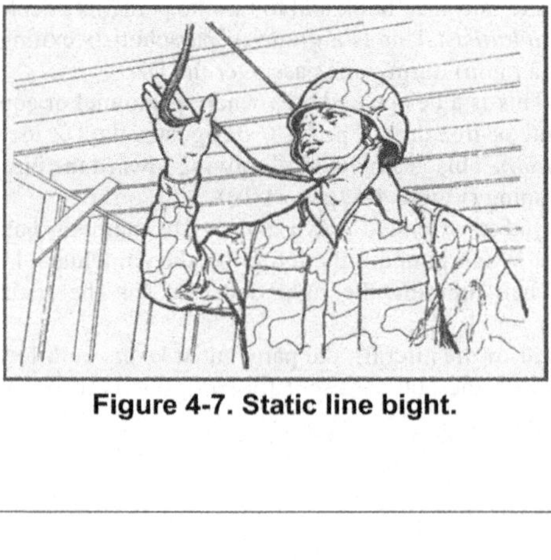

Figure 4-7. Static line bight.

(7) ***Jump Commands.*** The last two jump commands, STAND BY and GO, are used for each student when practicing exits.

(a) On the command STAND BY, the parachutists shuffle toward the jump door.

- When the first jumper is perpendicular to the jump door, he takes one more shuffle step and halts his movement about 2 feet from the center of the jump platform. He keeps his feet spread and legs slightly flexed so that his weight is equally distributed over both feet to maintain balance.
- He makes eye-to-eye contact with the safety and hands the static line to the safety.
- He executes a left or right turn to face the open jump door, ensures his arm is not entangled with the static line, holds his elbows firmly into his sides, and places the palms of his hands (fingers spread) on the ends of the reserve parachute with his right hand protecting the rip cord grip.
- At the command STAND BY, the number 2 jumper will be positioned about even with the leading edge of the jump door, 2 feet from the skin of the aircraft and facing to the rear, in the shuffle position with his feet spread and legs slightly flexed so that the weight is equally distributed over both feet to maintain balance.
- Follow-on jumpers close up behind the preceding jumper and keep the shuffle position with the feet spread and legs slightly flexed so that their weight is equally distributed over both feet to maintain balance.

(b) At the command GO, the number 1 jumper walks toward the door and onto the jump platform, focusing on the horizon. He pushes off with either foot and **vigorously** jumps **up** and **out** away from the jump platform, immediately snapping into a good tight body position.

(c) The number 2 jumper (and all following jumpers) performs the following:

- He shuffles toward the jump door, ensuring he is about 2 feet from the skin of the aircraft.
- As he begins to shuffle, he assumes an elbow-locked position with the arm that is controlling his static line. He places his static line control hand so that it is nearly touching the back of the pack tray of the jumper in front of him, which establishes the proper interval between jumpers. He does **not** place his static line control hand in a position so that it extends past the pack tray of the jumper in front of him.
- When about perpendicular to the jump door, he takes one more shuffle step, makes eye-to-eye contact with the safety, and hands the static line to the safety.
- He ensures his arm is not entangled with the static line, and he holds his elbows firmly into his sides.
- He places the palms of his hands (fingers spread) on the ends of the reserve parachute, with his right hand protecting the rip cord grip.
- He executes a left or right turn to face the open jump door.
- He walks toward the door and onto the jump platform, focusing on the horizon. He pushes off with either foot and **vigorously** jumps **up** and **out** away from the jump platform, immediately snapping into a good tight body position.

(8) ***Body Position.*** The student remains in the body position, is critiqued by the instructor, and is told to RECOVER and wait for further instruction.

4-7. ADVANCED PHASE

The training provided during the advanced phase is presented in the same manner as in the basic phase except that the entire sequence of time warnings and jump commands is given, and mass exits are substituted for individual exits.

a. **Personnel**. Extra instructors may be needed to ensure that all students in the mock door apparatus react properly to each of the jump commands.

b. **Execution**. When the mass exit technique is taught, each stick receives the commands STAND BY and GO. Each succeeding student shuffles to the door and exits the aircraft. A one-second interval must be maintained between students.

Section III. SUSPENDED HARNESS

The suspended parachute harness apparatus is a modified troop parachute harness suspended from a spreader bar assembly by four web risers (Figure 4-8). The spreader bars react to riser (T-10-series) or toggle (MC1-series) manipulation much the same as the canopy. The suspended harness simulates the third and fourth points of performance: canopy control and prepare to land.

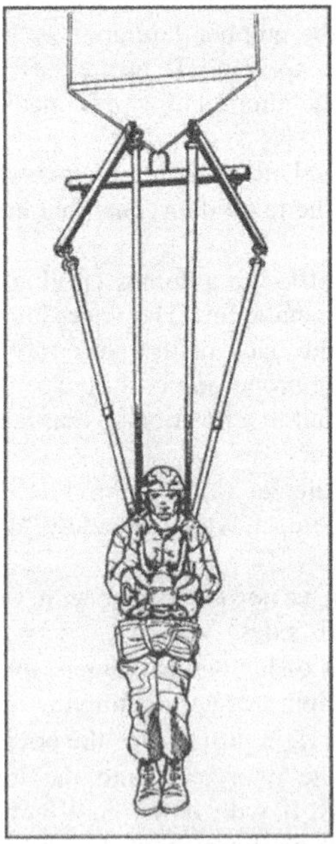

Figure 4-8. Suspended harness.

4-8. OBJECTIVES

The suspended harness apparatus is designed to teach the student to execute the following maneuvers:

- T-10-series slips (two-riser and one-riser).
- MC1-series turning, holding, running, and crabbing.
- T-10-/MC1-series landing attitude.
- T-10-/MC1-series emergency landings (tree, water, and wire).
- React to twists, collisions, and entanglements.

4-9. PERSONNEL AND EQUIPMENT REQUIREMENTS

The following personnel and equipment are needed to conduct training on the suspended harness.

a. One instructor is needed to give commands, and at least one assistant instructor is needed for each platform to control and supervise student performance. One troop harness is required for every two students. A modified MC1-series riser assembly is used, which facilitates training on a simulated T-10-series or MC1-series parachute.

b. A platoon is divided into teams of two students each. Number 1 is the parachutist, and number 2 is the coach.

(1) The parachutist obtains a troop parachute harness and moves to a specific point at the suspended harness apparatus.

(2) The coach mounts the platform and adjusts the risers so that the male fitting of the canopy release assembly is level with the parachutist's shoulders below. The coach dismounts the platform and helps the parachutist don the harness and attach it to the risers. The coach remounts the platform and takes up all the slack in the parachutist's risers.

(3) All parachutists face the instructor and await commands.

4-10. SEQUENCE OF COMMANDS

The instructor uses the following sequence of commands. (The let-up position is the starting position for all training on the suspended harness.)

a. **LET UP**. On this command, each parachutist:

- Has his head erect and his eyes on the horizon.
- Grasps all four risers (T-10-series) with his hands and locks his elbows.
- Grasps the steering toggles (MC1-series) with his hands (palms outward, eye level, elbows back) and applies moderate tension (enough to take out any slack in the control lines).
- Has his back straight.
- Has his feet and knees together. (The parachutist is standing flat-footed on the ground.)

b. **AT EASE IN THE HARNESS**. On this command, each parachutist comes to a modified position of parade rest in the harness. From the let-up position, the instructor directs the parachutist to practice appropriate parachute maneuvers.

c. **CHANGE OVER**. Number 2 switches places with number 1 and becomes the parachutist; number 1 becomes the coach. Instructors may give additional commands and instruction to ensure proper fitting, wearing, or removal of equipment and proper student performance on the apparatus.

Section IV. THE 34-FOOT TOWER

The 34-foot tower supports a replica of a section of a troop carrier aircraft (Figure 4-9). A jump door is on each side of the replica. Four steel cables are suspended parallel to the ground and slightly above each door. A trolley, which supports tow trolley risers, is attached to each cable. Each trolley riser has a ring attached to its free end. This ring is connected to a modified harness worn by students during training on the tower. The tower is a primary training apparatus to help teach basic jump techniques and points of performance. For ease in training, the instruction is divided into two phases: The **basic** training phase continues instruction presented on the mock door apparatus. The **advanced** training phase continues instruction on the mass exit technique, simulates parachute malfunctions, and familiarizes the students with jumping combat equipment.

Figure 4-9. The 34-foot tower.

4-11. BASIC TRAINING OBJECTIVES

Students practice the exit technique, the proper body position, and the 4000-count. The tower also gives each student experience to overcome his fear of height, and it simulates the opening shock.

a. **Training**. The initial period of instruction on the tower apparatus includes an orientation and demonstration of the apparatus, duties of instructor personnel (graders, JMs, harness checker), and duties of all detail personnel. Each student is given the commands STAND BY and GO by the JM in the tower. The student's exit, body position, 4000-count, and simulated canopy checks are critiqued and graded by the grader on the ground.

b. **Safety**. To prevent safety problems, all personnel must be safety conscious. The following measures are taken to prevent unsafe conditions:

(1) Tower JMs must wear safety belts.

(2) Students must have equipment adjusted properly, and instructors must check them to ensure a snug fit, the absence of frays or tears, the serviceability of snaps, and a properly fitted helmet.

(3) Qualified maintenance personnel must perform and record thorough inspections of the tower weekly. Inspections must include trolley risers, cables, trolleys, and JM safety belts.

(4) Students must lift their legs as they approach the mound at the end of the cable run to prevent injury.

4-12. PERSONNEL AND EQUIPMENT REQUIREMENTS

The following personnel manage, operate, and carry out 34-foot tower training.

a. **Instructors**. A minimum of seven instructors are required to operate this apparatus. Their positions are indicated in Figure 4-10 (page 4-16).

(1) Two JMs hook up students in the tower and issue the last two jump commands.

(2) One harness checker inspects all harnesses before students exit the tower.

(3) Two graders on the ground critique and grade the students' performance on the apparatus.

(4) Two mock door instructors correct students on problem areas and give them additional practice.

b. **Detail Personnel**. A minimum of 38 detail personnel are required.

(1) ***Mound Safety Officers/NCOs.*** A student stands at parade rest on top of the mound between and slightly to the rear of the moundmen assigned to his side of the tower. He is responsible for the safety and conduct of the moundmen. He should be the most senior officer or NCO on the mound on each side (2 total).

(2) ***Moundmen.*** Two students are assigned to each cable on the tower (16 total). They stand at parade rest on top of the mound, facing the tower. After the parachutists reach the mound and come into contact with the stop cable, the moundmen unhook the parachutist and then hook the rope to the trolley risers.

(3) ***Ropemen.*** One student at each point along the beaded path returns the trolley risers (with rope attached) to the tower after the parachutist is detached. The ropemen take commands from the ropeman safety and rope line safety/relay. The ropemen for each

two points rotate, and there are always two additional ropemen standing at the base of the pole supporting the tower (12 total).

(4) ***Riser Safeties.*** One student stands at parade rest off each side of the mound and outside of the path leading to the grader's booth (2 total). He is responsible for attaching the butterfly snaps on the risers to the D-rings of the harness.

(5) ***Ropeman Safeties.*** One ropeman safety is located between points 2 and 3 and the other between points 6 and 7 on the beaded path (2 total). They are responsible for controlling the flow of the ropemen from the mound to the tower.

(6) ***Rope Line Safeties/Relays.*** One student (officer or NCO) for each side of the tower (2 total) stands at parade rest near the base of the pole supporting the cables facing the rope line. He relays the commands of the jumpmaster to the ropemen and ensures students do not pass under the safety ropes and anchor cables.

(7) ***Base Safety Officers/NCOs.*** One student (officer or NCO) for each side of the tower (2 total) stands inside the base of the tower next to the stairs. He places the risers over the proper shoulder for the door (right or left) each parachutist will jump.

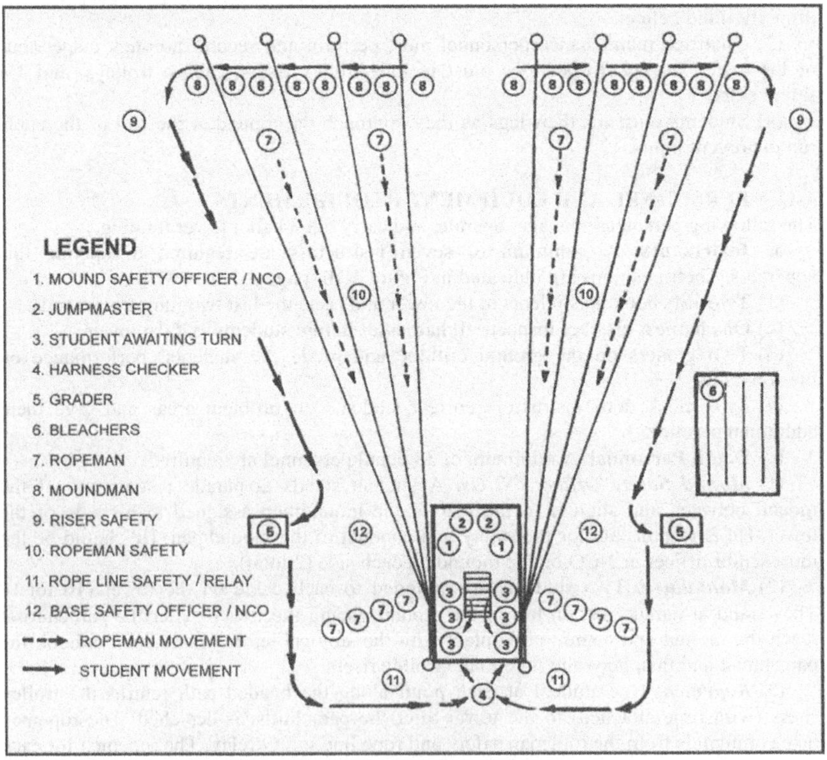

Figure 4-10. Tower personnel positions.

c. **Equipment**. The equipment includes one troop parachute harness assembly (with four risers) and a training reserve parachute for each student.

4-13. ADVANCED TRAINING OBJECTIVES

These objectives have the student practice exits using the mass exit technique with and without combat equipment. If exiting with combat equipment, he must also practice lowering the equipment. The student must practice simulating activation of the reserve parachute. He must do this instantly when told a malfunction has occurred.

a. **Training**. The initial period of instruction during this phase includes a lecture and demonstration on the mass exit technique.

b. **Common Student Errors**. The following errors are often made by students.

- Failing to maintain a 1-second interval between jumpers.
- Using improper body position (caused by rushing).
- Improper exit.
- Falling out of the door.
- Failing to count.

c. **Safety**. Safety considerations for the advanced phase are the same as in the basic phase.

d. **Personnel**. Personnel requirements are reduced slightly from the basic phase to a minimum of 29: 16 moundmen, 8 ropemen, 2 mound safety officers, 2 rope line safety officers, and a harness checker.

e. **Equipment**. Combat equipment is required for the students. (Actual packed combat equipment containers, or combat equipment containers which approximate packed loads, can be used to train students in the techniques of exiting with this type of load.)

Section V. METHODS OF RECOVERY

During methods of recovery training, the jumper learns the steps he must take to consolidate with his unit as a member of the fighting force. Once the jumper lands on the drop zone, he must quickly recover, correctly stow his air items, and quickly proceed to the unit assembly area.

4-14. TRAINING OBJECTIVES

This phase of training teaches the student to properly activate the canopy release assemblies on the parachute harnesses. He learns to perform the buddy-assist method of recovery and to react properly when using the various recovery training apparatuses. He is also taught how to recover the parachute and individual equipment from the DZ and to rapidly clear the DZ.

4-15. PERSONNEL AND EQUIPMENT REQUIREMENTS

Personnel required are one instructor, one assistant for every two dismount points, and four detail personnel for each dismount point. Sufficient harnesses and combat gear properly rigged to accommodate each student are also needed.

4-16. TRAINING APPARATUS

The hand-towed drag pad training apparatus is used in teaching students how to activate the canopy release assemblies. The hand-towed drag pad is made of two pieces of webbing attached to a metal bar with a loop on each end. Attached to the bar are two risers with the male fitting of the canopy release assemblies attached to the opposite ends (Figure 4-11). Students train in three-man teams as follows:

- Number 1 is the parachutist and wears the drag pad.
- Number 2 and number 3 pull the drag pad.

Figure 4-11. Hand-towed drag pad.

- Number 1 puts on the harness and passes the risers to numbers 2 and 3. Numbers 2 and 3 then attach the risers to the canopy release assemblies, and number 1 lies on his back.
- On the command PREPARE TO DRAG, number 1 places his chin on his chest, raises his feet 6 inches off the ground, and places his hands on the canopy release assemblies.
- On the command DRAG, numbers 2 and 3 drag number 1 across the ground.
- On the command RELEASE, number 1 reaches down and activates the canopy release assemblies using the prescribed method.
- The three students immediately return to the starting line.
- Numbers 2 and 3 hook up the parachutist to the drag pad again.
- At the command CHANGE OVER, number 2 becomes the parachutist (number 1), number 3 becomes number 2, and number 1 becomes number 3. Students rotate numbered positions until each student becomes proficient in the procedures.

4-17. CANOPY RELEASE ASSEMBLIES

Before the canopy release assemblies can be activated, the safety clip must be pulled down to expose the cable loop. There are two ways to activate the canopy release assembly. One is the hand-to-shoulder method and the other is the hand-assist method.

4-18. CANOPY RELEASE ASSEMBLY ACTIVATION

The instructor tells the student to pull the safety clip out and away from the body. The cable loop release does not require a great deal of strength to activate, and, if the parachutist is injured, the cable loop release can be easily activated by the thumb or fingers of either hand (Figure 4-12).

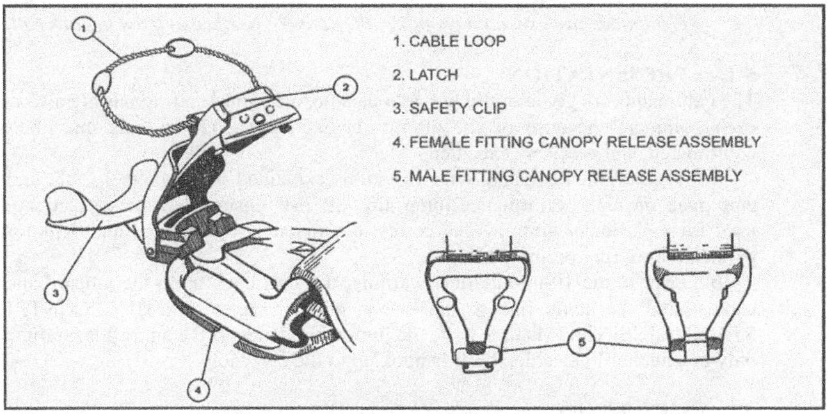

Figure 4-12. Cable loop release.

4-19. JUMP REFRESHER TRAINING

Before making a parachute jump, students make several refresher jumps from platforms. They are required to make a satisfactory PLF in each of the principal directions before they leave the platform area. (A detailed description of refresher training is in Appendix A.)

CHAPTER 5

JUMP COMMAND SEQUENCE AND JUMPER ACTIONS

The JM gives a sequence of nine jump commands to ensure positive control of parachutists inside the aircraft and immediately before exiting. Every command requires specific actions by each parachutist. When commands are executed properly, they ensure a safe exit from the aircraft.

5-1. PRESENTATION

The commands are given orally but, as a backup, arm-and-hand signals are also used with each command because of the aircraft engine noise. The signals must be smooth, coordinated, and precisely executed.

a. The commands listed, with variations explained in Chapters 17, 18, and 19, are employed on high-performance jump aircraft. JMs ensure that the correct sequence is used for a particular aircraft. The correct commands are explained and demonstrated to parachutists during prejump briefing.

b. Prior to the 10-minute time warning, the JMs hook up to the inboard anchor line cable, hand the static line to the safety, and announce, SAFETY, CONTROL MY STATIC LINE. The JMs then issue the jump commands. If the aircraft is configured with only one anchor line cable, the JMs hook up to the one cable.

5-2. GET READY

GET READY is the first jump command.

a. **Command.** This jump command alerts the parachutists seated in the aircraft and directs their complete attention to the JM.

(1) The JM starts the command with his arms at his sides and gives the arm-and-hand signal by extending both arms to the front at shoulder level with his palms facing the parachutists.

(2) He begins at shoulder level, fingers and thumbs extended and joined, palms facing toward the parachutists. He extends both arms forward until the elbows lock, with the palms toward the parachutists. He gives the oral command GET READY, then returns to the start position with arms at the sides (Figure 5-1).

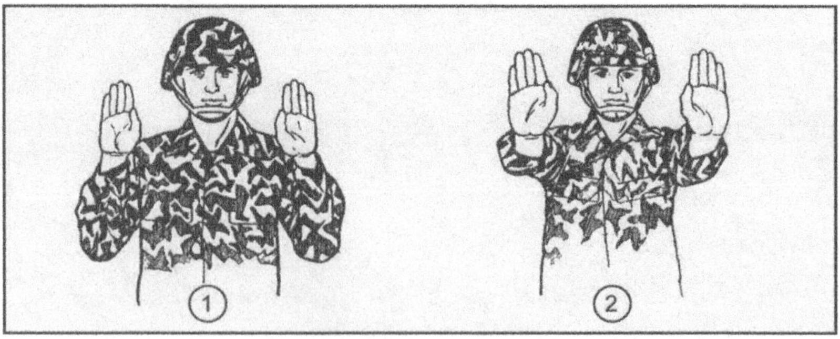

Figure 5-1. GET READY.

b. **Static Line**. The static line is over the appropriate shoulder and fastened to the top carrying handle of the reserve parachute. Parachutists do not remove the static line snap hook from the reserve parachute after the JM inspection or anytime before the command HOOK UP.

c. **Jumper Actions**. Each parachutist signifies alertness by leaning forward and placing both hands on his knees. Each parachutist positions his foot nearest the jump door under the seat and places his foot nearest the pilot's compartment in the aisle.

5-3. OUTBOARD PERSONNEL, STAND UP

OUTBOARD PERSONNEL, STAND UP is the second jump command. For this command, the arm-and-hand signal has two parts.

Part One. The JM starts at the shoulders, index and middle fingers extended and joined, with remaining fingers and thumbs curled to the palms. He gives the command OUTBOARD PERSONNEL, lowers the arms down to the sides at a 45-degree angle, and locks the elbows.

Part Two. The JM gives the command STAND UP. He extends and joins the fingers and thumb of each hand, rotates the hands so the palms face up, and then raises the arms straight overhead, keeping the elbows locked (Figure 5-2). At this command, parachutists sitting nearest the outboard side of the aircraft stand up, raise and secure the seats, face the jump doors, and assume the shuffle position.

NOTE: The method of releasing the seats from the floor varies, depending on the model and year of aircraft. Before takeoff, these devices are inspected and the method of release explained.

Figure 5-2. OUTBOARD PERSONNEL, STAND UP.

5-4. INBOARD PERSONNEL, STAND UP

INBOARD PERSONNEL, STAND UP is the third jump command. The arm-and-hand signal has two parts.

Part One. The JM starts with the hands centered on the chest at shoulder level, index and middle fingers extended and joined, remaining fingers and thumbs curled to the palms. He gives the command INBOARD PERSONNEL, extends the arms forward at a 45-degree angle, toward the inboard seats, and locks the elbows.

Part Two. The JM gives the command STAND UP. He first rotates his arms to the sides and down at a 45-degree angle. Then he extends and joins the fingers and thumb of each hand, rotates his hands so the palms face up, and raises his arms straight overhead, keeping the elbows locked (Figure 5-3).

The parachutists seated inboard react in the same manner as the outboard personnel described in the previous paragraph.

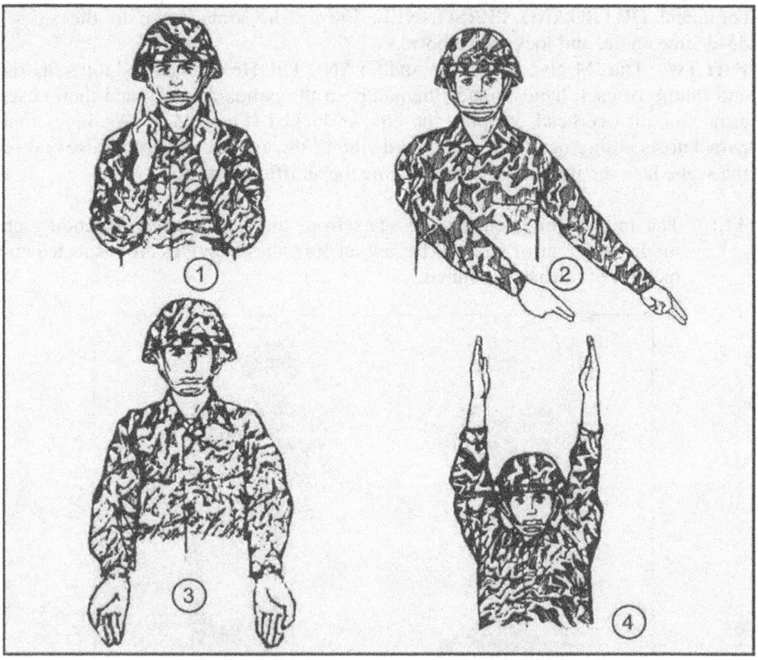

Figure 5-3. INBOARD PERSONNEL, STAND UP.

5-5. HOOK UP

HOOK UP is the fourth jump command.

a. **Command.**

(1) The JM begins with his arms either extended directly overhead with elbows locked or with arms bent, hands at shoulder level.

(2) He forms a hook with the index finger of each hand. He forms fists with the remaining fingers and thumb of each hand (Figure 5-4).

(3) As he gives the oral command, he moves his arms down and up in a pumping motion. He repeats the arm-and-hand signal at least three times.

Figure 5-4. HOOK UP.

b. **Jumper Actions.**

(1) At this command, each parachutist detaches the static line snap hook from the top carrying handle of the reserve parachute and hooks up to the appropriate anchor line cable, with the open portion of the snap hook toward the outboard side of the aircraft. Each parachutist must ensure that the snap hook locks properly.

(2) The safety wire is inserted in the hole and folded down. To protect the eyes, the wire is inserted by pointing it toward the rear of the aircraft. Then a bight is formed in the static line and held at eye level. The bight is not released until the parachutist moves into the door.

(3) Personnel jumping the left (right) door have the static line over the left (right) shoulder.

5-6. CHECK STATIC LINES

CHECK STATIC LINES is the fifth jump command.

a. **Command.**

(1) This is a plural command since there are several static lines attached to the anchor line cable. It begins at eye level, with the thumb and index finger of each hand forming an "O."

(2) The JM extends and joins his remaining fingers with the palms facing in. As he gives the oral command, he extends his arms to the front until the elbows are nearly locked, then returns to the starting position.

(3) He repeats the arm-and-hand signal at least three times, ensuring the knife edge of his hands are toward the parachutists and the palms face each other (Figure 5-5).

b. **Jumper Actions.**

(1) Upon receiving this command, each parachutist checks his static line and the static line of the parachutist to the front.

(2) Each parachutist checks visually and by feeling with his free hand. He does not release the bight for checks. He verifies the following items:

- Static line snap hook is properly attached to the anchor line cable with the safety wire properly inserted. Static line is free of frays and tears.
- Static line is not misrouted and is properly stowed on pack tray.
- All excess slack in the static line is taken up and stowed in the static line slack retainer.
- Pack closing tie is routed through the pack opening loop.
- Pack tray is intact.

NOTE: The last two jumpers in each stick face about. The next to last jumper inspects the last jumper's static line and gives him a sharp tap to indicate that the static line and pack tray have been checked and are safe for jumping.

(3) Each parachutist gives the parachutist to the front a sharp tap signifying that the static line and pack tray have been checked and are safe for jumping.

Figure 5-5. CHECK STATIC LINES.

5-7. CHECK EQUIPMENT

CHECK EQUIPMENT is the sixth jump command.

a. **Command.**

(1) The JM starts this arm-and-hand signal with the fingertips centered on his chest, palms facing the chest, and fingers and thumb of each hand extended and joined; or with his arms extended to the sides at shoulder level, fingers and thumbs extended and joined, and palms facing toward the parachutist.

(2) He gives the oral command, extends his arms to the sides at shoulder level, and then returns them to the chest; or bends his arms at the elbows, bringing the fingertips to the center of the chest, and then returns to the extended position (Figure 5-6).

(3) He repeats the arm-and-hand signal at least three times. (The JM must check his own equipment.)

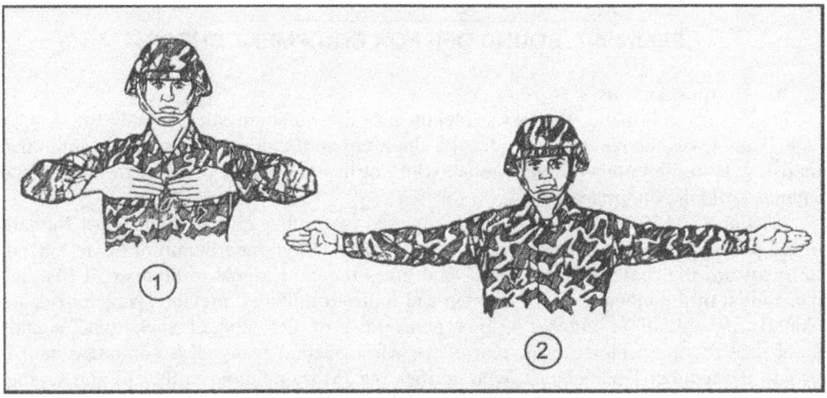

Figure 5-6. CHECK EQUIPMENT.

b. **Jumper Actions.**

(1) At this command, each parachutist checks his equipment, starting at the helmet, and ensures there are no sharp edges on the rim of the ballistic helmet and that the chin strap and parachutist retention straps are properly routed and secured. The parachutist then physically seats the activating lever of the chest strap ejector snap and the leg strap ejector snaps. If jumping combat equipment, the parachutist also ensures the ejector snap of the HPT lowering line is properly attached and seated.

(2) The parachutist completes these actions with the free hand while maintaining a firm grip on the static line bight with the other hand.

5-8. SOUND OFF FOR EQUIPMENT CHECK

This is the seventh jump command.

a. **Command.**

(1) The JM cups his hands and places the thumbs behind the ears.

(2) He gives the oral command SOUND OFF FOR EQUIPMENT CHECK (Figure 5-7).

Figure 5-7. SOUND OFF FOR EQUIPMENT CHECK.

b. **Jumper Actions**.

(1) At this command, the last parachutist in the outboard stick sounds off, saying "OK," and gives the parachutist in front a sharp tap on the thigh. The signal is continued until it gets to the number 1 parachutist, who notifies the JM by pointing to the JM and saying, "All OK, jumpmaster."

(2) For a C-130 aircraft, this signal is passed to number 25 parachutist (just forward of the wheel well), who forms a circle with his index finger and thumb of his free hand, turns toward the center of the aircraft, and gives the okay signal to number 24 (the last parachutist of the inboard stick). The tap and indication that all previous parachutists are okay is passed up to number 4, first parachutist of the inboard stick, who signals number 3, first parachutist to the rear of the wheel well. The signal is continued until it gets to the number 1 parachutist, who notifies the JM by pointing to the JM and saying, "All OK."

(3) A parachutist who has an equipment problem notifies the JM, AJM, or safety personnel by raising his outboard hand high above the anchor line cable, palm facing the JM. The parachutists do not pass this signal. The JM, AJM, or safety either corrects the deficiency or removes the parachutist from the stick.

NOTE: After the JM receives "All OK, jumpmaster," he regains control of his static line from the safety and takes the number 1 parachutist position.

5-9. STAND BY

STAND BY is the eighth jump command. This command (Figure 5-8, page 5-8) is given about 10 seconds before the aircraft reaches the release point and only after the aircraft has cleared all obstacles near the DZ.

a. **Command**.

(1) Starting at the shoulders, the JM extends and joins his index and middle fingers, curling the remaining fingers and thumb of each hand toward the palm.

(2) He extends his arms down to the sides at a 45-degree angle by locking the elbows, and points to both doors at the same time.

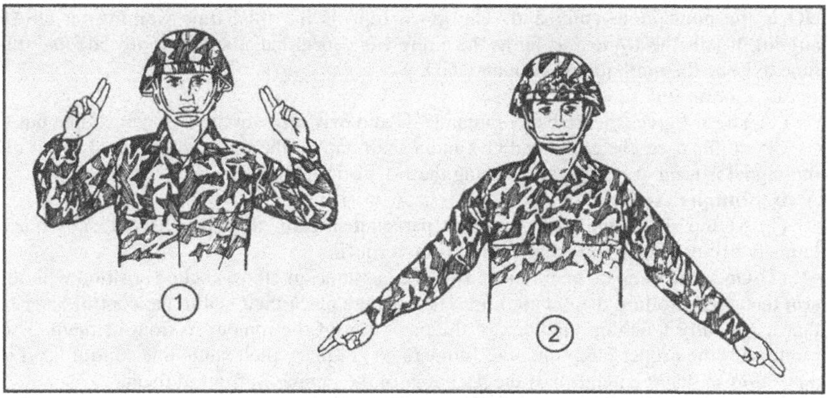

Figure 5-8. STAND BY.

b. **Jumper Actions**.

(1) At this command, parachutist number 1 shuffles toward the door, establishes eye-to-eye contact with the safety, hands the safety his static line, holds his elbows firmly into his sides with his palms on the end of the reserve, turns and centers himself in the open jump door, and awaits the command GO.

(2) All following parachutists maintain the static line bight and close up behind the preceding parachutist (Figure 5-9).

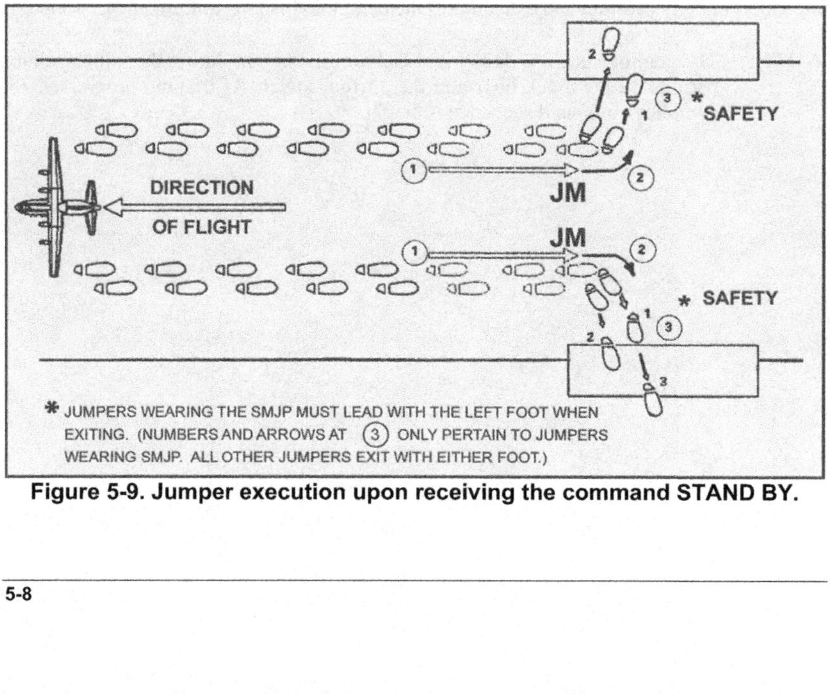

Figure 5-9. Jumper execution upon receiving the command STAND BY.

5-10. GO

GO is the ninth jump command. The green light is the final time warning on USAF aircraft. It tells the JM that as far as the aircrew is concerned, conditions are safe and it is time to issue the ninth jump command, GO.

a. **Command.**

(1) The JM gives the verbal command GO and may also tap the first parachutist out.

(2) In this case, the command GO and a sharp tap on the thigh is the signal to exit. If this signal is used, it is explained during the JM's briefing.

b. **Jumper Actions.**

(1) At the command GO, the first parachutist exits the aircraft. All subsequent jumpers begin moving toward the door using a shuffle.

(2) Once the jumpers begin to shuffle, they assume an elbow-locked position with the arm that is controlling their static line. The jumpers place their static line control hand so that it is nearly touching the back of the pack tray of the jumper in front of them. This establishes the proper jump interval. Jumpers do not place their static line control hand in a position so that it extends past the pack tray of the jumper in front of them.

(3) As each jumper approaches the door, he establishes eye-to-eye contact with the safety and hands his static line to the safety. Once the safety has control of the jumper's static line, the jumper returns his hand to the end of the reserve parachute with his fingers spread.

(4) After handing his static line to the safety (vicinity of the lead edge of the door), the jumper executes a left or right turn (as appropriate) and faces directly toward and centered on the door with both hands over the ends of the reserve parachute, fingers spread. He continues the momentum of his movement by walking toward the door, focusing on the horizon, and stepping on the jump platform. He pushes off with either foot and **vigorously** jumps up and **out** away from the aircraft. He immediately snaps into a good tight body position and exaggerates the bend into his hips to form an "L" shape.

NOTE: The commands STAND BY and GO are first taught during the initial training periods on the mock door and the 34-foot tower. As training progresses, the complete command sequence is taught.

CHAPTER 6

MAIN PARACHUTE MALFUNCTIONS AND EMPLOYMENT OF THE RESERVE PARACHUTE

A malfunction is any failure in the deployment or inflation of a parachute; or it is canopy damage, which can create a faulty, irregular, or abnormal condition that increases the jumper's rate of descent. The two classes of main parachute malfunction (total and partial) demand the jumper's immediate attention. Jumpers must be trained to identify the malfunction and take the appropriate action. Thorough training in what actions to take in the case of a malfunction is essential for parachutists. Practical exercises involving the activation of the reserve parachute are incorporated into all phases of training. Each type of malfunction is demonstrated so that jumpers can see exactly how each type of malfunction looks.

6-1. PULL-DROP METHOD (FOR MIRPS AND T-10 RESERVE)

When a total malfunction occurs, or the parachute provides no lift, the jumper must activate his reserve using the pull-drop method. Also, at the end of the 4000-count, if the jumper feels no opening shock, he should immediately activate his reserve using the pull-drop method. The jumper—

- Keeps a tight body position.
- Keeps his feet and knees together.
- Grasps the left carrying handle of the reserve parachute with his left hand.
- Turns his head left or right.
- Pulls the rip cord grip with his right hand and drops it to the ground.

When descending with only the reserve parachute inflated, the parachutist controls directional movement by slipping. The proper landing attitude is obtained by reaching up and grasping as many suspension lines in the opposite direction of drift as possible and slipping (as in the prepare-to-land attitude with the T-10-series parachute). Upon landing, the parachutist makes a quick recovery and collapses the canopy. In strong winds, if a quick recovery is impossible, the parachutist grasps one suspension line from either suspension line group and pulls the suspension line toward himself until the canopy collapses.

NOTE: If the MIRPS does not activate immediately, the jumper remains in a good tight body position while still grasping the left carrying handle. With the right hand forming a knife cutting edge, he sweeps the top panel of the reserve up and away, ensuring that the right hand does not interfere with the spring-loaded deployment assistance device.

6-2. DOWN-AND-AWAY METHOD (FOR T-10 RESERVE ONLY)

To activate the reserve parachute using the down-and-away method, the jumper—

- Returns to a tight modified body position with his feet and knees together.

- Places his hand on the middle of the reserve (over the rip cord protector flap). He exerts pressure on the reserve and grasps the rip cord grip with his right hand.

NOTE: Strong pressure must be maintained with the left hand to prevent the pilot chute and reserve canopy from springing out.

- Pulls the rip cord grip and drops it.
- With the right hand forming a knife edge, palm facing out, reaches between the pack assembly and canopy, and grabs as much canopy and reserve parachute suspension lines as possible.
- Throws the reserve parachute down and to the right (or left) side at about a 45-degree angle. (If the jumper is spinning, the canopy is thrown in the direction of the spin.) (If the reserve does not inflate, the jumper must retrieve the canopy and continue to throw it down and away until it inflates.)
- Uses the thumbs to clear all remaining suspension lines from the pack tray.

When the reserve parachute has activated, the jumper may have two inflated canopies. When descending with two inflated canopies, he has no direction control over the parachutes; all other jumpers remain clear.

6-3. TOTAL MALFUNCTION

A total malfunction is the failure of the parachute to open or to deploy.

a. **Causes of a Total Malfunction**. The failure of the parachute to deploy can be caused by a severed static line, a broken snap hook, or a broken anchor line cable. The jumper's failure to hook up also results in the failure of the parachute to deploy. Malfunctions of these types are rare.

b. **Streamer**. Although not defined as total malfunction, a deployed parachute with a "cigarette roll" or "streamer" provides little or no lift for the jumper. This malfunction must be treated as a total malfunction. This malfunction occurs when a portion of the skirt blows between two suspension lines and begins to roll with the opposite fabric. The heat generated by the friction of the fabric being rolled causes the nylon to fuze and blocks the air channel in the canopy. The jumper immediately activates his reserve using the pull-drop method.

c. **Towed Jumper**. Although not classed as a parachute malfunction, a parachutist can be towed behind the aircraft by a misrouted static line or by a piece of equipment that has snagged the aircraft during the jumper's exit. During the 4000-count, the jumper feels an excessive opening shock and then feels himself being towed by the aircraft. The jumper remains in a tight body position, protecting his rip cord grip until he is either retrieved inside the aircraft or is cut free by the loadmaster on the pilot's order. If the jumper is being towed by the static line and is cut free, the main parachute will not deploy, and the jumper will have to activate his reserve using the pull-drop method. The jumper's actions are as follows:

(1) ***Retrieving the Jumper Inside the Aircraft.*** The jumper remains in a tight body position until he is completely inside the aircraft. A towed jumper must not use his hands to assist the retrieving personnel. The most important action of a towed jumper is to protect his rip cord grip.

(2) ***Cutting the Jumper Away.*** Once the jumper is cut free of the aircraft, the main parachute may, or may not, deploy. If the jumper was towed by something other than the static line, the main parachute will deploy and inflate. There is no need to activate the reserve parachute. If the jumper was towed by the static line and is cut free, the jumper must immediately deploy his reserve using the pull-drop method.

WARNING

The towed jumper must remain in tight body position and protect the rip cord grip with his right hand. Accidental activation of the reserve while being towed may be fatal.

6-4. PARTIAL MALFUNCTION

The four types of partial malfunctions are **complete inversion**, **semi-inversion**, **blown section or gore**, and **broken suspension lines**.

a. **Complete Inversion**. This malfunction may occur when a portion of the skirt blows inward between a pair of suspension lines on the opposite side of the parachute. This portion of the skirt forms a secondary lobe that fills with air and enlarges at the expense of the rest of the canopy. The portion of the canopy forming the secondary lobe is inverted. The canopy turns inside out with no decrease in its lifting surface.

(1) It is difficult to detect if a complete inversion occurs during the initial deployment of the canopy. With a T-10-series, the rear risers control the front of the canopy and the front risers control the rear of the canopy. With an MC1-series, the orifice and control toggles are to the parachutist's front and maneuvering techniques are reversed.

(2) There may be no need for the parachutist to activate the reserve parachute unless the canopy was damaged during inversion. The parachute failed to function properly, but will support the parachutist. The inversion may increase the jumper's rate of descent. If the jumper's rate of descent is greater than that of other jumpers, he deploys his reserve. If he is using the T-10 reserve and his rate of descent is only slightly greater than other jumpers, he uses the down-and-away method. Otherwise, he uses the pull-drop method (the only method of activating the MIRPS).

b. **Semi-Inversion**. This malfunction may occur if development of the secondary lobe stops before completely inverting. This malfunction may remain stable, become completely inverted, or revert to normal during descent. The total lifting capability of the canopy is decreased by the formation of a secondary lobe. The fabric can be burned by friction and weakened during descent. With this malfunction, the parachutist must deploy his reserve. (If he is using the T-10 reserve, he uses the down-and-away method.) The T-10-series and MC1-series parachutes have anti-inversion nets that reduce the chance of this type of malfunction.

c. **Blown Section or Gore**. This malfunction occurs when the strain placed on the canopy during inflation is great and a panel, section, or gore is ripped or torn out, resulting in a hole(s) in the canopy. The jumper compares his rate of descent with that of other jumpers. If the jumper is falling faster than other jumpers around him, he must deploy the reserve parachute. (If he is using the T-10 reserve, he uses the down-and-away method.) Large holes in the canopy should be treated like a blown section or gore.

d. **Broken Suspension Lines**. This malfunction occurs when enough suspension lines break causing the jumper's rate of descent to be greater than fellow jumpers; the parachutist must activate his reserve using the down and away method. If control lines on an MC1-series canopy break, the parachutist controls the canopy by pulling only one of the rear risers in the direction that he wishes to turn. **He must use a rear riser**.

PART TWO

Duties and Functions of Key Personnel, Advanced Airborne Techniques and Training

CHAPTER 7

RESPONSIBILITIES AND QUALIFICATIONS OF KEY PERSONNEL

The initial training and follow-on refresher training of key personnel are of major concern to commanders. The proper training and supervision of key personnel ensure that correct procedures and operational safety measures are followed during airborne operations.

7-1. COMMANDER'S RESPONSIBILITIES

The airborne commander designates the key personnel for each airborne operation. These key personnel are the primary jumpmaster (JM), assistant jumpmasters (AJMs), safety personnel, departure airfield control officer (DACO), drop zone support team leader (DZSTL), drop zone safety officer (DZSO), and malfunction officer (MO).

a. Each aircraft has designated JM, AJM, and safety personnel. The airborne commander gives the designated JM command authority over, and responsibility for, all personnel onboard a jump aircraft. JM, AJM, and safety duties are described in Chapters 8, 9, and 10.

b. The DACO is located at the departure airfield and has coordination responsibility with the GLO and TALCE/aircrew for the loading of personnel, equipment, and supplies into the aircraft. Also, the DACO provides the JM with changes to station time and the overall operational plan, current DZ weather, airfield crossing procedures, and the aircraft parking plan. Complete DACO duties are discussed in Chapter 11.

c. Each DZ has a DZSO or DZSTL. The DZSO or DZSTL has command authority over the actions and safety of all personnel on the drop zone. DZSO and DZSTL procedures for DZ operations are described in Chapters 20, 21, and 22.

d. The MO, as a member of the DZST, is located on the DZ. Detailed MO duties are described in Chapter 23.

The JM, DACO, DZSTL, or DZSO can delegate their authority to subordinates, but they **cannot** delegate their responsibilities.

7-2. KEY PERSONNEL PREREQUISITES

The following minimum standards must be met before personnel will be allowed to perform JM, AJM, safety, DACO, DZSO, DZSTL, or MO duties for personnel and heavy equipment airdrop operations. Tables 7-1 and 7-2 (pages 7-3 and 7-4) indicate the duties that JM, DZSO, and DZSTL qualified and current personnel may perform, the airdrop method that will be used, and the type of airdrop mission that may be flown.

a. **JM, AJM, and Safety Personnel**. To be appointed as JM, AJM, or safety personnel, individuals must meet the following prerequisites:

(1) ***Primary Jumpmaster.***

(a) Commissioned officer, warrant officer, or NCO (US Army and US Navy: E5 or above; USMC and USAF: E4 or above).

(b) JM qualified. To be JM qualified, the JM must be a graduate from an authorized JM course (see Appendix B) at Fort Benning, GA; at a Fort Benning MTT; at Fort Bragg, NC; at a Fort Bragg MTT; or from an SOF JM course. (JMs qualified through SOF JM course must be JM refreshed prior to assuming JM duties outside SOF units.)

(c) JM current. To be JM current, the JM must have either performed JM or AJM duties within the preceding 180 days; or, if a senior or master rated parachutist, performed duty as a safety on military fixed-wing aircraft, utilizing door or ramp exit within the preceding 180 days; or completed a JM refresher course (see Appendix C) in the preceding 180 days and be a current jumper. (JM or safety duties performed on rotary-wing aircraft will apply for JM currency.)

(d) Prior experience. Perform duties twice as the AJM and perform duties twice as a safety. USAF personnel must complete the job qualification requirements for their Air Force specialty code (AFSC).

(2) ***Assistant Jumpmaster.***

(a) Commissioned officer, warrant officer, or NCO (US Army and US Navy: E5 or above; USMC and USAF: E4 or above).

(b) JM qualified and current.

(c) Prior experience. Perform duties twice as a safety.

(3) ***Safety Personnel.***

(a) Commissioned officer, warrant officer, or NCO (US Army: E5 or above; USMC, USN, USAF: E4 or above).

(b) JM qualified and current.

b. **DACO, DZSO, DZSTL, and MO**. To be appointed a DACO, DZSO, DZSTL, or MO, individuals must meet the following prerequisites:

(1) ***DACO.***

(a) Commissioned officer, warrant officer, or NCO (US Army: E5 or above; USMC, USAF: E4 or above).

(b) JM qualified and current.

(c) Prior experience. Perform duties as assistant DACO at least once.

(2) ***DZSO.***

(a) Commissioned officer, warrant officer, or NCO (US Army and US Navy: E5 or above; USMC, USAF: E4 or above).

(b) JM qualified and current or CCT/STS certified.

(c) Prior experience. Perform the duties of assistant DZSO in support of an airborne operation involving personnel or heavy equipment at least once.

NOTE: For combination airdrop operations the DZSO/DZSTL must follow the procedures for heavy drop operations, but observe the jumpers as they exit the aircraft.

(3) ***DZSTL.***

(a) Commissioned officer, warrant officer, or NCO (US Army: E5 or above; USMC, USAF: E4 or above), or civilian equivalent. (This position is not required for USN.)

(b) JM qualified and current for PE or HE.

(c) Certified as a DZSTL by having attended one of the following since 1988:

- USAIS Pathfinder Course.
- USAIS Jumpmaster Course (for CARP DZs only).
- USAIS DZSTL MTT.
- 82d Airborne Division Advanced Airborne School DZSTL Course.
- 82d Airborne Division Advanced Airborne School JM Course (for CARP DZs only).
- US Army Special Operations Command (USASOC) Jumpmaster Course.

(d) Prior experience. Observed and assisted a qualified and current DZSTL while performing his duties during an airdrop operation involving personnel or heavy equipment.

(e) DZSTL current. To be DZSTL current, the DZSTL must have performed the duties of DZSTL or assistant DZSTL within the preceding 180 days or completed a DZSTL refresher course taught by a current DZSTL within the preceding 180 days.

NOTE: DZSTLs in support of CDS airdrops are **not** required to be airborne qualified, on jump status, or JM qualified and current, but they must have attended an authorized Pathfinder or DZST course. For combination airdrop operations the DZSO/DZSTL must follow the procedures for heavy drop operations, but observe the jumpers as they exit the aircraft.

(4) ***Malfunction Officer.***

(a) Commissioned officer, warrant officer, or NCO (US Army and US Navy: E5 or above; USMC, USAF: E4 or above), or civilian equivalent. USAF: DZC/DCSO will fill the duties as the MO and be designated in writing as an MO. He will have a thorough understanding of the parachute equipment used for the operation. USAF combat control personnel are authorized to perform the duties of MO during unilateral operations.

(b) Qualified parachute rigger from the unit providing the air items used during the operation (IAW AR 59-4/MCO 13480.1B). (For the USMC only, the malfunction officer does not have to be from the organization providing the air items.)

	DUTY TO PERFORM	AIRDROP METHOD	TYPE AIRDROP
1. JM SCHOOL GRADUATE BEFORE SEP 1988:			
(A) JM (C)	JM, DZSO, DZSTL	CARP, VIRS, WSVC, GMRS	CDS, HE, PERS
(B) JM (NC)	NONE		
2. JM SCHOOL GRADUATE AFTER SEP 1988:			
(A) JM (C)	JM, DZSO	CARP	CDS, HE, PERS
(B) JM (NC)	DZSO	CARP	CDS
C = CURRENT NC = NOT CURRENT			

Table 7-1. Duties that Jumpmaster School certified personnel may perform.

	DUTY TO PERFORM	AIRDROP METHOD	TYPE AIRDROP
1. USAIS PATHFINDER SCHOOL GRADUATE AFTER SEP 1988:			
(A) JM (C) DZST (C)	JM, DZSTL	CARP, *GMRS, VIRS	CDS, HE, PERS
(B) JM (NC) DZST (C)	DZSTL	CARP	CDS
(C) JM (C) DZST (NC)	JM		
(D) NONAIRBORNE DZST (C)	DZSTL	CARP	CDS
C = CURRENT NC = NOT CURRENT * GMRS DZ NORMALLY RESERVED FOR SOCOM UNITS, AND REQUIRES DOCUMENTED TRAINING FROM DZSTL MTT			

Table 7-2. Duties that USAIS Pathfinder School graduates may perform.

CHAPTER 8

JUMPMASTER DUTIES AT THE UNIT AREA

The success of airborne operations depends mainly on how well the JM executes his duties. He must receive mission briefings, conduct prejump training, supervise rigging of equipment, and move to the departure airfield, all within a rigid time schedule.

Section I. ESSENTIAL INFORMATION

A key factor in the JM duties is the mission briefing. H-hour (TOT) is established at this time and the backward planning process begins.

NOTE: Jumpmaster duties begin immediately upon notification.

8-1. DESIGNATION NOTIFICATION

Upon notification that he has been designated as JM, the individual obtains or is provided the following information:

- Mission and ground tactical plan.
- Air movement plan to include time of flight, formations, route, direction of flight over drop zone, drop altitude, location and design of code letters, racetracks, and emergency call signs/frequencies.
- Names of AJM(s) and safety personnel, and time and place to brief them.
- Time and place of initial manifest call.
- Time and place of final manifest call.
- Time and place to conduct operations briefing.
- Time and place to conduct prejump training.
- Time and place to check and inspect parachutists' uniforms and equipment.
- Transportation (movement to marshaling area and departure airfield plan and times).
- Tactical cross-load plan.
- Time and place of parachute issue, including types of parachutes.
- Weather decision time(s).
- Time and place of troop safety briefing.
- Type of aircraft for the operation and special items of equipment being worn by jumpers (DMJP, AIRPAC, AT4JP, SMJP, or CWIE) or A-series containers aboard aircraft (door bundles or wedge).
- Aircraft tail numbers, chalk numbers, and parking spots.
- Load time.
- Time and place of aircrew/JM briefing.
- Station time.
- Takeoff time.
- Time on target.

NOTE: If, during the joint planning and preparation phase for airborne operations, it is decided host nation aircraft are to be used without navigational equipment, a detailed pilot, loadmaster, and JM brief must take place.

- Landing plan to include drop zones, drop times, delivery sequence, number/type of loads (PP, HD, CDS, LAPES), and types of drops (CARP, GMRS, VIRS, or WSVC).
- Mission and ground tactical plan.
- Air item turn-in plan.
- Medical support plan.

8-2. ASSISTANT'S BRIEFING

After he receives the initial operation briefing, the JM returns to the unit and briefs his AJM and safety personnel. The JM assigns them duties for the remainder of the operation. At this time, the JM determines who assumes responsibility for parachutists remaining onboard. The manifest of personnel scheduled to jump is prepared. The JM schedules a full rehearsal with the JM team before the JM team assembles the chalk (planeload).

a. Upon completion of the briefing, the JM organizes the chalk IAW the cross-loading plan and conducts the initial manifest call. AJM(s) and safety personnel check the identification card and identification tags of each parachutist.

b. Items for cross-loading include door bundles and large, bulky equipment carried by individual parachutists (CWIE, DMJP). The JM determines which chalk and informs parachutists which position and door they will jump.

c. The JM, aided by his AJM(s) and safety personnel, inspects each parachutist's equipment to ensure proper rigging. Parachutists pack and rig their equipment and containers before airborne operations.

NOTE: Primary jumpmaster can delegate authority, but not responsibility.

8-3. JUMPMASTER/SAFETY KIT BAG

The JM ensures aviator kit bags for use onboard the aircraft have been prepared to contain extra items that may be needed during any phase of the airborne operation. This is referred to as the JM/safety kit bag and is used by the JM, AJM, and safety personnel. Items to consider for use onboard the aircraft (depending on the type of airborne operation) are—

- Flashlight (night operations).
- Masking tape/cloth.
- Roll of 1/4-inch cotton webbing.
- Safety wires (with lanyards).
- Harness single point release (HSPR) complete, or H-harness complete.
- Hook-pile tape (HPT) lowering lines.
- Foam impact pads, chin straps, pull-the-dot fasteners with tabs, headbands with attaching clips, and parachutist retention straps.
- Quick-release snaps.
- Upper tie-downs for M1950 weapons cases.

- Retainer bands.
- Trash bags.
- Earplugs and airsickness bags.
- Two extra reserves and extra aviator kit bags (for static lines and deployment) (1 bag for every 15 deployment bags).
- A knife will be carried by the JM, AJM, and safety personnel. (Knife should not be carried on a point of contact.)

8-4. OPERATION BRIEFING

As soon as practical after the initial manifest call, the JM briefs personnel on the details of the operation. Following the operation briefing, prejump training is conducted in the unit area or at the departure airfield. It should be scheduled and conducted within 24 hours before takeoff and include the following:

- Drop zone.
- Type of aircraft.
- Chalk number(s).
- Type of parachute(s).
- Briefing on serials, container delivery system, heavy drop, and type of aircraft, if a part of a larger airborne operation.
- Weather decision time (for GO, NO-GO decision).
- Type of individual equipment and separate equipment that troops will be jumping (CWIE, AIRPAC, DMJP, ALICE pack, SMJP, M1950 weapons case).
- Time and place of parachute issue.
- Load time.
- Station time.
- Takeoff time.
- Length of flight.
- Time on target.
- Direction of flight over DZ.
- Drop altitude.
- Predicted winds on the DZ and direction.
- Route checkpoints.
- DZ assembly aids and area.
- Parachute turn-in point(s).
- Time and place of final manifest call.
- Medical support plan.
- Obstacles on or near the DZ.

NOTE: If the unit jumps nonstandard equipment containers, the rigging on these containers must be approved IAW individual service regulations. These rigging procedures must be forwarded for testing to US Army Airborne Special Operations Test Directorate (USAABNSOTD), Fort Bragg, NC, and then for approval to Commander, 1st Battalion (Abn), 507th Infantry, ATTN: Jumpmaster School, Fort Benning, GA 31905.

Section II. SUSTAINED AIRBORNE TRAINING

All personnel require sustained airborne training. The JM usually does not know the proficiency of all parachutists he is responsible for; therefore, basic airborne jump techniques are rehearsed so each parachutist can demonstrate his ability to perform them. Jumpmasters, safeties, and key leaders will make on-the-spot corrections of any jumper not properly performing the required training. (See paragraph 8-15 for a sample prejump training narrative.)

8-5. MINIMUM TRAINING

Sustained airborne training will consist of, at a minimum, prejump training to include PLFs and mock door training. Prejump training will include a review of the five points of performance, towed jumpers, collisions and entanglements, malfunctions, activation of the reserve, and emergency landings. Mock door training will include rehearsal of every detail involved with the airborne operation to include accidental activation of the reserve on board the aircraft. Jumpers must make a minimum of two exits from the mock door.

a. **Training Apparatuses**. Army personnel will use aircraft fuselage mock-ups to rehearse preflight and in-flight action. (Aircraft fuselage mock-ups can be complete airframes or may be field-expedient training devices as simple as rope strung to simulate an anchor line cable and ballistic helmets on the ground to mark the jump door. The use of field-expedient training devices is not the preferred method for conducting sustained airborne training.) Also, actions-in-the-aircraft training reminds parachutists what occurs in flight before jumping. The JM can use the mock door apparatus to show parachutists where their relative positions will be in the aircraft. If in-flight rigging is to be performed, the rigging station locations can be indicated also. The JM reviews and leads a rehearsal of all actions related to in-flight procedures so the jump mission will be smooth and safe.

b. **Execution**. Each parachutist should be seen by the JM and should hear him (a bullhorn should be used, if necessary). Performance-oriented training is conducted for emergency landings. AJM and safety personnel must make aggressive and positive on-the-spot corrections. Prejump training must be taught proficiently.

8-6. PREJUMP TRAINING

All personnel must attend prejump training. Jumpers must be positioned so that their actions can be viewed by the JM and so they can hear him. A bullhorn can be used, if necessary. Prejump training is performance-oriented and should be tailored to fit the mission. AJM and safety personnel must make aggressive and positive on-the-spot corrections. Prejump training must be taught proficiently.

8-7. FIVE POINTS OF PERFORMANCE

Training on the five points of performance must be attended by all parachutists and JMs.

a. **Proper Exit, Check Body Position and Count**.

(1) Keep chin on chest.

(2) Keep eyes open.

(3) Keep elbows into sides.

(4) Hands over the ends of the reserve parachute, fingers spread, right palm protecting the rip cord grip.

(5) Bend body forward at the waist.

(6) Keep feet and knees together.

(7) Lock knees.

(8) Count to four thousand (fixed-wing aircraft).

(9) Count to six thousand (helicopters).

(10) Immediately activate the reserve parachute using the proper method if opening shock is not felt.

b. **Check Canopy and Gain Canopy Control.**

(1) Reach up and grasp toggles (MC1-series); or grasp risers (T-10-series).

(2) Make a 360-degree check of the canopy.

(3) Remove twists, if any: grasp a set of risers in each hand, thumbs down with knuckles to the rear, and apply outward pressure while bicycling the legs in a vigorous motion.

(4) Immediately recheck canopy and gain canopy control.

c. **Keep a Sharp Lookout for All Jumpers During Descent.**

(1) Remember the three rules of the air.

(2) Always look before turning/slipping.

(3) Always slip/turn in the opposite direction from other jumpers.

(4) Always give lower parachutist the right-of-way.

(5) Avoid other parachutists all the way to the ground; maintain 25 feet of separation (T-10-series) or 50 feet of separation (MC1-series) between jumpers in the air.

(6) At the end of the third point of performance, release all appropriate equipment tiedowns.

d. **Slip/Turn into the Wind and Prepare to Land.**

(1) Check to ensure the air below and around is clear of other parachutists. At no higher than 200 feet, lower equipment. Immediately regain canopy control and continue to keep a sharp lookout for other parachutists.

(2) When 100 feet above ground level, slip into the wind (T-10-series), or turn and hold into the wind (MC1-series).

(3) Keep feet and knees together.

(4) Keep knees slightly bent and unlocked.

(5) Point balls of the feet toward the ground.

(6) Keep head and eyes toward the horizon. Before making contact with the ground, turn the lower portion of the body (below the waist) to a 45-degree angle (front or rear PLF), exposing the portion of the body that will come in contact with the ground.

e. **Land.**

(1) Make a PLF using the five points of contact.

(2) Make no stand-up landings.

(3) Remain on your back and activate one canopy release assembly using either the hand-assist or hand-to-shoulder method.

(4) Recover and turn in equipment.

- Remain in prone position; place weapon into operation and remove harness.
- Remove air items from D-rings on parachute harness.
- Elongate canopy into the wind, remove debris, and figure-eight roll.
- Insert canopy into aviator kit bag with bridle loop on top. Insert waistband through bridle loop.

- Snap aviator kit bag closed; do not zip.
- Attach reserve parachute to the carrying handle of the aviator's kit bag.
- Maintain noise and light discipline. Move rapidly to the nearest turn-in point.

8-8. FIVE POINTS OF CONTACT

Training on the five points of contact includes the following:

a. **Review of Five Points of Contact**.

- Balls of the feet.
- Calves.
- Thighs.
- Buttocks.
- Pull-up muscle (right or left side).

b. **Parachute Landing Falls**. All manifested jumpers must perform one satisfactory PLF in each of the four directions.

- Left side.
- Right side.
- Front.
- Rear.

8-9. TOTAL MALFUNCTIONS (NO LIFT CAPABILITY)

Parachutists must receive training on total malfunctions and control of the reserve parachute.

a. **Reserve Activation**. Activate the reserve using the pull-drop method.

- Remain in a tight body position.
- Keep feet and knees together.
- Grasp the left carrying handle of the reserve parachute with the left hand.
- T-10: Turn head to either side. With the right hand, pull the rip cord grip and drop it.
- MIRPS: Sweep up and away on rip cord protector flap.

b. **Reserve Parachute Control**. While descending under the reserve, control the canopy by slipping and assume the landing attitude by reaching up and grasping as many suspension lines in the opposite direction of drift as possible and slipping with both hands (as in the prepare-to-land attitude with the T-10-series parachute).

c. **MIRPS Sweep Method**. If the reserve parachute does not activate immediately, the jumper remains in a good tight body position while still grasping the left carrying handle. With the right hand forming a knife cutting edge, he sweeps up on the rip cord protector flap, ensuring that the right hand does not interfere with the spring-loaded deployment assistance device.

8-10. PARTIAL MALFUNCTIONS

Parachutists must receive training on partial malfunctions.

a. **Partial Malfunction Indicators**.

- Complete inversion.
- Semi-inversion.
- Blown section or gore.

- More than six broken suspension lines.
- Holes.

Should any of these malfunctions occur and cause the parachutist's rate of descent to increase in comparison to other parachutists, the parachutist activates his reserve parachute as follows:

(1) ***T-10 Reserve.***

(a) Snaps back into a tight body position.

(b) Places the left hand over the rip cord protector flap, fingers extended and spread. Applies pressure to prevent canopy from deploying.

(c) Pulls the rip cord grip with the right hand and drops it.

(d) Forms a knife cutting edge with the right hand to reach into the pack tray, at the same time releasing pressure with the right hand, and grasps as much canopy as possible.

(e) Keeps feet and knees together.

(f) Throws canopy down and away from the body in the same direction as the spin. (Spin is to right, throw right; spin is to left, throw left.) If the reserve does not inflate, pulls it back in and throws it in the opposite direction.

(g) Immediately frees all suspension lines using the thumbs in a downward raking motion.

(2) ***MIRPS.***

(a) Snaps back into a tight body position.

(b) With the left hand, grasps the left carrying handle.

(c) Pulls the rip cord grip with the right hand and drops it.

(d) If the reserve does not deploy, forms a knife cutting edge with the right hand and sweeps the rip cord protector flap up.

b. **Reserve Inflates**. When the reserve parachute inflates, and there are two inflated canopies, the parachutist has no directional control over his parachute. All other parachutists slip or steer clear. To assume the proper landing attitude, the parachutist reaches high on all four risers of the main parachute and maintains this attitude until making ground contact. Immediately upon landing, he releases the main parachute (using the canopy release assemblies) and collapses the reserve by using either the quick-recovery method or by detaching the connector snaps from the D-rings of the main lift web.

8-11. COLLISIONS

The parachutist must always attempt to slip or turn away. If unable to avoid a collision, he uses the spread-eagle method to bounce off another canopy or suspension lines. If a parachutist enters another parachutist's suspension lines, the entering parachutist assumes the modified position of attention with the right hand protecting (but not grasping) the rip cord grip, in hope that he will exit the same location without becoming entangled. If not, the entering parachutist may use his left hand to assist in exiting the other jumper's canopy and suspension lines the same way he entered.

8-12. ENTANGLEMENTS

If parachutists become entangled, their actions required to correct the problem depend upon the type parachute used. Reaction techniques are as follows:

a. **T-10-Series Parachute**. The higher parachutist moves hand under hand down to the lower parachutist. They attempt to establish eye-to-eye contact and hold onto each other by the left main lift web(s). They must **NOT** touch the other jumper's canopy release assemblies. They decide which parachute landing fall to execute upon contact with the ground and both parachutists execute the same PLF. If they are face-to-face, they will **NOT** execute a front PLF. If they are back-to-back, they will **NOT** execute a rear PLF. If one parachutist has a completely inflated canopy, neither parachutist activates the reserve parachute. If both parachutes lose lift capabilities, parachutists use the pull-drop method to activate their reserve parachutes.

b. **MC1-Series Parachute**. Both parachutists immediately activate their reserve parachutes for a partial malfunction. Neither parachutist attempts to climb to the other parachutist. The higher parachutist avoids the lower parachutist when landing.

8-13. EMERGENCY LANDINGS

Depending upon the type of emergency landing initiated, the parachutist performs specific actions.

a. **Tree Landing**. The parachutist—

(1) Attempts to avoid the trees.

(2) Retains combat equipment (if not already lowered).

(3) Checks below, then jettisons combat equipment (if already lowered). (Ballistic helmet remains on the head.)

(4) Maintains canopy control until making contact with trees.

(5) Rotates forearms in front of his face and chest when making contact with trees.

(6) Prepares to execute PLF if passing through trees.

(7) Considers the possibility of activating the reserve parachute and climbing down the outside of it if hung up in trees.

b. **Wire Landing**. The parachutist—

(1) Attempts to avoid the wires.

(2) Lowers combat equipment, checks below, then jettisons equipment.

(3) Raises both arms to the elbow-locked position and places the palms of the hands on the inside of the front set of risers before contact. The feet and knees are together (bend in knees is exaggerated), and chin is on the chest.

(4) Upon making contact with the wires, the jumper pushes forward on the front set of risers, arches his back, and kicks vigorously with the legs, initiating a rocking motion, and attempts to work through the wires.

(5) Prepares to do a PLF if he should pass through the wires.

c. **Water Landing without Life Preserver**. The parachutist—

(1) Attempts to avoid the water (lake or river).

(2) Checks below and then, if clear, jettisons headgear.

(3) Releases all equipment tie-downs and lowers equipment, but does **not** jettison it.

(4) Activates quick release on the waistband.

(5) Unsnaps the left connector snap on the reserve parachute.

(6) Rotates the reserve to the right side of the parachute harness.

(7) Seats himself well into the saddle.

(8) Activates the ejector snap on the chest snap.

(9) Places hands on the ejector snaps on the leg straps.

(10) Activates the ejector snaps on the leg straps, throws arms up, and arches out of the harness when entering the water.

(11) Prepares to execute a PLF if the water is shallow.

(12) Swims upstream, upwind, or away from the parachute to avoid becoming entangled with it.

d. **Water Landing With Life Preserver.**

(1) Emergency water landings could occur on a tactical training mission where the route to the DZ is over a large body of water. On such flights, life preservers are issued to parachutists. If the aircraft malfunctions, the parachutists may need to jump over the water.

(2) Emergency water landings require parachutists to leave any combat equipment onboard the aircraft so they do not become entangled with the equipment in the water.

(3) A parachutist may find himself over water with attached combat equipment due to his drifting off a DZ bordered by water, or during emergency bailout after being hooked up onboard the aircraft. If this occurs, the jumper must lower his combat equipment but **not** jettison it.

(4) Deliberate water landings are executed for training in selected water drop zones. Parachutists may wear combat equipment only after is has been waterproofed and float checked.

(5) The parachutist wears the inflatable life preserver under his harness with the inflatable portions under his armpits. Parachutists wearing a UDT vest will route the chest strap under the UDT vest to prevent crushing the chest if inflated.

(6) During the third point of performance, the parachutist inflates the life preserver by discharging the attached CO2 cartridges. If necessary, the life preserver can be inflated by blowing air into the inflation valve hose.

(7) Upon entering the water, the parachutist activates both canopy release assemblies and swims upstream or away from the canopy.

8-14. RESERVE ACTIVATION INSIDE AIRCRAFT

If the reserve parachute is activated inside an aircraft, the aircrew must follow certain procedures.

a. **Fixed-Wing Aircraft**. If the parachutist is aft of the wheel well, and the jump doors (or ramp) are open and the reserve canopy is in or going out the door (or ramp), the JM and safety personnel do not attempt to retain the parachutist inside the aircraft.

(1) The area is cleared, if possible, and the parachutist exits immediately. If the canopy is not in the door (or ramp), the deployed reserve canopy is secured as quickly as possible by anyone nearby. Then, the parachutist is moved to the forward section of the aircraft, the open reserve is removed, another reserve is attached, and the parachutist is returned to the stick to jump.

(2) If the parachutist is forward of the wheel well, and the jump doors (or ramp) are open, the parachutist either steps on or grasps the reserve canopy and traps it so it cannot inflate. Safety personnel move the parachutist to the forward section of the aircraft. They remove the reserve and attach another, and the parachutist jumps on the next pass over the DZ.

(3) If the jump doors (or ramp) are closed and a reserve parachute deploys, the JM or safety personnel move the parachutist to the forward section of the aircraft. The deployed

reserve is removed, and the loadmaster is told not to open the jump doors (or ramp). Another reserve is attached, and the parachutist is returned to the stick to jump.

b. **Rotary-Wing Aircraft**. If the reserve parachute is activated in an aircraft that requires the parachutist to sit in the door, no attempt is made to stop the parachutist from exiting the aircraft. The parachutist immediately exits. In an aircraft that requires parachutists to exit over the ramp, the procedures described for fixed-wing aircraft are followed.

8-15. TOWED PARACHUTIST PROCEDURES

A towed parachutist is retrieved or cut free as follows:

a. If the parachutist is being towed by the static line, he is retrieved back inside the aircraft. Once he nears the jump door, he does not reach for the JM but continues to protect the rip cord grip; the JM or safety reaches for him. If the jumper cannot be retrieved, he is cut free over the drop zone. Once he feels himself fall free from the aircraft, he must immediately activate his reserve parachute for a total malfunction.

b. If the parachutist is being towed by anything other than the static line, the JM or safety tries to jog him free from the aircraft. If the parachutist cannot be freed, he is retrieved. If he is freed from the aircraft, he does not need to activate his reserve since his main parachute will deploy.

c. Only the aircraft loadmaster and the JM (with assistance from the safety) will perform towed parachutist retrieval.

8-16. SAMPLE PREJUMP TRAINING NARRATIVE

Prior to beginning prejump training, jumpmaster personnel check ID card, ID tags, and ballistic helmets for serviceability and proper routing of parachutist helmet retention strap. Then, the JM organizes jumpers into either a half-moon or extended rectangular formation. The JM uses a half-moon formation for 30 or fewer jumpers and an extended rectangular formation for more than 30 jumpers. When in a half-moon formation, jumpers must be positioned so the JM can see them easily. The extended rectangular formation provides better control for on-the-spot corrections than does the half-moon formation. Once the jumpers are in the chosen formation, safety personnel check identification tags and cards. When finished, they make on-the-spot corrections on the jumpers as they conduct prejump training. Figure 8-1 shows a sample prejump training briefing that will help the JM conduct an effective prejump training session for his jumpers. The jumpmaster should give the prejump training narrative outlined in Figure 8-1 verbatim. The jumpmaster will not read the narrative to the jumpers.

NOTES: 1. The following items are to be covered during prejump training:

- Five points of performance.
- Recovery of equipment.
- Towed parachutist procedures.
- Malfunctions.
- Collisions and entanglements.
- Emergency landings (tree, wire, and water).
- Parachute landing falls.

2. Although the prejump briefing can be given by anyone in the jumpmaster team (JM, AJM, safety), the primary jumpmaster can delegate his authority but cannot delegate his responsibility.
3. Prejump training is performance-oriented training and should be tailored to each mission. The jumpmaster should refer to his unit SOP for additional guidance. During prejump training, the JM uses the command HIT IT as often as needed to keep the jumpers actively involved

Prior to prejump training, place your jumpers into a formation that allows you to easily control them and make on-the-spot corrections. The extended rectangular formation and the horseshoe formation are two preferred formations.

Prior to placing them in formation, ensure the jumpmaster team inspects ballistic helmets, ID tags, and ID cards. The jumpmasters or the safeties can accomplish this inspection.

Although prejump can be given by anyone on the jumpmaster team, the primary jumpmaster can delegate authority but not responsibility.

Holding, running, one riser slips, and other information can be inserted into the prejump if the airborne commander deems it necessary.

Although prejump training should be tailored to fit the mission, emergency landings should always be covered due to the many variables involved with emergency situations; for example, if jumpers have to conduct an emergency bailout over unfamiliar terrain.

Prejump training is performance-oriented training and the jumpmaster team must ensure that the jumpers are performing the actions as they are being covered. During prejump training, use the "HIT IT" exercise as often as needed to keep the jumpers actively involved. Jumpmasters will refer to their unit ASOPs for additional guidance.

When jumping rotary-wing aircraft, jumpers will extend their count to six thousand.

The first point of performance is PROPER EXIT, CHECK BODY POSITION, AND COUNT. Jumpers "HIT IT." Upon exiting the aircraft, snap into a good tight body position. Keep your eyes open, chin on your chest, elbows tight into your sides, hands on the end of the reserve with your fingers spread, right hand covering the rip cord grip. Bend forward at the waist keeping your feet and knees together, knees locked to the rear, and count to four thousand.

At the end of your four thousand count, immediately go into your second point of performance—CHECK CANOPY AND IMMEDIATELY GAIN CANOPY CONTROL. When jumping the T-10-series parachute, reach up to the elbow locked position and secure a set of risers in each hand; simultaneously conduct a 360-degree check of your canopy. When jumping the MC1-series parachute, secure a toggle in each hand and pull them down to eye level, simultaneously conducting a 360-degree check of your canopy. If, during your second point of performance, you find that you have twists, reach up with both hands and grab a set of risers, thumbs down, knuckles to the rear. Pull the risers apart and begin a vigorous bicycling motion. When the twist comes out, check canopy and immediately gain canopy control.

Figure 8-1. Sample prejump training briefing.

Your third point of performance is KEEP A SHARP LOOKOUT DURING YOUR ENTIRE DESCENT. Remember the three rules of the air and repeat them after me. Look before you slip/turn; slip/turn iin the opposite direction to avoid collisions, and the lower jumper has the right of way. Avoid all jumpers all the way to the ground and maintain a 25-foot separation when jumping the T-10-series parachute and a 50-foot separation when jumping the MC1-series parachute. Some time during your third point of performance, release all appropriate equipment tie downs.

This brings you to your fourth point of performance, which is PREPARE TO LAND. At 200 to 100 feet AGL, look below you to ensure there are no fellow jumpers, and lower your equipment. Regain canopy control. At approximately 100 feet AGL, slip/turn into the wind and assume a landing attitude. When jumping the T-10-series parachute and the wind is blowing from your left, reach up on the left set of risers and pull them deep into your chest. If the wind is blowing from your front, reach up on the front set of risers and pull them deep into your chest. If the wind is blowing from your right, reach up on your right set of risers and pull them deep into your chest. If the wind is blowing from your rear, reach up on your rear set of risers and pull them deep into your chest. When jumping the MC1-series parachute and the wind is blowing from your left, pull your left toggle down to the elbow locked position. When you are facing into the wind, let up slowly to prevent oscillation. If the wind is blowing from your rear, pull either toggle down to the elbow locked position. When you are facing into the wind, let up slowly to prevent oscillation. If the wind is blowing from your front, make minor corrections to remain facing into the wind. Once you are facing into the wind, assume a landing attitude by keeping your feet and knees together, knees slightly bent, with your head and eyes on the horizon.

When the balls of your feet make contact with the ground, you will go into your fifth point of performance, LAND. You will make a proper PLF by hitting all five points of contact. Touch them and repeat them after me. 1) BALLS OF THE FEET. 2) CALF. 3) THIGH. 4) BUTTOCKS. 5) PULL UP MUSCLE. You will never attempt the make a stand up landing. Remain on your back and activate one of your canopy release assemblies using either the hand-to-shoulder method or the hand assist method. To activate your canopy release assembly using the hand-to-shoulder method, with either hand reach up and secure a safety clip and pull it out and down exposing the cable loop. Insert the thumb from bottom to top through the cable loop, turn your head in the opposite direction, and pull out and down on the cable loop. To activate the canopy release using the hand assist method, with either hand reach up and secure a safety clip and pull it out and down exposing the cable loop. Insert the thumb from bottom to top. Reinforce that hand with the other hand, turn your head in the opposite direction, and pull out and down on the cable loop. If your canopy fails to deflate, activate the other canopy release assembly. Place your weapon into operation and remove the parachute harness.

I will now cover RECOVERY OF EQUIPMENT. Once out of the parachute harness, remove all air items from the parachute harness. Place the parachute harness inside the aviator's kit bag smooth side facing up, leaving the waistband exposed. Elongate the canopy and suspension lines removing all debris as you go. Once you reach the apex of the canopy, insert your thumb into the bridle loop and figure-eight roll the canopy and suspension lines all the way down the aviator's kit bag. Place the canopy and suspension lines into the aviator's kit bag. Route the waistband through the bridle loop leaving 6 to 8 inches of the waistband exposed. Snap the aviator's kit bag closed. Do not zip. Secure the reserve parachute to the aviator's kit bag, place it over your head, conduct a 360-degree police of your area, and move to your assembly area.

Figure 8-1. Sample prejump training briefing (continued).

I will now cover TOWED JUMPER PROCEDURES. "JUMPERS, HIT IT." If you become a towed jumper and are being towed by your static line and are unconscious, you will be retrieved inside the aircraft. If you are conscious, maintain a good tight body position with your right hand protecting your rip cord grip, and an attempt will made to retrieve you inside the aircraft. As you near the jump door, DO NOT REACH FOR US, continue to protect your rip cord grip. If you cannot be retrieved, you will be cut free. Once you feel yourself falling free from the aircraft, immediately activate your reserve parachute for a total malfunction.

If you are being towed by your equipment, regardless of whether you are conscious or unconscious, we will cut or jog your equipment free and your main parachute will deploy.

NOTE: If you are being towed from a rotary-wing aircraft, maintain a good tight body position and protect your rip cord grip. The aircraft will slowly descend to the DZ, come to a hove, and the jumpmaster will free you from the aircraft.

The next item I will cover is MALFUNCTIONS. There are two types of malfunctions: total and partial. A total malfunction provides no lift capability what-so-ever; therefore, you must activate your reserve using the PULL DROP METHOD.

There are several types of partial malfunctions and actions for each. If you have broken suspension lines or blown sections or gores, compare your rate of descent with fellow jumpers. If you are falling faster than fellow jumpers, activate your reserve for a partial malfunction. If you have a squid, semi-inversion, or a complete inversion with damage to the canopy or suspension lines, you must immediately activate your reserve for a partial malfunction. If you have a complete inversion with no damage to the canopy or suspension lines, do not activate your reserve.

I will not cover ACTIVATION OF THE MODIFIED IMPROVED RESERVE PARACHUTE SYSTEM. To activate the MIRPS, you will use the "PULL DROP METHOD." "JUMPERS, HIT IT." Maintain a good tight body position. Grasp the left carrying handle with your left hand; with right hand grasp the rip cord grip. Pull out on the rip cord grip and drop it. Your reserve will activate.

NOTE: If your reserve does not activate, maintain a good tight body position, and with your right hand form a knife cutting edge and sweep the rip cord protector flap up and away allowing the DADS to properly deploy.

To activate the T-10 reserve for a total malfunction, use the "PULL DROP METHOD." "JUMPERS, HIT IT." If you do not feel an opening shock at the end of your count, maintain a good tight body position. With your left hand, grasp the left carrying handle; with your right grasp the rip cord grip. Turn your head to the left or right; pull the rip cord grip and drop it. Your reserve will activate.

To activate the T-10 reserve for a partial malfunction, use the down-and-away method. "JUMPERS, HIT IT. CHECK CANOPY." Snap back into a good tight body position. With the left hand, cover the rip cord protector flap and with the right hand, grasp the rip cord grip. Apply inward pressure with the left hand and with the right hand pull the rip cord grip and drop it. Form a knife cutting edge with the right hand and insert it into the upper right hand corner of the reserve. Grasp as much canopy and suspension lines as possible and pull out and up over either shoulder and throw it down and away in the direction you are

Figure 8-1. Sample prejump training briefing (continued).

spinning. If the canopy fails to inflate, pull it back into your body and throw it down and away in the opposite direction. Once your reserve activates, with each hand form a fist, thumbs exposed, and with a sweeping motion, clear all of the suspension lines from the pack tray ensuring they are deployed.

NOTE: If you have to activate the reserve for a partial malfunction, any attempt to control either canopy will be useless as one canopy will act as a brake for the other.

The next items I will cover are COLLISIONS AND ENTANGLEMENTS. "JUMPERS, HIT IT. CHECK CANOPY." If you see another jumper approaching, immediately attempt to slip/turn away. If you cannot avoid the collision, assume a spread eagle position and attempt to bounce off the jumper's canopy and suspension lines and then slip/turn away.

If you become entangled and are jumping the T-10-series parachute, the higher jumper will climb down to the lower jumper using the hand-under-hand method. Once both jumpers are even, you will face each other and grasp each other's main lift web. Both jumpers will discuss which PLF to execute. Both jumpers will conduct a PLF with the direction of drift. Neither jumper will fall into the other jumper. Both jumpers will continue to observe their canopies. If one canopy collapses, neither jumper will activate their reserve as one T-10-series parachute can safely deliver two combat-equipped jumpers to the ground. If both canopies collapse, the jumpers will pull towards each other to create a clear path for the activation of their reserve parachutes, then activate their reserves using the PULL DROP METHOD.

If you are jumping the MC1-series parachute, both jumpers will remain where they are, obtain a clear path, and immediately activate their reserve using the PULL DROP METHOD.

The next items I will cover are EMERGENCY LANDINGS. The first emergency landing I will cover is the tree landing. If you are drifting towards the trees, immediately slip/turn away. If you cannot avoid the trees and have lowered your equipment, look below you to ensure there are no fellow jumpers and then jettison your equipment making a mental note of where it lands. If you have not lowered your equipment, keep it on you to provide extra protection while passing through the trees. At approximately 100 feet AGL, assume a landing attitude by keeping your feet and knees together, knees slightly bent, with your head and eyes on the horizon. When the balls of your feet make contact with the trees, rotate your hands in front of your face with your elbows high. Be prepared to execute a PLF if you pass through the trees.

If you get hung up in the trees, keep your ballistic helmet on and lower and jettison all unneeded equipment. Activate the chest strap ejector strap and activate the quick release in your waistband. Place your left hand over the rip cord protector flap and apply pressure. Grasp the rip cord grip with your right hand and pull it and drop it. Control the activation of the reserve parachute toward the ground ensuring that all suspension lines are completely deployed. Disconnect the left connector snap and rotate the reserve to the right. Grasp the main lift web with either hand below the canopy release assembly and with the other hand activate the leg strap ejector snaps and climb down the outside of the reserve.

Remember, when in doubt, stay where you are and wait for assistance.

Figure 8-1. Sample prejump training briefing (continued).

The next emergency landing I will cover is the wire landing. If you are drifting toward wires, immediately slip/turn away. If you cannot avoid the wires, look below you to ensure there are no fellow jumpers and then lower and jettison your equipment making a mental note of where it lands. Assume a landing attitude by placing your hands, fingers extended and joined, high on the inside of the front set of risers with the elbows locked. Place your chin on your chest, keep your feet and knees together, and exaggerate the bend in your knees. When the balls of your feet make contact with the wires, begin a vigorous rocking motion in an attempt to pass all the way through the wires. Be prepared to execute a PLF if you pass all the way through the wires. If you get hung up in the wires, stay where you are and wait for assistance.

The last emergency landing I will cover is the water landing. The water landing is the most dangerous emergency landing because it takes the most time to prepare for. If you are drifting towards a body of water, immediately slip/turn away. If you cannot avoid the water, look below you to ensure there are no fellow jumpers, and then lower, DO NOT JETTISON, your equipment. Next, jettison your ballistic helmet. Activate the quick release in your waistband, disconnect the left connector snap and rotate the reserve to the right. Seat yourself well into the saddle and activate the chest strap ejector snap. Regain canopy control. Before entering the water, assume a landing attitude by keeping your feet and knees together, knees slight bent, and place your hands on the leg strap ejector snaps. When the balls of your feet make contact with the water, activate the leg strap ejector snaps, arch your back, throw your arms above your head, and slide out of the parachute harness. Swim upwind or upstream away from the canopy. Be prepared to execute a PLF if the water is shallow. If the canopy comes down on top of you, locate a radial seam, follow it to the skirt of the canopy and swim upstream or upwind away from the canopy.

NOTE: If you are jumping the B-7 life preserver, activate it in the air. If you are jumping the B-5 life preserver, activate it in the water after you remove the parachute harness.

NIGHT JUMPS: When conducting night jumps, be sure to give your canopy an extra look, maintain noise and light discipline, and be prepared to execute a PLF because you will hit the ground 5 to 10 seconds before you think you will.

AWADS: When jumping under AWADS conditions, do not lower your equipment until you have passed through the clouds. Do not slip/turn unless you must to avoid a collision. If you have any type of malfunction, immediately activate your reserve using the pull drop method, because you cannot compare your rate of descent. Ensure you recheck your canopy once you pass through the clouds.

PLFs: We will now move to the PLF platform and conduct one satisfactory PLF in each of the four directions, ensuring you conduct a proper PLF.

Figure 8-1. Sample prejump training briefing (continued).

CHAPTER 9

JUMPMASTER AND SAFETY DUTIES AT THE DEPARTURE AIRFIELD

Time is a critical factor at the departure airfield. The following events occur at the same time to allow the unit to meet station time:

- *DACO/PJM update briefing.*
- *PJM aircraft inspection and coordination with aircrew.*
- *Control of parachute issue by AJM/safeties.*
- *Rigging/inspection of parachutists.*
- *Loading of aircraft.*

The PJM usually turns control of the chalk (planeload) over to the AJM and safeties while he accomplishes update briefings, aircraft inspection, and aircrew coordination. The AJM and safeties control parachute issue and prepare for rigging/inspection of the chalk. All JM team personnel must draw their own parachutes for wear during airdrop operations; the aircrew cannot provide parachutes for the JM team.

Section I. KEY PERSONNEL

The commander selects the best-qualified jumpers to perform the JM, AJM, DACO, and DZSO/DZSTL duties for a unit airdrop operation. Those selected key personnel must correctly perform their assigned duties to ensure mission success and jumper safety. This responsibility may be delegated to qualified personnel in accordance with individual service component regulations.

9-1. PRIMARY JUMPMASTER/ASSISTANT JUMPMASTER DUTIES

PJM/AJM duties are as follows:

a. **PJM/DACO Briefing**. Upon arrival at the airfield, the PJM reports to the DACO for an update briefing to include:

- Change in the station time.
- Change in the overall operation plan.
- Current weather and winds.
- Parking plan of aircraft (location and tail number of the assigned aircraft).
- Coordination with the USAF guide if wheeled vehicles are used for transport to aircraft.

b. **Manifest Distribution**. Normally the manifest (DA Form 1306) is distributed as follows:

- DACO—two copies (original plus one copy).
- PJM—one copy.
- Pilot or his representative—one copy.
- Parachute issue facility—one copy.
- Unit suspense file—one copy.

c. **PJM/Aircrew Initial Coordination**. After DACO coordination, the PJM should proceed to the aircraft for initial coordination. Normally, the aircraft is open with a crewmember onboard one hour before station time. Items to be discussed, verified, or agreed to include:

(1) Aircraft configuration IAW the unit mission. If the aircraft is incorrectly configured, the requesting unit has the option to accept or reject it. If the mission request asks for both doors to be open for a mass troop drop but one door is inoperative (in flight or on the ground), the requesting unit has the option to continue the mission using one door or to abort the mission. For single door operations, one door may be used for personnel airdrop. However, if the other paratroop door is open, it must be manned by a current JM. If only one door is being used, it must be manned by a current JM/safety team. A single-door aircraft operation must be conducted in a military controlled air space environment.

(2) Control of the jump doors.

(3) Drop altitude, speed, and heading.

(4) Racetracks.

(5) Towed parachutist procedures (in detail).

(6) Emergency actions onboard.

(7) Time warnings and checkpoints.

(8) Type of drop, for example, CARP, GMRS, VIRS.

(9) Load time.

(10) Station time.

(11) Takeoff time.

(12) Initial contact time with CCT or DZST for update on DZ conditions (if communications are being used).

(13) Drop time.

(14) Additional details:

(a) If a ground abort occurs, designate which key personnel onboard must be advised.

(b) If the PJM is not the last parachutist, designate who is in command of the troops on board in an emergency.

(c) Emphasize to the aircrew the importance of accurate direction and velocity of DZ winds (before the 1-minute time warning) and accurate time warnings.

(d) Ensure seats are stowed by releasing the support leg from the aircraft floor, rotating the support leg 270 degrees until it rests on the seat, folding the seat up until it is vertical to the aircraft floor, and securing it there by attaching the clip provided on the stowage strap to the upper seat back support.

d. **Aircraft Inspection**. The PJM, accompanied by a crew member (usually the loadmaster), inspects the aircraft and coordinates any activities related to the airborne operation. The PJM must check the exterior and interior portions of the aircraft directly related to the airborne operation.

9-2. SAFETY PERSONNEL

While the PJM inspects the aircraft, safety personnel control the chalk, making sure personnel remain in assigned stick at all times. Personnel must also be accounted for at all times.

a. **Parachute Issue**. AJM/safety personnel supervise the chalk during parachute and air item issue. They draw extra aviator kit bags (1 for every 15 jumpers) and at least two extra reserve parachutes. The extra kit bags are used to store the static lines and deployment bags after the jump. (The extra kit bags are placed in or with the PJM kit bag.)

b. **Departure Airfield Layout Inspection**. All air items and combat equipment should be displayed for inspection by AJM/safety personnel before donning or loading on the aircraft.

c. **Parachutes and Equipment**. AJM/safety personnel ensure that all parachutists use the buddy system when donning parachutes and equipment. (Personnel should not start donning parachutes and equipment earlier than one hour before load time to avoid unnecessary time in the harness.)

d. **Final DACO Coordination**. If directed by the JM, AJM/safety personnel report to the DACO for any special or last-minute instructions that must be passed to the JM.

e. **JMPI**. AJM/safety personnel assist in rigging, inspecting, and correcting deficiencies as directed by the PJM. The PJM's role during JMPI is to observe and supervise. The PJM should only perform JMPI to facilitate meeting station time.

NOTE: All current and qualified JMs assist during the JMPI.

Section II. JUMPMASTER PERSONNEL INSPECTION

The PJM is responsible for the inspection of his parachutists before an airborne operation. Only by a complete and systematic equipment inspection of each parachutist can the PJM ensure that personnel aboard his aircraft are safe to jump.

9-3. HANDS-ON INSPECTION

During JMPI, the PJM uses a control hand and a working hand. During the course of the inspection, the control hand and working hand may change. The control hand remains stationary, while the eyes follow the working hand. The principle is to look at what is touched by the working hand. When locations are described in this sequence (top, right, left), it is in relation to the parachutist—**not** the JM. The word "trace" describes the working hand moving along the item being inspected to ensure that it is not twisted, cut, frayed, or misrouted. When tracing metal items, the JM inspects for sharp edges and proper assembly. The inspection sequence described in this section is followed for a typical combat parachutist rigged with LBE (exposed), combat pack (ALICE pack/FPLIF), M1950 weapons case, and main and reserve parachutes. (Other items and combinations of equipment are prescribed elsewhere in this manual.) When performing JMPI on these items, the JM starts at the point of attachment to the parachute harness and completely inspects that item before proceeding with the remainder of the inspection sequence.

9-4. BALLISTIC HELMET (FRONT)

The jumper moves to the JM and the JM tells him, "Open your rip cord protector flap." The JM then proceeds as follows:

a. Place both hands on the rim of the ballistic helmet, on the extreme right side, fingers extended and joined, palms facing the parachutist. Using the right hand as the working hand and the left hand as the control hand, trace across the rim of the ballistic helmet with the working hand to the opposite side of the ballistic helmet, inspecting for any sharp or protruding edges that may damage or cut the static line during aircraft exit. Once the hands are parallel, insert the thumb of each hand under the rim and feel for the locking nuts to ensure they are present and secured.

b. Gently tilt the parachutist's head back and look at the headband. Ensure that the smooth leather portion is toward the parachutist's head and the openings of the attaching clips are down, visible, and secured. Place the right index finger on the pull-the-dot fastener with tab. Ensure that it is a serviceable pull-the-dot fastener in that there are four plies of nylon, three of which must run through the snap portion, and it is secured.

c. Bypass the pull-the-dot fastener with tab and trace down to the point of attachment for the chin strap. Ensure the chin strap is properly routed through the adjusting buckle and that the parachutist retention strap is routed around the long continuous portion of the chin strap and under the adjusting buckle, with the pile portion of the parachutist retention strap away from the jumper's face.

d. Continue to trace the long continuous portion of the chin strap down under the parachutist's chin and back up to the point of attachment on the opposite side, and conduct the same inspection.

e. Finally, trace the inside of the nylon portion of the adjusting buckle until you make contact with the left thumb, which is still on the locking nut. Trace the short, sewn portion of the chin strap across the front of the parachutist's chin. Drop both hands.

9-5. ADVANCED COMBAT HELMET (FRONT)

The ACH is the programmed replacement for the ballistic helmet. It is compatible with NVDs and chemical defense equipment, allows maximum sensory and situational awareness, provides an unobstructed field of view, and increases ambient hearing capabilities. With proper sizing, the ACH retention and suspension system provides balance, stability, comfort, fit, ventilation, and superior impact protection throughout all static line airborne operational scenarios. The JM inspects the ACH as follows:

a. Move to the jumper's front and issue the command "OPEN YOUR RIP CORD PROTECTOR FLAP." Place your hands on the extreme right hand side of the helmet, fingers and thumb extended and joined, palms facing the jumper. The left hand is the control hand and the right hand is the working hand. With the working hand, trace across the rim of the ACH and feel for any sharp or protruding edges.

b. Place the right index finger on the ladder locks to ensure the chinstrap is properly routed through the ladder locks and free of all cracked components and is serviceable. Trace the chinstrap down to the fastener and ensure that it is free of all cracked components and is properly secured. Continue to trace the long continuous portion of the chinstrap under the jumper's chin to its point of attachment on the opposite side and conduct the same inspection. Place your index finger on the short sewn portion of the chinstrap and trace it across the jumper's chin.

9-6. CANOPY RELEASE ASSEMBLIES

The JM inspects the right canopy release assembly.

a. Look at the right canopy release assembly. Tap the canopy release assembly with the knuckles of the left hand and ensure that it sounds solid. Turn the right canopy release assembly one-quarter turn out and look at it closely. Ensure that it is properly assembled, has no cracked components, and is free of all foreign matter. Look at the left canopy release assembly and, with the right hand, conduct the same inspection. If the static line is resting against either canopy release assembly, move it so that it does not obstruct the view of the assembly.

b. When inspecting a parachute harness equipped with the modified D-rings, use the thumb of the left hand to pull down on the right upper D-ring to ensure it is secured to the jumper's right main lift web with gutted Type II or Type III nylon cord, and that it is routed under the jumper's chest strap. With the right hand, conduct the same inspection on the jumper's left upper D-ring.

9-7. CHEST STRAP

The JM inspects the chest strap next.

a. Simultaneously slide both hands down the main lift web until the little fingers make contact with either the D-rings or the L-bar connector link.

b. When two sets of D-rings are present, look at the left existing D-ring verifying it is rotated in the up position and safetied behind the main lift web using either Type II or Type III nylon cord (core threads removed).

c. Move the left hand below the "new" D-ring and rotate the main lift web inward, verifying the D-ring is positioned immediately below the existing D-ring and directly above the confluence wrap.

d. Visually verify the screw pin is fully seated. It will be slightly recessed below the screw pinhead housing and staked. (Any sign of screw pin misalignment immediately renders the harness unserviceable. It will not be jumped and must be removed from service.)

e. Repeat this inspection procedure for the right main lift web.

f. Look at the chest strap and ensure that it is not misrouted around the main lift web.

g. Insert the right hand (fingers and thumb extended and joined, fingers pointed skyward, palm facing the jumpmaster) from bottom to top behind the chest strap (next to where it is sewn into the main lift web)

h. Trace the chest strap until the right hand is behind the ejector snap, ensuring it is not twisted, cut, or frayed. The ejector snap pad will not come between the right hand and the ejector snap.

i. With the thumb of the right hand, press in on the activating lever for the ejector snap to ensure that it is properly seated over the ball detents and is free of all foreign matter. Leave the right hand and thumb in place and move to the right side of the jumper.

9-8. WAISTBAND

The JM begins the inspection by moving to the right side of the parachutist.

a. Insert the left hand (fingers and thumb extended and joined, fingers pointed skyward, palm facing the jumpmaster) from the bottom to the top behind the waistband (next to where it is sewn into the pack tray). Look at the waistband where it is sewn into the pack tray and ensure that at least 50 percent of one row of stitching is present.

b. Trace the waistband forward to ensure that it is not twisted, cut, or frayed and that it has not been misrouted behind the horizontal back strap. Trace the waistband forward until the left hand makes contact with the right D-ring. Look at the waistband to ensure that it is routed over the right main lift web and under the right D-ring.

c. Rotate the right hand down and grasp the top carrying handle of the reserve parachute, palm facing the reserve. Simultaneously lift up and out on the reserve parachute and place the left hand, palm facing the jumper, into the jumper's chest. Look at the waistband where it is routed behind the reserve parachute to ensure that it is routed through both waistband retainers and is not twisted, cut, or frayed.

d. Withdraw the left hand from the jumper's chest, reach under the right forearm, and grasp the left carrying handle of the reserve parachute (palm facing away from the reserve, fingers spread). Release the top carrying handle of the reserve with the right hand and move to the left side of the parachutist. Insert the right hand (fingers and thumb extended and joined, pointing up, and the palm facing away from the parachutist) from bottom to top behind the waistband as close to the left D-ring as possible. Look at the waistband to ensure that it is routed over the left main lift web and under the left D-ring.

e. Trace waistband back to the metal adjuster, ensuring that it is not twisted, cut, or frayed. Leave the right hand in place behind the metal adjuster. Remove the left hand from the left carrying handle and insert the index and middle fingers of the left hand into the quick release of the waistband, ensuring that it is a two- to three-finger quick release, and not a false quick release. Remove the index and middle fingers from the quick release and, with the index finger and thumb of the left hand, pinch off the free-running end of the waistband where it comes out of the metal adjuster. Trace the free-running end of the waistband until the fingers fall off the end. Ensure the free-running end is not cut, torn, or frayed and that it is easily accessible to the jumper.

f. Reinsert the left hand into the left carrying handle of the reserve, with the palm facing away from the reserve and the fingers spread. Look at the right hand and trace the waistband adjuster panel back to the pack tray, ensuring it is not twisted, cut, frayed, or misrouted under the horizontal back strap. Look at the waistband adjuster panel where it is sewn to the pack tray and ensure that at least 50 percent of one row of stitching is present. Drop both hands and move to the jumper's front.

9-9. RESERVE PARACHUTE

The JM begins by grasping the top carrying handle with the left hand (with the palm facing the reserve) and pulling up and out slightly. He looks at the left connector snap and, with the index finger or thumb of the right hand, checks the left connector snap to ensure that it is secured to the left D-ring, that it has spring tension, and is not safetied.

a. With the left hand, grasp the top carrying handle of the reserve, palm facing the reserve parachute, and lift up and out. Look at the left connector snap and, with the index finger of the right hand, finger the left connector snap one time to ensure that it is properly secured to the left D-ring, has spring tension, and has not been safetied. Grasp the top carrying handle of the reserve parachute with the right hand, palm facing the reserve, and lift up and out. Look at the right connector snap and, with the index finger of the left hand, finger the right connector snap one time to ensure that it is properly secured to the D-ring, does not have spring tension, and has been safetied. To further ensure that the right connector snap has been safetied, with the left index finger, pull down and out on the safety wire and lanyard to ensure that the safety wire has been routed through the drilled hole in the right connector snap. Insert the left index finger from the top to the bottom on the inside of the right connector snap to ensure the safety wire has been bent down toward the ground.

b. Place the right hand on the left end panel of the reserve parachute. Form a knife edge with the left hand, palm facing the jumpmaster, and trace from top to bottom between the rip cord grip and the right end panel of the reserve parachute to ensure that the rip cord grip is not winterized (unless the mission calls for it) and that the right pack opening spring band is not misrouted over the rip cord grip. Insert the index finger of the left hand into the rip cord grip stow pocket and feel for the steel swaged ball. With the left index finger and thumb, pinch off the first locking pin where it comes out of the rip cord grip stow pocket. Apply inward pressure on the left end panel with the right hand and physically seat the first locking pin from the jumper's right to the jumper's left through its cone. Trace the cable over to the second locking pin and continue tracing until the fingers fall off end, ensuring that the locking pins are not bent, cracked, or corroded and that the cable is not kinked or frayed.

c. Grasp the right end panel of the reserve parachute with the left hand and look at the first set of locking pins, cones, and grommets. Rotate the reserve parachute 360 degrees ensuring there is no exposed canopy or suspension lines and no rust or corrosion, and that the locking pins are properly assembled. Look at the second set of locking pins, cones, and grommets and repeat the same inspection. With the index finger and thumb of either hand, pinch off the rip cord protector flap, visually and physically ensure the log record book is present, then close the rip cord protector flap. When inspecting the MIRPS, ensure that a 1/4-inch strip of yellow binding tape has been permanently sewn across the top of the rip cord protector flap. The jumpmaster then places either hand on the bulge created by the deployment assistance device to ensure it is centered behind the rip cord protector flap.

d. Inspect the pack opening spring bands for exposed metal, spring tension, and proper routing. Form a knife edge with the left hand (fingers and thumb extended and joined, palm facing the jumpmaster) and sweep the top carrying handle and universal static line snap hook back toward the jumper (this will be the control hand). Begin the inspection with the top right pack opening spring band.

(1) With the index finger and thumb of the right hand, pinch off the tab portion of the top right pack opening spring band and pull it down toward the rip cord protector flap. Look at the pack opening spring band to ensure that it is routed through the reinforced nylon webbing at the back of the reserve, it is properly routed under the top carrying handle, and that there is no exposed metal on the pack opening spring band. When the tab portion of the pack opening spring band is released, the pack opening spring band should snap back into place. Repeat the same inspection for the top left pack opening spring bands. Place the control hand on the bottom right corner of the reserve parachute, lifting it up and out, so that the bottom left and bottom right pack opening spring bands can be inspected in the same manner.

(2) Lower the reserve parachute back to its original position, leaving the control hand on the bottom right corner of the reserve, and inspect the right pack opening spring band.

e. Conduct an overall inspection of the reserve parachute to ensure that it is free of grease, oil, dirt, mud, tears, and exposed canopy. Place both hands on the top right corner of the reserve parachute, palms facing the reserve. The left hand is the control hand and the right hand is the working hand. With the head and eyes 6 to 8 inches away from the working hand, trace across the top panel and down the left end panel. Move the control hand down to the bottom right corner of the reserve and lift the reserve parachute up and out, exposing the bottom panel. Trace the bottom panel across until contact is made with the control hand. Lower the reserve parachute back to its original position, returning the control hand to the top right corner. Flip the right hand over and trace up the right end panel until contact is made with the control hand. Raise the control hand up out of the way and trace across the top right corner where the control hand had been. Raise the reserve parachute to the jumper and command, HOLD, SQUAT.

9-10. LEG STRAPS

Insert the index and middle fingers of both hands, from outside to inside, under the leg straps right below the aviator's kit bag where the natural pocket is formed. Simultaneously slide both hands rearward on the leg straps back to the saddle, ensuring that the leg straps are not crossed.

a. With the left hand, trace the right leg strap forward to the quick-fit V-ring. Ensure that the leg strap is not twisted, cut, or frayed. With the left thumb, push in on the activating lever of the ejector snap to ensure it is properly seated over the ball detents and is free of all foreign matter.

b. Leave the left hand in place and look at the left leg strap. With the right hand, trace the left leg strap up to the quick-fit V-ring, ensuring that it is not twisted, cut, or frayed, and is properly routed through the exposed carrying handle of the aviator's kit bag, over the bottom and under the top. With the right thumb, push in on the activating lever of the ejector snap to ensure it is properly seated over the ball detents and is free of all foreign matter. Look at the aviator's kit bag and ensure that it has not been reversed and that the sewn, reinforced portion is facing away from the jumper. Tap the jumper on the thigh and command, RECOVER.

9-11. STATIC LINES

If the static line has not been routed over the parachutist's shoulder, the JM routes it at this time. The following explanation is for the static line when routed over the right shoulder. For the left shoulder, the hands are reversed.

a. **Static Line.**

(1) With the right hand, secure the static line snap hook. Pull up on the static line snap hook and ensure that it is attached to the top carrying handle of the reserve parachute. Open the palm and look at the static line snap hook. Place the index finger of the left hand on the static line snap hook next to the drilled hole and ensure the hole for the safety wire is present and free of all obstructions.

(2) Regrasp the static line snap hook and, with the left thumb and index finger, pinch the safety wire and lanyard at its point of attachment; look at the safety wire and lanyard. Ensure that the lanyard is attached to the upper loop portion of the static line and not to the cut-away portion of the static line snap hook. Trace the safety wire and lanyard to ensure it is not too long, too short, or excessively bent. Drop the safety wire and lanyard.

(3) Grasp the static line above the static line snap hook with the left hand, palm facing the jumper, thumb down just above the upper loop portion. Rotate it to the parachutist's right and push it toward the static line snap hook, exposing the inside of the looped portion of the static line. Look inside the looped portion of the static line to ensure that it is not cut, frayed, or burned. Rotate the static line snap hook 180 degrees and inspect it again in the same manner. Drop the left hand.

(4) Use either hand to form a bight in the static line, and inspect the static line retainer to ensure it is not cut, torn, or frayed more than 50 percent. Insert the static line bight from top to bottom through the static line slack retainer and pull the excess through. Flip the bight on top of the pack tray and place either hand on it. The hand that controls the bite becomes the control hand. With the index finger and thumb of working hand pinch off the first stow and pull it one or two inches toward the center of the pack tray. Look behind the stow to ensure that the static line has not been misrouted around the static line stow bar and that it is free of cuts, frays, or burns. Release the first stow and let it pop back into place.

(5) Insert the index finger of the working hand from bottom to top behind the first strand of static line as close as possible to the first stow. Trace the first strand static line over to the second stow to ensure that it is free of all cuts, frays, and burns. Once contact is made with the second stow, pinch it off with the index finger and thumb of the working hand, pull it one to two inches toward the center of the pack tray, and conduct the same inspection. Place the index finger or thumb of the working hand behind the second strand of static line and trace it away from you ensuring it is not cut, frayed, or burned. Continue to inspect the static line in the same manner down to the pack opening loop ensuring that you inspect the last strand of static line with the index finger only. Ensure that the last strand of universal static line is routed from the right outer static line stow bar.

(6) Once contact is made with the pack opening loop ensure that it is situated between the pack closing loops at the 6 and 9 o'clock position. Insert the index finger of the working hand from bottom to top into the pack opening loop. Pull down and out on the pack opening loop, look inside to ensure the pack closing tie has been routed through the pack opening loop, and that the pack opening loop is not torn or frayed. Let the pack opening loop pop off your finger.

(a) Place the index finer of the working hand on the pack closing loop at the 6 o'clock position. Look at the pack closing loop to ensure the pack closing tie is routed through the pack closing loop and the pack closing loop is not torn or frayed more than 50 percent. Inspect the remaining pack closing loops in the same manner using a clockwise motion (9 o'clock, 12 o'clock, 3 o'clock).

(b) Look at the pack closing tie and the surgeon's knot locking knot. Ensure the surgeon's knot locking knot is properly positioned between the pack closing loops at the 3 o'clock and 6 o'clock position. Insert the index finger of the working hand from bottom to top behind the surgeon's knot locking knot. Pull down and out to ensure it is secure and that the pack closing tie has been properly constructed of one turn **only** of 1/4-inch cotton webbing. Let the pack closing tie pop off the end of your finger, stand up directly behind the jumper, and issue the command TILT YOUR HEAD FORWARD.

b. **Universal Static Line.**

(1) With the right hand, grasp the USL snap hook and give it a slight tug, ensuring that the USL snap hook is secured to the top carrying handle of the reserve parachute. Let the USL snap hook rest in the palm of the right hand. Place the index finger of the left hand on the girth hitch to ensure that it is routed around the narrow portion of the static line snap hook and that the green marking stitching is present on the static line. Place your index finger on the rivet ensuring that it is present, then place the index finger on the opening gate of the USL snap hook and finger it one time ensuring that it has spring tension and that the opening gate is facing toward the jumper. Regrasp the USL snap

hook and position it perpendicular to the reserve. With the left hand palm facing the jumper, thumb pointing down, grasp the USL just above the girth hitch and rotate it down into the jumper's right. Push the static line toward the girth hitch and inspect the inside upper loop portion of the static line ensuring it is free of all cuts, frays, or burns. With the right thumb or index finger, push in on the upper loop portion of the static line and conduct the same inspection. Redress the static line around the narrow portion of the USL snap hook.

(2) Since the static line is routed over the jumper's right shoulder, form an "O" with the index finger and thumb of the right hand around the USL snap hook. The right hand now becomes the working hand. Raise the hand, simultaneously inspecting the static line as it runs through the "O" ensuring the static line is free of all cuts, frays, or burns. When the working hand is as high as it will go, issue the command TURN. Once the jumper has completed the turn, the right hand should have been raised high enough to pull all of the slack from the static line slack retainer. Keep the USL tight between the control hand and the first stow and place the index finger (or index finger and middle finger) of the working hand behind the USL below the control hand so there is skin-to-skin contact. Trace the USL down to the first stow ensuring that it is free of all cuts, frays, and burns and it has not been misrouted under or through either riser assembly.

(3) The JM will then use either hand to form a bight in the static line. The JM will now inspect the static line retainer to ensure it is not cut, torn, or frayed more than 50 percent. Insert the static line bight from top to bottom through the static line slack retainer and pull the excess through. Flip the bight on top of the pack tray and place either hand on it. The hand that controls the bite becomes the control hand. With the index finger and thumb of the working hand, pinch off the first stow and pull it one or two inches toward the center of the pack tray. Look behind the stow to ensure that the USL has not bee misrouted around the static line stow bar and it is free of cuts, frays, or burns. Release the first stow and let it pop back into place.

(4) Insert the index finger of the working hand from bottom to top behind the first strand of USL as close as possible to the first stow. Trace the first strand of USL over to the second stow to ensure that it is free of all cuts, frays, and burns. Once contact is made with the second stow, pinch it off with the index finger and thumb of the working hand, pull it one to two inches toward the center of the pack tray, and conduct the same inspection. Place the index finger or thumb of the working hand behind the second strand of static line and trace it away from you ensuring it is not cut, frayed, or burned. Continue to inspect the USL in the same manner down to the pack opening loop ensuring that you inspect the last strand of static line with the index finger only, and the last strand of USL is routed from the right outer static line stow bar.

(5) Once contact is made with the pack opening loop ensure that it is situated between the pack closing loops at the 6 and 9 o'clock position. Insert the index finger of the working hand from bottom to top into the pack opening loop. Pull down and out on the pack opening loop and look inside to ensure the pack closing tie has been routed through the pack opening loop and that the pack opening loop is not torn or frayed. Let the pack opening loop pop off your finger. Place the index finger of the working hand on the pack closing loop at the 6 o'clock position.

(a) Look at the pack closing loop to ensure the pack closing tie is routed through the pack closing loop and the pack closing loop is not torn or frayed more than 50 percent.

Inspect the remaining pack closing loops in the same manner using a clockwise motion (9 o'clock, 12 o'clock, 3 o'clock).

(b) Look at the pack closing tie and the surgeon's knot locking knot. Ensure the surgeon's knot locking knot is properly positioned between the pack closing loops at the 3 o'clock and 6 o'clock position. Insert the index finger of the working hand from bottom to top behind the surgeon's knot locking knot and pull down and out. Ensure it is secure and that the pack closing tie has been properly constructed of one turn **only** of 1/4-inch cotton webbing. Let the pack closing tie pop off the end of your finger, stand up directly behind the jumper, and issue the command TILT YOUR HEAD FORWARD.

9-12. BALLISTIC HELMET (BACK)

Place both hands on the rim of the ballistic helmet on the extreme left-hand side, fingers and thumb extended and joined, fingers pointing skyward, and palms facing the jumper.

a. The left hand is the control hand and the right hand is the working hand. With the working hand, trace the rim of the ballistic helmet across to the opposite side, ensuring there are no sharp or protruding edges that could cut or damage the jumper's static line upon exiting the aircraft.

b. Once the hands are parallel, place the thumb of each hand on the rim of the ballistic helmet and tilt the jumper's head forward. Look at the parachutist's retention strap to ensure that it is properly routed between the helmet shell and the foam impact pad and that it is not twisted.

c. With the index finger and thumb of either hand, pinch off the foam impact pad and pull down slightly to ensure that it is secured to the shell of the ballistic helmet and that it is serviceable.

9-13. ADVANCED COMBAT HELMET (BACK)

Place both hands on the left side of the ACH with the fingers extended and joined, fingers pointing skyward, with the palms facing the jumper. The left hand is the control hand and the right hand is the working hand.

a. With the working hand, trace the rear of the ACH checking for sharp or protruding edges that could cut or fray the static line. Once both hands are parallel with each other, tilt the jumper's head forward.

b. Place the right index finger on the ladder lock to ensure the back adjustment strap is properly routed through the ladder lock, is free of all cracked components, and is serviceable. Trace the back adjustment strap forward until contact is made with the chin strap to ensure the back strap is not twisted, cut, or frayed. Place the left index finger on the left ladder lock and conduct the same inspection.

c. Conduct a visual inspection of the nape pad to ensure it is present, secure, and serviceable.

NOTES:
1. There are four different configurations for the individual pads inside the ACH.
2. Ensure there is a minimum of seven pads inside the ACH and four pads are covering the locking nuts.
3. The pads in the rear should extend slightly past the bottom of the ACH.

9-14. RISER ASSEMBLIES

Reach as far forward over the jumper's shoulders as possible and, with each hand, grasp a riser assembly just above the canopy release assemblies.

a. Give the left riser assembly a sharp tug to the rear. Open the hand to form an "L." Apply upward pressure with the left thumb and trace the riser assembly rearward to where it disappears into the main pack tray, ensuring that it is not twisted, cut, torn, or frayed. Leave your left hand in place and repeat the same procedure for the right riser assembly.

b. Ensure a log record book is present in either riser assembly.

9-15. PACK TRAY

An overall inspection of the pack tray is conducted to ensure the pack tray is free of grease, oil, dirt, mud, or tears. Place both hands on the top left corner of the pack tray, palms facing the pack tray. The left hand is the control hand and the right hand is the working hand.

a. With the head and eyes 6 to 8 inches away from the working hand, trace across the top pack closing flap, down the right pack closing flap, across the bottom pack closing flap, flip the right hand over, and trace up the left pack closing flap.

b. When the working hand makes contact with the control, raise the control hand out of the way and trace across the top left corner of the pack tray where the control hand had been. Form knife edges with both hands, palms facing the JM, and issue the command ARCH YOUR BACK.

9-16. DIAGONAL BACK STRAPS

Insert each hand under the "X" formed by the diagonal back strap.

a. Look at the diagonal back straps to ensure they have been properly routed over the appropriate shoulder and that the top diagonal back strap has one more row of exposed stitching than the one on the bottom

b. Look at the diagonal back strap retainers to ensure that they are routed through the sizing channels in the diagonal back straps. The diagonal back strap retainers are routed around the diagonal back strap keepers and the pull-the-dot fasteners are secured.

c. With both thumbs, apply upward pressure to the pull-the-dot fasteners to further ensure that they are secure. Focus attention on the left hand and the left side of the jumper.

d. With the left hand, trace down the diagonal back strap to the back strap adjuster, ensuring it is not twisted, cut, or frayed. Grasp the back strap adjuster with the left hand and focus attention on the jumper's right side.

e. With the right hand, trace the diagonal back strap, ensuring that it is not twisted, cut, or frayed; bypass the back strap adjuster; and continue the inspection of the horizontal back strap.

9-17. HORIZONTAL BACK STRAP

The JM picks up the horizontal back strap.

a. Trace the horizontal back strap down to where it disappears into the main lift web. Withdraw the right hand from under the horizontal back strap and reinsert it (fingers and thumb extended and joined, fingers pointed skyward, palms facing the JM) from bottom to top behind the horizontal back strap where it reemerges from the main left web. Issue the jumper the command, BEND FORWARD AT THE WAIST.

b. Release the back strap adjuster and, with the left hand, reach across the pack tray and grasp the center of the bottom pack closing flap. With the head and eyes 6 to 8 inches from the working hand, trace the horizontal back strap across the jumper's back, ensuring that it is not twisted, cut, or frayed. Ensure that the horizontal back strap is properly routed through both horizontal back strap retainers, the horizontal back strap retainers are routed around the horizontal back strap keepers, and the pull-the-dot fasteners are secured.

c. Continue tracing the horizontal back strap to where it disappears into the main lift web on the left side of the jumper. With the left hand, palm facing skyward, regrasp the back strap adjuster on the jumper's left side.

d. Withdraw the right hand from behind the horizontal back strap and reinsert it (fingers and thumb extended and joined, palm facing the JM) from top to bottom or bottom to top behind the horizontal back strap where it reemerges from the main lift web. Trace the horizontal back strap up to where contact is made with the control hand still in place on the back strap adjuster. Withdraw the right hand from behind the horizontal back strap and get left hip to left hip with the jumper.

9-18. SADDLE

Place the fingertips of the right hand (fingers and thumb extended and joined, fingers pointed down, palm facing the jumper) on the accessory attaching ring of the lowering line adapter web or the single "X" boxed stitch on the left main left web.

a. Trace the saddle across the jumper's buttocks, ensuring that the saddle is not inverted, twisted, cut, or frayed, and that neither leg strap has been misrouted around the saddle. Trace the saddle until contact is made with the single "X" boxed stitching on the right main lift web.

b. Give the jumper the good seal of approval by lightly tapping him on the buttocks and issuing the command RECOVER.

9-19. WEAPONS CASE, M1950

The M1950 weapons case will be inspected in its entirety before inspecting the reserve parachute.

a. The inspection of the M1950 weapons case begins with its point of attachment on the left D-ring. Look at the opening gate of the quick-release snap to ensure that the opening gate is facing the jumper's body and is the outermost item on the left D-ring (unless the harness is not equipped with either the lowering line adapter web or triangle links). With the right index finger, finger the opening gate one time to ensure that it is properly attached to the left D-ring, it has spring tension, and it has not been safetied. With the heel of the right hand, press up on the activating arm of the quick-release snap to ensure that it is seated between the ball detents. With the index finger of the right hand, trace down until contact is made with the V-ring. Ensure the quick-release link is routed through the V-ring and is secured by the rotating claw.

b. Continue to trace down the inside of the M1950 weapons case until contact is made with the adjusting strap. Ensure the adjusting strap is routed through the appropriate set of adjusting strap connectors, secured by means of a half hitch, and is not twisted, cut or frayed. Continue tracing down the inside of the M1950 weapons case until the fingers fall off the bottom.

c. Form a knife-edge with the right hand, palm facing skyward, and trace from front to rear along the bottom of the M1950 weapons case to ensure the muzzle of the weapon is not protruding.

d. Place the index finger of the right hand on the slide fastener at the bottom of the closing flap. Ensure the slide fastener is secure by tracing up the outside of the M1950 weapons case to the vicinity of the lift fastener. With the index finger and thumb of the right hand, pinch off the slide fastener and tab thong. Pull down and out to ensure the slide fastener and tab thong are secured by the upper tie-down tape or have been separated over the lift fastener.

e. Drop the right hand down 10 to 12 inches from the top of the M1950 weapons case and give it a sharp slap, feeling for the forward assist of the M16-series rifle or the charging handle of the M249.

f. With the index finger and thumb of the right hand, pinch off the bow knot of the upper tie-down tape on the front of the M1950 weapons case. Visually inspect the upper tie-down tape to ensure it is properly routed behind the M1950 weapons case, around the main lift web, above the chest strap, and secured by a bow knot. This concludes the inspection of the M1950 weapons case.

g. With the left hand, grasp the top carrying handle of the reserve parachute, palm facing the reserve parachute, and lift up and out. Place the index finger of the right hand on the inner mounting screw of the L-bar connector link to ensure that it is present and tightly secured. Rotate the right index finger around to the outer mounting screw of the L-bar connector link and conduct the same inspection.

h. Place the right index finger behind the main lift web from the outside to the inside and feel for the exposed metal of the L-bar connector link. Inspect the reserve parachute in the same manner as if it were on a Hollywood jumper and issue the jumper the command HOLD.

9-20. ALICE PACK WITH HARNESS, SINGLE-POINT RELEASE, AND HOOK-PILE TAPE LOWERING LINE

After the JM has completed the overall inspection of the reserve parachute, he then inspects the ALICE pack.

a. With the left hand, secure the snap hook for the right adjustable D-ring attaching strap. With the right hand, secure the snap hook for the left adjustable D-ring attaching strap. Begin the inspection with the right adjustable D-ring attaching strap. Look at the snap hook to ensure that the opening gate is facing the jumper's body and it is attached to the outside of the right connector snap. With the index finger of the left hand, finger the opening gate one time to ensure that it is properly secured to the right D-ring and has spring tension. With the left thumb, flip the free-running end of the right adjustable D-ring attaching strap out of the way.

b. Place the index of the left hand on the front of the right adjustable D-ring attaching strap just below the snap hook. Trace down the right adjustable D-ring attaching strap until contact is made with the white attaching loop, ensuring that it is not twisted, cut, or frayed. Once contact is made with the white attaching loop, bypass the triangle link and pick up the inspection of the white attaching loop in front of the triangle link.

c. With the left index finger, trace down the attaching loops to ensure that the white attaching loop is routed from bottom to top through the triangle link, the green attaching loop is routed from bottom to top through the white attaching loop, the red attaching loop is routed from bottom to top through the green attaching loop, and all are routed from bottom to top through the grommet in the female portion leg strap release assembly. Place the index finger of the left hand on the single "X" boxed stitch on the release handle cross strap.

d. Look at the release handle cable where it emerges from the release handle cross strap. Ensure the release handle cable is properly routed through the red attaching loop and secured by the cable loop retainer. Leave the left index finger in place and, with the right hand; conduct the same inspection on the left adjustable D-ring attaching strap. After inspecting the left adjustable D-ring attaching strap, focus on the release handle.

e. With the right hand, lift up on the release handle. Ensure the release handle is properly routed between the two plies of the release handle cross strap and secured by the hook-pile tabs.

f. Form knife edges with each hand, palms facing each other, and place them on the equipment retainer straps. Trace the equipment retainer straps down between the external cargo compartments of the ALICE pack until you make contact with the adjustable cross strap.

g. Leave the left hand in place and with the right hand grasp the free-running end of the adjustable cross strap and give it a tug to the jumper's left, ensuring that all the slack has been removed from the adjustable cross strap. Raise the ALICE pack to the jumper and issue the command HOLD.

h. Reform the knife edges with each hand, palms facing each other, and place them on the equipment retainer straps where they come out of the cushioned envelope at the top of the ALICE pack. Trace the equipment retainer straps down, bypassing the girth hitch formed by the hook-pile tape lowering line, until contact is made with the friction adapters. Look at the friction adapters to ensure the equipment retainer straps are properly routed through the friction adapters.

i. Simultaneously, place the index finger and middle finger of each hand on top of the quick releases to ensure they are no more than three fingers and no less than two. With the thumbs of each hand, simultaneously pick up the free-running ends of the equipment retainer straps. Look at them to ensure that they are "S"-folded or accordion folded and not rolled and are secured by two turns of retainer band.

j. With the left hand, palm facing the ALICE pack, grasp the hook-pile tape lowering line in the vicinity of the girth hitch. Look at and pull up on the girth hitch to ensure it is properly routed from north to south and not east to west to the "X" formed by the equipment retainer straps.

k. Place the index finger of the right hand on the hook-pile tape lowering line so there is skin-to-skin contact with the control hand. Begin tracing the hook-pile tape lowering line ensuring that it is routed over the left adjustable shoulder-carrying strap. Continue to trace the hook-pile tape lowering line until contact is made with the first set of hook-pile tabs. Ensure the hook-pile tabs are secured and no folds of the hook-pile tape lowering line are protruding from the retainer flap.

l. Continue to trace down the hook-pile tape lowering line ensuring that it is secured to the tubular portion of the frame by two retainer bands. Continue tracing the hook-pile tape lowering line until contact is made with the second set of hook-pile tabs. Ensure they are properly secured and no folds of the hook-pile tape lowering line are protruding from the retainer flap. Continue to trace the hook-pile tape lowering line until it disappears behind the nylon chaff of the M1950 weapons case. Leave the right hand in place and, with the left hand, release the hook-pile tape lowering line.

m. With the left hand, grasp the trail edge of the M1950 weapons case and pull it forward. With the right index finger, pick up the hook-pile tape lowering line on the backside of the nylon chaff and trace it up to its point of attachment. Once the hook-pile tape lowering line has been traced to its point of attachment, look at the ejector snap to ensure the opening gate is facing the jumper's body.

n. With the right thumb, press in on the activating lever to ensure that it is properly seated over the ball detent and free of all foreign matter. Turn the ejector snap 1/4-turn out to ensure the small tooth is present.

o. Visually inspect the yellow safety lanyard to ensure that it is serviceable and has not been wired, tied, or taped down.

p. Place the index finger of the right hand on the surgeon's knot locking knot to ensure that the lowering line adapter web has been tied off to the main lift web.

q. With the right index finger, trace up the lowering line adapter web until it disappears behind the waistband, ensuring that it is not twisted, cut, or frayed, or misrouted in any manner.

r. Move back to the front of the jumper and issue the command SQUAT. Continue to inspect the remainder of the jumper in the same manner as a Hollywood jumper.

9-21. JMPI OPTIONS WITH COMBAT EQUIPMENT

The jumpmaster is responsible for inspecting each jumper and item of equipment as outlined in paragraphs 9-3 through 9-20. Long flights, heavy loads, extreme heat, tactical updates, and special items of equipment (DMJP, STINGER, and so on) may warrant delaying the attachment of equipment until a predesignated time later in the mission timeline. Under such conditions, the airborne commander may decide to conduct the JMPI without equipment. The commander **must** consider individual and unit proficiency when conducting the command risk assessment. When equipment is attached after the JMPI, a jumpmaster should attach and inspect the equipment.

9-22. MOLLE

Rigging and JMPI procedures for the MOLLE are contained in this paragraph.

a. **Rigging Procedures.** Rigging procedures include preparing the MOLLE for rigging and securing the HSPR to the MOLLE.

(1) ***Preparation.*** Prepare the MOLLE for rigging as follows:

(a) Remove the excess slack from both adjustable shoulder straps.

(b) Secure all excess webbing (roll and tape).

(c) Attach two retainer bands to the left side of the frame.

(d) When jumping special items of equipment (60-mm modified M1950 weapons case, AT4JP, DMJP, SMJP, or side-mount AIRPAC) or any other equipment that exceeds

30 inches in width, either empty the outer pockets or remove them and place them inside the MOLLE.

(2) ***Secure the HSPR to the MOLLE.*** Both sides of the HSPR are prepared in the same manner. Secure the HSPR to the MOLLE as follows:

(a) Place the HSPR on a smooth flat surface with the attaching loops facing up and the equipment retainer straps toward the jumper. Remove all twists or turns from the equipment retainer straps and the male portion leg strap release assembly.

(b) Route the release handle assembly between the two plies of the release handle cross strap from bottom to top (release handle cable first). Secure the release handle assembly in place with the hook-pile tape.

(c) Place the adjustable D-ring attaching strap alongside the HSPR so the opening gate of the snap hook is facing downward and the triangle link is over the white attaching loop.

(d) Route the white attaching loop from bottom to top through the triangle link; route the green attaching loop from bottom to top through the white attaching loop; route the red attaching loop from bottom to top through the green attaching loop and then through the grommet on the female portion of the leg strap release assembly. Ensure the cable loop retainer is facing upward.

(e) Route the release handle cable through the red attaching loop and the cable loop retainer.

(f) Rotate the HSPR over (so the bottom is on top) and remove all twists and turns. Ensure all running ends of webbing on the MOLLE are properly secured with an elastic keeper or retainer band.

(g) With the frame side facing up, place the MOLLE upside down with the bottom of the pack toward the adjustable D-ring attaching straps on the HSPR.

(h) Route the equipment retainer straps through the carrying strap on the top of the pack (bottom, as rigged), under the top horizontal support of the frame (between the shoulder straps), and over the back pad. Cross the retainer straps forming an "X" pattern at the back of the MOLLE.

(i) From the bottom of the MOLLE (top, as rigged), route the two friction adapters between the frame and the waist belt. Secure the retainer straps to their appropriate friction adapter. Form a quick release in each equipment retainer strap.

(j) Adjust the HSPR before tightening. On the front of the MOLLE, ensure the equipment retainer straps are routed on each side of the pouch. Ensure the white attaching loops are in line with the center of the bottom portion of the MOLLE (top, as rigged).

(k) Tighten the equipment retainer straps by pulling on the lower looped portion of the quick releases using a see-saw method. After tightening, ensure the quick releases are accessible for the JM to inspect during JMPI.

(l) Adjust the length of the quick releases so they are no shorter than two fingers width and no longer than three fingers width.

(m)S-fold and secure the equipment retainer straps with masking tape **or** retainer bands—always one, never both. Do not secure the quick releases to the S-folds.

(n) Route the male portion of the leg strap release assembly from the point where it is sewn to the equipment retainer strap by the most direct route down the side of the MOLLE and attach it to the female portion of the leg strap release assembly forming the adjustable leg strap.

(o) Tighten and S-fold, or roll, the excess webbing and secure in the webbing retainer. Tighten both adjustable shoulder-carrying straps and secure excess webbing.

(p) Secure any exposed items with 1/4-inch cotton webbing, or tape.

(r) Bend each side of the waist belt toward the frame and secure to the frame with a retainer band or 1/4-inch cotton webbing.

b. **JMPI Procedures.** After the JM has completed the inspection of the reserve parachute, he lifts up on the reserve parachute and issues the command HOLD. He then inspects the MOLLE as follows:

(1) With the left hand, secure the snap hook on the right adjustable D-ring attaching strap. With the right hand, secure the snap hook on the left adjustable D-ring attaching strap. Begin the inspection with the right adjustable D-ring attaching strap.

(a) Look at the snap hook to ensure that the opening gate is facing the jumper's body and is the outermost item on the right D-ring. With the left index finger, finger the opening gate to ensure that it is properly secured to the right D-ring, and it has spring tension.

(b) With the left thumb, flip the free-running end of the right adjustable D-ring attaching strap out of the way.

(c) With the index and middle finger of the left hand behind the right adjustable D-ring attaching strap, trace down the right adjustable D-ring attaching strap until contact is made with the white attaching loop, ensuring that strap is not twisted, cut, or frayed.

(d) Once contact is made with the white attaching loop, bypass the triangle link and pick up the inspection of the white attaching loop in front of the triangle link. With the left index finger, trace down the attaching loops to ensure that the white attaching loop is routed from bottom to top through the triangle link, the green attaching loop is routed from bottom to top through the white attaching loop, and the red attaching loop is routed from the bottom to the top through the green attaching loop and then through the grommet in the female portion of the leg strap release assembly. The only way to ensure that the red attaching loop has been routed properly is to pull back on the female portion of the leg strap release assembly and look behind it.

(e) Place the left index finger on the single "X" boxed stitch on the release handle cross strap just below the female portion of the leg strap release assembly. Look at the release handle cable where it comes out of the release handle cross strap to ensure that it is routed through the red attaching loop and is secured in the cable loop retainer.

(f) Leave the left hand in place and focus attention on the right hand and left adjustable D-ring attaching strap. Inspect it in the same manner.

(2) Once the inspection of the left adjustable D-ring attaching strap has been completed, focus attention on the release handle. With the right hand, gently pick up on the release handle assembly and ensure that it has been routed through the two plies of the release handle cross strap and is secured by the hook-pile tab provided.

(a) Form a knife edge with each hand, palms facing each other, and place them on the single "X" boxed stitch. Simultaneously trace down the equipment retainer straps to the outside of the pouch of the MOLLE pack until contact is made with the adjustable cross strap.

(b) Leave the left hand in place. With the right hand grasp the free-running end of the adjustable cross strap and give it a sharp tug to the jumper's left, ensuring that all of the slack has been taken out of the adjustable cross strap.

(c) Continue to trace down the equipment retainer straps until your fingers fall off the bottom.

(3) Raise the MOLLE pack up and visually inspect the equipment retainer straps to ensure they are routed through the top carrying strap, under the top horizontal support frame, and over the back pad. Then issue the next command of HOLD.

(a) Reform knife edges with each hand, palms facing each other, and place them on the equipment retainer straps where they reemerge under the support frame on top of the back pad. Simultaneously trace down the equipment retainer straps bypassing the girth hitch formed by the hook-pile tape lowering line, and trace all the way down to the friction adapters. Look at the friction adapters to ensure that the equipment retainer straps have been properly routed through them.

(b) Simultaneously place the index and middle finger of each hand on top of the two- to three-finger width quick releases in the equipment retainer straps to ensure that they are no more than three fingers width and no less than two fingers width.

(c) Simultaneously pick up the free-running ends of the equipment retainer straps with your thumbs. Look at them to ensure that they are S-folded (or accordion folded) and secured by at least two turns of a retainer band **or** masking tape—always one, never both.

(4) Next, secure the free-running ends of the adjustable shoulder-carrying straps and conduct a visual inspection ensuring that they are S-folded (or accordion folded) and secured with a retainer band **or** masking tape—always one, never both.

c. **Hook-Pile Tape Lowering Line.** Inspect the hook-pile tape lowering line as follows:

(1) With the left hand (palm facing the MOLLE pack, thumb down), grasp the hook-pile tape lowering line just above the girth hitch. This is the control hand. Look at and pull up on the girth hitch to ensure it is properly secured to the "X" formed by the equipment retainer straps, from north to south, and is centered on the back pad.

(2) Place the index finger of the right hand on the hook-pile tape lowering line directly beside the control hand making skin-to-skin contact. Begin tracing the hook-pile tape lowering line to ensure that it is routed over the left adjustable shoulder-carrying strap. Continue to trace the hook-pile tape lowering line until contact is made with the first hook tab.

(3) Ensure that the first hook tab is properly secured and no folds of the hook-pile tape lowering line are protruding out of the retainer flaps. Continue to trace down the hook-pile tape lowering line ensuring that it is secured to the frame of the MOLLE by a retainer band on each end of the retainer flap. Continue tracing the hook-pile tape lowering line until contact is made with the second hook tab to ensure that it is properly secured and no folds of the hook-pile tape lowering line are protruding out of the retainer flaps. Continue to trace the hook-pile tape lowering line until it disappears behind the nylon chafe on the M1950 weapons case.

(4) Release the hook-pile tape lowering line with the left hand and grasp the trail edge of the M1950 weapons case and pull it forward.

(5) With the right index finger, continue the inspection of the hook-pile tape lowering line on the backside of the nylon chafe and trace it up to its point of attachment.

(6) Once the hook-pile tape lowering line has been traced to its point of attachment, look at the ejector snap to ensure the opening gate is facing the jumper's body. With the

right thumb, press in on the activating lever to ensure that it is properly seated over the ball detent and free of all foreign matter. With the right index finger and thumb, pinch off the yellow safety lanyard and trace down to ensure that it is serviceable and has not been wired, tied, or taped down.

(7) With the right index finger, pull out on the accessory attaching ring of the lowering line adapter web to ensure that it has been tied off to the main lift web. With the right index finger, trace up the lowering line adapter web until it disappears behind the waistband ensuring that it is not twisted, cut, frayed, or misrouted in any manner.

(8) Move back to the front of the jumper, issue the command SQUAT, and inspect the remainder of the jumper in the same manner as a Hollywood jumper.

9-23. JMPI SEQUENCE FOR AIRPAC

The jumpmaster inspects the AIRPAC as follows:

NOTE: This JMPI sequence is for the front-mounted AIRPAC when jumped with a M1950 weapons case.

a. Place the left hand on the snap hook for the right D-ring attaching strap and the right hand on the snap hook for the left D-ring attaching strap. Leave the right hand in place and focus attention on the left hand and the right D-ring attaching strap.

(1) With the index finger of the left hand, finger the opening gate to ensure it has spring tension and is properly attached to the right D-ring to the outside of the connector snap. Place the index and middle finger of the left hand from outside to inside behind the nylon portion of the D-ring attaching strap and trace down until the middle finger makes contact with the circular ring, ensuring the strap is not twisted, cut, or frayed.

(2) Bypass the circular ring and place the index finger on the red attaching loop. Visually inspect the circular release ring to ensure it has been routed through the circular ring and folded back against the nylon portion of the D-ring attaching strap. Ensure the red attaching loop has been routed bottom to top through the circular release ring and then through the grommet on the female portion of the leg strap release assembly. With the thumb and index finger, peel back on either the top or bottom corner of the female portion of the leg strap release assembly to further ensure the red attaching loop has been properly routed.

(3) Place the index finger of the left hand on the single "X" boxed stitch just below the circular ring. Conduct a visual inspection of the release handle cable to ensure it is properly routed under both plies of the release handle cross strap and through the red attaching loop. Conduct the same inspection on the left D-ring attaching strap.

b. Leave the left index finger in place and, with the right hand, gently lift up on the release handle to ensure it has been properly routed under both plies of the release handle cross strap and has been secured by the hook-pile tabs.

c. Form a knife edge with both hands, palms facing each other, and place them on the outer edge of the equipment retainer straps. Simultaneously trace down the equipment retainer straps until contact is made with the friction adapters. Leaving both hands in place, look at the friction adapters, ensuring the equipment retainer straps have been properly routed through them and the free-running ends are secured by means of the webbing retainers.

d. With either hand, carefully open the circular closing flap and hold it up and out of the way. This hand is now the control hand. Place the index finger of the working hand on the single "X" boxed stitch on the 6-o'clock set of equipment retainer straps. Visually inspect both friction adapters to ensure the free-running ends of the equipment retainer straps are properly routed through them and are secured by means of the webbing retainers.

e. Conduct the same inspection for the 9-o'clock and 3-o'clock set of friction adapters. Place the index finger of the working hand on the eyelet of the cotter pin.

f. Ensure that the white grommet securing loop has been routed through all three grommets, is not cut or frayed, and has been secured by the cotter pin.

g. Close the circular closing flap, lift the AIRPAC to the jumper, and issue the command HOLD.

h. With the left hand, grasp the hook-pile tape lowering line, palm facing the jumpmaster. Pull up if the short bridal is attached horizontally, or to the jumper's left if it is attached vertically.

i. Insert the index finger and middle finger of the right hand, from the jumper's left to right, behind the locking D-ring. Ensure that it is properly attached to the triangle link and is locked into place.

j. Trace up the short bridal, ensuring it is not twisted, cut, or frayed, until contact is made with the girth hitch of the hook-pile tape lowering line.

k. Place the index finger of the right hand on the girth hitch, ensuring the cotton buffer is present.

l. Place the index finger and middle finger of the right hand from the jumper's left to right behind the short bridal on the opposite side of the girth hitch. Trace the short bridal until contact is made with the triangle link, ensuring it is not twisted, cut, or frayed, paying special attention to the point of attachment at the triangle link.

m. Leave the left hand in place and put the index finger of the right hand on the hook-pile tape lowering line where it emerges from the lowering line stow pocket. Ensure that no excess of the hook-pile tape lowering line is protruding from the retainer flap. Trace the hook-pile tape lowering line until the index finger disappears behind the nylon chaff portion of the M1950 weapons case ensuring it is not cut, frayed, or burned.

n. Leave the right index finger in place and, with the left hand, grasp the trail edge of the M1950 weapons case and pull it toward him. Remove the right index finger and rotate it around to the backside of the M1950 weapons case and continue the inspection of the hook-pile tape lowering line where it reemerges from the nylon chaffing.

o. Trace the hook-pile tape lowering line to its point of attachment.

p. With the thumb of the right hand, push in on the activating lever of the ejector snap to ensure it is properly seated over the ball detent. Ensure it is free of any foreign matter, and the opening gate is facing the jumper's body.

q. With the thumb and index finger of the right hand, pinch off the yellow safety lanyard and trace it until his fingers fall off the end, ensuring it is not wired, tied, or taped.

r. Move to the front of the jumper and issue the command SQUAT.

9-24. TECHNICAL INSPECTION FOR SIDE-MOUNT AIRPAC

The side-mount container is technically inspected and attached at the 20-minute time warning.

a. Ensure all visible container retaining straps are properly secured to their appropriate friction adapter. (All visible container retaining straps have a five- to six-finger quick release.)

NOTE: Five- to six-finger quick releases will depend on the item of equipment inside the side-mount AIRPAC.

b. Ensure the free-running ends of the container retaining straps are S-folded and secured to the quick-release loop by two turns of a retainer band.

c. Ensure the slide fastener is secure.

d. Ensure there is no excess in the fabric flap.

e. Ensure packing material is present, if required for particular weapon system.

9-25. M82, MEDIC JUMP PACK

Inspect the M82, medic jump pack (MJP) as follows:

a. With the left hand, secure the snap hook for the right adjustable D-ring attaching strap. With the right hand, secure the snap hook for the left adjustable D-ring attaching strap.

b. Begin the inspection with the right adjustable D-ring attaching strap. Look at the snap hook to ensure that the opening gate is facing the jumper's body and is the outermost item on the right D-ring. With the left index finger, finger the opening gate to ensure it is properly secured to the right D-ring and has spring tension. With the left thumb, flip the free-running end of the right adjustable D-ring attaching strap out of the way.

(1) With the index and middle finger of the left hand behind the right adjustable D-ring attaching strap, trace down the right adjustable D-ring attaching strap until contact is made with the white attaching loop, ensuring it is not twisted, cut, or frayed.

(2) Once contact is made with the white attaching loop, bypass the triangle link and continue the inspection of the white attaching loop in front of the triangle link. With the left index finger, trace down the attaching loops to ensure the white attaching loop is routed from bottom to top through the triangle link; the green attaching loop is routed from bottom to top through the white attaching loop; and the red attaching loop is routed from the bottom to top through the green attaching loop and then through the grommet in the female portion of the leg strap release assembly.

(3) Pull back on the female portion of the leg strap release assembly and look behind it to further ensure the red attaching loop is routed properly. Place the index finger of the left hand on the single "X" boxed stitch on the release handle cross strap just below the female portion of the leg strap release assembly.

(4) Look at the release handle cable where it comes out of the release handle cross strap to ensure it is routed through the red attaching loop protruding out of the grommet in the female portion of the leg strap release assembly and is secured in the cable loop retainer.

(5) Leave the left hand in place and focus attention on the right hand and the left adjustable D-ring attaching strap and inspect it in the same manner.

c. After inspecting the left adjustable D-ring attaching strap, focus attention on the release handle. With the right hand, gently pick up on the release handle and ensure that it has been routed through the two plies of the release handle cross strap and is secured by the hook-pile tab provided.

d. Form a knife edge with each hand, palms facing each other, and place them on the equipment retainer straps. Simultaneously trace down the equipment retainer straps until contact is made with the adjustable cross strap. Once contact has been made with the adjustable cross strap, leave the left hand in place.

e. With the right hand, grasp the free-running end of the adjustable cross strap and give it a sharp tug to the jumper's left ensuring that all slack has been taken out the adjustable cross strap.

f. Raise the M82 to the jumper and issue the command HOLD.

g. Re-form knife edges with each hand, palms facing each other, and place them on the equipment retainer straps where they come out of the cushioned envelope on the top of the M82. Simultaneously, with both hands, trace down the equipment retainer straps, bypassing the girth hitch formed by the hook-pile tape lowering line and trace all the way down to the friction adapters.

h. Look at the friction adapters to ensure the equipment retainer straps have been properly routed through them. Place the index and middle fingers of each hand on top of the two- to three-finger quick releases in the equipment retainer straps to ensure they are not more than three fingers and no less than two. Simultaneously pick up the free-running ends of the equipment retainer straps and ensure they are S-folded (or accordion folded) and have not been rolled. Ensure they are secured by two turns of a retainer band.

i. With the left hand (palm facing the M82, thumb down), grasp the hook-pile tape lowering line just above the girth hitch.

j. Look at, and pull up on, the girth hitch to ensure it is properly routed south to north through the "X" formed by the equipment retainer straps.

k. Place the right hand index finger on the hook-pile tape lowering line directly beside the control hand. Begin tracing the hook-pile tape lowering line ensuring it is routed over the left adjustable shoulder-carrying strap. Trace the hook-pile tape lowering line until contact is made with the first hook tab. Ensure the first hook tab is properly secured and no folds of the hook-pile tape lowering line protrude from the retainer flap.

l. Continue to trace down the hook-pile tape lowering line ensuring it is secured to the tubular portion of the frame by retainer bands. Continue tracing the hook-pile tape lowering line until contact is made with the second hook tab. Ensure the second hook tab is properly secured and no folds of the hook-pile tape lowering line protrude from the retainer flap. Continue to trace the hook-pile tape lowering line until the right index finger disappears behind all three plies of nylon chaffing on the M1950 weapons case.

m. Release the hook-pile tape lowering line with the left hand and grasp the trailing edge of the M1950 weapons case and pull it forward.

n. With the right index finger, pick up the hook-pile tape lowering line on the back side of the three plies of nylon chaffing and trace it to its point of attachment. Look at the ejector snap to ensure the opening gate faces the jumper's body.

o. With the right thumb, press in on the activating lever to ensure it is properly seated over the ball detent and free of foreign matter.

p. With the right index finger and thumb, pinch off the yellow safety lanyard and trace down to ensure it is serviceable and has not been wired, tied, or taped down.

q. Move back to the front of the jumper and inspect the remainder of the jumper in the same manner as a Hollywood jumper.

Section III. MOVEMENT ON THE AIRFIELD

After personnel inspection, safety personnel load the parachutists aboard the aircraft. Load time is the time agreed on by the Army and Air Force for loading the aircraft. Station time is the time the aircrew, parachutists, and equipment are inside the aircraft and are prepared for takeoff (everyone seated with seat belts fastened and helmets on). (See Table 9-1.)

9-26. AIRFIELD MOVEMENT PROCEDURES

These procedures must be followed when moving parachutists on or across an active airfield:

a. Coordinate with DACO for permission to cross the airfield.

b. Keep parachutists in closed formation.

c. Cross active runways at authorized crossing points.

d. Cross only on light signals from airfield control tower:

- Green (GO).
- Red (STOP, DO NOT PROCEED).
- Flashing red (CLEAR TAXI STRIPS AND RUNWAYS).
- Flashing red and green (EMERGENCY WARNING, BE ALERT).
- Flashing white (RETURN TO STARTING POINT).

e. Display checkered flags on the first and last escort vehicles.

f. Keep vehicles in low gear while crossing runways.

g. Do not raise radio antennas within 50 feet of any aircraft.

h. Do not smoke in the vicinity of any aircraft.

i. Avoid aircraft propellers.

j. Avoid jet engine intakes/exhausts. Stay about 50 feet from intakes and 200 feet from exhausts.

9-27. LOADING AIRCRAFT

Parachutists are loaded in the aircraft in reverse order. During loading, safety personnel move forward in the aircraft ahead of the chalk and supervise seating of the chalk to ensure that all seats are filled, seat belts are fastened, and that personnel are in proper stick order. They also assist in loading equipment aboard the aircraft. The aircrew briefing may be given before or after loading the aircraft but must be completed before takeoff.

NOTE: The actions taken are according to type of aircraft used during the airdrop.

9-28. IN-FLIGHT EMERGENCY PROCEDURES

Jumpers must be briefed on in-flight emergency procedures described in Table 9-1.

NOTE: Rotary-wing aircraft in-flight emergency procedures are different from those shown in Table 9-1. They are explained by either the PJM or pilot before boarding the aircraft.

SITUATION	SIGNAL	ACTIONS IN USAF AIRCRAFT	ACTIONS IN ARMY AIRCRAFT
Crash landing during takeoff	Continuous ringing of alarm bell or oral warnings	Remain seated until aircraft stops, then exit.	Pull legs inside aircraft; remain in position; cover head with arms.
Crash landing during flight	Six short rings of alarm bell or oral warning	Jump if time permits. If not, secure seat belts; brace for impact on continuous ring; remain seated until aircraft has stopped.	As directed by pilot.
Emergency bailout	Three short bells of alarm bell or oral warning	Stand up and hook up on continuous ring; exit aircraft under direction of PJM.	Exit aircraft under direction of PJM.
Ditching over water with insufficient drop attitude	Six short rings and oral warning	Use available padding, remain seated, and secure seat belts.	Pull legs inside aircraft; remain in position; cover head with arms.
Lighten load	Oral warning	Designated parachutist may assist PJM or loadmaster in jettisoning cargo.	As directed by pilot.
Fire in flight	Oral warning	Remove parachutists from fire area and extinguish fire.	As directed by pilot.

Table 9-1. In-flight emergency procedures.

CHAPTER 10

JUMPMASTER AND SAFETY DUTIES IN FLIGHT

This chapter provides an overview of the duties of the jumpmaster, assistant jumpmaster, and safety personnel during flight. For aircraft-specific personnel duties, see Part Four, Aircraft Used in Airborne Operations.

Section I. PRIMARY JUMPMASTER, SAFETY PERSONNEL, AND ASSISTANT JUMPMASTER

After aircraft takeoff, the JM must remain oriented to aircraft position at all times and keep the parachutists informed of any deviations from flight plan. He may coordinate with the navigator or use strip maps and checkpoints. He also remains in communication with the pilot. This is performed by relay through the loadmaster over the interphone. On Army aircraft, the JM or safety personnel should wear a flight helmet or headset for direct communication with the pilot and to monitor the ground control element. If the JM/safety cannot wear a flight helmet or headset, communication can be made through the crew chief.

10-1. PRIMARY JUMPMASTER

General rules stress that the JM must—

- Not sacrifice safety for any reason.
- Rehearse JM procedures on the ground.
- Hook up before opening jump door(s) or ramp.
- Face the open jump door or tailgate when in flight.
- Maintain a firm handhold on the aircraft when working in or close to the open jump door or ramp.
- Not allow anyone in or near the open jump door without a helmet (or equivalent) and safety harness or parachute.

The JM must remember that jumpers' eyes require 20 minutes to adjust to lower light levels before night operations. Therefore, only red compartment lights or night vision imaging systems (NVIS) lighting will be used 20 minutes before drop time. This allows for physical adjustment and minimizes the impairment of a jumper's night vision during the critical first seconds of a night drop. If the white light is substituted for the red light before drop time, the jumpers' eyes must readjust by turning on the red compartment lights for 20 minutes.

10-2. SAFETY PERSONNEL

During flight, safety personnel constantly check on the condition of all parachutists and distribute airsickness bags where needed. They also assist the JM in relocating personnel who are too sick to jump, jump refusals, and other types of no-jumps. If a jump refusal occurs, safety personnel move the parachutist forward in the cargo compartment to be seated. During in-flight rigging missions, safety personnel assist in parachute issue. They operate rigging, JMPI, and correction stations as directed by the JM. The safety double checks to ensure the pilot and loadmaster have designated a specific individual to cut away a towed jumper.

10-3. PRIMARY JUMPMASTER/ASSISTANT JUMPMASTER DUTIES

JM/AJM duties are as follows:

a. Enforce flight rules and regulations.
b. Issue time warnings.
c. Issue jump commands.
d. Perform door safety checks.
e. Perform outside air safety checks.
f. Control exit of all parachutists.
g. Eject door bundles.
h. Perform in-flight rigging mission.

Section II. DOOR PROCEDURES AND DOOR BUNDLE EJECTION

Time warnings, door procedures, and door bundle ejection are events that commonly occur during each airdrop operation. Time warnings and door procedures are completed during each airdrop operation, but door bundles are not always ejected. The time warnings and door procedures are considered critical aspects of JM operations during all airdrop operations.

NOTE: The 20-minute time warning may be increased to 30 minutes aboard the C-141 or C-17 if the JMs need more time to accomplish their duties.

10-4. THE 20-MINUTE TIME WARNING

JMs unbuckle their seat belts and stand up. They move to the rear of the aircraft, turn, and face the parachutists. (All time warnings begin and end at the shoulders with closed fists.) They extend both arms straight forward, extending and spreading the fingers and thumb of each hand. This is repeated twice while the oral command is given.

a. All CWIEs, AT4JPs, and DMJPs are attached to designated parachutists and are inspected. Safety personnel assist the JMs with attaching equipment and ensuring the tie-downs are secure and the lower lines are attached and secure. JMs and safeties unlash door bundles (if any) and move them near the jump door(s). Once positioned, the static line of each cargo parachute is attached to the outboard anchor line cable and the following inspection sequence is completed:

- Static line and clevis (safety wire must be bent so that it has metal-to-metal contact).
- Pack closing tie.
- Drogue device (one for C-130; two for C-141).
- Point of attachment to the bundles (risers).
- Tie-down tapes (one on each corner).
- Pack tray and bundle (for any loose or excess webbing).

b. When the inspection is completed, the JM slaps the smooth side of the bundle and ensures it is facing the trailing edge of the door(s).

NOTE: The 10-minute time warning may be increased to 15 minutes aboard the C-141 or C-17 if the JMs need more time to accomplish their duties.

10-5. THE 10-MINUTE TIME WARNING

When the loadmaster informs the JMs that the aircraft is 10 minutes from the DZ, the JMs unbuckle their seat belts and stand up (if seated). They hook up to the inboard anchor line cable, move to the rear of the aircraft near the jump door, and transfer control of their static lines to safety personnel. They announce, SAFETY, CONTROL MY STATIC LINE, and ensure the safety has positive control before turning to face the parachutists. The JMs begin jump commands at the 10-minute time warning. During this process, the safeties help the JMs hook up, watch for excess static line, and stow any excess static line in the static line slack retainer. After inspection of the JMs' static lines, safeties move out to inspect the sticks, as indicated in the following paragraph.

10-6. FIRST SEVEN JUMP COMMANDS

The JM is now ready to issue the first seven jump commands.

a. After the parachutists are standing, safety personnel inspect the following items on each parachutist while moving forward in the aircraft:

- Waistband for proper quick release.
- Ejector snap on the HPT lowering line for proper attachment.
- Quick-release assembly on the weapons case for proper attachment.
- Adjustable leg straps on harness, single-point release.

NOTE: Safeties must be alert for and correct any excess webbing or loose lowering line.

b. Once the safeties have checked the last parachutist, and after the command HOOK UP, they return to the aft end of the aircraft. While moving to the aft end, safeties check the full length of each parachutist's static line to ensure proper routing and attachment to the anchor line cable. They check with both eyes and hands and stow any excess static line in the static line slack retainer so that each static line is taut from the anchor line cable to the main pack tray. As the safety checks each static line, he reminds each jumper to "make eye-to-eye contact and hand the safety your static line." After he checks the first parachutist and the JM's static line, the safety resumes control of the JM's static line.

c. After the JM receives "All okay, JM," he regains control of his static line from the safety, takes the number 1 parachutist position, and waits for slow-down procedures.

d. Once the aircrew has completed their slow-down checklist (slow aircraft to drop speed, open jump doors, deploy air deflectors, and position jump platforms), the loadmaster relinquishes control of the jump door to the JM by giving the JM a thumbs-up signal. He then moves to the ramp area.

10-7. DOOR SAFETY CHECK

Safety personnel position themselves near the trail edge of the jump door and control the static line for the JM as he performs the door safety check and outside air safety check. He grasps the lead edge of the door with his lead hand and transfers control of his static line back to the safety (Figure 10-1). Then, he grasps the trail edge of the door with his trail hand. Using only the lead hand, he ensures the pip pin is in place, securing the jump door in the open position on the C-130, or that the door-latch catch operates properly on the C-141. (See Chapter 16 for door procedures for the C-17A.)

a. **Jump Platform**. The JM checks the jump platform by—

(1) Tapping the lead down lock with the lead foot and then the trail down lock with the trail foot and visually checking to ensure they are engaged.

(2) Placing his trail foot in the center of the jump platform and shifting his weight to the platform, ensuring the platform can support his body weight. The trail foot remains in place. (If the JM removes his foot from the platform, he must again check the down locks and ensure the platform can support his body weight.) From this rest position, the JM performs all remaining duties in the door until it is time to issue the eighth jump command (or place bundles in the door).

b. **Trail Edge of Door**. With the trail hand, the JM starts at the top of the trail edge of the door, tracing all the way down to the trail down lock, then back to the top to ensure no sharp or protruding edges exist.

c. **Air Deflector**. The JM looks at the air deflector, ensuring it is fully deployed.

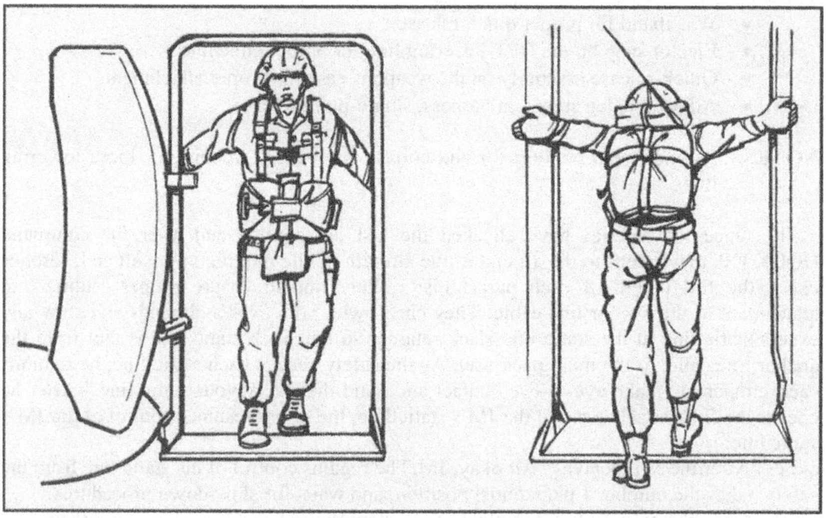

Figure 10-1. JM door position.

10-8. INITIAL OUTSIDE AIR SAFETY CHECK AND CHECKPOINTS

The JM must make outside air safety checks to ensure there are no unsafe conditions outside the aircraft (aircraft in the formation to the rear that are below drop altitude or other low-flying aircraft). From the same rest position, the JM leans far enough outside the aircraft to make a proper air safety check, and visually checks—

- Direction of flight.
- Overhead.

- Rear.
- Straight down.
- Straight to his front.
- Direction of flight.

The JM continues observing outside the aircraft and spotting for checkpoints en route to the DZ. In the absence of checkpoints, he uses additional time advisories from the air crew, which must have been identified during the pilot/JM briefing. The navigator and primary loadmaster should also be present for this briefing. The JM relays checkpoints or time advisories to the parachutists by leaning back inside the aircraft, keeping his foot centered on the platform, facing the parachutists, and issuing the checkpoint or time advisories.

NOTES: 1. When the JM leans back inside the aircraft to assume the rest position, he will not collapse his trail elbow with the lead arm locked. Allowing his trail elbow to collapse could cause accidental MIRPS activation resulting in JM extraction from the aircraft. Such an extraction can result in serious injury or death.

2. The JM may lock his elbows, if deemed necessary, to properly conduct outside air safety checks, or when checking for towed jumpers. However, the JM is not required to lock his elbows when performing those actions.

3. The loadmaster should relay time warnings to the jumpmaster or safety by using hand signals. The exact hand signals must be coordinated between the jumpmaster and loadmaster during the aircrew brief. However, suggested hand signals are as follows: 1-minute time warning--the index finger extended vertically; 30-second time advisory—the index finger and thumb held closely together.

10-9. THE 1-MINUTE TIME WARNING

Approximately one minute from the green light, the safety passes along loadmaster time warnings if the jumpmaster fails to hear them. The JM relays the 1-minute warning to the parachutists by leaning back, keeping his foot centered on the platform, facing the jumpers, and extending the index finger of his lead hand. He announces, ONE MINUTE. The jumpers relay the announcement to the rear. The JM continues observing outside for the 30-second checkpoint, or until he receives a 30-second time advisory from the loadmaster.

10-10. FINAL OUTSIDE AIR SAFETY CHECK

The JM announces, THIRTY SECONDS. The jumpers relay the announcement to the rear. After the JM has given the 30-second checkpoint or relayed a 30-second time advisory, he makes his final outside safety check and, based on his checkpoint or USAF/other service aircraft advisory, determines when the aircraft is about 10 seconds (20 seconds for bundle drops) from the release point. Once again, he leans out and visually checks—

- Direction of flight.

- Overhead.
- Rear (ensures no aircraft in the formation have dropped below jump altitude).
- Straight down.
- Straight to the front.
- Direction of flight (the JM may be able to see the DZ at this time, depending on the aircraft axis of approach).

The safety, positioned at the aft end of the door, controls the JM's static line during all outside safety checks.

NOTE: For a CARP/VIRS DZ, the JM has no responsibility to positively identify the DZ, code letter, or color of smoke. The responsibility to positively identify the DZ, code letter, or color of smoke rests with the USAF/other service aircraft.

a. **Unsafe Condition Outside Aircraft**. If the JM observes any unsafe conditions outside the aircraft, he notifies the loadmaster by a preplanned arm-and-hand signal (for example, a cutting motion across the throat) that a NO DROP situation exists for this pass.

b. **Bundle Drops, 20 Seconds from Release Point**. The JM moves back inside the aircraft and positions the first door bundle on the jump platform so that it is on its balance point with its longest dimension vertically in the door and the parachute on the top or inboard side of the bundle. Then the JM—

(1) Maintains a firm grasp on the bundle with the lead hand and a firm grasp on the trail edge of the jump door with the trail hand. He ensures that the cargo parachute static line is routed above the trail arm.

(2) Keeps his eyes on the jump caution light. When the light turns green, he ejects the door bundle(s), ensuring that it goes straight out and does not tumble.

(3) Maintains a firm grasp with his trail hand and turns toward the cargo area. The JM and AJM make eye-to-eye contact and give each other the thumbs-up signal, indicating that the door bundle has been ejected, that they are not aware of any unsafe conditions, and that they are ready to exit personnel. (Parachutists are not positioned in the door until this is accomplished.) This procedure is followed when ejecting door bundles from one door or both doors, as long as both doors are open.

(4) Moves on line with the lead edge of the jump door and issues the eighth jump command IAW the exit procedures.

c. **No Bundle Drops, 10 Seconds from Release Point**. When door bundles are **not** used on an airborne operation, the JM makes his final check to determine that the aircraft is about 10 seconds out from the RP, based on his checkpoint or USAF/other service aircraft advisory. Then he—

(1) Maintains a firm grasp on the trail edge of the door with the trail hand and turns as far as possible toward the cargo area. The JM and AJM make eye-to-eye contact and give each other the thumbs-up signal, indicating that there are no unsafe conditions outside the aircraft and that they are ready to exit personnel. This procedure is followed when exiting personnel from one or both doors, as long as both doors are open.

(2) Releases the trail edge of the door and issues the eight jump command (keeping his body centered on the trail edge of the door).

10-11. EIGHTH JUMP COMMAND

The JM issues the eighth jump command, STAND BY. The safety positions himself near the trail edge of the door, ensuring he does not impede the flow of jumpers.

a. **Computed Air Release Point**. After issuing the eighth jump command, the JM immediately regains control of his static line from the safety. He backs away from the door toward the middle of the aircraft to allow the safety room to control static lines. The safety receives the first parachutist's static line with his forward hand and secures it with his aft hand by pinning it against the aft anchor line support.

b. **Ground Marking Release System**. After issuing the eighth jump command, the JM continues to spot for the DZ and DZ markings. He backs away from the door toward the middle of the aircraft to allow the safety room to control static lines. When the line of the panels break the trailing edge of the jump door, the JM issues the command, GO. The safety receives the first parachutist's static line with his forward hand and secures it with his aft hand by pinning it against the aft anchor line support. For USAF operations, a safety is not used and either the JM or loadmaster fulfills the duty after coordination.

10-12. NINTH JUMP COMMAND

When the JM issues the ninth jump command, GO, he ensures that he has backed away from the door where he can best control his stick and away from the safety, to avoid confusion or congestion. Safety personnel take static lines while the JM controls the flow of parachutists. The safety grabs each static line with his forward hand, passes it back to his aft hand, and ensures the line is firmly seated against the aft anchor line cable support. After all parachutists have exited the aircraft, the JM and AJM hand off their static lines to the safeties and exit the aircraft. The safety immediately makes a visual check for toward jumpers. He then looks directly at the loadmaster, passes him a thumbs-up signal, and states YOUR DOOR, AIR FORCE.

a. **Red Light Procedures**. If a JM sees a red jump caution light, he sounds off with "RED LIGHT, RED LIGHT, RED LIGHT," and then moves forward to block the flow of the stick. If any parachutist tries to exit on the red light, he will be allowed to exit, except in the case of an unsafe parachutist. No one touches or physically tries to stop a parachutist who is past the leading edge of the door. Parachutists will be stopped only by oral command; this reduces the risk of accidentally activating the jumper's reserve.

b. **Jump Refusals**. When removing a jump refusal from the door, the JM tells the refusal, "GREEN LIGHT, GO; GREEN LIGHT, GO; GREEN LIGHT, GO." If the jumper does not exit, he tells the refusal, "YOU ARE A JUMP REFUSAL AND I AM TAKING YOU OUT OF THE DOOR." Then, the JM should approach the parachutist from the rear and grasp the sides of his parachute. The JM **never** puts his hand in front of the parachutist's face or grabs the back of his ballistic helmet.

(1) The safety warns the refusal that he is being brought back inside the aircraft. Under **NO** condition should the JM reach around in front of the refusal to pull him back into the aircraft; the JM could accidentally hit or grab the reserve rip cord grip.

(2) After the jump refusal is brought inside the aircraft, he is guided as far forward in the aircraft as is possible, seated, buckled up, unhooked, and directed to not touch his equipment. During training, a jump refusal will stop the jump for that pass.

c. **Unsafe Parachutist**. If a parachutist has passed the leading edge of the door and must be stopped because of a misrouted static line or other critical deficiency, the safety approaches the jumper from the back to grab the sides of his parachute. The safety never reaches around in front of the parachutist, puts his hands in front of the parachutist's face, or grabs the back of his helmet.

10-13. TOWED PARACHUTIST (FIXED-WING AIRCRAFT)

Actions to retrieve or cut free a towed parachutist are as follows:

a. **JM Actions**. If the JM observes a towed parachutist, he takes the following actions. The JM—

(1) Stops the stick of parachutists (if applicable).

(2) Notifies the loadmaster, who then notifies the pilot and requests that drop altitude be maintained.

(3) Identifies how the parachutist is being towed. If the parachutist is being towed by anything other than the lowering line or static line, the JM attempts to free him. If the parachutist is being towed by a lowering line, the JM immediately cuts the lowering line, thus freeing the parachutist. If the parachutist is being towed by the static line, the JM initiates recovery procedures.

b. **JM Recommendations**. The JM observes the towed parachutist and recommends whether to retrieve or cut the parachutist free. The recommendation is relayed by the loadmaster to the pilot. The pilot makes the decision. If the decision is to cut the jumper free, the loadmaster will cut the static line on command of the pilot. Priority is as follows:

(1) ***Door***. Safety personnel move the remainder of the stick toward the front of the aircraft. If the parachutist is to be retrieved, the loadmaster uses the Towed Parachutist Retrieval System (TPRS) or installs the retrieval bar on the C-141 or the CGU1-B cargo strap on the C-130, retracts or folds in the jump platform, and initiates retrieval. All personnel stay clear of the door and line of travel of the static line retriever cable. When the parachutist has been retrieved to the door, the JM and safety personnel gain physical control of the parachutist. The loadmaster relieves tension from the static line retriever so that the parachutist can be brought inside the aircraft. The retrieved parachutist is moved all the way forward and is seated. He does not jump again. If the retrieval is unsuccessful and the parachutist must be cut free, the loadmaster cuts the static line.

(2) ***Ramp***. If a parachutist is towed following a ramp exit and is to be cut free, the loadmaster partly retrieves the static lines to reach the towed parachutist's static line in order to cut it. If the parachutist is to be retrieved, the loadmaster uses the Towed Parachutist Retrieval System (TPRS) or installs a CGU1-B cargo strap (C-130) about 5 1/2 feet above the ramp. The static lines are retrieved over the CGU1-B strap. As the parachutist is retrieved to the ramp area, the JM and safety personnel gain physical control of him. The parachutist is pulled into the aircraft (under the strap) as the loadmaster relieves tension from the static line retriever cable. The parachutist is moved all the way forward and is seated; he does not jump again.

c. **Modifications to Towed Jumper Procedures**. Occasionally, the above towed jumper procedures must be modified. JMs are responsible for the safety of the parachutists' equipment checks and deployment safety as long as they are onboard the deployment aircraft. The loadmaster(s) is trained and responsible for retrieving towed parachutists. No additional JMs or safeties are required to remain onboard the aircraft unless specified in the jumping unit's ASOP. If all parachutists have exited, those actions described as JM responsibilities are accomplished by the loadmaster(s). The aircrew is responsible for all equipment left onboard the aircraft by the jump unit until it can be retrieved or turned over to the unit concerned.

NOTE: Towed jumper procedures for rotary-wing aircraft are addressed in Chapter 17.

d. **Safety Duties**. After all parachutists have exited (including JMs), the safety makes a visual check for towed parachutists to the rear of the jump door, gives the USAF loadmaster a thumbs-up signal, and an oral YOUR DOOR, AIR FORCE. This indicates that all parachutists are free and clear of the aircraft.

e. **Static Line and Deployment Bag Retrieval**. Safety personnel and the loadmaster retrieve the static lines and deployment bags. Once the static lines and deployment bags are inside the aircraft, safety personnel detach the static lines and store them with the deployment bags in extra aviator kit bags.

f. **During Return to Departure Airfield**. While en route to the departure airfield, safety personnel obtain the name, rank, social security number, unit, and reason (that is, sickness, equipment malfunction, jump refusal) for any manifested parachutist remaining onboard. They also check the aircraft for any Army equipment that was left onboard for turn-in to the DACO, collect trash and airsick bags, and reinstall seats and seat belts if subsequent lifts are planned.

g. **At Departure Airfield**. On return to the departure airfield, safety personnel turn in all air items (reserves, deployment bags, and kit bags) left onboard the aircraft to the storage facility (**obtain a receipt**). They also turn over any unit or personal equipment left aboard the aircraft to the DACO. All jump refusals and personnel left aboard the aircraft are immediately turned over to the DACO with a full account of the circumstances for each. Any jump refusal's personal equipment remains under DACO control until relieved of authority by competent unit personnel.

CHAPTER 11

DEPARTURE AIRFIELD CONTROL OFFICER

The DACO is appointed by the airborne commander and is responsible for coordination and control of the loading of personnel, equipment, and supplies into aircraft. The DACO is located at the departure airfield.

11-1. INITIAL COORDINATION

When advised of appointment, the DACO is furnished the following information:

- The unit(s) that is jumping.
- Type of jump and number of personnel.
- Type and number of aircraft.
- Number of lifts.
- Load time.
- Station time.
- Weather decision time.
- Drop zone.
- Names of DZSO, assistant DZSO, DZSTL, JM, AJM, and safety personnel.

11-2. TANKER/AIRLIFT CONTROL ELEMENT COORDINATION

At the departure airfield, the DACO makes contact with the tanker/airlift control element(TALCE)/aircrew to discuss, coordinate, and confirm the following:

- Aircraft parking plan.
- Aircraft tail number(s).
- Time of pilot/jumpmaster briefing.
- Time aircraft will be available for jumpmaster inspection.
- Weather data.
- Flight line safety measures.
- Appointment and identification of the GLO.
- Current safety regulations (TALCE/aircrew advises DACO).
- Aircraft loading procedure and time.
- Return of jump refusals (how they are to be managed).
- Manifest distribution to TALCE/aircrew.
- Inspection of aircraft for air items and other airborne unit equipment left onboard upon its return to departure airfield.

11-3. DROP ZONE SAFETY OFFICER/DROP ZONE SUPPORT TEAM COORDINATION

The DACO—

- Contacts the DZSO/DZSTL one hour before drop time.
- Informs the DZSO/DZSTL of aircraft aborts.
- Updates the DZSO/DZSTL on changes.

- Requests a closure report from the DZSO/DZSTL.
- Requests timely wind readings from the DZSO/DZSTL.

11-4. ADDITIONAL RESPONSIBILITIES OF THE DEPARTURE AIRFIELD CONTROL OFFICER

The DACO—

- Maintains radio and telephone contact with TALCE/aircrew and higher headquarters.
- Controls movement of all vehicles.
- Coordinates with TALCE/aircrew/GLO for airfield lighting to assist night airdrop departure airfield operations.
- Ensures all units police their areas before enplaning.
- Reports all serious incidents.
- Is physically present for all weather briefings and decisions.
- Ensures parachute maintenance personnel are present for issue and technical assistance.
- Becomes familiar with regulations and the SOP of the unit involved in the operation.

11-5. AIRFIELD AND RUNWAY SAFETY

The DACO—

- Is responsible for the movement in and around aircraft, taxiways, and runways.
- Briefs JM/AJM on airfield procedures.

PART THREE
Equipment

CHAPTER 12
INDIVIDUAL COMBAT EQUIPMENT JUMP LOADS

Individual equipment and weapons are attached or worn by the parachutist in several configurations--for example, exposed, placed inside containers, or a mix of the two. Unit airborne SOPs specify ways of packing equipment and other mission-essential items consistent with safety requirements and this manual. However, to ensure personnel safety, all small arms must be on "SAFE" when jumping.

Section I. LOAD PLACEMENT

Fragile items, such as weapon sights, are padded. Crushable items are not placed directly under the attaching harness. Exposed weapons or equipment, snap hooks, and projections on containers are potential safety hazards and must be taped.

12-1. LOAD DISTRIBUTION

Individual equipment attached to the equipment belt is placed on the front or sides of the body away from the PLF points of contact. The medium or large ALICE pack is attached to the front of the parachutist using an H-harness or HSPR. A lowering line is attached if the pack exceeds 35 pounds or has a frame attached. Hard, bulky, or irregularly shaped items are not placed to the rear of the parachutist or on the thighs. Jumpmasters and unit leaders must inspect parachutists to ensure weight and distribution of the load does not interfere with the parachutist's ability to perform a strong exit from the aircraft door.

12-2. CONSIDERATIONS

In accordance with AFI 11-2-C-130-Vol III, commanders will not load parachutists with equipment to exceed 350 pounds total weight and will reduce this weight by 12.5 pounds for every thousand feet of exit altitude above 3,000 feet MSL for fixed- and rotary-wing jumps. Total weight will not exceed 325 pounds for C-130 ramp jumps. Failure to comply could result in D-ring failure if the reserve is deployed at terminal velocity and or the JM is unable to retrieve the towed parachutist. The variety and weight of equipment or weapons attached to a jumper may exceed the parachute load limits and a safe descent rate. Also, the jumper's actions or time available to release tie-down straps and lower equipment may interfere with control of the parachute close to the ground (Tables 12-1 and 12-2, page 12-2). Commanders should use all the equipment methods of entry available to them (CDS, door bundles, wedge, and so on) to prevent overloading any individual parachutist.

TWO EQUIPMENT LOADS AND BASIC LOAD (POUNDS)								
Soldier Load Type	Load-Bearing Equipment with Two Canteens (Water)	ALICE Pack with Sleeping Bag (Winter)	Battle Dress Uniform Helmet, and Boots (1)	M1950 Weapons Container	Weapons Loaded with Ammunition (2)	Suspended Weight of T-10-/ MC1-Series Parachute and Reserve	Soldier Weight	Total Suspended Weight (Pounds) (3)
M16 Rifleman	11.5	32	15	7.3	31.0	23	205	324.8
M60 Machine Gunner	11.5	32	15	7.3	54.4	23	205	348.2
M203 Gunner	11.5	32	15	7.3	40.0	23	205	333.8
Dragon Gunner	11.5	32	15	6.5 (weapons only)	55.9	23	205	348.9
Radio Operator	11.5	32	15	7.3	71.6	23	205	365.4

(1) Weight of uniform does not include winter gear (for example, parka, liners, underwear).
(2) Includes basic load of ammunition, grenades, Claymore, bayonet, and cleaning kit.
(3) Does not include arctic gear.

Table 12-1. Weight of parachutist.

DESCRIPTION	WEIGHT (POUNDS)	REFERENCE	REMARKS
Maximum Load-Bearing Capacity T-10-Series/MC1-Series Main Canopy	360	Natick Research, Development, and Engineering Center	For every 1,000 feet above 3,000 feet MSL subtract 12.5 pounds from the jumper's total weight, which includes the parachute assembly. Failure to comply may result in D-ring or reserve failure if the reserve is deployed at terminal velocity.
Maximum Load-Bearing Capacity to Achieve 22 Feet Per Second or Less Descent Rate, T-10-Series/ MC1-Series Main Canopy	360	Natick Research, Development, and Engineering Center	Total suspended weight
Air Movement Planning Weight of Combat-Equipped Parachutist	260		Parachutist with one equipment container

Table 12-2. Parachute load limits.

Section II. LIFE PRESERVERS

Jumpers wear life preservers whenever a flight is conducted over water, a water obstacle is on the DZ, or the intended DZ is close to a water obstacle. (See Appendix D.) The B-7, the B-5, the LPU-10/P, and the life preserver inflatable (NSN 4220-0027-6892-9050) life preservers are currently certified for use by parachutists. Other life preservers may be used if approved by the individual service components.

12-3. B-7 LIFE PRESERVER

The B-7 is worn **under** the parachute harness. To fit the B-7 life preserver, the parachutist places one flotation packet under each arm so that the packet flaps are to the outside and the toggle cords are down and to the front. He runs the shoulder strap from front to rear over the left shoulder, under the back strap, then from rear to front over the right shoulder, and attaches it to the ring on the right flotation packet. He adjusts the shoulder strap so that the flotation packets fit snugly against his armpits. The parachutist then attaches the chest strap to the attachment rings on the left flotation packet, forming a quick release.

> **WARNING**
> **Ensure that the B-7 life preserver is worn so that the inflatable portion is *not* between the parachute harness and the body. Serious injury may result if it is in this position when inflated.**

12-4. B-5 LIFE PRESERVER

The B-5 is worn **under** the parachute harness. The parachutist inflates the vest by pulling two toggle cords (at the bottom of the vest) which activate CO2 cartridges that fill the vest with gas. An alternate provision for inflating the vest is by blowing into the manual inflation valve rubber hoses located on the upper right side of the vest. Manual inflation should be used only if the CO2 inflation valves fail to operate. The flotation vest is placed over the neck so that the inflatable vest is on the parachutist's chest. The manual inflating valves should be completely closed when donning the life vest. The back strap and leg straps are then adjusted.

> **WARNING**
> **Do not inflate the B-5 life preserver until the parachute harness is removed. The B-5 can crush an individual's chest if inflated beneath a parachute harness.**

12-5. LPU-10/P LIFE PRESERVER

The LPU-10/P is a standard USAF carbon-dioxide cartridge-activated life preserver assembly worn during flights over water or during airdrops when water obstacles are near or on the intended DZ. The LPU-10/P has an adjustable harness and underarm inflation bladders. It is designed to keep the wearer's head above water at weights up to 250 pounds for up to 10 minutes. The LPU-10/P is compatible with the USAF C-9, T-10, and MC-4 parachute harness assemblies. The LPU-10/P must be maintained IAW USAF TO 14S-1-102.

a. **Donning the LPU-10/P**. The LPU-10/P is worn **under** the parachute harness. The harness is worn so that the inflatable pockets are under the parachutist's arms. The manual inflating valves should be completely closed when donning the life vest. The shoulder and waist straps are then adjusted to ensure the inflation container is one hand width beneath the armpit and not constrained by the parachute harness.

WARNING

The inflation wings must be one hand-width beneath the jumper's armpit and clear of the harness straps. If the inflation pockets are too snug under the armpit, or if they are between the harness and the jumper's body, the jumper can experience severe pain or crushed ribs during inflation.

b. **Inflating the LPU-10/P.** The parachutist inflates the flotation bladders by pulling two toggle cords (at the bottom of the vest) which activate CO2 cartridges that fill the bladders with gas. An alternate provision for inflating the vest is by blowing into the manual inflation valve rubber hoses located on the bottom side of the wings. Manual inflation should be used only if the CO2 inflation valves fail to operate.

Section III. HARNESSES AND LOWERING LINE

All load carriers (ALICE packs and weapons cases) are attached to the parachutist by harnesses and, if lowered, rigged with the lowering line. Two types of harnesses are used--the older H-harness (modified) and the new Standard A, HSPR assembly. (The HSPR replaces the modified H-harness through attrition.)

12-6. H-HARNESS

The H-harness (modified) (Figure 12-1) consists of two equipment retainer straps.

a. These straps are connected by two cross straps. Each equipment retainer strap has two friction adapters 3 inches apart. Two D-ring attaching straps terminate in a free-running end on one end and a snap hook on the other end.

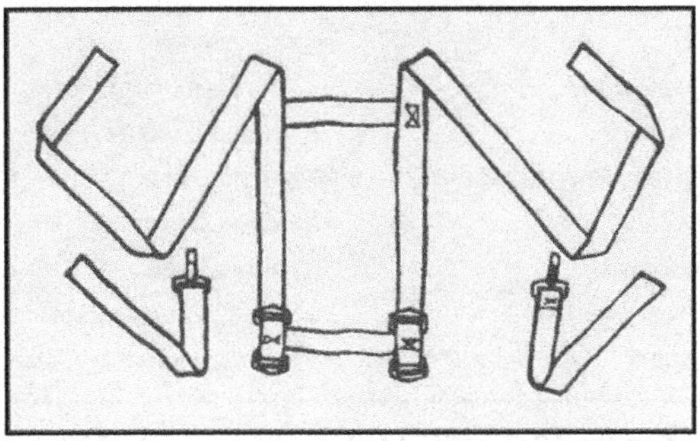

Figure 12-1. H-harness (modified).

b. The H-harness is used to rig the ALICE pack to the parachute harness. When rigging the H-harness, the parachutist ensures that equipment fits snugly under the reserve parachute and that the D-ring attaching strap snap hooks are spaced a four-finger distance from the H-harness friction adapters.

12-7. HARNESS, SINGLE-POINT RELEASE

The HSPR is an H-type design. It is made of nylon webbing with friction adapters to secure it around the load, and it has two adjustable D-ring attaching straps. To stabilize the pack to the parachutist during movement in the aircraft, exit, and main parachute deployment, two adjustable leg straps are provided to secure the pack to the parachutist's right and left legs. The leg straps are equipped with the male portion of the leg strap release assembly. The harness has a single-point release assembly that simultaneously releases the load and leg straps from the parachutist and parachute harness (Figure 12-2).

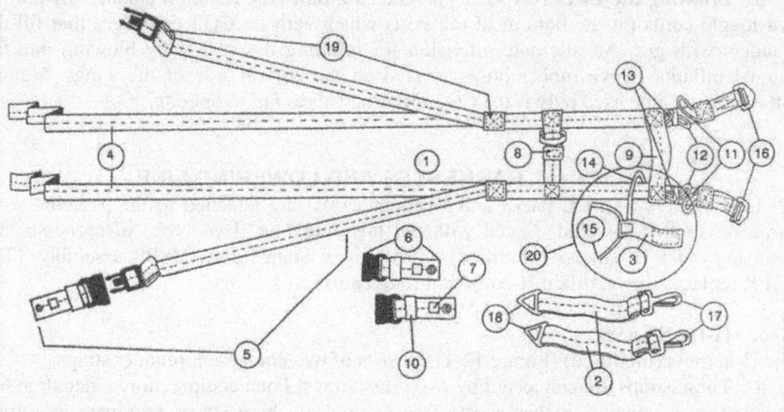

1. HARNESS, SINGLE-POINT RELEASE	11. WHITE ATTACHING LOOP
2. ADJUSTABLE D-RING ATTACHING STRAP	12. GREEN ATTACHING LOOP
3. RELEASE HANDLE	13. RED ATTACHING LOOP
4. EQUIPMENT RETAINER STRAP	14. RELEASE HANDLE LANYARD
5. ADJUSTABLE LEG STRAP	15. RELEASE HANDLE CABLE
6. FEMALE PORTION, LEG STRAP RELEASE ASSEMBLY	16. FRICTION ADAPTER
7. CABLE LOOP RETAINER	17. SNAP HOOK
8. ADJUSTABLE CROSS STRAP	18. TRIANGLE LINK
9. RELEASE HANDLE CROSS STRAP	19. MALE PORTION, LEG STRAP RELEASE ASSEMBLY
10. WEBBING RETAINER	20. RELEASE HANDLE CABLE ASSEMBLY

Figure 12-2. HSPR (NSN 1670-01-227-7992).

12-8. HOOK-PILE TAPE LOWERING LINE

The HPT lowering line (Figure 12-3) is used to lower all equipment attached to the parachutist. The HPT has been modified to accommodate the DMJP and AT4JP when lowered as a tandem load.

a. The standard 15-foot lowering line is made of tubular nylon (1 inch wide) with two retainer flaps sewn on. The retainer flaps have HPT sewn to the edges.

b. Two-inch tabs are sewn on the lowering line and, when the line is stowed, the tabs are secured to prevent line spillage. The ejector snap has a yellow safety lanyard (1 inch by 8 inches) attached.

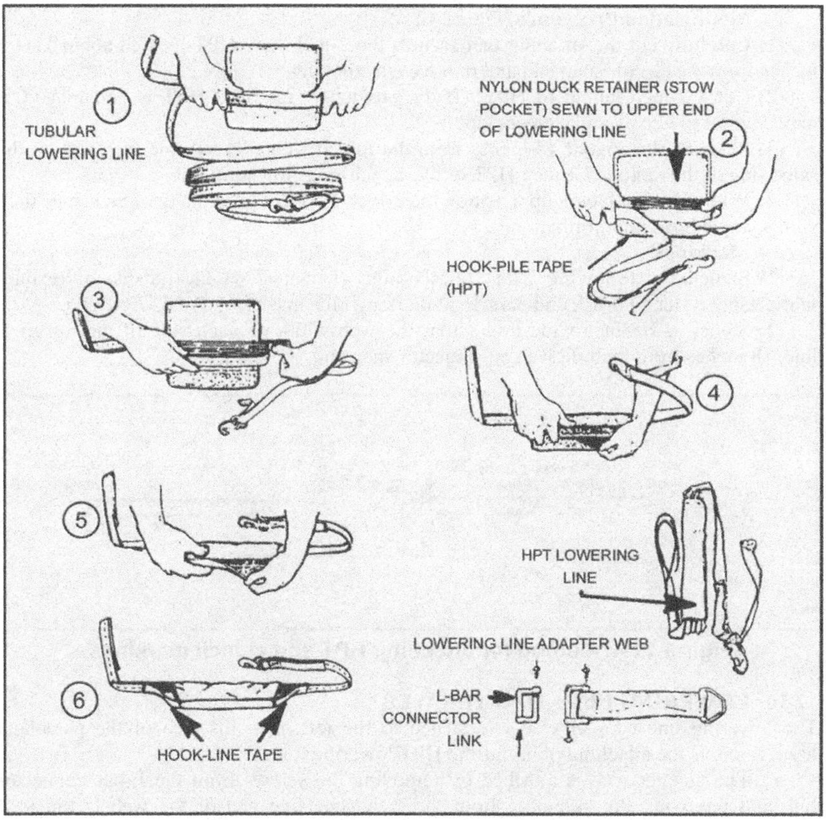

Figure 12-3. Hook-pile tape lowering line.

12-9. HOOK-PILE TAPE LOWERING LINE (MODIFIED)

The modified HPT lowering line (NSN 1670-01-067-6838) must be used when the DMJP and AT4JP are lowered as a tandem load. The modification is accomplished in the field.

a. **Materials Required.**

(1) Fastener tape, pile, color OD 106, 1 inch wide, Class 1, MIL-F-21840, NSN 8315-00-106-5974.

(2) Thread, nylon, color OD-S1, Type II, Class A, size E, V-T-295, 2,500-yard tube, NSN 8310-00-244-0609.

(3) Ink, marking, parachute, color strata blue, Type IV of MIL-I-6903 (or AA-59291), NSN 7510-00-286-5362.

b. **Stitching Requirements.**

(1) Stitching will conform to FED-STD-751, Type 301, 7 to 9 stitches per inch.

(2) Ends of stitching will be over-stitched not less than 1/2 inch.

c. **Modification Procedure** (Figure 12-4).

(1) Carefully cut the stitching that secures the 2-inch-long HPT located about 11 3/4 inches from the ejector snap end and remove cut stitching.

(2) Cut a 2-inch length of HPT. (If the previously removed HPT is undamaged, it may be used in lieu of replacement tape.)

(3) Place marks 46 and 48 inches from the folded web edge ejector snap end on the same side of the removed 2-inch HPT of the 1-inch-wide lowering line.

(4) With pile side facing up, position the 2-inch HPT between the markings and stitch with a single box stitch formation.

d. **Markings.**

(1) Stencil the following with 1/2-inch-high characters on the outside of retainer fabric using a stencil brush and parachute-marking blue ink: "DMJP/AT4JP MOD."

(2) Stencil a 1/8-inch-wide line across the web width on each side of the lowering line, 18 inches from the fold web edge ejector snap end.

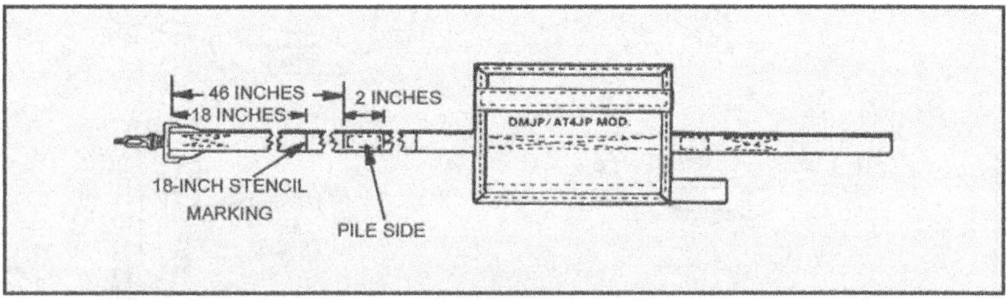

Figure 12-4. Method for attaching HPT and stencil markings.

12-10. LOWERING LINE ADAPTER WEB

The lowering line adapter web is attached to the left main lift web of the parachute harness and is the attaching point for the HPT lowering line.

a. The adapter web is attached by removing the screws from the L-bar connector link and removing the assembly from the web. The free end of the web is threaded through the left D-ring of the parachute harness and between the L-bar and main lift web. The link assembly is reattached through the loop on the free end of the adapter web and around the main lift web, and the screws are securely reset (Figure 12-5).

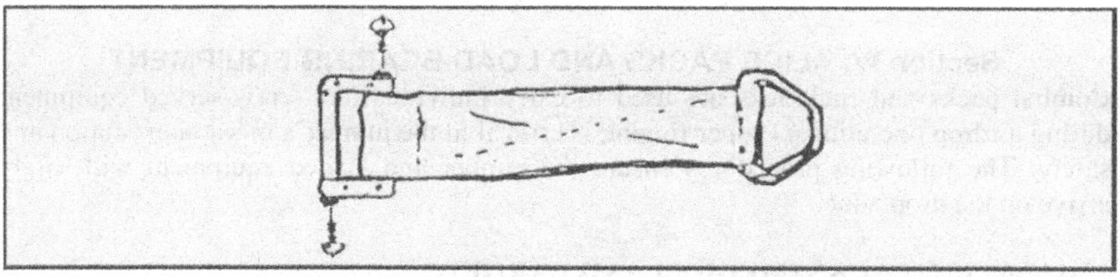

Figure 12-5. Lowering line adapter web.

b. When attaching equipment to be lowered, the web must be routed under the waistband. Type II or III nylon cord (gutted) is used to eliminate slack between the accessory attaching ring and the parachute harness main lift web. This is done by making one turn through the accessory attaching ring loop in the adapter web, and one turn around the main lift web of the parachute harness and the lowering line adapter web (Figure 12-6). The loose ends must be tied together using a suitable joining knot such as a square knot followed by two overhand knots or a surgeon's knot.

NOTE: The lowering line adapter web will become obsolete when the T-10-series modified parachute harness with triangle links is received by units.

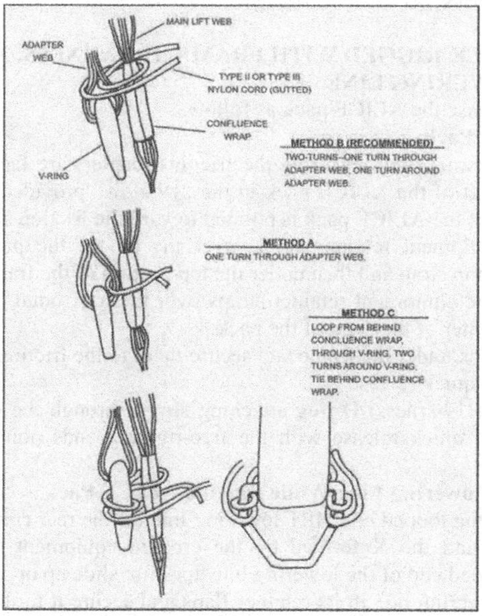

Figure 12-6. Securing lowering line adapter web with Type II or Type III nylon cord.

Section IV. ALICE PACKS AND LOAD-BEARING EQUIPMENT

Combat packs and rucksacks are used to carry individual and crew-served equipment during airdrop operations. Proper rigging is critical to the jumper's mission execution and safety. The following procedures ensure the jumper and loaded equipment will safely arrive on the drop zone.

12-11. ALICE PACKS (MEDIUM AND LARGE)

Medium and large ALICE packs are attached to the parachutist using the H-harness or the HSPR. They can be lowered during descent by attaching the HPT lowering line.

a. Items of equipment are inserted, and padding is placed between the load and the front portion of the pack. The outside pockets are filled with nonfragile items (full pockets help position the H-harness on the pack). The pack is closed by engaging the drawstrings and tie-down straps. The adjustable shoulder carrying straps are adjusted snugly. The excess webbing on the adjustable shoulder carrying straps is rolled and taped. The running ends of the waist straps are routed around the frame opposite the lower back pad, tightened, and secured in place by taping. This secures the adjustable shoulder carrying straps and reduces the possibility of entanglement onboard the aircraft.

b. If carried, the sleeping mat is rolled tightly to reduce its size. The mat is placed between the two top vertical tie-down straps and the top cover of the pack. It is secured by tightening the straps. The H-harness or HSPR encompasses the mat when it is routed around the ALICE pack.

12-12. ALICE PACK RIGGED WITH FRAME, H-HARNESS, AND HOOK-PILE TAPE LOWERING LINE

Rig, attach, and release the ALICE pack as follows:

a. **Rigging the Pack.**

(1) Lay the H-harness flat, ensuring the friction adapters are facing down. Place the center outside pocket of the ALICE pack in the "window" provided by the cross straps. Ensure the bottom of the ALICE pack is pointed toward the friction adapters.

(2) Run the equipment retainer straps over the top of the pack load outside the shoulder carrying loop strap and then under the top portion of the frame.

(3) Then, run the equipment retainer straps over the horizontal bar of the frame and cross them at the center of the back of the pack.

(4) Run the straps under the frame and secure them to the friction adapters, forming a two- to three-finger quick release.

(5) Thread the (H-harness) D-ring attaching straps through the intermediate friction adapters, forming a quick release with the free-running ends pointing away from the parachutist.

b. **Attaching Lowering Line While Rigging ALICE Pack.**

(1) Girth-hitch the looped end HPT lowering line (at the rear center of the pack from top to bottom) around the X formed by the crossed equipment retainer straps. This ensures that the looped end of the lowering line does not slide up or down the H-harness.

(2) Stow the lowering line in its retainer flaps and secure it to the left side (as worn) of the pack frame with two retainer bands.

NOTE: It is strongly discouraged that any additional items, such as canteens and entrenching tools, be attached to the outer portion of the ALICE pack. However, mission requirements may dictate that some items be rigged to the outside of the ALICE pack. Items such as canteens and entrenching tools may be attached to the outside of the ALICE pack in accordance with individual unit SOPs, providing they are secured by point of attachment (clips) and further secured with 1/4-inch cotton webbing (two turns) or Type II or III nylon cord.

c. **Attaching the ALICE Pack.**

(1) Attach the snap hooks of the D-ring attaching straps to the D-rings on the parachute main lift web outside the reserve parachute connector snaps. Secure the ALICE pack in one of two ways:

NOTE: This load must be snug under the reserve parachute.

(a) When jumping from the right door, route the lower tie-down tape on the weapons case around the left leg and frame of the ALICE pack. Tie it with a bowknot on the leading edge of the weapons case.

(b) When jumping from the left door, attach 1/4-inch cotton webbing to the ALICE frame on the right side (as worn) by means of a girth hitch, secure it around the right leg, and tie in a bowknot.

(2) Attach the HPT lowering line ejector snap to the accessory attaching ring on the lowering line adapter web (Figure 12-7).

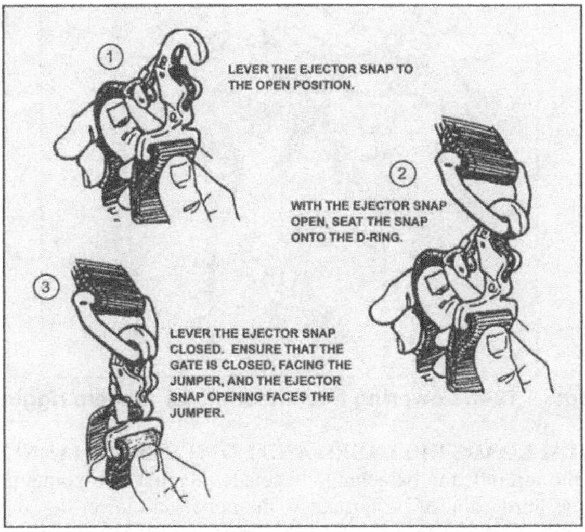

Figure 12-7. Ejector snap on HPT lowering line attached to accessory attaching ring on lowering line adapter web.

d. **Releasing the Pack.**

(1) Upon exiting the aircraft, execute the first two points of performance and, during the third point of performance, release all tie-downs.

(2) At 100 to 200 feet above the ground (during the fourth point of performance), check below for other parachutists, then pull both free-running ends of the D-ring attaching straps, allowing the ALICE pack to fall the length of the lowering line.

12-13. TANDEM LOAD AND LOWERING LINE

After the pack is rigged to the H-harness and frame, the lowering line is attached to the H-harness and stowed. The lowering line is secured to the left vertical bar of the frame (as worn) with two retainer bands. The ALICE pack and M1950 weapons case, when jumped together, are rigged as a tandem load and lowered on the same lowering line. After the ALICE pack and weapons case are attached to the parachute harness D-rings, the lowering line ejector snap is passed between the weapons case and the nylon chafe material of the case. The lowering line ejector snap is then attached to the accessory attaching ring on the lowering line adapter web (Figure 12-8).

Figure 12-8. Lowering line attached for tandem rigging.

12-14. TANDEM LOADS RELEASED AND LOWERED (H-HARNESS)

Upon exiting the aircraft, the parachutist executes the first two points of performance. Then, during the third point of performance, the upper and lower tie-down tapes on the weapons case are untied. The right leg tie-down on the ALICE pack is released (if jumping the left door). During the fourth point of performance at an altitude of about 100 to 200 feet above the ground, the ALICE pack is lowered by pulling at the same time on the free-running ends of the D-ring attaching straps, allowing the pack to fall to the end of the line. The activating arm of the quick-release assembly on the weapons case is activated, and the case slides down the lowering line to rest on top of the (lowered) pack.

NOTE: The D-ring attaching straps are removed from the parachute harness and secured to the H-harness before the parachute is returned to parachute maintenance after a jump exercise.

12-15. ALICE PACK RIGGED WITH FRAME USING HARNESS, SINGLE-POINT RELEASE AND HOOK-PILE TAPE LOWERING LINE

Before attaching the HSPR to the pack, the release handle and adjustable D-ring attaching straps are attached to the HSPR (Figure 12-9).

a. Route the two release handle cables between the two plies of the release handle cross strap. Attach the pile tape of the release handle to the hook tape attaching tab located between the plies of the release handle cross strap. Ensure that the release handle lanyard is not misrouted. Place the triangle links of the adjustable D-ring attaching straps on top of the white attaching loops. Route the white attaching loop up through the triangle link.

b. Route the green attaching loop up through the white attaching loop. Route the red attaching loop up through the green attaching loop. Route the red attaching loop through the grommet on the female portion leg strap release assembly. Ensure that the cable loop retainer on the female portion of the leg strap release assembly is facing up. Route the release handle cable through the red attaching loop and then through the cable loop retainer.

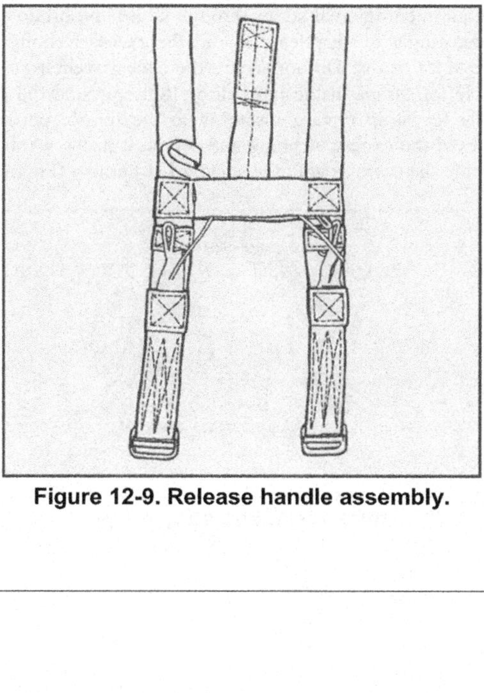

Figure 12-9. Release handle assembly.

c. Repeat the process for the other strap. Turn the harness over so that the adjustable D-ring attaching straps are on the bottom. Place the ALICE pack on top of the harness so that the middle outer cargo pocket is placed between the release handle cross strap and the adjustable cross strap. Ensure the top of the pack is facing the equipment retainer straps. Route the equipment retainer straps underneath the top of the frame, cross them on the back of the pack to form an X, then route them underneath the frame and the backrest of the pack (Figure 12-10).

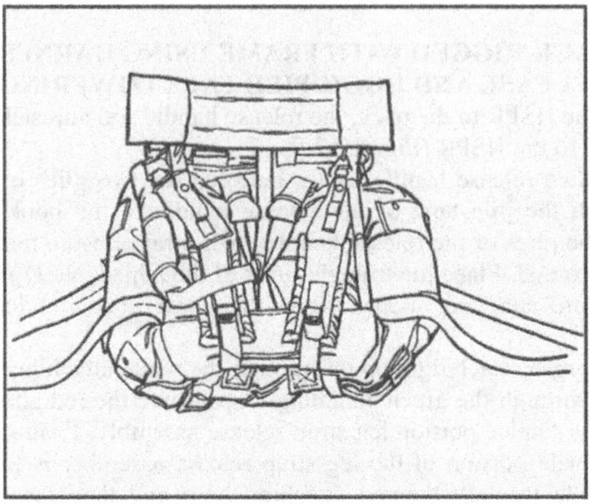

Figure 12-10. Harness routed.

d. Route the equipment retainer straps through their appropriate friction adapters and form a two- or three-finger quick release. S-roll the excess webbing and secure it with two turns of the retainer bands. Do not secure the excess webbing to the quick release. Route the adjustable leg straps in the most direct route around the pack and attach the male portion of the leg strap release assembly to the female portion of the leg strap release assembly. Fold the excess webbing and secure it in the webbing retainer. Attach the HPT lowering line the same as with the standard H-harness (Figure 12-11).

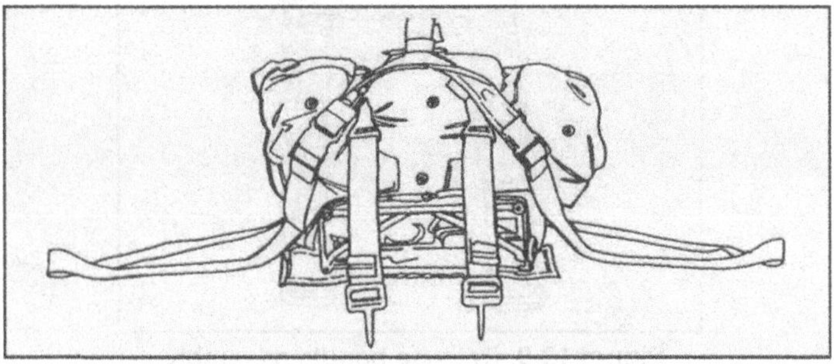

Figure 12-11. Leg straps routed.

12-16. ATTACHMENT OF HARNESS, SINGLE-POINT RELEASE AND ALICE PACK TO PARACHUTIST

When completely rigged, the HSPR is attached to the parachutist in the following sequence.

a. Using the buddy system, the parachutist stands in front of the HSPR-rigged ALICE pack. He grasps the harness by the two adjustable D-ring attaching straps and secures them to the D-rings on the main lift web outside of the connector snaps of the MIRPS/T-10 reserve (Figure 12-12).

b. The parachutist pulls on the free-running ends of the adjustable D-ring attaching straps and snugs the pack under the reserve. He then routes the ejector snap of the HPT lowering line through the nylon chafe material on the M1950 weapons case, from front to rear, then attaches the M1950 weapons case. He attaches the ejector snap on the HPT lowering line to the accessory attaching ring or triangle line as the outermost item on the D-ring. To attach the HPTL to the left D-ring, first open the ejector snap.

c. After the JM completes his JMPI, he routes the male portion leg strap release assemblies around the parachutist's legs and the M1950 weapons case, and attaches them to the female portion leg strap release. Adjustable leg straps may be tightly secured, but **must not** be so tight that they prevent the parachutist from executing a strong exit.

Figure 12-12. Parachutist adjusts leg straps.

12-17. TANDEM LOAD ATTACHED TO PARACHUTIST

The attachment of the rigged pack to the parachutist is identical to the procedure described in paragraph 12-16, with the following exceptions (Figure 12-13).

a. The lower tie-down tape is removed from the weapons case. If using the modified weapons case (with the HPT leg tie-down strap), the tie-down is routed around the case, the HPT is pressed together, and excess webbing is secured.

b. After the weapons case is attached to the parachutist, the leg strap of the HSPR is routed around the jumper's leg and the outside of the weapons case. Then the jumper tightens the adjustable leg strap after hooking the static line snap hook to the anchor line cable.

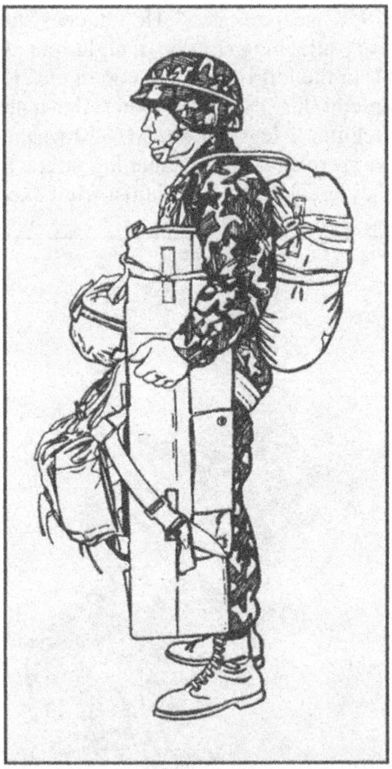

Figure 12-13. Tandem load attached.

c. The long end of the upper tie-down tape on the weapons case is routed around the case and main lift web directly above the chest strap of the parachute harness. It is tied with a bowknot on the front leading edge of the weapons case.

12-18. TANDEM LOADS RELEASED AND LOWERED (HARNESS, SINGLE-POINT RELEASE)

Upon exiting the aircraft, the parachutist executes the first two points of performance and then, during the third point of performance, unties the upper tie-down tape on the weapons case.

a. About 100 to 200 feet above the ground, the parachutist grasps the release handle and pulls up and out quickly, while at the same time releasing the load (and leg straps) and allowing it to drop the length of the lowering line. The activating arm of the quick-release snap on the weapons case is activated, and the case slides down the lowering line to rest on top of the (lowered) pack.

b. The release handle is released immediately following separation of the load from the jumper. The handle for releasing the load is secured to the HSPR with a release handle lanyard and stays with the HSPR to prevent its loss or separation.

c. To jettison the HSPR in an emergency, the parachutist lowers the pack, lowers the M1950 weapons case, then pulls out on the yellow safety lanyard (attached to the ejector snap on the HPT lowering line), which allows the pack to fall free.

12-19. JUMPING OF EXPOSED LOAD-BEARING EQUIPMENT

The procedures for rigging and wearing the protective mask carrier and the exposed load-bearing equipment during airdrop operations are as follows (Figure 12-14).

a. **Protective Mask Carrier**. The exposed protective mask carrier is worn in a different manner when airdrop operations are conducted.

(1) Wrap the narrow waist strap around the carrier, underneath the flap of the lower outside pocket, through the large O-ring, and hook it to the small D-ring to secure the quick opening flap securely closed.

CAUTION
Do not place masking tape on the mask carrier.

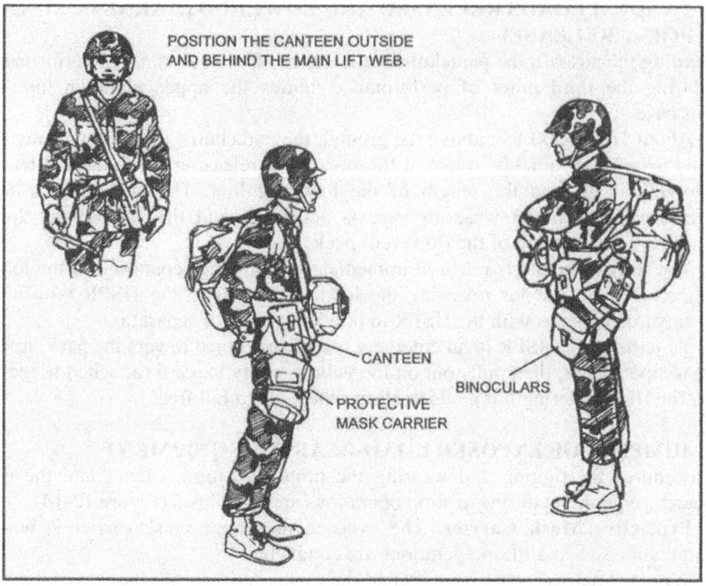

Figure 12-14. Binoculars and protective mask carrier rigged.

(2) Route the wide waist strap around the jumper's waist and hook it to the large D-ring, before donning the parachute and LBE. Don LBE and parachute, if LBE is to be jumped exposed.

(3) Place the aviator kit bag under the protective mask carrier. Route the parachute harness left leg strap through the carrying handle of the aviator kit bag and connect it to the left ejector snap. Route the parachute harness right leg strap directly to the right ejector snap. Do not attempt to secure the mask carrier with the leg straps.

(4) Adjust the length of the wide waist strap of the mask carrier so the entire mask carrier rests below the hip pad of the ALICE pack, but not so low that it interferes with the parachutist's ability to walk.

b. **Load-Bearing Equipment**. The LBE is worn in a standard manner when airdrop operations are conducted.

(1) Position the LBE just below the bottom of the parachute pack tray so that no portion of the parachute harness (for example, waistband, horizontal back strap, diagonal back straps, and so forth) rests on any major accessory item placed on the LBE (for example, canteen or ammunition pouch).

(2) Ensure the LBE is not covering any point of contact (for example, thighs, buttocks, or pull-up muscle) and is not buckled in front. JMs and jumpers must not wear knives on a point of contact.

(3) If a field pack is worn (for example, buttpack), make sure it does not contain any fragile items or equipment that is in a hard case.

(4) Tape any sharp edged piece of equipment, such as the wire cutter on the M9 bayonet sheath, with 100-mph tape to prevent snagging on other equipment.

(5) Ensure proper wear of the LBE by fitting it snugly against the jumper's body to prevent it from becoming entangled in risers, suspension lines, or equipment inside the aircraft.

(6) Route the LBE through the parachute harness before donning, or wear it before donning the parachute.

12-20. ENHANCED TACTICAL LOAD-BEARING VEST

The ETLBV must be in its full extended position. Both upper and lower chest buckles and the pistol belt remain unfastened. The jumper removes all slack from the chest buckle adjusting strap, S-folds or rolls the excess webbing, and secures the tape. After the jumper dons his parachute, the pistol belt and canteens are routed below the horizontal back strap and over the saddle. The ETLBV chest-mounted sewn-in magazine pouches are routed in the V formed by the horizontal back straps.

Section V. ADJUSTABLE INDIVIDUAL WEAPONS CASE (M1950)

The parachutist's M1950 individual weapons case is designed to carry individual or crew-served weapons. The M1950 may also be modified for other configurations. It is 10 inches wide and is adjustable in length from 33 1/2 inches to 50 1/2 inches. It is secured vertically by a quick-release snap attached to the left D-ring on the parachute harness.

12-21. M1950 SECURED TO PARACHUTIST

To prevent the M1950 from swaying during the opening shock of the parachute, two tie-downs fasten the case to the parachutist.

a. The upper tie-down tape is tied around the main lift web of the harness, and the lower tie-down tape goes around the parachutist's leg. If the weapons case has been modified, the lower tie-down strap (HPT) is secured around the leg.

b. Upon landing, the parachutist can secure his weapon quickly by opening the slide fastener, which is protected by a closing flap.

c. The slide fastener is designed as a quick release. To activate it (when the case is closed), a sharp tug on the slide fastener and tab thong (in the same direction as when zipping the container closed) causes the slide fastener to come apart.

12-22. M1950 ATTACHED TO PARACHUTIST

The M1950 is attached after the parachutist dons (and adjusts) the main and reserve parachutes.

a. The quick-release snap is attached to the quick-release link on the case. The opening gate of the quick-release snap faces the parachutist and is attached to the parachutist's left D-ring (to the outside of the reserve parachute connector snap).

b. The long end of the lower tie-down tape is passed around the outside of the case and in back of the left leg above the knee. Using a bowknot, the ends of this tape are tied together on the front leading edge of the weapons case. (The knot is untied before landing.) If the case has been modified with HPT leg straps, the piles are pressed together.

c. The long end of the upper tie-down tape is passed (from left to right) around the weapons case and main lift web of the parachute harness above the chest strap. Using a bowknot, the ends of this tape are tied together on the front leading edge of the weapons case.

Section VI. M16 RIFLE/M203 GRENADE LAUNCHER, EXPOSED AND PACKED

The M16 rifle or M203 grenade launcher can be jumped exposed or inside the M1950 weapons case. If the field-expedient method (weapon exposed) is used, the commander must consider that this may increase risk of injury to the parachutists and, therefore, hinder success of the mission. Inherent hazards of the exposed weapon include—

- The weapon becoming entangled with another jumper's parachute if a midair collision occurs.
- Possible injury to the parachutist during the PLF.
- Damage to the weapon during landing, which may cause a failure to fire.

12-23. M16 RIFLE/M203 GRENADE LAUNCHER EXPOSED

The sling is extended all the way, and the keeper is taped in place. The padding is secured over the side-mounted bolt forward assist and charging handle.

a. Place a plastic muzzle cap on the M16 muzzle, or pad and tape the muzzle and sight. This prevents the weapon from becoming entangled with the parachute suspension lines, or dirt from clogging the weapon during landing. Insert a magazine into the weapon and tape the magazine to the receiver, including the ejector port cover, to prevent loss of the magazine and debris from entering the bolt area. (Ensure that a round is **not** chambered and the weapon is placed on SAFE). Tape the hand guards to prevent loss upon impact when landing. (To aid in the removal of the padding and tape, fold and press the adhesive side on the running end of the tape together to form a quick-release pull tab.)

b. Further secure the weapon with two tie-downs of 1/4-inch cotton webbing (or a like item) with bowknots (Figure 12-15). Secure the sling to the diagonal back strap with the upper tie-down, which is a 12-inch tie strap. Secure the barrel of the rifle to the leg with a 24-inch strap to prevent possible entanglement with suspension lines. Remove the lower tie-down before landing to avoid personal injury.

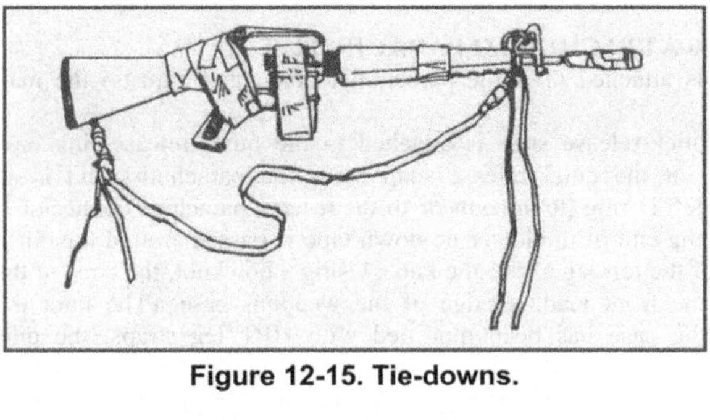

Figure 12-15. Tie-downs.

c. Sling the weapon over the left shoulder, with the muzzle down, and rotate it so that the pistol grip is facing the (parachutist's) rear (Figure 12-16). Place the sling from the lower keeper (stock) on the outside of the stock and over the left shoulder. Then run it under the chest strap of the main lift web. Thread the waistband through the carrying handle and into the metal adjuster on the waistband adjuster panel. Tighten the waistband securely so the weapon lies snugly against the parachutist's side.

Figure 12-16. Rifle positioned with muzzle down.

d. Prepare the M203 grenade launcher as outlined above, but place additional padding and tape around it (Figure 12-17).

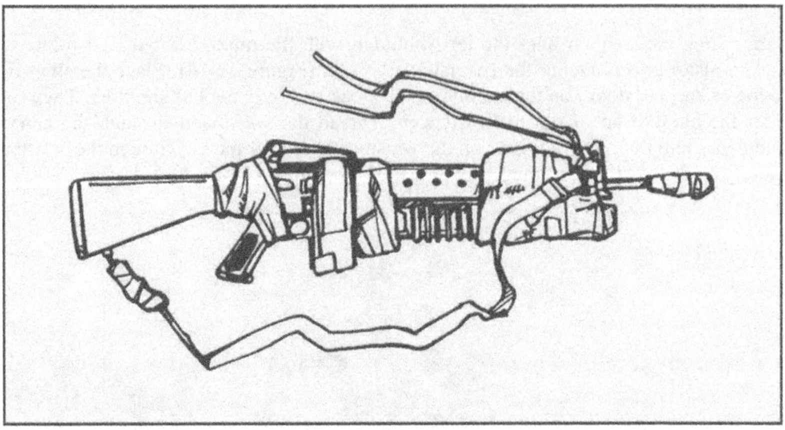

Figure 12-17. M203 grenade launcher, padded and taped.

12-24. M16 RIFLE/M203 GRENADE LAUNCHER PACKED IN M1950

The M1950 weapons case is laid flat with the closing flap facing up and opened. The weapon is inserted muzzle end first into the case with the forward assist facing up. A 20-round magazine is placed in the magazine well, if desired, but the parachutist ensures that a round is not chambered.

a. Close the case by using the flap thong, slide fastener and tab thong, and slide fastener. Leaving the slide fastener 1 to 2 inches from the top, separate the slide fastener and tab thong over the male portion lift fastener, and secure the female portion lift fastener to the male portion. If the lift fastener (or post) is unserviceable, route the upper tie-down tape through the slide fastener and tab thong.

b. Stand the case on its end so the weapon's muzzle is pointed up. Fold the excess case over the back side of the case and route the adjusting strap through the top set of adjusting strap connectors and secure it with a half hitch.

c. The M16/M203 may be jumped inside the case with a LAW or starlight scope. To do this, place the LAW or starlight scope on top of the M16/M203 and add padding (cellulose wadding) between the two items. It is usually rigged to the parachutist as a tandem load to be lowered on a single lowering line with the ALICE pack (paragraph 12-13).

Section VII. M60 MACHINE GUN

The M60 machine gun can be rigged on the individual parachutist or dropped as an accompanying load. If jumped on the individual, it is jumped as a team load.

12-25. M60 PACKED ASSEMBLED

The M1950 weapons case is laid down with the closing flap facing up. The muzzle of the M60 is placed into the lower-right corner of the weapons case. The weapon is pressed down until it is seated inside the case with the operating bolt up.

12-26. M60 PACKED DISASSEMBLED

The M1950 weapons case is shortened to 36 inches and laid down with the closing flap facing up (required when characteristics of a particular aircraft dictate length restrictions). It is packed as follows:

a. Disassemble the two groups by removing the barrel group.

b. Place the receiver group in the weapons case with the forearm assembly to the right and the cover facing down.

c. Place the barrel group in the case with the front sight to the left and pointing down.

d. Fit in the barrel group by sliding it to the right as far as possible so that the bipod-leg feet are not opposite the trigger housing group.

e. Add padding between the two groups.

NOTE: When the M60 is team loaded, the accessory bag, spare barrel, and tripod are placed in a separate weapons case and jumped by the assistant gunner. Both cases are usually tandem rigged and lowered (paragraph 12-13).

Section VIII. M249 SQUAD AUTOMATIC WEAPON

The SAW can be rigged for lowering as a tandem load or individual load. Tandem rigging is the same as for the M1950 weapons case and ALICE pack.

12-27. SAW MOD M1950 WEAPONS CASE

A modified M1950 weapons case is used for the SAW. The case is marked “SAW MOD” on the outside and has an extended 6 1/2-inch closing flap, which allows the weapon to fit in the case. A piece of cellulose wadding, about 20 inches long and 10 inches wide, is folded to form a pad about 10 inches long and 5 inches wide. This pad is placed in the muzzle end of the case. A 30-round magazine may be jumped with the SAW. The magazine is taped to the left side of the buttstock. The SAW is placed inside the case muzzle first. The pistol grip is toward the inside, and the carrying handle is facing up and away from the parachutist’s body.

12-28. ATTACHMENT TO PARACHUTIST

When rigging the SAW as a single item to be lowered (Figure 12-18), the parachutist attaches the quick-release snap to the quick-release link. This ensures that the opening gate quick-release snap is facing away from the main body of the SAW MOD M1950 weapons case when the quick-release link is pointed toward the top of the case. The parachutist **does not** route the quick-release link through the metal V-ring. No safety tie is used.

a. Prepare the HPT lowering line by folding it in the normal manner. No HPT lowering line should protrude from the retainer flaps. Secure each end of the HPT retainer flap with the HPT tab.

b. Route the looped end HPT lowering line from top to bottom through the V-ring of the SAW MOD M1950 weapons case. Route the entire HPT lowering line through the looped end HPT lowering line, forming a tight girth hitch around the V-ring.

c. Place the HPT lowering line against the main body of the SAW MOD M1950 weapons case to the left of the nylon chafe material, which secures the V-ring and quick-release link. The lowering line ejector snap points toward the top of the SAW MOD M1950 weapons case. The 1-inch tubular nylon webbing, which forms the looped end HPT lowering line, should rest between the retainer flap of the HPT lowering line and the nylon chafe material.

d. Secure the HPT lowering line with four turns of 1-inch-wide masking tape--two turns are routed around the main body directly below the quick-release link, and two turns are routed around the main body directly above the upper set of the adjusting strap connectors.

e. To attach the SAW MOD M1950 weapons case to the jumper's parachute harness, attach the quick-release assembly to the left D-ring of the parachute harness as the outermost item of equipment. Route the lowering line ejector snap below the quick-release assembly and attach it to the accessory attaching ring on the lowering line adapter web, with the opening gate facing toward the parachutist. Secure the upper tie-down tape and lower tie-down strap.

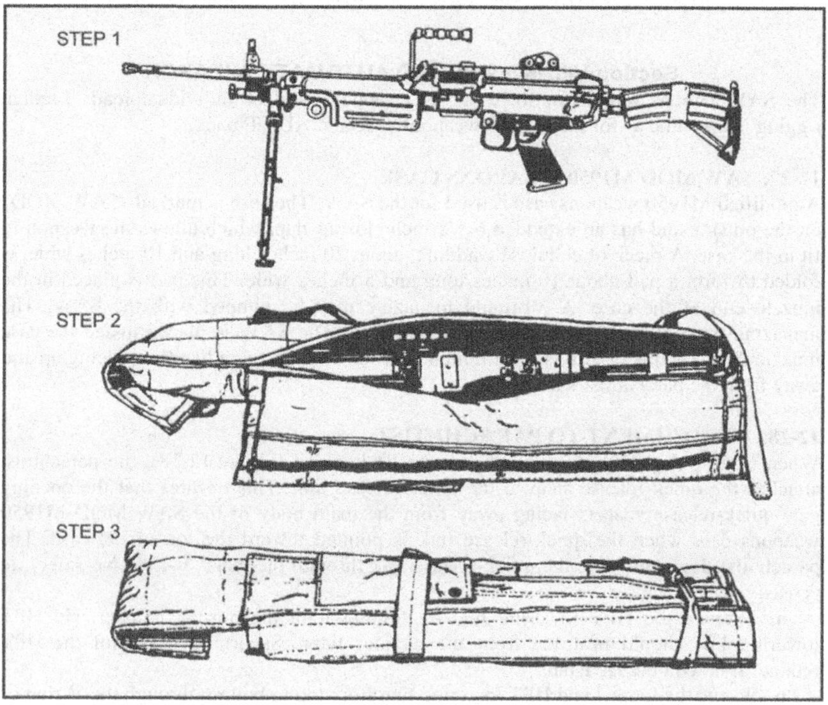

Figure 12-18. SAW rigged as a single item to be lowered.

Section IX. M224, 60-MM MORTAR

A modified M1950 weapons case is used for the M224 60-mm mortar. It is marked "60-mm MOD" on the outside and has an extended 11 3/4-inch closing flap.

12-29. MAJOR COMPONENTS

The components of the M224 are as follows (Figure 12-19).

- Aiming posts with case.
- M8 baseplate (small).
- M64 sight unit.
- M225 barrel.
- M170 bipod assembly.
- M7 baseplate (large).

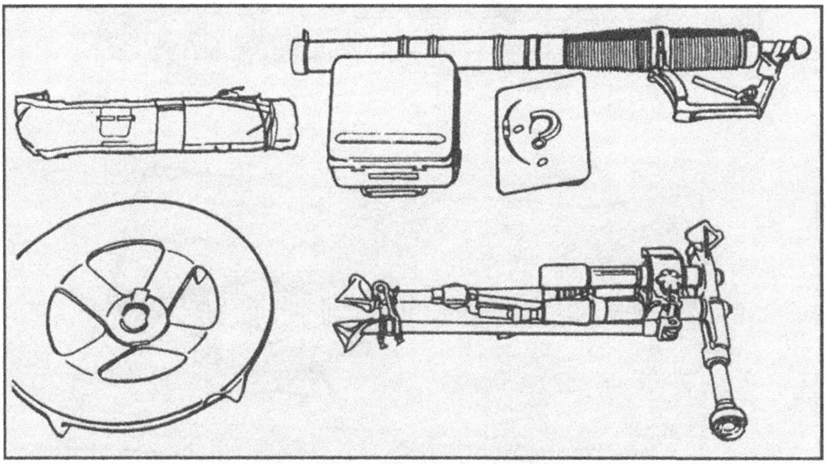

Figure 12-19. Components, M224 mortar.

12-30. LOAD DISTRIBUTION

The M224 crew members jump with the following components.

a. **Gunner**.
   - M224 barrel, in M1950 modified weapons case marked "60-mm MOD."
   - M8 baseplate, in ALICE pack.
   - M64 sight unit, centered in ALICE pack.

b. **Assistant Gunner**.
   - M170 bipod, in M1950 modified weapons gunner assembly case marked "60-mm MOD."

c. **Ammunition Bearer**.
   - Ammunition, in ALICE pack.

- M7 baseplate, in ALICE pack (outside of ALICE pack).
- Aiming posts, in ALICE pack (on top) with case.

12-31. INSTRUCTIONS FOR RIGGING

Following are instructions for rigging the M224 60-mm mortar.

a. **Gunner**. Place the barrel inside the case, muzzle down. Pack the small baseplate and sight unit inside the ALICE pack. Pad the sight unit with clothing or cellulose wadding to absorb the impact shock. Tandem-rig the pack and case for lowering (Figure 12-20).

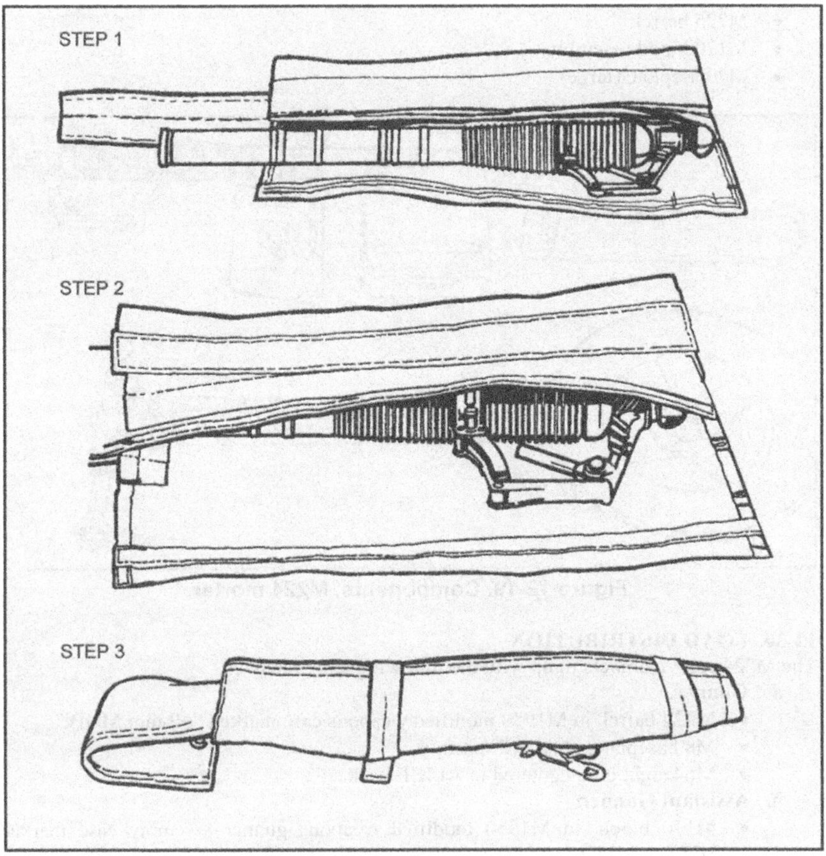

Figure 12-20. Barrel packed.

b. **Assistant Gunner**. Place the bipod assembly inside the case (Figures 12-21 and 12-22, page 12-26), ensuring that the traversing mechanism is in the middle of the case.

Secure the closing flap. Tandem-rig the case and pack for lowering (Figure 12-23 and paragraph 12-13).

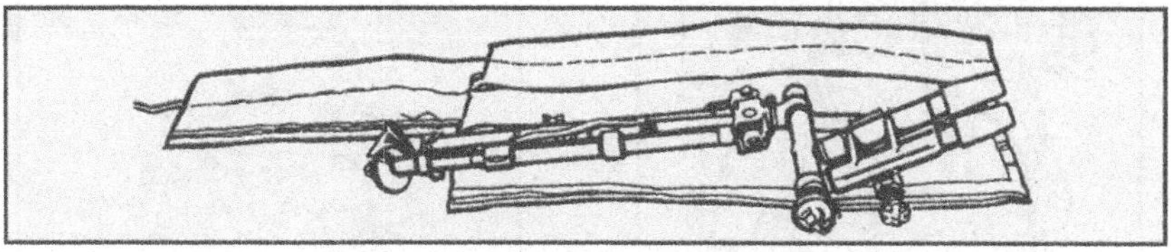

Figure 12-21. Bipod with case.

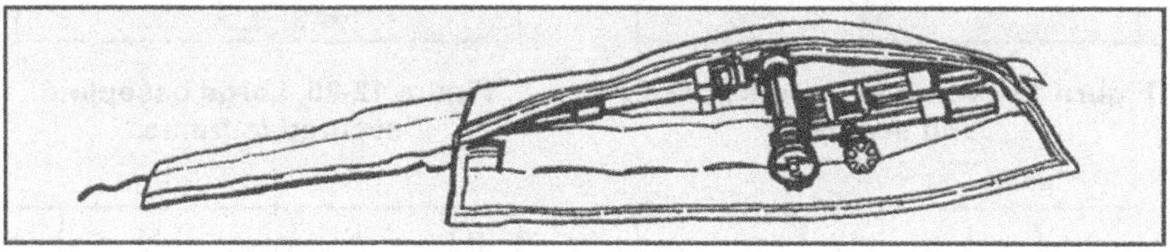

Figure 12-22. Bipod in case.

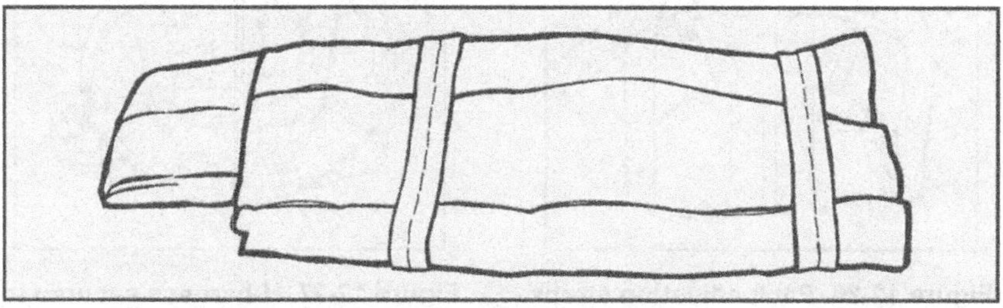

Figure 12-23. Assistant gunner's case completed for jumping.

c. **Ammunition Bearer**. Place the aiming posts under the top flap of the pack and secure with 1/4-inch cotton webbing to the top of the frame (Figure 12-24). Place the large baseplate over the outer accessory pouches and secure to the top of the frame with 1/2-inch tubular nylon webbing (Figure 12-25). Route the free-running ends of the pack adjusting straps through the baseplate (over, under, and over) and secure (Figure 12-26). Secure the H-harness to the pack and baseplate, ensuring that the equipment retainer straps are routed under the baseplate (top of the ALICE pack) (Figure 12-27) and over the baseplate (bottom of the ALICE pack).

NOTES:
1. Protect ammunition with cellulose wadding or clothing in ALICE pack.
2. Do not exceed weight limitations of the ALICE packs (large, 95 pounds; medium, 70 pounds).
3. Do not exceed weight limitation of parachute (360 pounds rigged weight for the T-10-series or the MC1-series).

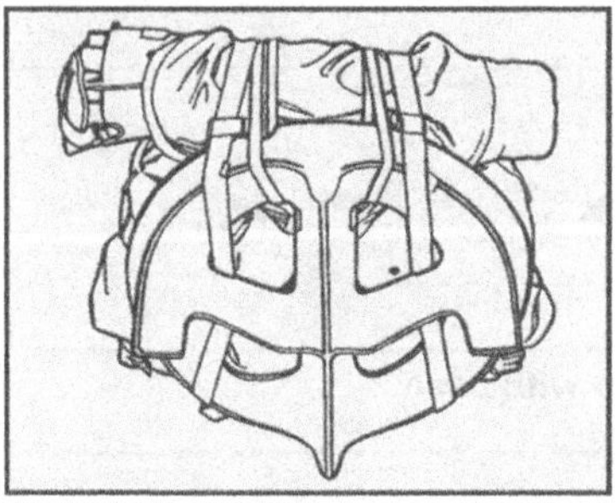

Figure 12-24. Aiming posts packed and secure.

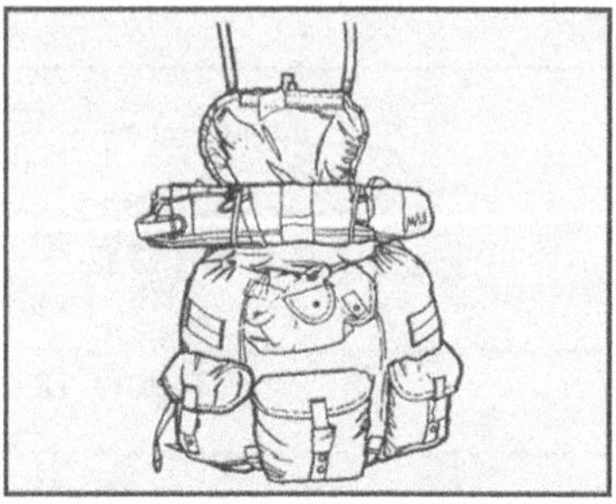

Figure 12-25. Large baseplate secured to frame.

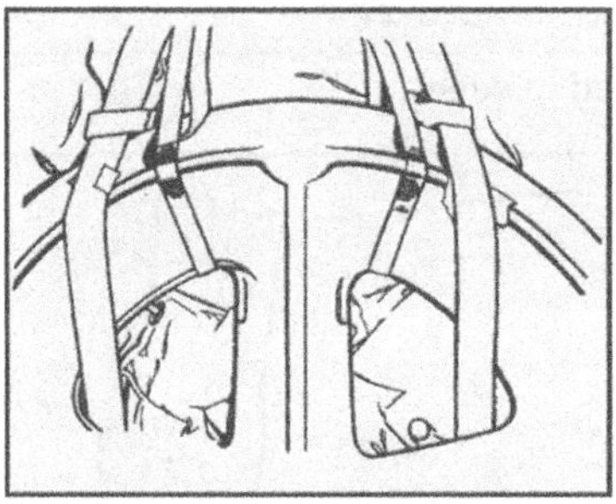

Figure 12-26. Pack adjusting straps routed through baseplate.

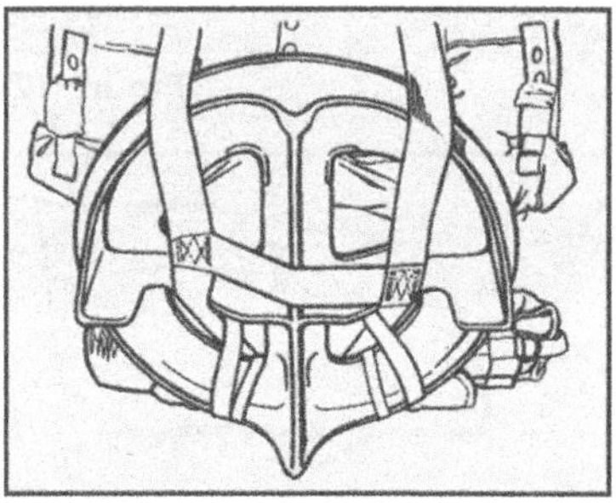

Figure 12-27. H-harness secured to pack and baseplate.

Section X. CONTAINER, WEAPON, INDIVIDUAL EQUIPMENT AND M202A1 ROCKET PACK

The CWIE is a general-purpose item used to carry designated combat equipment. It consists of a container and harness assembly, and when rigged for jumping, the container is attached to a harness assembly. The maximum dimensions for the CWIE are 12 inches wide, 12 inches deep, and 36 inches long. It can be adjusted to a minimum size of 12 inches wide, 6 inches deep, and 18 inches long. The CWIE weighs 16 1/2 pounds and, when packed, the container and contents must **not** exceed 95 pounds gross weight.

12-32. PREPARATION OF CONTAINER FOR PACKING

The container is laid down with the open portion facing up. A check is made to ensure that all securing straps and quick-fit adapters are present and serviceable.

a. **Packing Container.**

(1) Enough padding is used between items to prevent metal-to-metal contact. Special care is given to fragile items such as radios.

(2) Weapons that are too long to fit into the container are disassembled, and the parts are wrapped in padding.

(3) A combat pack or sleeping roll is placed in the bottom, which will hit the ground first to cushion the load.

(4) Related items are packed as one load.

(5) The heaviest items of equipment are loaded at the bottom of the container.

b. **Closing Container.**

(1) The side panels are folded over the contents.

(2) The bottom of the container is folded in an S-pattern to the desired length.

(3) The running ends of the three horizontal securing straps are passed over the container and fastened to the appropriate quick-fit adapters (using quick-release folds).

(4) The running ends of the two vertical securing straps are passed over the S-fold at the bottom of the container and fastened in a like manner.

12-33. HARNESS ASSEMBLY ATTACHED TO CONTAINER

The harness assembly is laid down with the inside portion facing up. All straps are straightened. The container is placed on the harness assembly. (The word TOP faces up. The arrow points toward the top of the harness assembly.)

a. The top and middle horizontal securing straps are placed around the container. The parachutist ensures that they are routed through (not over) the carrying straps. Each securing strap is fastened to the appropriate quick-fit adapters with a quick-release fold.

b. The bottom horizontal securing strap is placed around the container and secured to its quick-fit adapter. The two vertical securing straps are placed around the container. The parachutist ensures the securing straps are routed through (not over) the carrying straps and fastened to the appropriate quick-fit adapters with quick-release folds. The quick-fit adapters pass under the top carrying handle and the cable and conduit assembly.

c. The side securing strap is placed around the bottom of the container and over the two vertical securing straps. It is fastened to the appropriate quick-fit adapter with a quick-release fold.

d. The parachutist stands the entire assembly on end, top up. He tightens all securing straps and tapes any excess webbing. He then pulls out the release knob (red ball attached to the harness assembly).

e. The quick-release, quick-fit connecting links are inserted into the female part of the side fasteners on the harness assembly. The release knob, when pushed in, allows the cable to engage the claws in the hole of the quick-release, quick-fit connecting links.

f. The safety pin is inserted into the recess in the release knob from either side.

g. The looped end of the lowering line is threaded through the V-ring and under the carrying handle on top of the harness assembly. The entire length of the lowering line is pulled back through the loop and then pulled tight (Figure 12-28).

h. The HPT lowering line is folded, stowed in its container, and attached to the top carrying handle with two retainer bands.

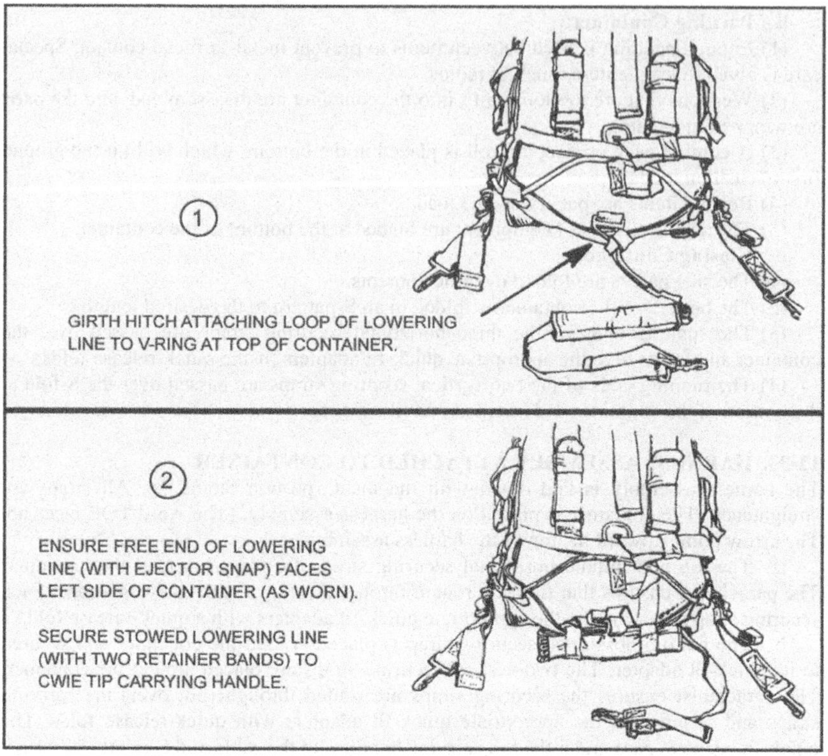

Figure 12-28. HPT lowering line attached to CWIE.

12-34. CONTAINER AND ASSEMBLY ATTACHED TO PARACHUTIST

The snap hooks of the quick-release straps are attached to the D-rings on the parachute harness. The parachutist ensures that the snap hooks are positioned to the outside of the connector snaps of the reserve parachute. The leg retaining strap is fastened around the leg (left leg, right door; right leg, left door), through the friction adapter, and a quick-release fold is made. The ejector snap on the HPT lowering line is attached to the lowering line adapter web on the left side of the parachute harness.

12-35. CONTAINER RELEASED

After exiting the aircraft, the parachutist executes the first three points of performance. After he executes the third point of performance, he releases the leg retaining strap and checks below to ensure that no other parachutists are in the way. The parachutist removes the safety pin from the red release knob when about 200 feet above the ground. He pulls the red release knob up and out, allowing the container to drop the full length of the lowering line.

Section XI. DRAGON MISSILE JUMP PACK

The Dragon (M47) missile jump pack (DMJP) is designed to carry one missile system and the M16 or M203 (Figure 12-29). The jump pack consists of a pack body constructed of nylon duck material with 1/4-inch felt material permanently sewn inside, nylon securing straps, lowering line, and quick-release assembly.

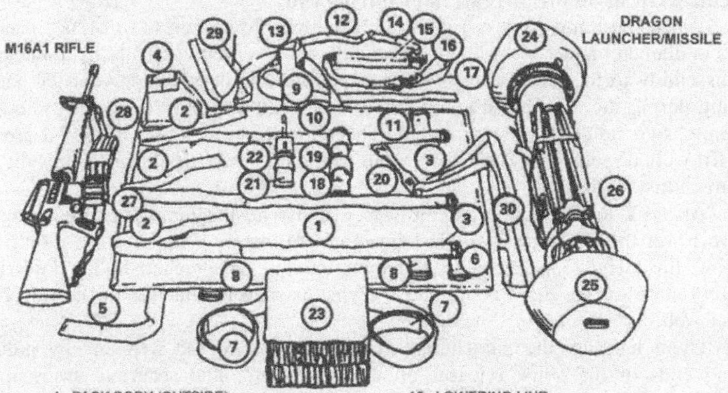

1. PACK BODY (OUTSIDE)
2. HORIZONTAL SECURING STRAPS
3. HORIZONTAL SECURING STRAPS QUICK-FIT ADAPTERS
4. RIFLE MUZZLE STOW POCKET
5. RIFLE BUTT SECURING STRAP
6. RIFLE BUTT SECURING STRAP QUICK-FIT ADAPTER
7. LAUNCHER FORWARD END SECURING STRAPS
8. LAUNCHER FORWARD END SECURING STRAP QUICK-FIT ADAPTERS
9. LAUNCHER AFT END BRIDLE
10. LAUNCHER AFT END BRIDLE SECURING STRAP QUICK-FIT ADAPTER
11. LAUNCHER AFT END BRIDLE SECURING STRAP QUICK-FIT ADAPTER
12. LOWERING LINE ATTACHING STRAP
13. LOWERING LINE ATTACHING STRAP LARGE LOOP END
14. LOWERING LINE ATTACHING STRAP SMALL LOOP END
15. LOWERING LINE
16. LOWERING LINE QUICK-EJECTOR SNAP
17. LOWERING LINE EJECTOR SNAP YELLOW SAFETY LANYARD
18. LOWERING LINE STOW POCKET
19. SECURING TABS
20. CARRYING HANDLE
21. QUICK-RELEASE LINK
22. QUICK-RELEASE SNAP
23. CARDBOARD HONEYCOMB ENERGY DISSIPATER
24. AFT END SHOCK ABSORBER (LARGE END CAP)
25. FORWARD END SHOCK ABSORBER (SMALL END CAP)
26. BIPOD
27. 20-ROUND MAGAZINE
28. 30-ROUND MAGAZINE
29. UPPER TIE-DOWN
30. LOWER TIE-DOWN

Figure 12-29. Nomenclature, DMJP.

12-36. MISSILE AND TRACKER

The missile and tracker can be jumped in two configurations. They can be rigged to be lowered as individual items, requiring two lowering lines, or tandem rigged on a single lowering line. The tandem rig requires the DMJP/AT4JP MOD HPT lowering line and modifications to the jump pack.

a. Due to the length of the missile and the difficulties in handling and moving the equipment around in the aircraft, the DMJP is restricted to parachutists who are at least 5 feet 6 inches tall. **The DMJP cannot be jumped from any aircraft that requires the parachutist to sit on the aircraft floor during exit.**

b. When the jump pack is rigged with the missile and an M16/M203, it is 11 1/2 inches in diameter and weighs about 40 pounds. It is secured vertically by a quick-release snap assembly to the left D-ring of the parachutist's harness. To prevent the pack from swaying during the parachutist's exit from the aircraft or from the opening shock of the parachute, two tie-downs are provided. The upper tie-down tape is routed around the main lift web, directly below the chest strap, while the lower tie-down secures the pack to the parachutist's left leg.

c. An HPT lowering line assembly is issued with the jump pack and can only be used to lower the DMJP as a single item. Tandem lowering requires the modified HPT lowering line. The looped end of the lowering line is attached to the lowering line attaching strap on the jump pack, and the ejector snap is attached to the lowering line adapter web.

d. Upon landing, the parachutist can rapidly get to the weapons by pulling the running ends of the quick releases on the three horizontal securing straps and three adjustable end straps.

12-37. DRAGON MISSILE JUMP PACK RIGGED

A detailed explanation of rigging the jump pack follows.

a. **Position the Dragon missile**.

(1) Place the pack with the felt side of the pack body facing up and extend all straps.

(2) Position the missile in the pack with the bipod facing up (Figure 12-30).

(3) Fit the aft shock absorber (large end) into the launcher aft end bridle of the jump pack.

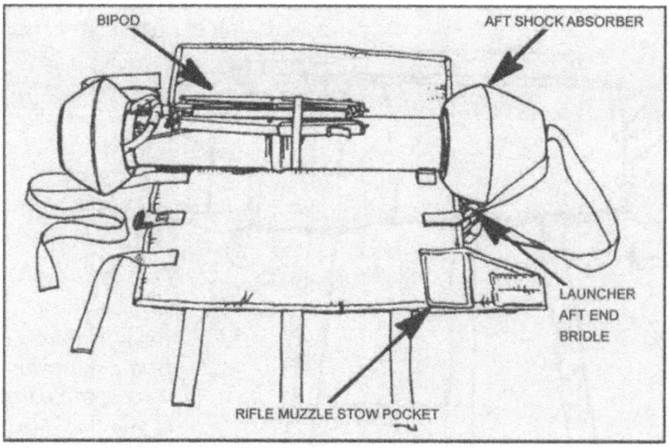

Figure 12-30. Missile being positioned.

b. **Position the weapon**.

(1) Insert a 20-round magazine into the magazine well (optional).

(2) Secure a 30-round magazine to the weapon sling, using adhesive tape (optional).

(3) Position the rifle on top of the pack, on top of the missile bipod, by inserting the rifle muzzle into the rifle muzzle stow pocket. The butt of the rifle pistol grip lies along the inner edge of the jump pack (Figures 12-31 and 12-32).

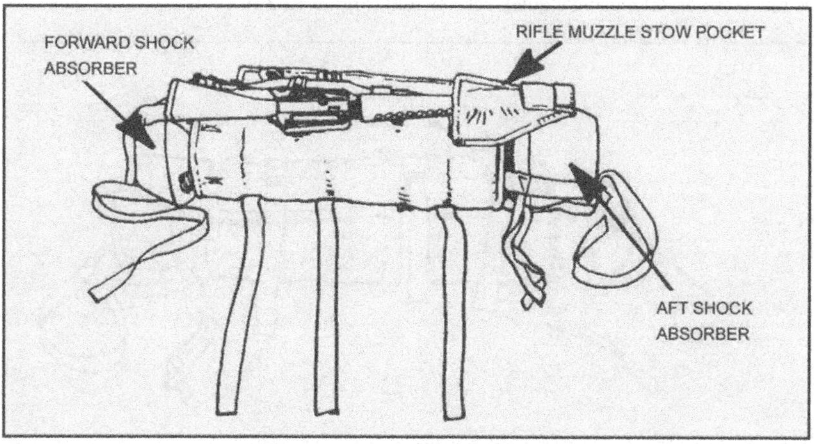

Figure 12-31. Rifle positioned.

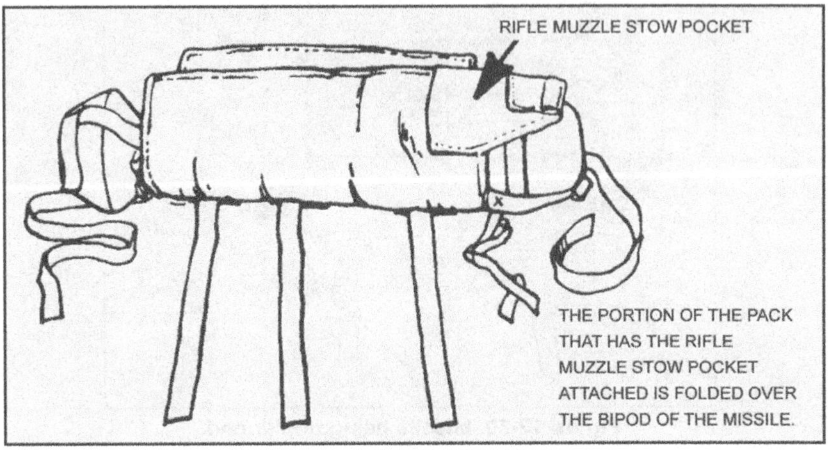

Figure 12-32. Pack folded.

c. **Close the pack**.

(1) Fold the outer flap of the jump pack over the weapon (if packed).

(2) Route the three horizontal securing straps through the quick-fit adapters using a quick-release fold, but do not tighten.

(3) Route the rifle butt securing strap through the quick-fit adapter. Tighten the strap as much as possible, using a quick-release fold (Figure 12-33).

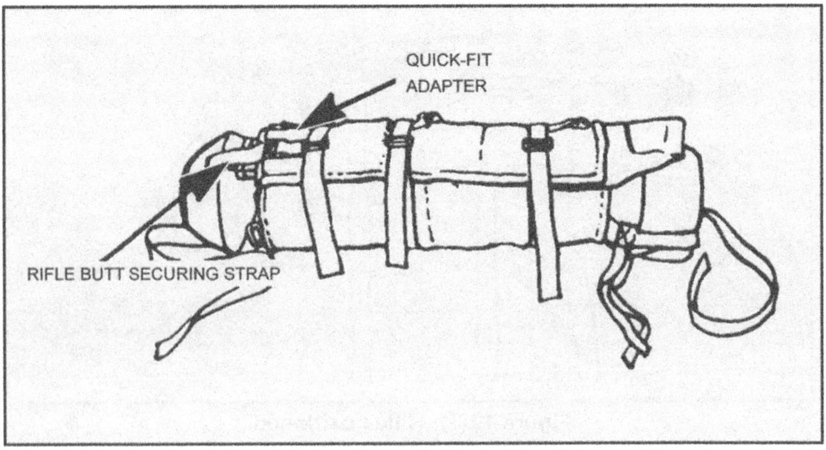

Figure 12-33. Pack closed.

d. **Secure honeycomb squares.**

(1) Route the missile launcher aft end bridle securing strap through its quick-fit adapter and tighten, using a quick-release fold.

(2) Secure two 9-inch-square blocks of cardboard honeycomb energy dissipater across the forward shock absorber of the missile, using the launcher forward end securing straps. Tighten the straps and form an X over the center of the honeycomb. Secure the straps using a quick-release fold (Figure 12-34). Tie one turn of 1/4-inch cotton webbing around the X formed by the launcher forward end, securing straps in a surgeon's knot and locking knot. Tie another one turn of 1/4-inch cotton webbing around the launcher forward end, securing straps below the cardboard honeycomb energy dissipator in a surgeon's knot and locking knot. If the missile forward end shock absorber is missing from the Dragon missile training round, it should be replaced with a 9- by 9-inch square of cardboard honeycomb.

WARNING

Never jump a live missile without this honeycomb in place and securely fastened. Place parachutists with DMJPs or CWIEs to the front of the stick. Staggered exits are difficult to control if individuals carrying either the DMJP or CWIE are not near the front of the sticks.

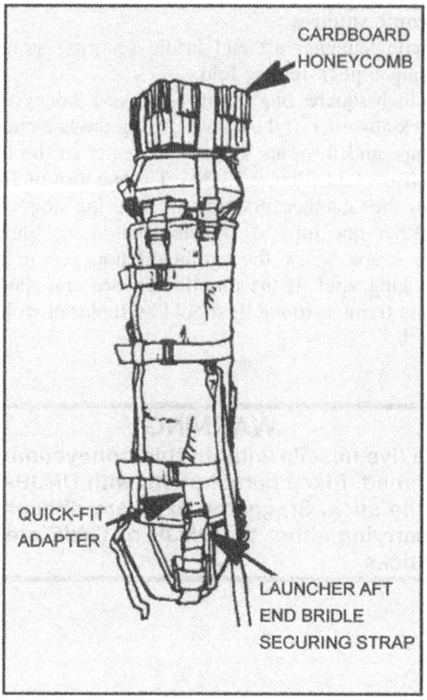

Figure 12-34. Honeycomb squares secured.

(3) If not already installed, attach the lowering line attaching strap to the launcher aft end bridle by means of a girth hitch with the large loop end (Figure 12-35, page 12-36). Run the small loop end through the large loop end and tighten (run the small loop end toward the lowering line stow pocket).

(4) Position the DMJP on the aft end shock absorber and retighten the launcher forward end securing straps using quick-release folds. All excess on the securing straps are folded back under the corresponding quick-fit adapter. Retighten one turn of 1/4-inch, 80-pound test cotton webbing in an X pattern around the launcher forward end securing straps, and tie securely with a surgeon's knot and locking knot (Figure 12-36, page 12-36).

(5) Tie one turn of 1/4-inch, 80-pound test cotton webbing around the launcher forward end, securing straps below the 9-inch square honeycomb. Tie in a surgeon's knot and locking knot to prevent the honeycomb from shifting.

(6) Position the DMJP on the honeycomb and tighten the launcher aft end bridle securing strap. Secure its running end by folding it back under the corresponding quick-fit adapter.

(7) Tighten the three horizontal securing straps as evenly as possible around the weapon(s). Incorporate a quick-release fold and secure with a retainer band or two turns of masking tape.

(8) Attach the quick-release assembly to the pack (Figure 12-37).

Figure 12-35.
Lowering line attaching strap installed.

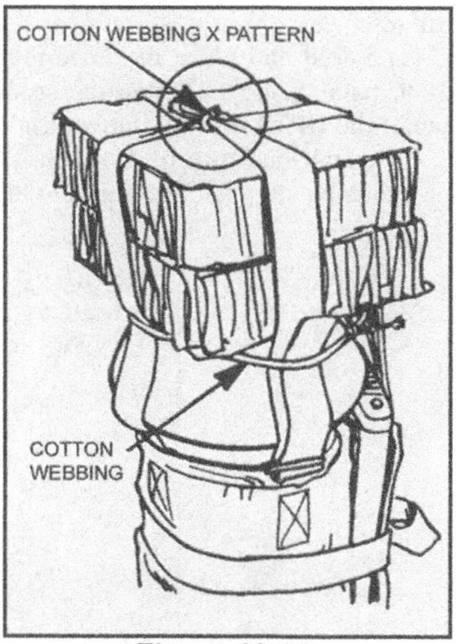

Figure 12-36.
X pattern.

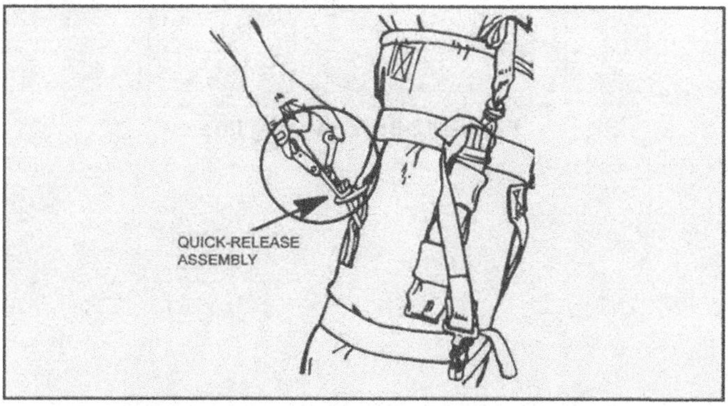

Figure 12-37. Quick-release assembly.

NOTE: Apply more tension to the end strap by tightening the three horizontal securing straps evenly. Uneven or insufficient tension can cause misalignment of the overlap portion of the pack and result in improper retention and protection of the rifle.

e. **Attach lowering line to DMJP.**

(1) Attach the lowering line to the lowering line attaching strap by routing the looped end of the HPT lowering line through the looped end of the lowering line attaching strap (Figure 12-38). Route the lowering line through its own loop and pull tight, forming a girth hitch.

(2) S-fold and place the lowering line inside the lowering line stow pocket (Figure 12-39, page 12-40). Tighten the securing tabs (Figure 12-40, page 12-41). The ejector snap of the HPT lowering line protrudes from the lowering line stow pocket.

(3) Install one turn of 1/4-inch, 80-pound test cotton webbing around the lowering line attaching strap and the adjacent launcher aft end bridle strap, and tie securely.

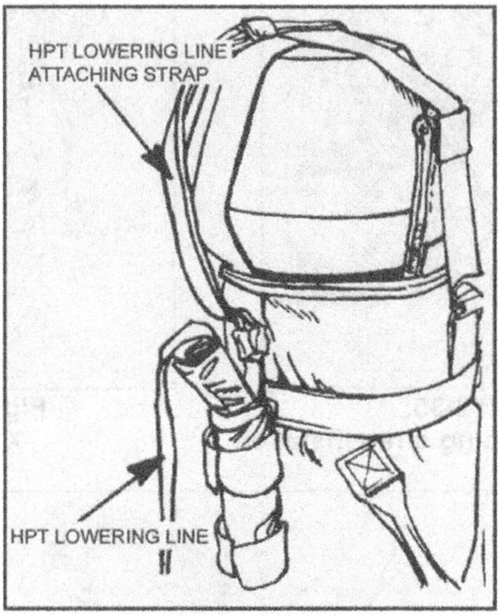

Figure 12-38. Lowering line.

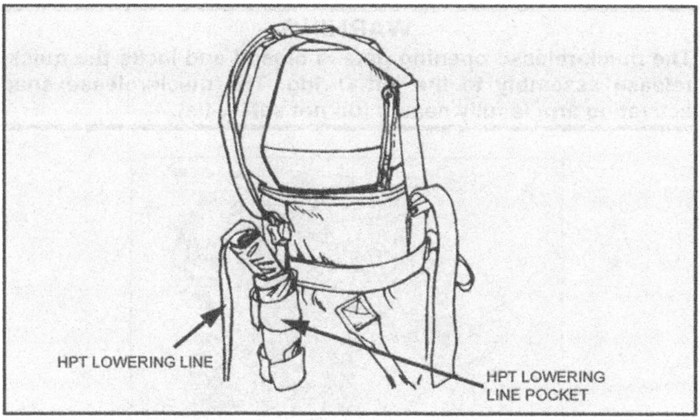

Figure 12-39. Lowering line inserted in pocket.

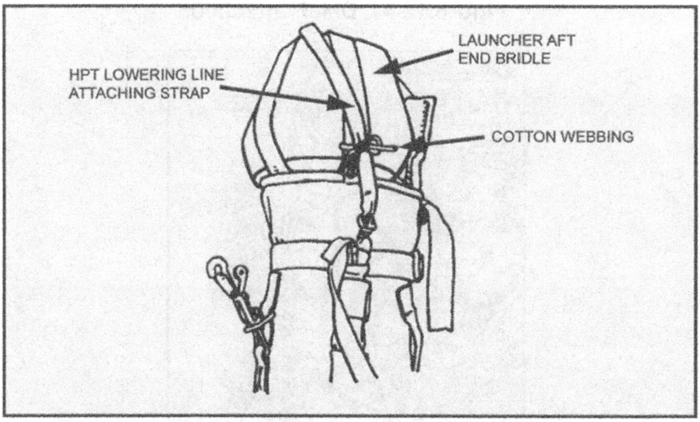

Figure 12-40. Lowering line attaching strap secured.

12-38. DRAGON MISSILE JUMP PACK ATTACHED TO PARACHUTIST

The DMJP is the last item attached to the parachutist.

a. The DMJP is secured to the parachutist by attaching the quick-release snap to the parachutist's left D-ring as the outermost item (Figure 12-41).

b. The lower tie-down tape is routed around the DMJP and the parachutist's left leg (Figure 12-42). A single-loop bowknot is tied on the front of the DMJP where it is easy for the parachutist to reach.

c. The upper tie-down tape is routed around the left main lift web directly below the chest strap. It is tied snugly with a single-loop bowknot (Figure 12-43, page 12-42).

WARNING

The quick-release opening gate is closed and locks the quick-release assembly to the left D-ring. The quick-release snap activating arm is fully seated (do not safety tie).

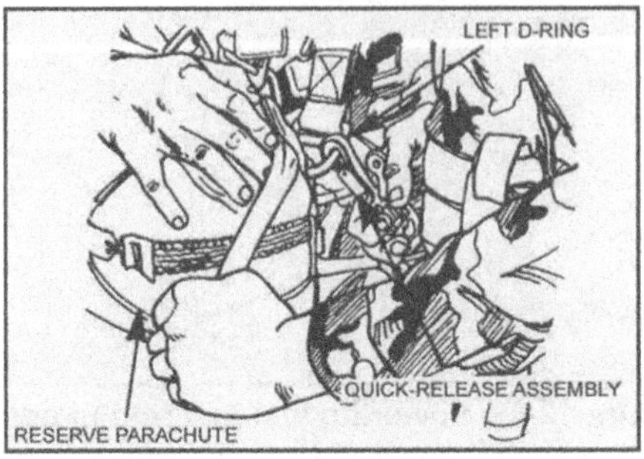

Figure 12-41. DMJP attached.

Figure 12-42. HPT lowering line attached.

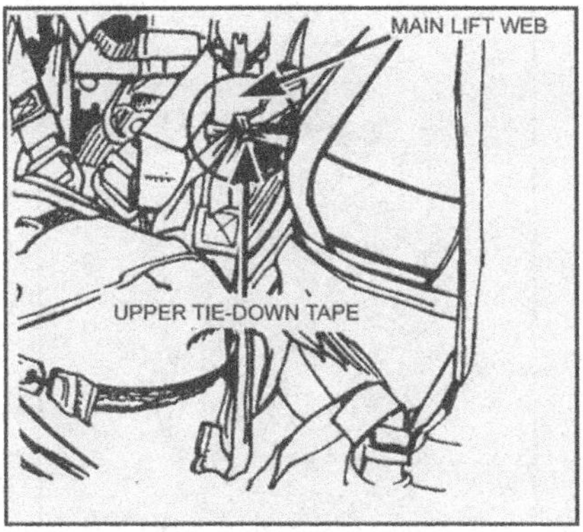

Figure 12-43. Upper tie-down.

12-39. INDIVIDUAL JUMP PROCEDURES

The DMJP is jumped with a front-mounted large ALICE pack (Figure 12-44).

a. At the 20-minute time warning, one of the safety personnel attaches the DMJP to the parachutist and inspects it.

b. After leaving the aircraft, the parachutist—

(1) Releases the lower tie-down tape, then the upper tie-down tape, in sequence. If either tie malfunctions and the parachutist must ride the missile down, it is important that the missile be vertical, not hanging directly below the feet.

(2) Looks to see that the area below is clear (about 100 to 200 feet above ground), and activates the DMJP quick-release assembly by using the left hand to push down and away on the activating arm, allowing it to fall the length of the lowering line.

NOTE: To jettison the DMJP in an emergency, the parachutist lowers the pack, then pulls out on the yellow safety lanyard attached to the ejector snap, allowing the missile pack to fall free.

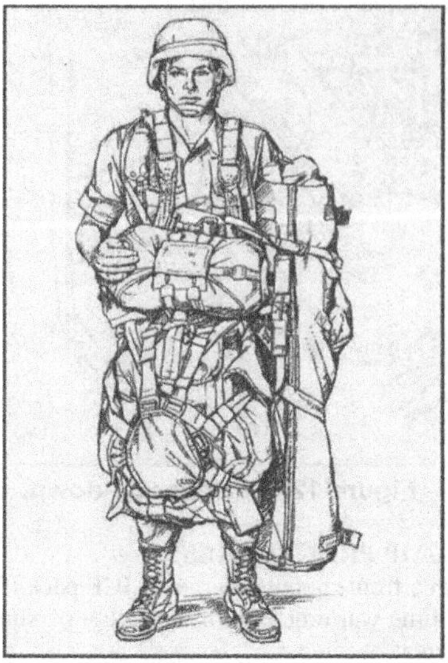

Figure 12-44. DMJP attached to parachutist.

12-40. DRAGON TRACKER

The tracker (sighting system) is placed inside the large ALICE pack and is attached to the parachutist. The system must remain in its protective case during the jump.

12-41. DRAGON MISSILE JUMP PACK AND ALICE PACK RIGGED AS A TANDEM LOAD

To rig the DMJP and ALICE pack for lowering, modifications are required along with the DMJP/AT4JP modified HPT lowering line.

a. **Materials**.

- Fastener tape, pile, color OD 106, 1-inch width, Class 1, MIL-F-21840, NSN 8315-00-106-5974.
- Fastener tape, hook, color OD 106, 1-inch width, Class 1, Type II, MIL-F-21840, NSN 8315-00-106-5973.
- Webbing, textile, nylon, Type X, Class 1, 1 23/32-inch width, color olive drab, MIL-W-4088, and Class R, MIL-W-27265, NSN 8305-00-261-8584.
- Thread, nylon, color OD-S1, conforming to Type II, Class A, Size 3, 800-yard tube of V-T-295, NSN 8310-00-559-5212.
- D-ring, parachute harness, drawing, No. 11-1-485.

b. **Stitching Requirements**.

(1) Stitching must conform to FED-STD-751, Type 301, 5 to 8 stitches per inch.

(2) Ends of stitching must be over-stitched not less than 1/2 inch.

c. **Modification Procedure**.

(1) ***Replacement of the V-ring with D-ring, 11-1-485.***

(a) Place the jump pack on a repair table, positioning it so the outside faces upward.

(b) Cut the box-X stitching on the chafe web end, which secures the V-ring to the pack. Remove the V-ring and cut the stitching.

(c) Pass the end of the chafe through replacement D-ring 11-1-485 and stitch with the box-X stitch formation to pack as in the original construction.

(2) ***Attachment of the D-ring 11-1-485 and web chafe to aft shock absorber harness cross adjustable strap.***

(a) If installed, remove the lowering line attaching web, which is girth-hitched to the aft shock absorber harness cross straps.

(b) Cut a 7 1/4-inch length of Type X nylon webbing and sear cut ends to prevent fraying.

(c) Place marks 3 inches from each end and in the center of the webbing.

(d) Place marks 3 5/8 inches from each side of the center of the aft harness cross adjustable strap.

(e) Pass the end of 7 1/4-inch Type X nylon webbing through the D-ring 11-1-485 loop and position it between the markings of the aft harness cross adjustable strap. Stitch with a 3-inch, 4-point WW stitch formation at each end.

(3) ***Attachment of hook-and-pile lowering line retainer.***

(a) Cut a 3 1/2-inch length of hook tape.

(b) Cut a 4-inch length of pile tape.

(c) With hook and pile facing each other, overlap tabs 1 3/4 inches and press together.

(d) Place marks 1 and 2 inches (on the inside) from the binding edge of the pack body on the aft cross adjustable strap.

(e) With pile side facing toward outside of pack and hook facing to inside of pack, position edges of tape overlap even with cross strap edges between markings and stitch to cross strap with a single box stitch formation.

12-42. ALICE PACK (LARGE) JUMPED WITH DRAGON MISSILE JUMP PACK

The large ALICE pack with frame must be jumped with the DMJP to stabilize the DMJP and to accommodate the Dragon tracker assembly.

a. Completely pad the tracker assembly with cellulose wadding to prevent damage.

b. Place the padded tracker assembly in the inside pouch with soft articles of clothing or equipment packed around it. Place additional items in the pouch, with hard and sharp items padded.

c. Fill the outside ALICE pockets with nonfragile items to capacity, since the filled pockets aid the positioning and prevent slippage of the H-harness during parachute opening and lowering line deployment.

d. Close the pack by use of the drawstring closure, engaging the securing straps and attachment of pocket snap fasteners.

e. Adjust the adjustable shoulder carrying straps snugly. S-fold and tape the excess webbing on the adjustable shoulder carrying straps. Route the running ends of the waist straps around the frame opposite the lower back pad, tighten, and secure in place by taping. This secures the adjustable should carrying straps and reduces the possibility of entanglement on board the aircraft.

f. To stow the Dragon modified lowering line, S-fold the lowering line neatly on top of the web inside the retainer, ensuring that ends are stacked evenly with the retainer outer edges. Secure the pile tab on web located 46 inches from the ejector snap end to hook the extension on the retainer. Fold the hook side of flap tightly over the S-folded lowering line and, holding it firmly, fold pile side of the flap over and secure hook and pile together. Secure the pile extension on the retainer flap to the hook tab at the looped end of the lowering line.

12-43. ALICE PACK RIGGED WITH FRAME

The pack is turned upside down.

a. Place the modified H-harness on the pack so that the cross straps are in front of the pack and the friction adapters are touching (or near) the bottom of the pack frame.

(1) Route the equipment retainer straps over the top of the pack and then under the top portion of the frame.

(2) Route the equipment retainer straps over the horizontal bar of the frame and cross at the center of the back of the pack.

(3) Route the straps under the frame and secure them to the friction adapters, forming a quick release.

b. To attach the lowering line after rigging pack, girth-hitch the lowering line loop at the intersection of the crossed equipment retainer straps (at the rear center of the pack).

(1) Secure the stowed lowering line to the left side (as worn on the parachutist) on the keepers of the pack with two turns of masking tape or two sets of retainer bands. The lowering line ejector snap faces up and to the right or left.

(2) Thread the (H-harness) D-ring attaching straps through the intermediate friction adapters, forming a quick release with the running ends that are pointing away from the parachutist.

12-44. DRAGON MISSILE JUMP PACK RIGGED

The DMJP is rigged as described above, omitting installation of the lowering line attaching strap.

12-45. ALICE PACK ATTACHED TO PARACHUTIST

The D-ring attaching strap snap hooks (on rigged ALICE) are attached to the harness D-rings outside of the reserve connector snaps.

12-46. DRAGON MISSILE JUMP PACK RIGGED FOR TANDEM LOAD

The rigged DMJP is placed next to the parachutist's left side.

a. To secure the 15-foot HPT lowering line, grasp the ejector snap of the HPT lowering line. Route it (from bottom to top) through the D-ring openings at the pack body and aft end locations. Position the 18-inch strata blue marking on the HPT lowering line even with the D-ring on top of the DMJP (Figure 12-45).

b. Secure the lowering line to the aft cross strap at the pack body end by folding the hook tab over the lowering line and fastening pile together.

c. Attach the quick-release snap to the quick-release link, which has been routed through the D-ring. The snap opening must face away from the pack.

12-47. DRAGON MISSILE JUMP PACK ATTACHED TO PARACHUTIST

The DMJP quick-release assembly is attached to the left harness D-ring outside of the reserve connector snap and H-harness or HSPR snap hook (Figure 12-46).

a. Attach the HPT lowering line ejector snap to the lowering line adapter web.

b. Route the lower tie-down tape around the DMJP and the parachutist's left leg.

c. Tie with a bowknot on the front of the DMJP so it is easy for the parachutist to reach.

d. Route the upper tie-down tape under and around the left main lift web directly below the chest strap and tie snugly with a bowknot.

e. Use the adjustable leg strap of the HSPR as a lower tie-down tape.

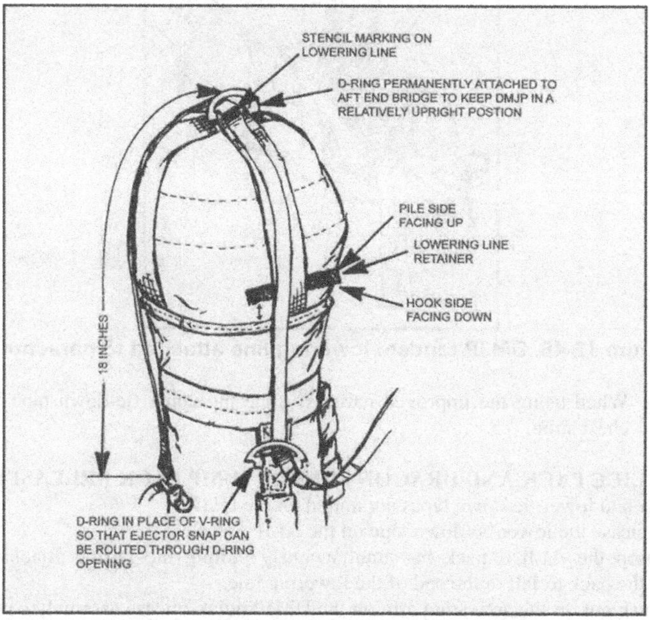

Figure 12-45. DMJP modified with D-ring.

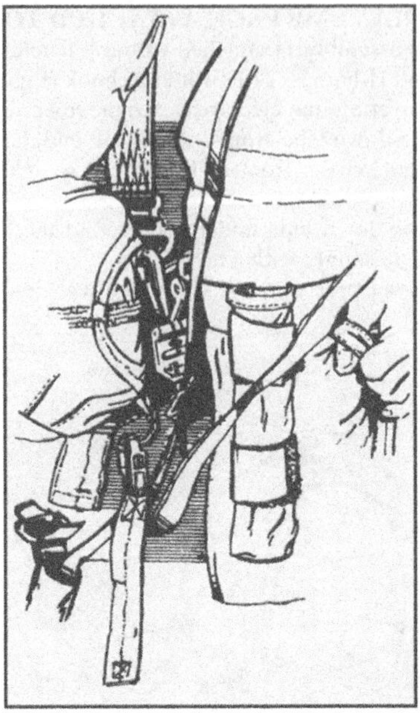

Figure 12-46. DMJP tandem lowering line attached to parachutist.

NOTE: When using the improved harness, route the upper tie-down tape below the chest strap.

12-48. ALICE PACK AND DRAGON MISSILE JUMP PACK RELEASED

The upper and lower tie-down tapes are untied on the DMJP.

a. Release the lower tie-down tape on the ALICE pack.

b. Drop the ALICE pack by simultaneously pulling the D-ring attaching straps, allowing the pack to fall to the end of the lowering line.

c. Push out on the activating arm on the DMJP quick-release assembly so the DMJP slides down the lowering line to the end.

NOTE: To jettison the ALICE pack and DMJP in an emergency (after performing the above), lift the reserve up and pull the yellow safety lanyard on the lowering line ejector snap, allowing the ALICE pack/DMJP to fall free.

WARNING
Under no circumstances will the parachutist release the DMJP before releasing the ALICE pack.

12-49. REMOVAL OF LOWERING LINE

Upon landing, the parachutist removes the lowering line by pulling the line through the two D-rings and releasing the girth hitch on the H-harness. The DMJP MOD lowering line is stored with the DMJP for reuse.

Section XII. AT4 JUMP PACK

The parachutist's AT4JP is designed to carry one AT4 weapon round (SM-136) and one M16 rifle. The jump pack consists of a pack body constructed of nylon and 1/4-inch thick felt material. When the jump pack is rigged with the AT4 weapon round, M16, and shock absorber, it is 47 inches long, 9 inches in diameter, and weighs about 29 pounds. It is secured vertically by a quick-release snap to the left D-ring of the parachutist's harness.

12-50. COMPONENTS

To prevent the jump pack from swaying during the parachutist's exit from the aircraft or the opening shock of the parachute, two tie-down tapes are provided. The upper tie-down secures the pack to the main lift web, while the lower tie-down secures the pack to the parachutist's left leg.

a. A lowering line stow pocket is attached to the exterior of the pack for retaining the HPT lowering line. A 24-inch lowering line attaching strap is issued with the jump pack and is required for assembling the pack and lowering line when rigged for delivery of the AT4JP only. (When the AT4JP and ALICE are rigged for tandem load, a modified [DMJP/AT4] lowering line is required.) The looped end of the lowering line or lowering line attaching strap (based upon the configuration to be lowered) is attached to the nonadjustable cross D-ring strap on the jump pack so it will be suspended below the parachutist and impact in a vertical position. The ejector snap on the opposite end is attached to the lowering line adapter web on the parachutist's harness.

b. The shock absorber (polyurethane/honeycomb) configuration at the aft end of the jump pack is intended to prevent damage to the AT4 weapon round during ground impact. Upon landing, the parachutist can rapidly gain access to the weapon round and M16 by pulling the quick releases rigged into the jump pack securing straps. The total assembly is shown in Figure 12-47.

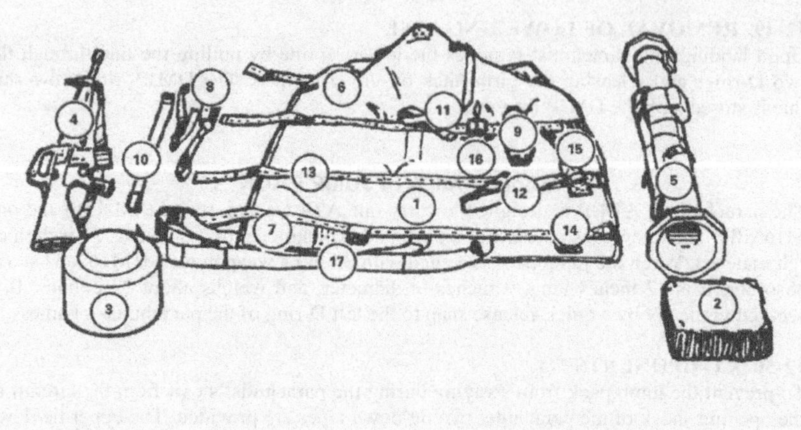

1. AT4 JUMP PACK
2. CARDBOARD HONEYCOMB
3. LAUNCHER AFT END SHOCK ABSORBER
4. M16 RIFLE
5. AT4 WEAPON
6. RIFLE MUZZLE STOW POCKET
7. RIFLE BUTT STOW POCKET
8. HPT LOWERING LINE
9. LOWERING LINE POCKET
10. LOWERING LINE ATTACHING STRAP
11. UPPER TIE-DOWN TAPE
12. LOWER TIE-DOWN TAPE
13. SIDE SECURING STRAPS
14. QUICK-FIT ADAPTER
15. CARRYING HANDLE
16. QUICK RELEASE ASSEMBLY
17. LAUNCHER AFT END SECURING STRAPS

Figure 12-47. AT4 nomenclature.

12-51. AT4 JUMP PACK RIGGED

To prepare the pack, the parachutist lays the pack down with the felt side of the pack facing up and extends all securing straps.

a. **Positioning the AT4 Weapon Round.**

(1) Position the weapon round on the pack with the carrying sling facing down, the launcher forward (small) end fitted into the forward end securing strap, and the launcher aft (large) end centered on the middle launcher aft end securing strap (Figure 12-48, page 12-48).

(2) Ensure that the launcher forward (small) end fits snugly into the nonadjustable cross D-ring strap.

(3) Fold the portion of the pack that has the rifle muzzle and butt stow pockets attached over the weapon round.

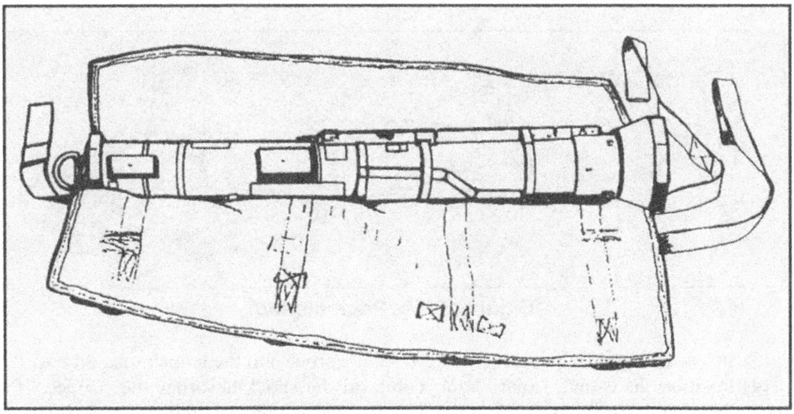

Figure 12-48. AT4 weapon round positioned.

b. **Positioning the M16.**

(1) Insert a 20-round magazine into the rifle (optional).

(2) Secure a 30-round magazine to the sling of the rifle, using adhesive tape (optional).

(3) Insert the rifle muzzle into the muzzle stow pocket and place the butt in the rifle butt stow pocket as shown in Figure 12-49. Ensure that the magazine well and pistol grip point toward the side securing straps, and the rifle muzzle is inserted into the muzzle pocket as far as possible.

Figure 12-49. Positioning the M16.

c. **Closing the Pack.**

(1) Rotate the unfolded portion over the rifle and weapon round snugly.

(2) Route the four side securing straps through the quick-fit adapters and tighten, then thread the end through the adapter, forming a quick-release fold. Fold the strap excess back under its corresponding quick-fit adapter (Figure 12-50).

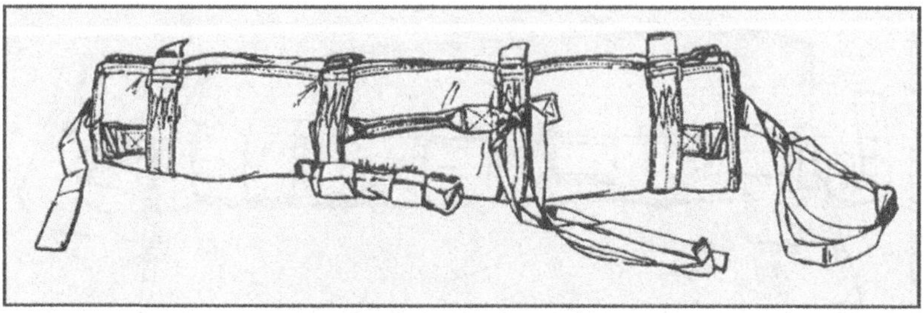

Figure 12-50. Pack closed.

(3) Insert the cone-shaped end of the shock absorber into the launch tube aft end.

(4) Position the 8-inch square honeycomb on the shock absorber flat surface. If the shock absorber is missing from the training round, replace it with two 8- by 8-inch cardboard honeycomb energy dissipaters. Thread the end of the securing strap through the launcher aft end securing strap keeper. Route the two aft end securing straps over the cardboard honeycomb through each quick-fit adapter and tighten, then thread the end through the adapter, forming a quick-release fold. Fold excess securing straps back under the corresponding quick-fit adapter (Figure 12-51).

(5) Route the launcher forward end securing strap under the nonadjustable cross D-ring strap through the quick-fit adapter and tighten. Then thread the end through the adapter, forming a quick-release fold. Fold excess securing straps back under the corresponding quick-fit adapter.

Figure 12-51. Running ends secured.

NOTE: The procedures for rigging the AT4JP vary at this point, depending on the method used to lower the parachutist's equipment. Follow paragraphs 12-51 and 12-52 if a single lowering line (tandem) is used.

12-52. AT4 AND ALICE PACK RIGGED

To attach the quick-release snap, the parachutist attaches it to the quick-release link that is routed through the D-ring. The opening gate should face away from the pack (Figure 12-52).

a. **Stowing the Standard Lowering Line.**

(1) S-fold the lowering line neatly on top of the web inside the retainer, ensuring that the ends are stacked evenly with the retainer flap outer edges. Secure the pile tab on the web located at the ejector snap end to hook extension on retainer.

(2) Fold the hook side of the flap tightly over the S-folded lowering line and, holding it firmly, fold the pile side of flap over and secure hook and pile together.

(3) Secure the pile extension on the flap retainer to the hook tab at the loop end of the lowering line (Figure 12-53).

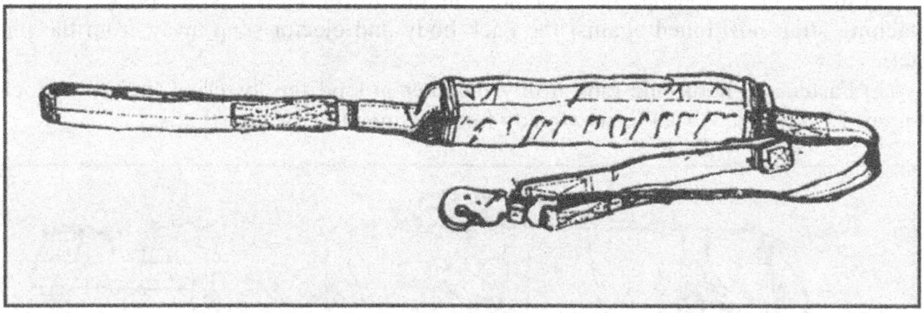

Figure 12-52. HPT lowering line stowed.

NOTE: The hook-pile tape located on the retainer flap and at each end of the stowed lowering line assembly must be secured firmly and without slack to prevent line spillage. To provide a secure closure of the hook and pile, press them firmly, evenly, and smoothly together without puckers.

b. **Securing the Attaching Web to the Standard Lowering Line.**

(1) Attach the lowering line to the lowering line attaching strap by routing the loop of the lowering line through the small loop end of the lowering line attaching strap.

(2) Route the lowering line through its own loop and pull tight (Figure 12-53).

c. **Fastening Attaching Strap to Nonadjustable Cross Strap D-Ring.**

(1) If not already installed, attach the lowering line attaching strap to the nonadjustable cross strap D-ring by routing the large loop around the D-ring.

(2) Route the lowering line through the attaching strap loop and pull tight (Figure 12-53).

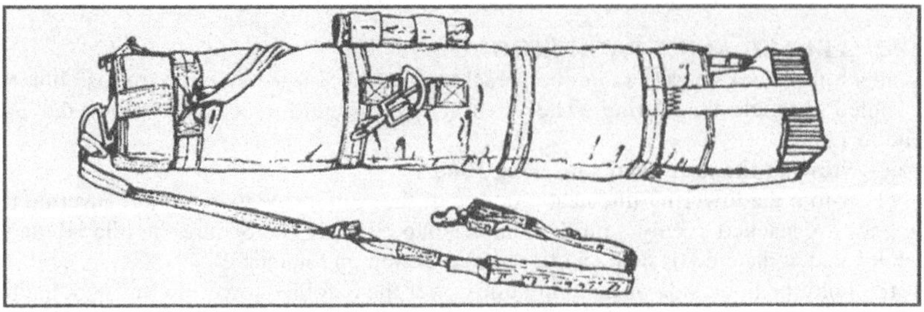

Figure 12-53. Attaching strap secured.

d. **Stowing the Standard Lowering Line in Stow Pocket**.

(1) Insert the lowering line assembly in the lowering line stow pocket with the attaching strap positioned against the pack body and ejector snap away from the jump pack.

(2) Fasten the hook-pile tabs firmly together around the lowering line stow pocket (Figure 12-54). The AT4JP is now ready for attachment to the parachutist.

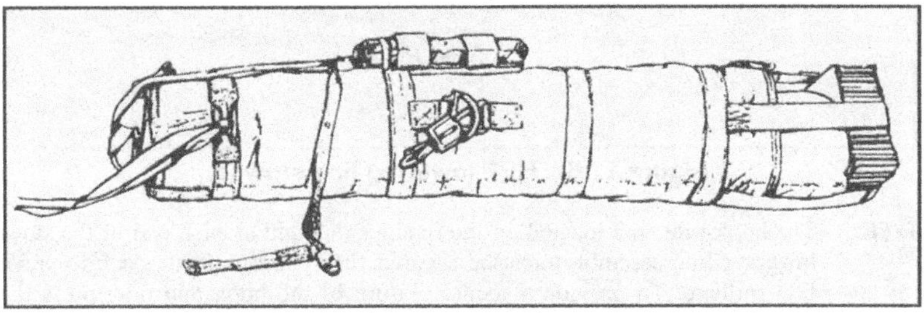

Figure 12-54. Standard lowering line stowed.

NOTE: Before attachment of the reserve parachute, individual equipment, and AT4JP, ensure that the canopy release assemblies of the T-10-series harness are in the hollows of the parachutist's shoulders for a proper fit. Adjust an improperly fitted harness by repositioning the diagonal back straps to the size channel that correctly fits the parachutist. A properly adjusted T-10-series harness on the parachutist ensures that the reserve parachute, individual equipment, and AT4JP are in the proper position. An improperly fitted harness with a full equipment and weapon load is uncomfortable and could result in problems during ground movement prior to enplaning, movement in the aircraft, aircraft exit, and main parachute deployment.

12-53. EQUIPMENT ATTACHED TO PARACHUTIST (STOWED LOWERING LINE)

To rig the ALICE pack with frame, the parachutist's ALICE pack with frame is prepared following procedures listed in paragraph 12-43, with the following exception: secure the stowed lowering line to the right side (as worn on the parachutist) on the vertical pack frame bar in two places with two turns of masking tape or two sets of retainer bands. The lowering line ejector snap faces up and to the right or left.

a. Attach the D-ring attaching strap snap hooks (on rigged ALICE pack) to the harness D-rings outside of the reserve connector snaps.

b. Secure ALICE pack to the parachutist's leg.

c. Attach the HPT lowering line ejector snap from the ALICE pack to the right D-ring on the harness.

d. Secure the AT4JP by attaching the quick-release snap to the left D-ring outside of the H-harness snap hook (Figure 12-55). Ensure that the quick-release safety gate is closed and the quick-release snap is locked to the D-ring, and that the quick-release assembly release attachment sequence is inboard to outboard as follows: reserve, H-harness, and AT4JP quick-release snap.

Figure 12-55. AT4JP secured.

e. Route the upper tie-down tape under and around the left main lift web directly below the chest strap. Tie snugly with a double-loop bowknot on the front of the AT4JP where it is easy for the parachutist to reach.

f. Route the lower tie-down tape around the AT4JP and the left leg. Tie with a double-loop bowknot on the front of the AT4JP where it is easy to reach. (If the HSPR is used, remove the lower tie-down on the AT4JP and route the left leg strap of the HSPR around the outside of the AT4JP, then retighten.)

12-54. ALICE PACK WITH FRAME AND AT4JP RIGGED

The quick-release snap is not attached at this point. This will be accomplished during the equipment attaching sequence. To stow the DMJP/AT4JP modified lowering line for tandem load, the following must be performed:

a. S-fold the lowering line neatly on top of the web inside the retainer, ensuring that ends are stacked evenly with the retainer outer edges.

b. Secure the pile tab on the web that is located 46 inches from the ejector snap end to hook extension on retainer.

c. Fold the hook side of the flap tightly over S-folded lowering line and, holding it firmly, fold the pile side of the flap over and secure hook and pile together.

d. Secure the pile extension on the retainer flap to the hook tab at the looped end of the lowering line (Figure 12-56, page 12-56).

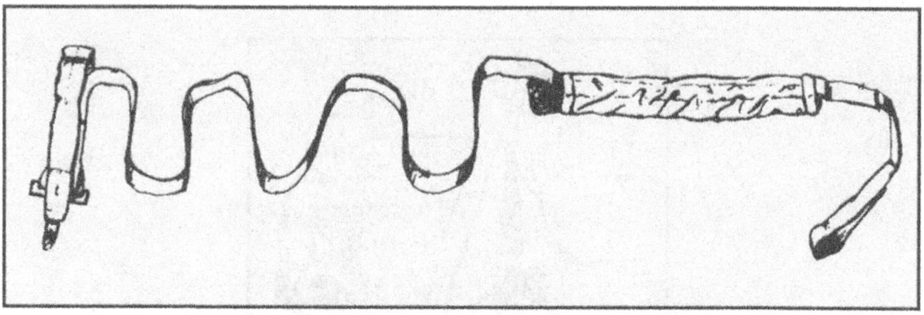

Figure 12-56. HPT lowering line stowed for tandem load.

12-55. EQUIPMENT ATTACHED TO PARACHUTIST (MODIFIED STOWED LOWERING LINE)

To rig the ALICE pack with frame, the parachutist's ALICE pack with frame is prepared following procedures listed in paragraph 12-43 with the following exception: secure the modified stowed lowering line to the left side (as worn on the parachutist) on the vertical pack frame bar in two places with two turns of masking tape or two sets of retainer bands. The lowering line ejector snap faces up and to the right or left.

a. Attach the D-ring attaching strap snap hooks (on rigged ALICE pack) to the harness D-rings outside of the reserve connector snaps.

b. Secure the ALICE pack to the parachutist's leg.

c. Stand the rigged AT4JP next to the parachutist's left side.

d. Grasp the ejector snap of the modified HPT lowering line already attached to the ALICE pack and route (from bottom to top) through the D-ring opening at the center of the pack, upper tie-down

Figure 12-57. Attaching equipment to parachutist.

chafe loop, and top D-ring located at the aft end. Position the 18-inch strata blue marking on the modified HPT lowering line even with the D-ring on top of the AT4JP (Figure 12-57).

e. Attach the quick-release snap to the AT4JP. The snap opening faces away from the pack.

f. Attach the modified HPT lowering line ejector snap to the lowering line adapter web.

g. Secure the AT4JP by attaching the quick-release assembly to the left D-ring outboard of the H-harness snap. Ensure that the quick-release safety gate is closed, the quick-release snap is locked to the D-ring, and the quick-release assembly release lever is fully seated (do not safety tie). The harness left D-ring snap attachment sequence is inboard to outboard as follows: reserve, H-harness, and AT4JP quick-release snap.

h. Route the upper tie-down tape under and around the left main lift web directly below the chest strap. Tie snugly with a double-loop bowknot on the front of the AT4JP where it is easy to reach (Figure 12-58).

i. Route the lower tie-down tape around the AT4JP and the left leg. Tie with a double-loop bowknot on the front of the AT4JP where it is easy to reach. If the HSPR is used, remove the lower tie-down on the AT4JP and route the left leg strap of the HSPR around the outside of the AT4JP, then tighten (Figure 12-59).

Figure 12-58. Routing upper tie-down.

Figure 12-59. Lower tie-down tape routed around AT4JP.

12-56. ALICE PACK AND AT4JP RELEASED

The lower and upper tie-downs on the AT4JP are released.

a. Release the ALICE pack as follows:

(1) *H-harness (modified)*. Release the lower tie-down tape on the ALICE pack. About 100 to 200 feet above ground, drop the ALICE pack by simultaneously pulling the quick-release folds on the D-ring attaching straps, allowing the pack to fall to the end of the line.

(2) *HSPR*. About 200 feet above the ground, grasp the release handle and pull up and out quickly, simultaneously releasing the load from the harness and legs, allowing the load to drop the length of the lowering line. **Release the handle immediately following separation of the load from the parachutist.** Secure the handle for releasing the load to the HSPR with a length of tubular webbing and stay with the HSPR to prevent its loss and separation.

b. Push out on the activating arm of the quick-release assembly of the AT4JP to lower the AT4JP. The AT4JP will either fall to the end of its lowering line or slide down the tandem lowering line to the ALICE pack.

NOTE: To jettison the ALICE pack and AT4JP in an emergency, the parachutist (after performing the above) lifts the reserve up and pulls the yellow safety lanyard on the lowering line ejector snap, allowing the ALICE pack and AT4JP to fall free.

WARNING

Under no circumstances will the parachutist release the AT4JP before releasing the ALICE pack.

c. Upon landing, remove the HPT lowering line by pulling the line through the two D-rings and disassembling the girth hitch on the H-harness.

d. Store the AT4JP modified HPT lowering line and polyurethane shock absorber with the AT4JP for reuse.

Section XIII. ALL-PURPOSE WEAPONS AND EQUIPMENT CONTAINER SYSTEM (AIRPAC)

The AIRPAC is a lightweight nylon equipment pack system used to enclose a jump load. The AIRPAC front-mount container can accommodate all size packs, M47 tracker, radios, and 60-mm mortar with baseplate or bipod. The side-mount container can hold one M47 round of ammunition, the AT4, M4, M24, SAW, M60 machine gun, and other ancillary equipment.

12-57. COMPONENTS

The AIRPAC consists of the front-mount container, the side-mount container, and the parachutist's individual equipment rapid release (PIE/R2) release mechanism.

12-58. RIGGING LOADS IN THE FRONT-MOUNT CONTAINER

Loads in the front-mount container are rigged as follows:

a. **Rigging the ALICE Pack and the Field Pack, Large, Internal Frame (FPLIF).**

(1) Extend all retaining straps outward to their limits. (The retainer strap with the circular closing flap is the top of the container.)

(2) Place the container so that the retaining straps face the ground and the parachute recovery bag faces up.

(3) Fold the parachute recovery bag (PRB) so that no portions protrude outside the container limits.

(4) Place pack on the PRB. Ensure the circular release rings on the bottom of the container are not constrained by the pack. The ALICE pack is jumped upside down with the waist pad against the front of the thighs. The FPLIF is jumped right side up.

(5) Bring the right side of the container material up first, then the left side and the bottom flap, and then the top flap.

(6) Place the circular closing flap on top of the pack and open the circular closing flap by pulling the tab. Bring the left, right, and bottom retaining straps to the center of the circular closing flap and route the white grommet securing loop through each grommet of the retaining straps. No specific order is required. Bring the top flap material up last. Place the cotter pin through the white grommet securing loop.

(7) Tighten all retaining straps, ensuring that the circular closing flap remains centered on the ALICE pack. Adjust the protective flaps to stow excess fabric. Close the circular closing flap using three hook-pile tabs.

(8) Underhand fold the free-running ends of the retaining straps toward the friction adapters and secure them under the webbing retainers.

NOTE: The circular release rings should be exposed on the top flap of the front-mount container. If the rigged pack is above the rings and restricts the rings' movement, the load must be adjusted.

b. **Rigging the Release Mechanism of the Front-Mount Container.**

(1) Stand the rigged pack in an upright position with the bottom of the AIRPAC facing the jumper.

(2) Route release handle cables through the cross strap on top of the container and secure the HPT.

(3) With the attaching strap snap hook facing down and towards the jumper, route the circular ring of the attaching strap (from bottom to top) through the circular release ring on top of the container. Fold the circular ring of the attaching strap back over and route the red attaching loop through the attaching ring on the attaching strap. With the cable loop retainer on the female portion of the leg strap release assembly facing up, route the red attaching loop from bottom to top through the grommet on the female portion of the leg strap release assembly. Route the release handle cable through the red attaching loop and then through the cable loop retainer. Repeat this sequence for the other strap.

c. **Rigging the HPT Lowering Line.**

(1) Select the desired landing orientation of the front-mount container (horizontally or vertically) and attach the short bridle to the appropriate triangle link.

(2) Route the looped end of the HPT lowering line through the unsewn portion of the short bridle and form a girth hitch. Place a retainer band around the nylon tab located inside the closure flaps. Route the HPT lowering line through the retainer band and stow in the retention pocket. Secure the HPT.

(3) For transportation, attach the ejector snap to the primary short bridle triangle link.

12-59. RIGGING LOADS IN THE SIDE-MOUNT CONTAINER

Rigging procedures for the side-mount container are as follows:

a. Place the side-mount container on the ground with the friction adapters facing down and the free-running ends of the internal securing strap to the right. Then, position packing material flush with the bottom edge of the fabric flaps and place the weapon on the container in the following orientation (Table 12-3):

WEAPON	PACKED ORIENTATION	PACKING MATERIAL
M60 MG/SAW	Muzzle up	None required
AT4	Muzzle up	1 piece 6" by 6" honeycomb on bottom
60-mm Mortar	Muzzle down	1 piece 6" by 6" honeycomb on bottom
Dragon	Muzzle down	2 pieces 8" by 8" honeycomb on bottom

Table 12-3. Packed weapon orientation and packing material required for side-mount container.

b. Tape and pad exposed or protruding parts (that is, sight posts, grip stocks, or bipods) with a soft material such as cellulose wadding or an adequate substitute.

c. Center load on the internal securing strap and pull lower fabric flap over packing material. Route free-running end of internal securing strap through the friction adapter on the lower fabric cap. Place floating V-ring of internal securing strap on the top center of the load. Form a 5- to 6-inch quick release in the strap and tighten. From the top, fold excess weapons container material downward until flush with the top of the load.

d. Wrap unpadded fabric flap over weapon. Bring padded fabric flap over load and loosely secure container retaining straps to friction adapters; form a 5- to 6-inch quick release in the strap. Do not tighten strap.

NOTE: If jumping an individual weapon, place weapon muzzle down in container slide-fastener compartment. Secure the slide fastener.

e. Tighten the container retaining straps (with quick releases). S-fold excess webbing and secure to the quick-release loop with two turns of a retainer band, leaving a 3-inch tail. Secure drawstring on top of container with a quick-release knot, if necessary.

12-60. RIGGING AIRPAC AS TANDEM LOAD WITH HOOK-PILE TAPE LOWERING LINE

The AIRPAC is rigged as a tandem load with HPT lowering line as follows:

a. If the HPT lowering line is attached to the front-mount container, detach and place to the side. Using the extended bridle, place the looped end through unsewn portion of short bridle and secure with a girth hitch. Route looped end of HPT lowering line through floating V-ring of extended bridle and form a girth hitch.

b. Stow the HPT lowering line in the retention pocket. Stow extended bridle excess in the retention pocket, allowing for a 2-foot running end. Secure the retention pocket and attach the ejector snap of the HPT lowering line to the accessory attaching ring, triangle link, or left D-ring on the parachutist's harness.

c. After the side-mount container is attached to the jumper, the extended bridle snap hook is attached to the V-ring on the top of the side-mount container.

12-61. AIRPAC ATTACHED TO PARACHUTIST USING PIE/R2 RELEASE MECHANISM

AIRPAC is attached to parachutist using the PIE/R2 release mechanism as follows:

a. Place the front-mount container attaching straps under the reserve parachute and attach the front-mount container snap hooks to the enlarged D-rings on the outside of the reserve connector snaps (same as a single-point release system).

b. Attach the HPT lowering line ejector snap to the parachutist harness lowering line adapter web.

NOTE: When using the PIE/R2 release mechanism, ensure the harness has a lowering line adapter web or new harness triangle link.

c. Attach the PIE/R2 release mechanism to the side-mount container (located on jumper's left side) by placing the hook portion of the snap shackle with the hook through the quick-release link from top to bottom. The spring-loaded gate should face away from the container. Remember to hook the gate toward jumper's groin muscle.

d. Open the snap shackle with hook, attach to the D-ring on the outside of the attaching strap snap hook from back to front, and close the snap shackle.

e. Take the extended bridle snap hook and attach to floating V-ring on top of side container. Secure extended bridle under retention flap on container.

f. On the PIE/R2 release mechanism, route the upper attaching strap around the back of the side-mount container, through the main lift web, and above the chest strap. With orange release handle facing forward on the side-mount container, route the grommet tab through the looped end of the upper attachment strap, over the red attaching loop, and secure with cable. Tighten the upper strap, ensuring that the orange release handle stays on the front portion of the side-mount container.

g. Connect the white release lanyard snap hook to the short bridle attaching the V-ring on the top of the front-mount container. Listen for a metallic click and pull to ensure the snap hook is secure.

h. Attach the leg straps, ensuring the left leg strap is also routed around the side container. Tighten and secure. Stow excess webbing under webbing retainers.

Section XIV. STINGER MISSILE JUMP PACK

The Stinger missile jump pack provides a means for the air delivery of the Stinger missile system when it is attached to a jumper during airborne operations. The SMJP is an overpack body constructed of nylon duck material, shock-absorbing material, nylon securing straps, an HPT lowering line, and a quick-release snap shackle.

> **WARNING**
> **Parachutists carrying the SMJP will jump from the first two positions in the stick only. Parachutists who are less than 5 feet 8 inches in height are not permitted to jump the SMJP. The SMJP is authorized to be jumped from C-130 and C-141 aircraft only.**

12-62. COMPONENTS

Components of the Stinger missile jump pack are as follows:

a. The SMJP consists of a pack body constructed of nylon duck material, shock-absorbing material, nylon securing straps, an HPT lowering line, and a quick-release snap shackle.

b. When rigged with a Stinger missile round, the SMJP is 12 inches in diameter, 68 inches long, and weighs approximately 38 pounds. It is secured vertically by the quick-release snap shackle to the left D-ring on the parachutist's harness. To prevent the SMJP from swaying during exit and deployment of the parachute, the parachutist uses the nylon webbing tie-downs to secure the SMJP to the left leg and to the parachute harness at chest level.

c. The HPT lowering line with a 6-foot extension is part of the SMJP system. The web loop end of the lowering line extension is attached to the SMJP with the HPT lowering line ejector snap routed through the D-ring on the other end of the extension sling and to the lowering line adapter web on the parachute harness. The HPT lowering line is attached to a front-mount container (ALICE pack, large, or the FPLIF) which contains the detachable Stinger weapon round components (gripstock, IFF, and the three BCUs). The parachutist lowers the SMJP and ALICE pack by pulling the SMJP quick-release assembly. Upon landing, the parachutist can rapidly gain access to the Stinger.

12-63. RIGGING PROCEDURES

Rigging procedures for the Stinger missile jump pack are as follows:

NOTES:
1. Throughout the rigging procedure, duct tape and masking tape are used to tape various parts of the webbing. The jumper can use duct tape for masking tape, **except** where the procedures state "masking tape only."
2. Throughout the instructions, the phrases "forward end" and "aft end" are used to describe the SMJP, Stinger missile round, and FHT. The forward

end is nearest the gripstock and sight assembly; the aft end is furthest from the gripstock and sight assembly.

a. **Stinger Round**. Before rigging the SMJP, inspect the Stinger round or FHT IAW TM 9-1425-429-12. A Stinger round exhibiting damage must not be jumped under any circumstances. It must be returned to the ASP for repair. Field handling trainers that exhibit signs of general minor damage and no major ruptures or cracks may be jumped if the damage does not exceed safe or serviceable limits. Pad the Stinger round with a 6-inch-wide piece of cellulose wadding placed around the aft end of the round near, but not on, the blowout disk. Tape the wadding with **masking tape only**.

b. **Aft Foam End Cap**. Place the foam end cap on the aft end of the missile round, one half at a time, ensuring the foam posts slide into the holes in the opposite end cap half. The end cap should fit snugly. If not, check to see that both halves fit together without any gaps and there is enough cellulose wadding about the missile round. Tape the foam end cap halves together using duct tape.

c. **Front Cover**. Make sure the front cover is on firmly. If loose, use a pen (or similar object) to press in the three rubber spacers on the cover. Push each spacer in evenly to achieve a firm fit between the cover and the missile round. The cover should slide on snugly when properly adjusted. Make sure the tab on the cover aligns with the notch on the missile round. When the cover is properly fitted, the "lollipop" should be visible when looking through the sight.

d. **Forward Foam End Cap**. Place the foam end cap on the forward end of the missile round in the same fashion, ensuring that the "lollipop" on the front cover is positioned in the cutout of the foam end cap. Tape the foam end caps together using duct tape, ensuring that there are no gaps.

e. **Sight**. Flip up the sight on the missile round.

(1) Fold a piece of cellulose wadding, making a piece that is 6 inches by 24 inches. Place the wadding under the sight and lower the sight down onto the wadding. Make sure that the wadding is not placed over the clip. Firmly press down on the sight to make the wadding fill the gap between the sight and the missile round and lock the sight into position.

(2) Loosely wrap cellulose wadding around the entire sight area until the wadding measures about 11 to 12 inches in diameter, about the same as the diameter of the foam end cap. Tape the wadding in place using **masking tape only**.

f. **Stinger Round Positioning**. Place the SMJP on the ground, felt side up. Make sure all straps are extended. Position the missile in the SMJP with the aft foam end cap fitted in the SMJP nonadjustable end strap bridle and with the missile round carrying strap facing up.

g. **SMJP Closing**. Fold the pack around the missile, keeping the securing strap quick-fit adjusters on the outside of the pack.

(1) Route the side securing straps through the quick-fit adjusters using a quick-release fold. Loosely tighten the straps to form the SMJP around the missile. Do not tighten the straps.

(2) Place the SMJP on the aft end so the free-running end straps are on top. Route the end straps through the quick-fit adapter opposite of each strap. Sit on the ground, place

your foot on the forward end cap, and tighten the adjustable straps. Tighten the end straps. Do **not** use a quick-release.

(3) Place the pack on the side and tighten the side securing straps, using a quick release. Tighten the side securing straps so the pack overlaps the entire length of the round. S-fold excess webbing and tape, using **masking tape only**.

NOTE: If there is a gap near the sight assembly, remove cellulose wadding until the gap is closed.

(4) Stand the SMJP on the aft end cap. Center a 6-inch by 6-inch piece of honeycomb on the forward end cap. Route the free-running end straps over the honeycomb, through the opposite quick-fit adapters, and tighten to form an X over the honeycomb.

(5) Secure straps with a quick release.

(6) S-fold excess webbing and tape with duct tape.

h. **Attachment of the Quick-Release Assembly**. Secure the quick-release web and SMJP HPT. Route the coated wires into the fabric guides. Hook the quick-release shackle onto the lowest SMJP D-ring which, when the SMJP is attached to the jumper, keeps the SMJP from dragging the ground.

NOTE: The quick-release handle must face toward the jumper's rear.

WARNING

A parachutist harness with too much slack, or with the incorrect SMJP D-ring, will cause the SMJP to drag on the aircraft floor and can result in jumper injury when he exits the aircraft.

i. **Weapon Components Rigging**. The gripstock, IFF, and the three BCUs are packed in the ALICE pack or FPLIF where they can be quickly recovered on the DZ.

(1) To protect the weapon components, load hard items in the lower portion of the pack, place soft items on top of the hard items, and pack the Stinger weapon components on top of the soft items near the top of the pack.

(2) Wrap each component with one wrap of cellulose wadding and secure with **masking tape only**.

(a) Gripstock antenna. The gripstock antenna must be in the folded position. Secure the gripstock antenna attaching clip in the closed position, with **masking tape only**.

(b) IFF. The interconnecting cable must remain attached to the IFF and wrapped separately to avoid metal-to-metal contact between the cable and IFF. Secure the wrapped cable to the wrapped IFF. Ensure the electrical outlet protecting cap is in place.

12-64. STINGER MISSILE JUMP PACK ATTACHED TO PARACHUTIST

The SMJP is the last item attached to the parachutist harness.

a. Secure the SMJP to the parachutist harness by attaching the quick-release snap shackle to the left D-ring, ensuring it opens away from the jumper.

NOTE: Failure to place the snap shackle away from the jumper may cause the SMJP to hang when the release is activated.

b. Route the lowering line ejector snap hook through the SMJP lowering line extension sling D-ring and connect to the lowering line adapter web.

c. Secure the lowering line and SMJP HPT.

d. Secure the lowering line extension to the SMJP and tape with **masking tape only**. Make sure there is no slack in the lowering line extension.

e. Route the chest tie-down around the left main lift web directly below the chest strap and around the SMJP. Route the chest tie-down strap with grommet through the chest tie-down loop. Route the red fabric loop through the grommet and place the coated wire through the red loop.

f. Route the leg tie-down around the parachutist's left leg, through the frame on the ALICE pack or through the nearest equipment retaining ban on the FPLIF, and around the SMJP. Secure in the same manner as the chest tie-down.

g. Attach a 6- to 8-foot doubled piece of 1/4-inch, 80-pound cotton webbing to the SMJP carrying handle. Route the cotton webbing through a secure loop on or near the bottom of the front-mount ALICE pack and tie to the handle of the HSPR.

NOTE: The jumper must use the HSPR when jumping the SMJP.

12-65. INDIVIDUAL JUMP PROCEDURES

The SMJP must be jumped with a front-mount container.

a. At the 20-minute warning, a safety attaches the SMJP to the parachutist and inspects it.

b. At exit, the jumper must **not** assume a tight body position. A tight body position causes severe twists and may flip the jumper through his risers. It also causes entanglement in suspension lines.

c. About 200 to 100 feet above the ground, the jumper ensures the area below is clear and activates the release by pulling the orange release tab down and away from his body with the left hand. The SMJP falls the length of the lowering line.

NOTE: To jettison the SMJP, the parachutist lowers the SMJP and pulls out on the yellow lanyard attached to the quick-ejector snap.

Section XV. RANGER ANTIARMOR/ANTIPERSONNEL WEAPON SYSTEM PACKED IN AT4JP AND DRAGON MISSILE JUMP PACK

The AT4JP and Dragon missile jump pack can be used to secure the RAAWS during airborne operations.

12-66. COMPONENTS AND CONTAINER DESCRIPTION

The RAAWS consists of three main components plus the cleaning kit and gun-mount.

a. **Main Components**. The main components are the weapon system, the telescopic sighting system, and various types of loads.

(1) *Weapon and telescoping sighting system.* When carried by the parachutist, they are packed in the AT4JP.

(2) *Rounds of ammunition.* The missile round containers are constructed of molded polyethylene. The rounds come in two container sizes: 27.6 inches (TP and HEAT) and 21 inches (illumination, smoke, and HE). The TP and HEAT round containers must be rigged in a DMJP (modified). Illumination, smoke, or HE missile rounds are packed under the top pack flap of the large ALICE pack or FPLIF.

b. **Cleaning Kit and Gun-Mount**. The cleaning kit and gun-mount are packed separately near the top of the jumper's ALICE pack or FPLIF.

c. **Container Description**. When configured for RAAWS rigging, the AT4JP cannot be rigged with the M16/M203. The AT4JP is about 49 inches long and 9 inches in diameter, and it weighs about 26 pounds. It is secured vertically to the left D-ring on the parachutist harness. To prevent the AT4JP from swaying, two tie-down tapes are attached. The upper tie secures the AT4JP to the main lift web below the chest strap, and the lower tape secures the jump pack to the parachutist's left leg (provided the jumper is not using the HSPR). A 15-foot lowering line and 24-inch lowering line attachment web are used. Two 9-inch by 9-inch pieces of honeycomb are used at the venturi end of the weapon (on the bottom, when rigged) to cushion the weapon during impact. After landing, the parachutist rapidly gains weapon access by pulling seven quick releases.

12-67. RIGGING PROCEDURES

Rigging procedures for RAAWS components are as follows:

a. Rig the RAAWS weapon and sighting system.

(1) Lay the jump pack on the ground (felt side up) and fully extend all securing webs.

(2) Place the venturi and muzzle covers on the weapon.

(3) Position the telescopic sight package on the weapon (between the front grip and the shoulder pad) and alongside the firing mechanism. Secure the telescopic sight package to the weapon with two lengths of 1/4-inch cotton webbing.

(4) Place the weapon on the pack with the front grip pointed straight up, the muzzle (small end of weapon) fitted into the nonadjustable cross D-ring strap, and the venturi (large end of weapon) centered on the middle launcher aft end securing strap.

(5) Ensure the muzzle fits snugly into the nonadjustable cross D-ring strap.

(6) Fold the rifle and muzzle pockets portion of the pack over the weapon.

(7) Rotate the unfolded flap portion over the other flap and route the four side securing straps through the adapters and form quick releases. S-fold the running ends and secure by folding excess back under corresponding quick-fit adapter.

(8) Position two 9-inch by 9-inch pieces of honeycomb onto the venturi of the weapon. Route the two aft end securing straps over the honeycomb, through the quick-fit adapters, and tighten securely. Form a quick release with the running ends, S-fold the excess, and secure by folding excess back under the corresponding quick-fit adapter.

(9) Route one turn of 1/4-inch cotton webbing around the securing straps between the honeycomb and the pack; tie securely. (This tie will prevent the honeycomb from shifting.)

(10) Attach the HPT lowering line to the AT4JP using procedures described in paragraph 12-8. If the AT4JP lowering line is to be employed, a 24-inch lowering line adapter web must be attached to the parachutist harness.

b. Rig the RAAWS 84-mm TP and HEAT rounds in the DMJP MOD.

NOTE: Due to the length of the TP and HEAT twin-tube containers, the containers must be jumped inside the DMJP MOD. DMJP modification procedures are described in paragraph 12-70.

(1) Lay the DMJP MOD on the ground felt side up.

(2) Center the long twin tube containers on the DMJP MOD.

(3) Route the center vertical and horizontal securing straps over the container and secure with a quick release. Ensure the tubes are oriented so the container caps are at the upper portion of the pack.

(4) Fold the flap containing the M16/M203 rifle muzzle stow pocket over the tubes.

(5) Place the M16/M203 rifle muzzle into the stow pocket, ensuring the carrying handle is facing away from the tubes.

(6) Fold the unfolded flap over the M16/M203 and secure the three horizontal securing straps. Ensure the center securing strap is routed through the rifle carrying handle. Form quick releases in the running ends, S-fold, and fold the excess back under the corresponding quick-fit adapters.

(7) Route the rifle butt securing strap over the butt and through the quick-fit adapter. Pull as tight as possible and form a quick release.

(8) Cut two pieces of honeycomb 5 inches by 11 inches. Place the two pieces onto the end of the container with the exposed rifle butt. Route the two aft end securing straps over the honeycomb, through the quick-fit adapters, and secure as tight as possible. S-fold the excess back under the corresponding quick-fit adapters.

(9) Tie the two securing straps together at the point where they cross with one turn of 1/4-inch cotton webbing to prevent them from shifting.

(10) Cut two pieces of 5 1/2-inch by 9 1/2-inch honeycomb. Place the honeycomb on the other end (top) portion of the container.

(11) Route the center securing strap through a D-ring, over the honeycomb, and through the quick-fit adapter. Tighten securely, S-fold, and fold the excess back under the corresponding quick-fit adapters.

NOTE: The nonadjustable strap is not used. It is pushed into the top of the pack before placing the honeycomb on top.

(12) Secure the D-ring to the center of the strap by running a length of 1/4-inch cotton webbing through the lower portion of the D-ring, around the strap, and tying. This will prevent the D-ring from shifting on the strap.

(13) Attach the lowering line to the DMJP MOD.

NOTE: The DMJP MOD with the long tubes must be rigged with its own lowering line.

(a) Attach the lowering line to the D-ring on top of the DMJP MOD by routing the lowering line through the D-ring and then girth-hitch the lowering line through itself.

(b) Stow the lowering line into the stow pocket on the DMJP MOD by pushing it as far as possible into the pocket. It will be halfway into the pocket. Secure the upper HPT flaps around the stowed lowering line.

NOTE: Do not tighten the lower HPT. This may bind the lowering line and prevent the DMJP MOD from lowering.

c. Rig the RAAWS 84-mm (illumination, smoke, or HE) rounds (short twin tubes) in the large ALICE pack with frame.

(1) Place the short tubes on the top of the ALICE pack and under the top flap. Secure all straps.

(2) Route one turn single Type I or III nylon cord through the equipment attaching points on one side of the ALICE pack, over the tubes, and through the equipment attaching points on the other side of the ALICE pack, then secure the running ends together.

d. Attach the HSPR to the ALICE pack.

12-68. EQUIPMENT ATTACHED TO PARACHUTIST

The AT4JP is secured to the parachutist after the reserve parachute and ALICE pack or FPLIF have been attached to the parachute harness D-rings.

a. Secure the AT4JP to the parachutist by attaching the quick-release assembly to the left D-ring. Ensure that the safety gate is completely closed, secure, and the activating arm is fully seated.

b. Ensure the harness left D-ring snap attachment sequence is inboard to outboard as follows:

(1) Reserve parachute.

(2) H-harness or other equipment.

(3) AT4JP quick-release snap.

c. Secure lowering line to parachutist by attaching the ejector snap to the lowering line adapter web. Ensure activating lever is fully seated. Ensure the yellow safety lanyard is not misrouted or entangled and is immediately accessible to the parachutist.

d. Route the upper tie-down tape around the main lift web at a point below the chest strap. Secure the lower tie-down tape around the left leg, or use the HSPR leg strap.

12-69. MODIFICATION PROCEDURES FOR THE DRAGON MISSILE JUMP PACK

Equipment needed to modify a DMJP for RAAWS missile round rigging are as follows: DMJP, two each A-7A straps, markers, a cutting tool, suitable needle, and No. 6 nylon thread.

a. Lay the DMJP for modification flat, felt side up. Using a suitable marker, make a mark at the center of the pack and in line with the aft end bridle securing strap at the top of the DMJP (Figure 12-60).

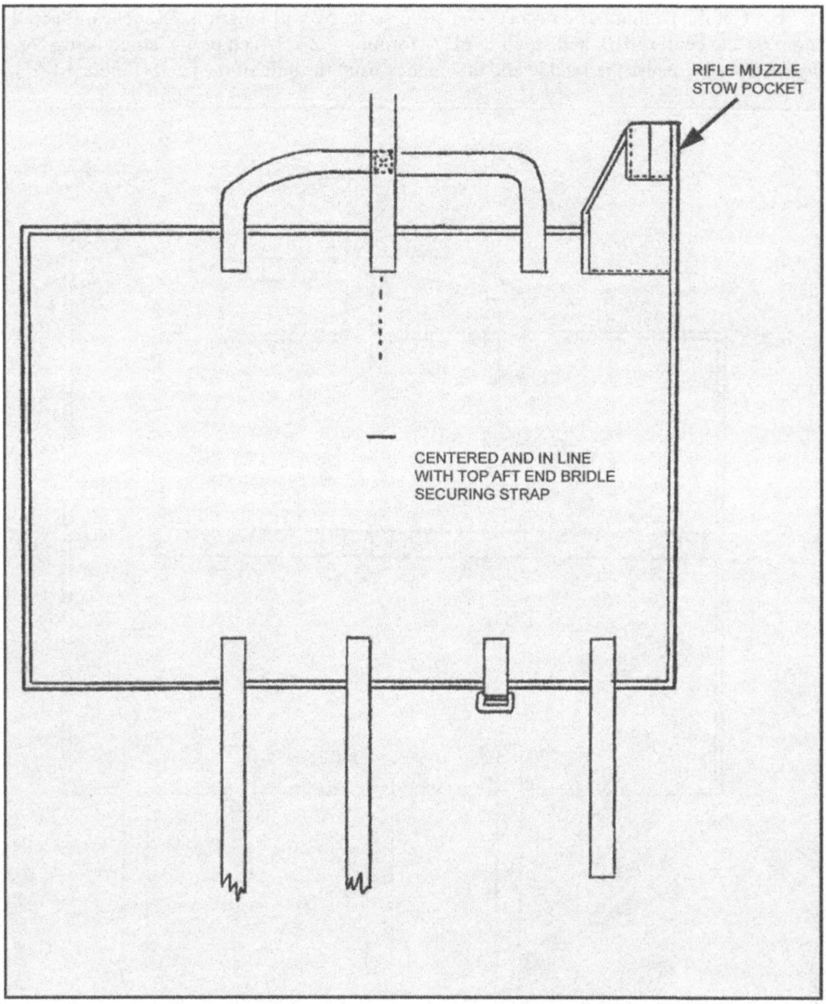

Figure 12-60. Marking the center of the DMJP.

b. Cut the running end of an A-7A strap so its overall length is 38 inches. Place the strap on the center mark and stitch in place forming a 2 1/2-inch box-X stitch using No. 6 nylon thread. Ensure the buckle end is 5 inches from the side of the pack (Figure 12-61).

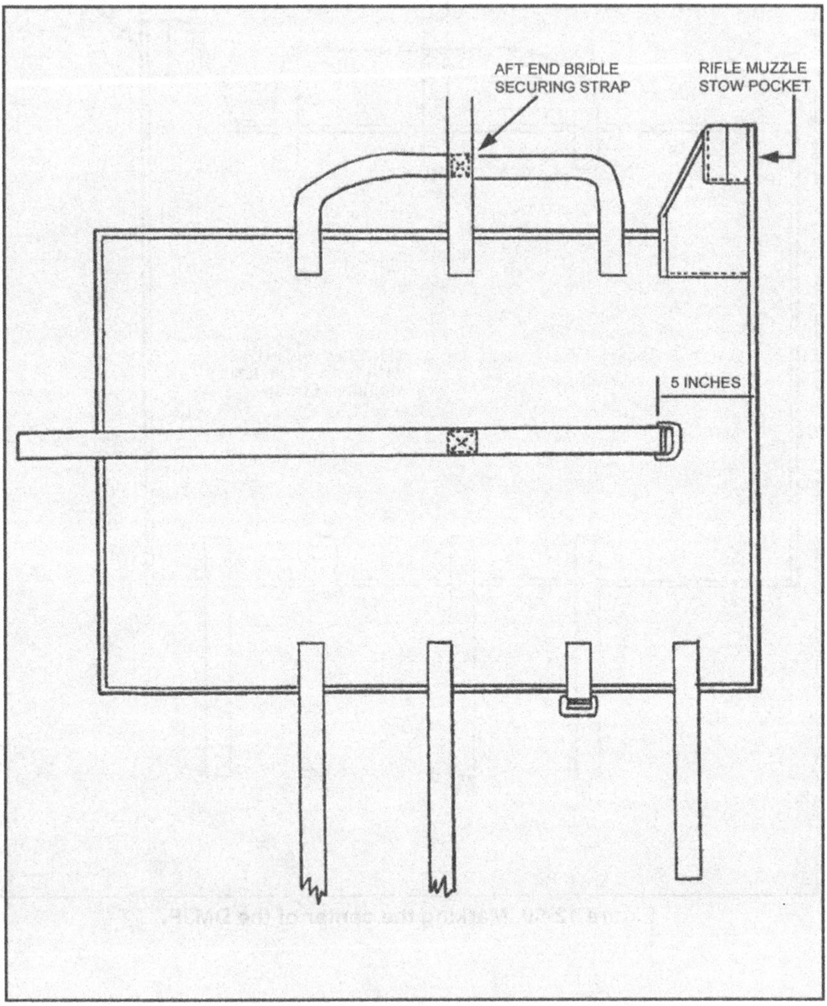

Figure 12-61. A-7A strap centered and stitched.

c. Locate the point where the aft end bridle securing strap crosses the U-shaped portion and is sewn. Cut and remove this stitching, being careful to not damage the straps (Figure 12-62).

d. From this side of the DMJP, measure and make two marks at 3 1/2 inches and 6 inches. From the other side of the DMJP, measure and mark at 6 inches and 8 1/2 inches (Figure 12-62).

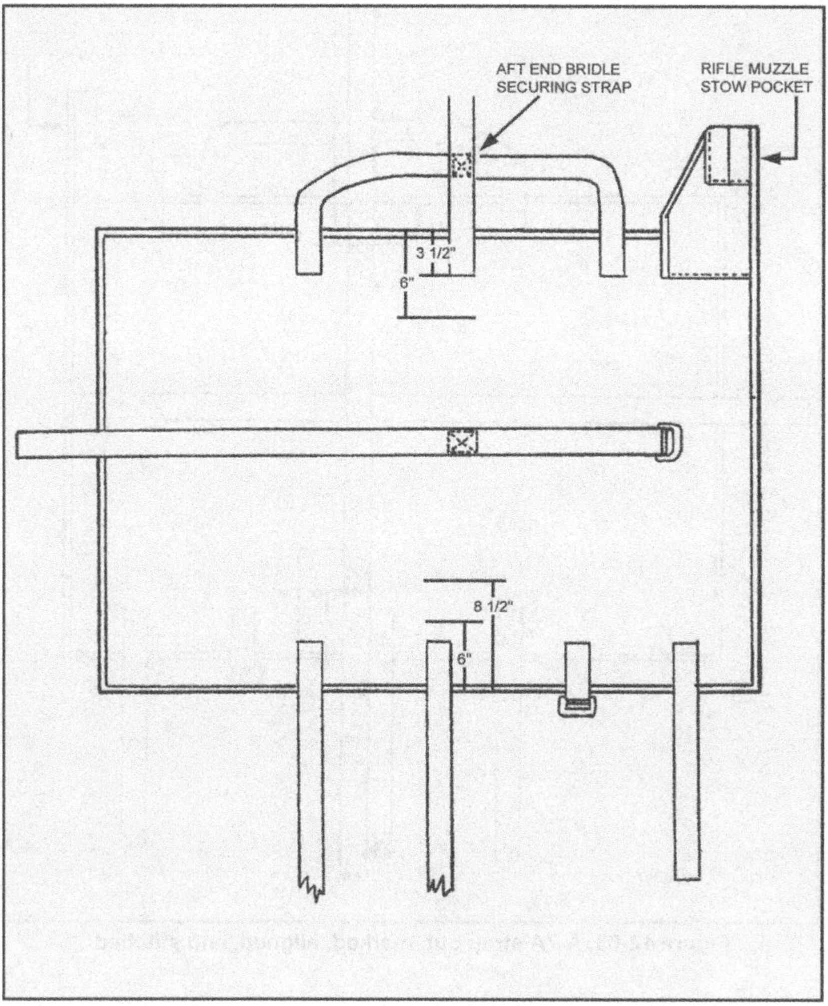

Figure 12-62. Stitching removed and DMJP markings.

e. Cut a second A-7A strap to an overall length of 68 inches. Mark the A-7A strap at a point 17 inches from the buckle. Align this mark with the 3 1/2-inch mark and stitch with two box-X stitch formations (Figure 12-63).

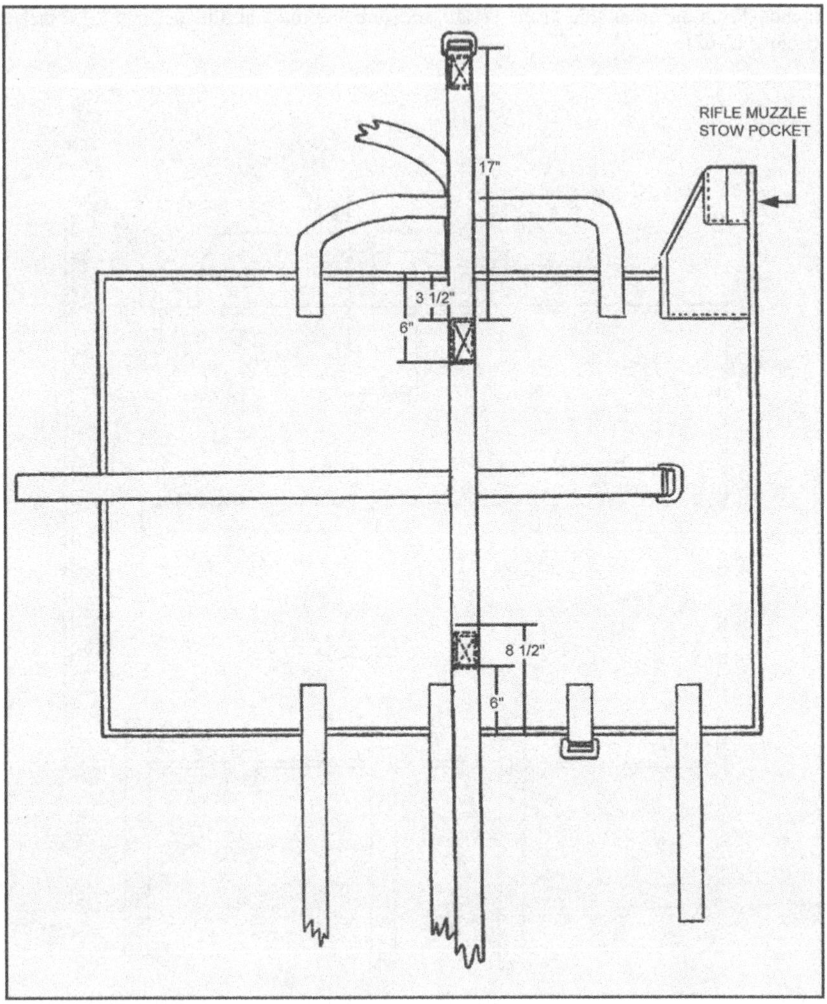

Figure 12-63. A-7A strap cut, marked, aligned, and stitched.

Section XVI. FIELD PACK, LARGE, INTERNAL FRAME

The FPLIF, with or without the patrol pack, can be rigged in the HSPR during airborne operations. However, the FPLIF is **not** recommended for jumping; substantial damage can occur to the pack and cause it to be unserviceable for further use.

12-70. RIGGING THE FIELD PACK, LARGE, INTERNAL FRAME WITHOUT PATROL PACK

Rigging instructions for the FPLIF without patrol pack, using the HSPR, are as follows:

a. Before attaching the harness to the FPLIF, attach the release handle assembly, adjustable D-ring attaching straps, and leg strap release assembly to the harness. S-fold, or accordion fold, all excess on the equipment retainer straps and secure with two turns of retainer band. No additional equipment will be attached to the outside of the FPLIF.

WARNING

The packed FPLIF can be no longer than 30 inches when rigged with the standard HSPR. The maximum recommended total rigged weight is 75 pounds, including the patrol pack. because of its internal frame, the FPLIF should be jumped in its fully extended position. THE H-harness *cannot* be used with the FPLIF.

b. Shorten the shoulder straps to their minimum length, roll and tape the free-running ends, route the cross-chest strap around the shoulder straps, and then fasten the cross-chest strap buckle. Stow upper patrol pack attaching straps inside the FPLIF by tucking them through the antenna closure flaps.

c. Extend the waistband straps, turn the FPLIF over with the shoulder straps down, route the waistband straps through the bottom loops formed by the lower patrol pack attaching straps, and buckle the waistband fastener.

d. Lay the HSPR on top of the FPLIF with the release handle assembly toward the top of the FPLIF, the adjustable cross-strap facing up, and the leg straps above the equipment retainer straps. Route the running ends of the HSPR equipment retainer straps from the top of the FPLIF down through the FPLIF's second and third equipment attaching loops. Route the adjustable leg straps through the second equipment attaching loops only. Route the HSPR equipment retainer straps under the waistband. Extend the HSPR adjustable cross-strap halfway.

e. Turn the FPLIF over with the shoulder straps facing up. Route the HSPR equipment retainer straps over the bottom of the FPLIF, under the waistband (through the outer waistband frame loops), and up over the comfort pad; form an X with the running ends. Ensuring that the release handle assembly is positioned on the top of the FPLIF, continue to route the HSPR equipment retainer straps under the shoulder straps, through the loops sewn on the back of the FPLIF near the top of the frame staves, and connect the free-running ends to the friction adapters, forming quick releases. S-fold the free-running ends and secure with retainer bands.

f. To attach the HPT lowering line to the HSPR, route the looped end of the HPT lowering line under the X formed by the HSPR equipment retainer straps, from top to bottom. Route the ejector snap through the loop, forming a girth hitch. Route the HPT lowering line over the top of the right FPLIF shoulder strap. Secure the HPT lowering line on the right side of the FPLIF to the horizontal FPLIF compression straps with two retainer bands, routing the ejector snap from bottom to top so that the HPT lowering line ejector snap is at the top.

g. Temporarily attach the HPT lowering line ejector snap to the triangle link of the left attaching harness strap. Attach the adjustable leg straps to the leg strap release assemblies, take up the slack in the adjustable leg straps, and fold and stow the excess using the webbing retainers. Attach the adjustable D-ring attaching strap snap hooks to the top carrying handle on the FPLIF.

12-71. RIGGING THE FIELD PACK, LARGE, INTERNAL FRAME WITH PATROL PACK

Rigging instructions for the FPLIF with patrol pack, using the HSPR, are as follows: (See Figure 12-64, page 12-75 and Figure 12-65, page 12-76.)

a. Before attaching the harness to the FPLIF, attach the release handle assembly, adjustable D-ring attaching straps, and leg strap release assembly to the harness. Fold and tape the running ends of all straps. No additional equipment will be attached to the outside of the FPLIF.

b. Shorten the shoulder straps to their minimum length, roll and tape the free-running ends, route the cross-chest strap around the shoulder straps, and then fasten the cross-chest strap buckle.

c. Extend the waistband straps, turn the FPLIF over with the shoulder straps down, route the waistband straps through the bottom loops formed by the lower patrol pack attaching straps, and buckle the waistband fastener.

d. Place the patrol pack on the outside front of the FPLIF and fasten the upper patrol pack attaching straps. Lay the HSPR on top of the patrol pack with the release handle assembly toward the top of the FPLIF, with the adjustable cross-strap facing up, and with the adjustable leg straps above the equipment retainer straps. Route the running ends of the HSPR equipment retainer straps and adjustable leg straps over the top of the patrol pack and down through the bottom equipment attaching loops of the patrol pack. Then route the equipment retainer straps through the second and third equipment attaching loops on the FPLIF. Route the adjustable leg straps through the second equipment attaching loops, only. Route the HSPR equipment retainer straps under the waistband. Route the lower patrol pack attaching straps under the third FPLIF equipment retainer loops and fasten them to the bottom of the patrol pack.

e. Turn the FPLIF over with the shoulder straps facing up. Route the HSPR equipment retainer straps over the bottom of the FPLIF, under the waistband (through the outer waistband frame loops), and up over the comfort pad; form an X with the running ends. Ensuring that the release handle assembly is positioned on the top of the FPLIF, continue to route the HSPR equipment retainer straps under the shoulder straps, through the loops sewn on the back of the FPLIF near the top of the frame staves, and connect the free-running ends to the friction adapters, forming quick releases. S-fold the free-running ends and secure with retainer bands.

f. To attach the HPT lowering line to the HSPR, route the looped end of the HPT lowering line under the X formed by the HSPR equipment retainer straps, from top to bottom. Route the ejector snap through the loop, forming a girth hitch. Route the HPT lowering line over the top of the right FPLIF shoulder strap. Secure the HPT lowering line on the right side of the FPLIF to the horizontal FPLIF compression straps with two retainer bands, routing the ejector snap from bottom to top so that the HPT lowering line quick-ejector snap is at the top.

g. Temporarily attach the HPT lowering line ejector snap to the triangle link of the left attaching harness strap. Attach the adjustable leg straps to the leg strap release assemblies, take up the slack in the adjustable leg straps, and fold and stow the excess using the webbing retainers. Attach the adjustable D-ring attaching strap snap hooks to the top carrying handle on the FPLIF.

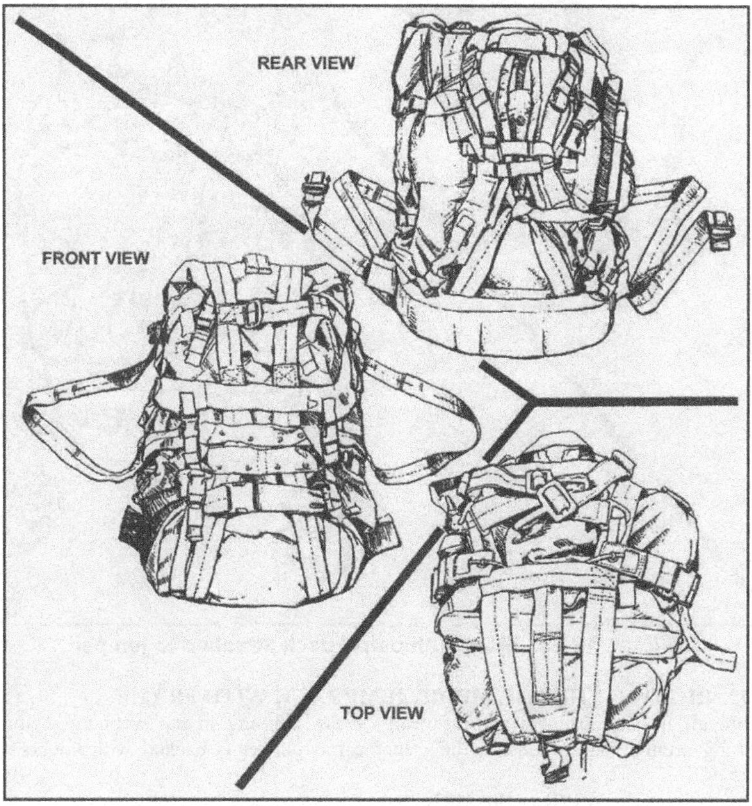

Figure 12-64. FPLIF with patrol pack on front.

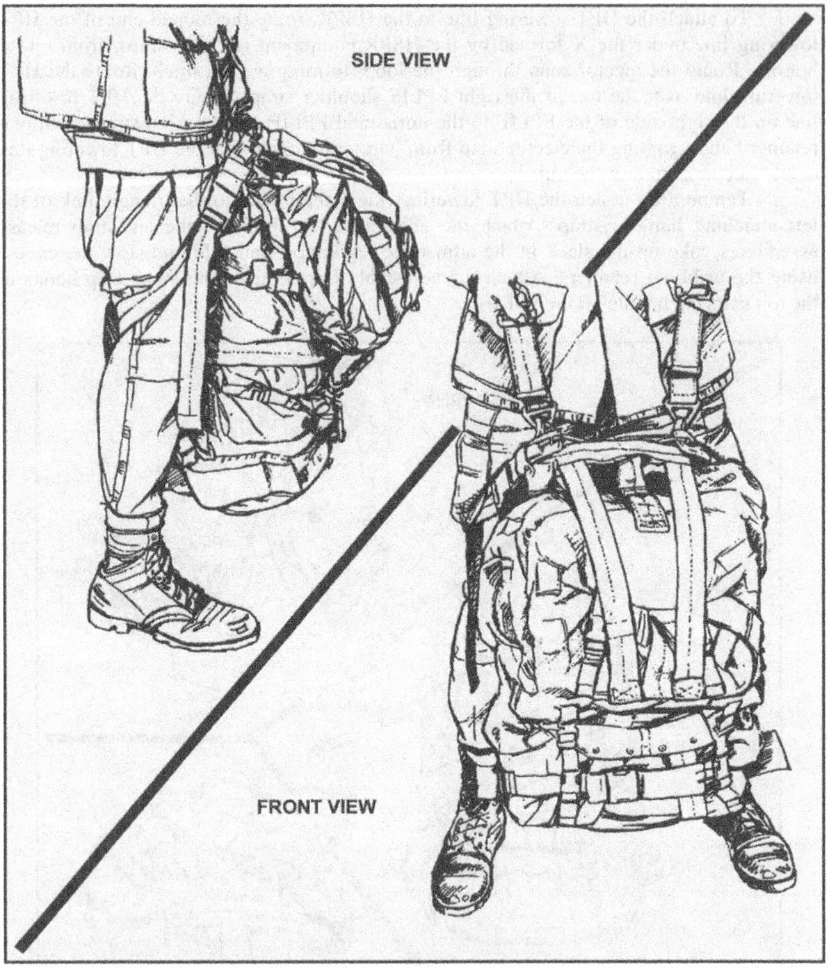

Figure 12-65. FPLIF with patrol pack attached to jumper.

12-72. RIGGING THE M82, MEDIC JUMP PACK WITH FRAME

Secure all Fastex fasteners and stow all excess webbing in the webbing retainer or webbing retainer bands. Ensure the center cargo pocket is packed with nonbreakable items.

a. **Secure the HSPR to the M82.**

(1) Lay out the HSPR with the attaching loops facing up.

(2) Route the two release handle cables between the two plies of the release handle cross strap. Attach the pile tape of the release handle assembly to the hook tab located between the plies of the release handle cross strap. Ensure the release handle lanyard is not misrouted.

(3) Place the triangle links of the adjustable D-ring attaching straps on top of the white attaching loops, with the opening gates of the snap hooks facing down. Route the white attaching loop through the triangle link; the green loop through the white attaching loop; the red attaching loop through the green attaching loop and then through the grommet on the female portion of the leg strap assembly. Ensure the cable loop retainer on the female portion of the leg strap assembly is facing up. Route the release handle cable through the red attaching loop and then through the cable loop retainer. Repeat this procedure for the other adjustable D-ring attaching strap.

(4) Turn the HSPR over so that the opening gates on the snap hooks of the adjustable D-ring attaching straps are facing up.

(5) Place the M82 upside down on the HSPR so that the center cargo pocket is in the window formed by the adjustable cross strap and the release handle cross strap.

(6) Route the equipment retainer straps under the frame and cross them in the center forming an "X", then route them to the appropriate friction adapter. Form a two- to three-finger quick release and tighten. S-roll the free-running ends and secure with two turns of retainer band (do not secure the excess to the quick releases).

(7) Route the male portion of the leg strap release assembly around the M82. Ensure the leg straps are fully extended with no twists and then attach them to the female portion of the leg strap release assembly.

NOTE: When stowing the HPT lowering line within the retainer flap, ensure stows are even with the retainer flap. No excess should exceed either end of the retainer flap.

b. **Attach the HPT Lowering Line to the HSPR.**

(1) Route the looped end of the HPT lowering line under the "X" formed by the equipment retainer straps from top to bottom (bottom to top), north to south (south to north).

(2) Route the remainder of the HPT lowering line through the loop end of the HPT lowering line, then pull it tight forming a girth hitch around the "X" formed by the equipment retainer straps.

(3) Secure the HPT lowering line to the left side of the frame by routing the ejector snap and HPT lowering line within the closed retainer flap through two retainer bands from bottom to top as the M82 is worn by the jumper.

CHAPTER 13
ARCTIC RIGGING

When using arctic rigging, the number of personnel who can be parachuted from a single aircraft is reduced by the bulk of equipment and cold weather clothing. When computing weight factors, the cold weather-equipped parachutist is estimated to weigh 310 pounds.

Section I. ARCTIC EQUIPMENT SPACE CONSIDERATIONS

Exiting interval between each parachutist is increased to two seconds when using arctic rigging. Aircraft compartment space required for a parachutist is 1 1/2 times more in cold regions than in temperate climates. Commanders must be familiar with the airborne operations portion of FM 31-71 for successful arctic rigging operations. However, plane-side parachute issue and rigging are impossible during winter months due to harsh temperatures.

13-1. WEIGHT FACTORS

Aircraft must be within 200 meters of the parachute rigging facility to keep rigged jumpers from walking through deep snow or over ice during winter months when temperatures are low and the individual parachutist's equipment is the heaviest. The serviceability of the activating lever on the ejector snap of the HPT lowering line should be checked, since there is an increased risk of the lever malfunctioning due to the heavy loads.

13-2. MODIFICATIONS

Modifications of standard equipment must be made for airborne operations under cold weather conditions.

a. **Waistband**. A modified waistband strap is used in lieu of the standard waistband when parachuting with snowshoes or skis. The strap consists of two pieces: a 6-foot A-7A strap and a 16-inch strap with a buckle at each end. The skis or snowshoes are attached to the side of the jumper opposite the static line (to prevent fouling). The modified waistband allows the buckle for the quick-release fold to be located on the same (either) side with the snowshoes or skis (Figure 13-1, page 13-2). Tandem loads are dropped on a single lowering line. Rigging or lowering procedures are contained in this chapter. Under arctic conditions, most individual equipment is lowered during descent due to the weight.

b. **Reserve Parachute**. Deployment of the reserve parachute under arctic or extreme cold weather conditions may be hindered by the bulkiness of the gloves worn. To overcome this obstacle, the rip cord is inverted as an optional requirement. Commanders requesting the MIRPS or T-10 reserve parachute for an arctic airborne operation must allow enough time for the reserves to be modified.

c. **Mittens**. Arctic mittens are not worn during the parachute jump; their bulkiness interferes with deployment of the reserve parachute and the lowering of equipment. The mittens are tucked inside the front of the jacket or under the parachute harness. They are not attached to, or packed in, a container. Trigger-finger mittens are stowed inside the jacket for wear as soon as the jumper is on the ground.

d. **Arctic Canteen**. The arctic canteen poses a hazard due to its long neck and metal body, which can injure a jumper if the PLF is executed on top of the canteen. Commanders should consider packing it in the ALICE pack to prevent personal injury or damage to the canteen.

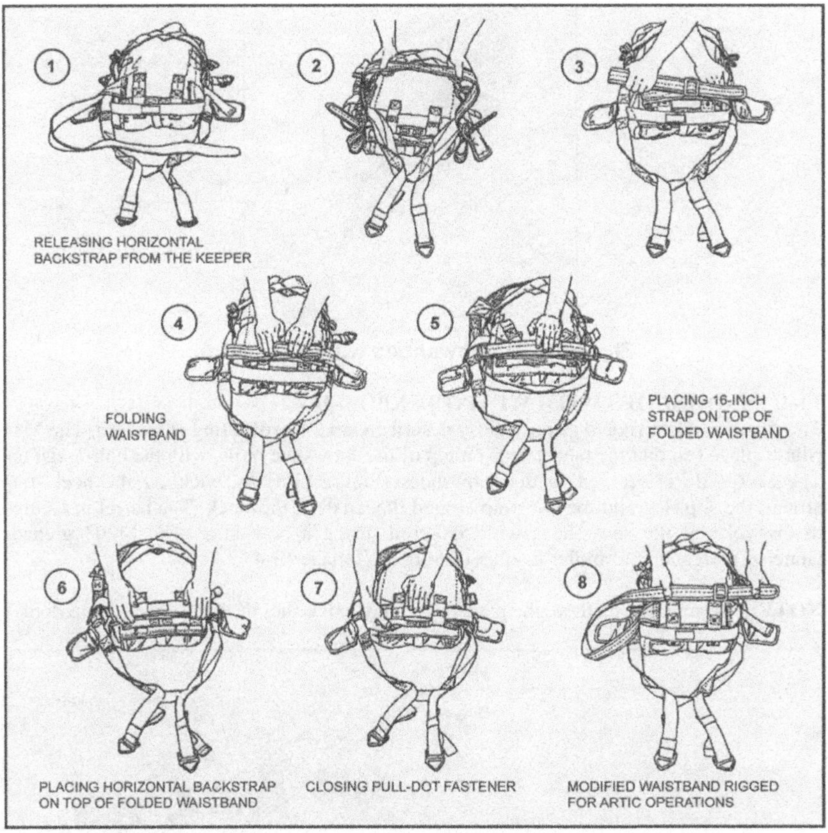

Figure 13-1. Modified waistband routed on parachute harness.

Section II. SNOWSHOES AND INDIVIDUAL WEAPON

Snowshoes are usually rigged on the parachutist to allow for immediate access.

13-3. SNOWSHOES WITHOUT WEAPON

The prefitted snowshoes are placed one on top of the other. The heel strap of the lower snowshoe is run underneath the lower shoe and up between the frame and webbing of both snowshoes. The heel strap buckle is brought up similarly on the other side of the snowshoes. A sling may be fabricated using 550-cord (or other suitable material). The snowshoes are secured with an additional tie-down, using 550-cord at the toe (Figure 13-2).

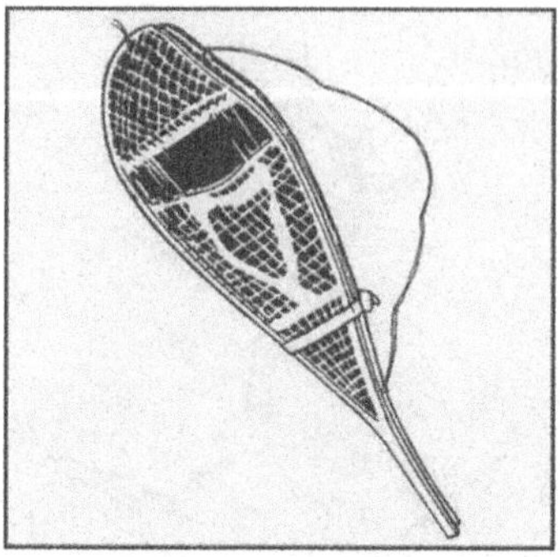

Figure 13-2. Snowshoes without weapon.

13-4. SNOWSHOES WITH WEAPON EXPOSED

The snowshoes are rigged as previously described, and the rifle sling is secured. The M16 rifle is placed so that the barrel rests on top of the snowshoe trails, with the bolt-assist up. The M16 rifle is secured to the snowshoes (Figure 13-3) by buckling the heel strap around the slip ring and the toe strap around the small of the stock. The barrel is secured to the yoke of the snowshoes with 550-cord, using a bow knot. The M203 grenade launcher is rigged in a similar manner (Figure 13-4, page 13-4).

NOTE: Using this method, the parachutist may exit either the right or left jump door.

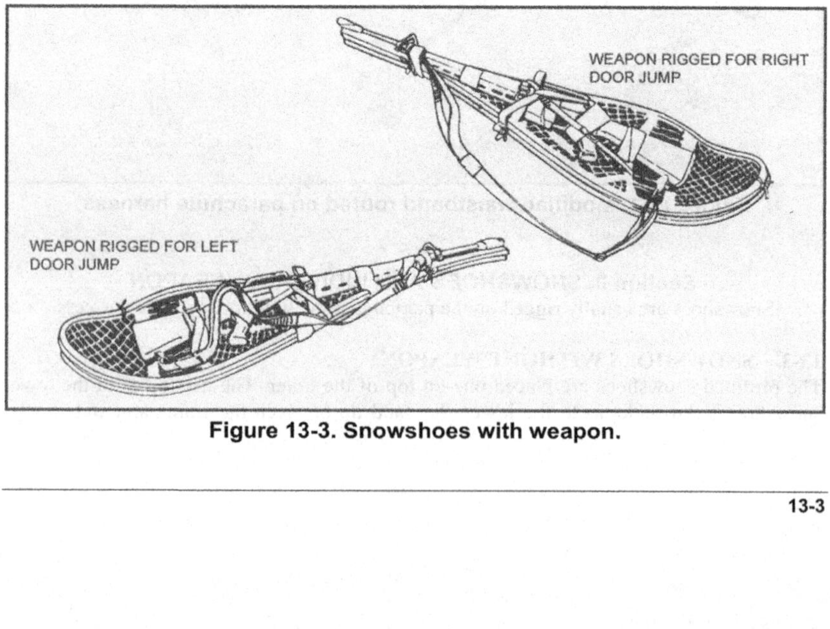

Figure 13-3. Snowshoes with weapon.

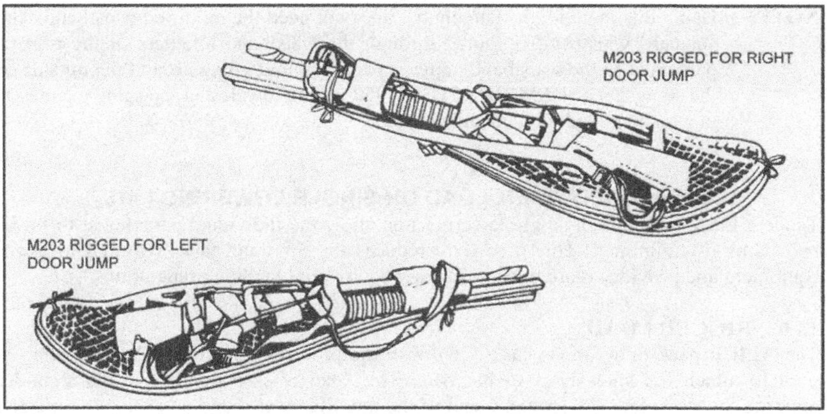

Figure 13-4. Snowshoes with M203.

13-5. JUMPING SNOWSHOES WITH M1950 WEAPONS CASE

Snowshoes are attached to the outside of the M1950 with the tails down and the tips of the snowshoes facing the jumper. The 550-cord is used to secure the snowshoes to the M1950 through both the upper and lower tie-down tape retaining bars. The running ends of the upper tie-down tape are routed through the toe window on the snowshoes and then around the M1950 to the left side of the main lift web. The M1950 is then attached as the outermost item to the left D-ring (Figure 13-5).

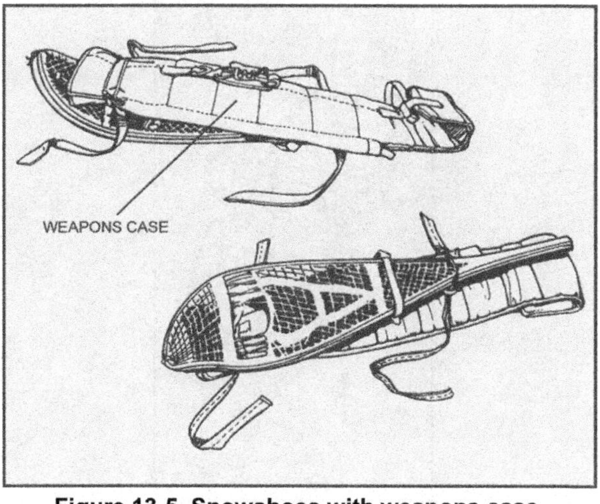

Figure 13-5. Snowshoes with weapons case.

NOTE: Using this method, the parachutist does not need the modified waistband. The standard waistband is routed through the waistband retainers of the reserve parachute to the waistband adjuster panel without going around the outside of the snowshoes and M1950. The M1950 is then lowered in the same manner as in a tandem load without snowshoes.

Section III. TANDEM LOAD ON SINGLE LOWERING LINE

Tandem loads rigged on a single lowering line allow the individual parachutist to lower two items of equipment. This procedure reduces the time and tasks required to lower equipment and provides more time for canopy control and landing preparations.

13-6. RIGGED LOAD

The ALICE pack or weapons case is worn in the prescribed manner. The rifle sling is used to attach the snowshoes to the rifle. This load is suspended over the shoulder opposite the static line. The running end of the modified waistband extension is threaded through the waistband retainers of the reserve parachute and around the snowshoes, and the end is made into a quick-release fold (Figure 13-6).

NOTE: The rifle sling is adjusted to fit the parachutist snugly. The sling adjustment is also small enough so that it does not come off the load when released to slide down the lowering line.

Figure 13-6. Parachutist rigged to jump left door.

13-7. HOOK-PILE TAPE LOWERING LINE

The lowering line is threaded through the rifle sling and attached to the main lift web of the parachutist (on the side the snowshoes are attached).

a. A length of 80-pound test tape is attached around the sling and the main lift web with a bow knot (just below the canopy release assembly). The lowering line adapter web is attached (Figure 13-7) to the left (right) side, corresponding to the side to which the snowshoes and rifle are to be attached.

b. When jumping with the ALICE pack or weapons case, the parachutist routes the lower tie-down on the case through the metal frame and around the leg. A separate tie-down tape for each is not necessary (Figure 13-8, page 13-8).

c. The stowed lowering line is secured with two retainer bands to the left (right) side of the vertical bar of the combat pack. After the pack, snowshoes, and rifle are attached to the parachutist, the lowering line ejector snap is passed through the rifle sling and attached to the accessory attaching ring of the adapter web. If the adapter web is not used, the ejector snap is attached directly to the D-ring on the harness. If the weapons case is attached to the snowshoes (Figure 13-9, page 13-8), the lowering line ejector snap is passed between the case and the cotton chafe material, which is attached to the case V-ring.

d. The upper tie-down tape (above the chest strap and below the canopy release assembly) and the lower tie-down tape around the parachutist's leg are untied, and the load is dropped by pulling the white release handle located on the cross strap of the HSPR. This ensures that the pack load falls the length of the line. The modified waistband quick-release fold is pulled and releases the snowshoes, which slide down the lowering line on top of the pack (Figure 13-10, page 13-9).

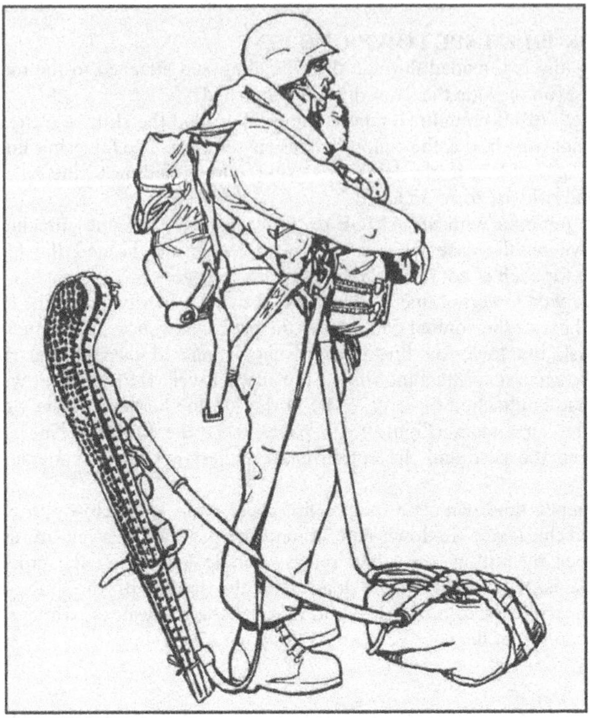

Figure 13-7. Lowering line assembly attached for a tandem load.

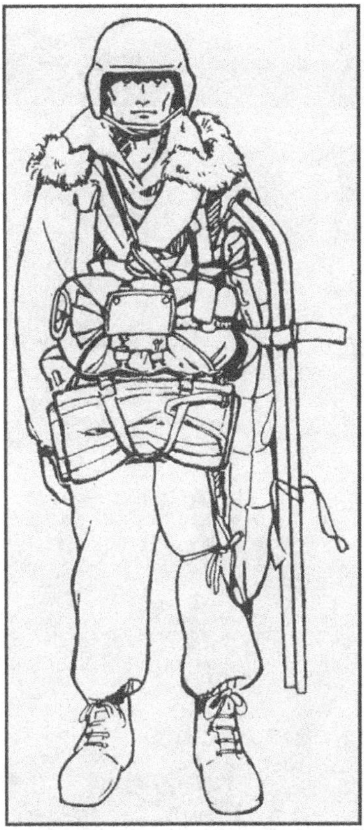

Figure 13-8. Upper and lower tie-downs.

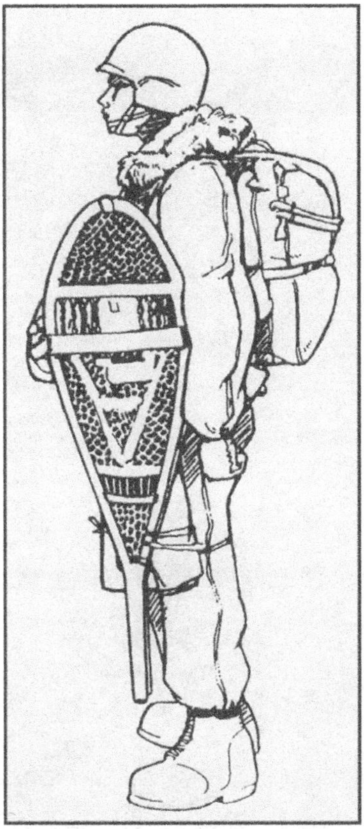

Figure 13-9. Left side view with weapons case.

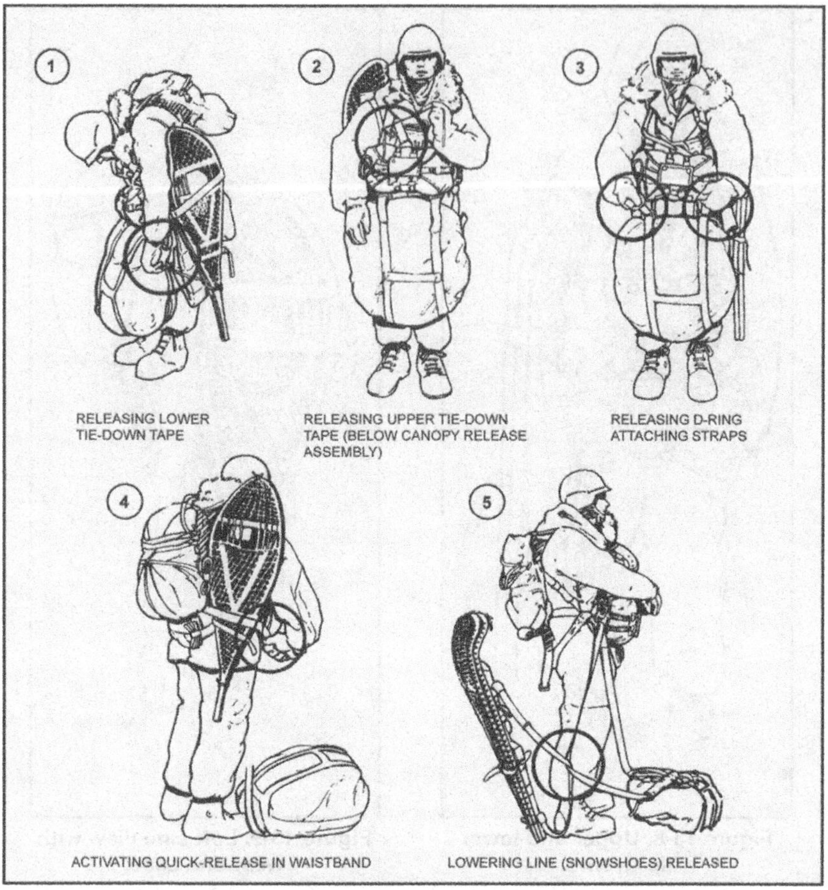

Figure 13-10. Arctic equipment released and lowered.

Section IV. SKIS JUMPED WITH RIFLE OR ALICE PACK

Skis, with the rifle attached, can be jumped using the procedures outlined.

13-8. SKIS AND RIFLE

When the parachutist jumps with skis, or with skis as part of the individual jump load, the jump must be from a rear platform or ramp (Figure 13-11).

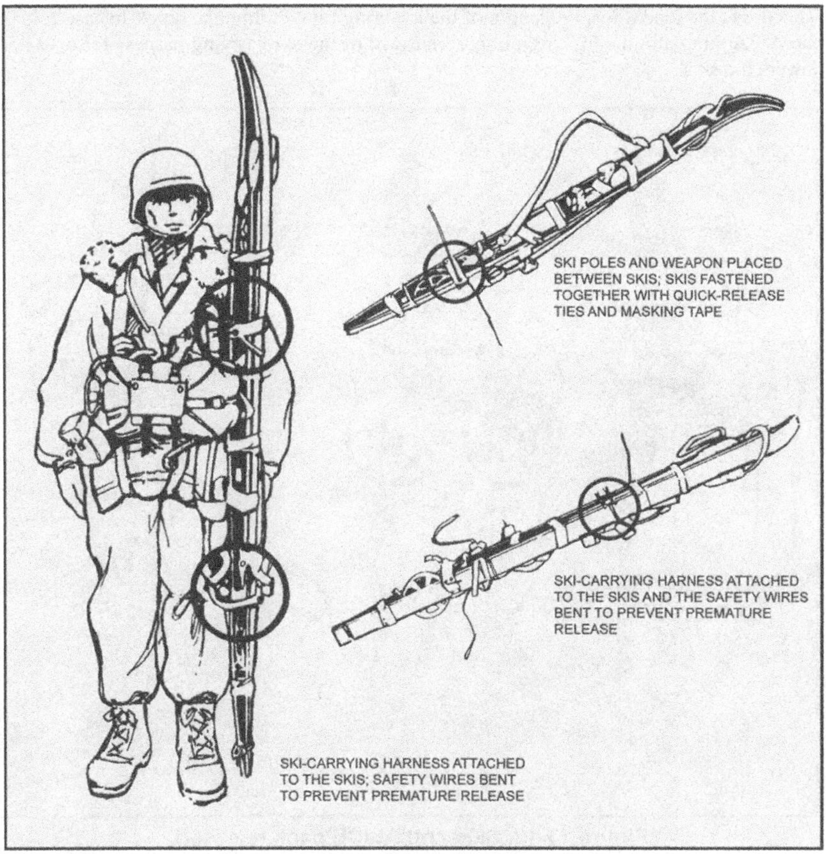

Figure 13-11. Skis and rifle rigged.

13-9. SKIS AND ALICE PACK OR WEAPONS CASE

The ALICE pack (or weapons case) lowering line is threaded between the skis and attached to the main lift web on the side to which the skis are attached. The lowering line is routed the same way, and the ejector snap is fastened to the lowering line adapter web or to the D-ring on the parachute harness (Figure 13-12). To secure and stabilize the skis, a length of 80-pound test is attached to the top and bottom of the skis with a bow knot. The skis and ALICE pack (or weapons case) are lowered by pulling the release handle. This drops the load down the length of the lowering line. Pulling the quick-release fold of the A-7A strap and the locking-pin cord lanyard of the ski-carrying harness releases and lowers the skis.

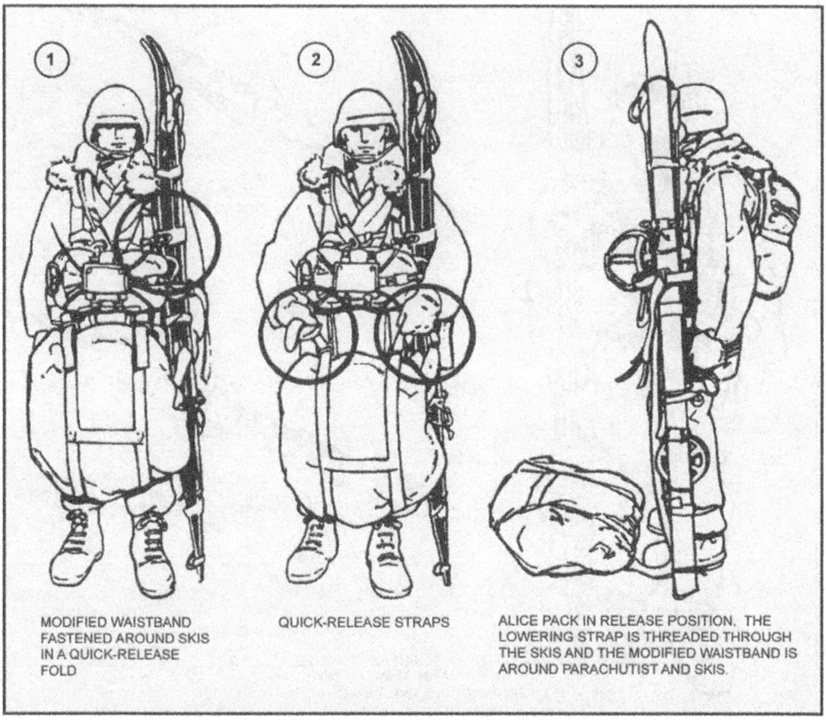

Figure 13-12. Skis and ALICE pack released.

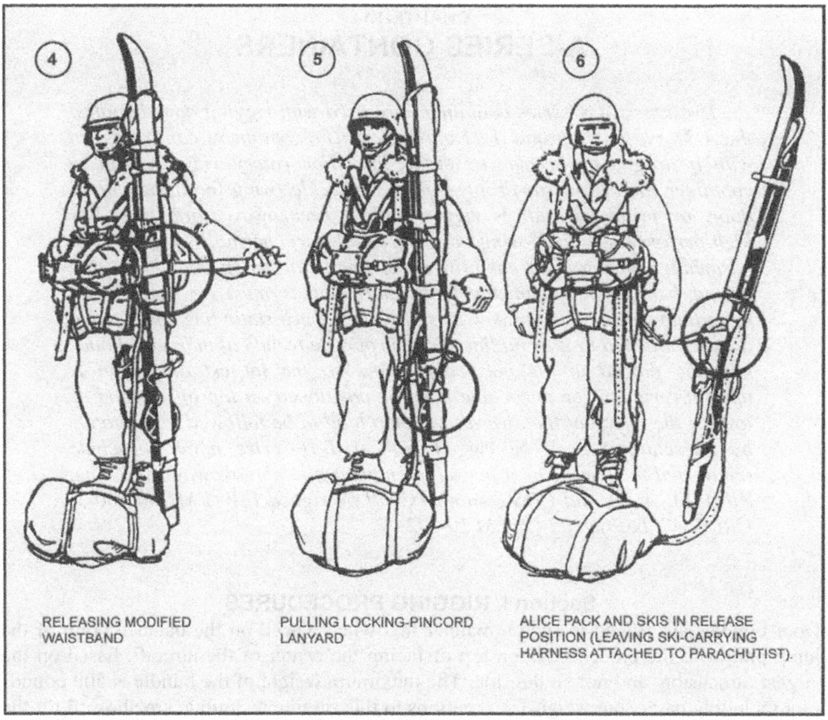

Figure 13-12. Skis and ALICE pack released (continued).

CHAPTER 14

A-SERIES CONTAINERS

Two types of A-series containers are used with rigging door bundles: the A-7A cargo sling and A-21 cargo bag. The containers can be rigged with a drogue or breakaway static line. For rotary-wing aircraft, a container load to be air-dropped from a shackle (wing load), helicopter door, or utility aircraft is rigged with a breakaway static line. For high-performance fixed-wing aircraft, loads are normally rigged with parachutes having nonbreakaway static lines. Paratroop door loads that are to be followed immediately by parachutists must be rigged with parachutes having nonbreakaway static lines. Each static line must have a drogue attached to it as outlined in appropriate technical manuals. Loads must be placed in the paratroop doors so the largest dimension is upright/vertical. The parachute must be positioned on top of the load or toward the inside of the aircraft. A ramp load to be followed immediately by parachutists must be rigged with a T-10-series main parachute (converted for cargo) or a parachute having a nonbreakaway static line. All A-7A, A-21, and CDS containers will be rigged IAW FM 100-500-3; CRRC will be rigged IAW FM 10-542.

Section I. RIGGING PROCEDURES

Door bundles are rigged in such a manner that when placed on the balance point of the jump platform, the parachute is on top or facing the center of the aircraft, based on the largest dimension, and not on the side. The maximum weight of the bundle is 500 pounds (not including parachute weight). Exceptions to this rigging technique are allowed for the 90-mm recoilless rifle and the Stinger missile. In both cases, the bundle is placed upright with the parachute facing the center of the aircraft. Both are rigged using the A-21 container. The skid board on the Stinger is placed inside the canvas cover.

14-1. ASSEMBLIES

When rigging an item, all components needed for its assembly must be packed in the same airdrop bundle. (For example, a radio and battery are packed in the same bundle.) When items such as radio equipment are rigged, each item is individually wrapped. Padding or honeycomb is placed under the item being prepared and inserted between the items comprising the load to prevent contact. Cellulose wadding, felt, or other suitable material must be used to avoid metal-to-metal or metal-to-wood contact.

14-2. WEBBING

All excess lengths of webbing are rolled and tied with 1/4-inch cotton webbing in a surgeon's knot and locking knot. This reduces the danger of bundles becoming snagged when ejected or released from the aircraft.

14-3. HAZARDOUS MATERIALS

If hazardous materials are placed inside bundles, they must have a shipper's certificate completed IAW AFJMAN 24-204/TM 38-250.

NOTE: The shipper's certificate is attached to the manifest, not the bundle.

Section II. A-7A CARGO SLING

The A-7A consists of the following components:

- Four straps, 188 inches long, constructed of Type X cotton or Type VII nylon.
- A strap fastener at one end of each strap.
- Four D-rings.

14-4. CHARACTERISTICS

Container components weigh 8 pounds with a maximum weight of 500 pounds (not including the parachute). The minimum weight depends on the parachute used. The dimensions are a maximum 30 inches wide by 48 inches deep by 66 inches high (to include cargo parachute).

14-5. TWO-STRAP BUNDLE

The jumpmaster lays out one strap perpendicular (lengthwise) to the bundle with the thick lip portion of the friction bar on the strap fastener facing down. He lays out one strap parallel (width) to the bundle with the thick lip portion of the friction bar on the strap fastener facing down and over the top of the perpendicular strap. When the straps are in place, they are ready to receive the bundle.

a. Center the bundle on the perpendicular strap. Route the perpendicular strap over the top of the bundle and through the single D-ring (through the rectangular portion of the D-ring), fold, and secure. When using the G-14 or T-10-series, two D-rings are used on the perpendicular strap. They are tied in place with Type II or Type III nylon cord gutted.

b. Route the parallel strap through the D-ring (through the rectangular portion of the D-ring), roll, and secure. Tie all excess webbing onto itself using one turn of 1/4-inch cotton webbing tied in a surgeon's knot and locking knot.

c. Tighten all straps. Tie off the excess above the strap fastener; ensure that the excess webbing is not above the top of the bundle. The bundle has one smooth side for ease in ejecting from the aircraft.

14-6. THREE-STRAP BUNDLE

The jumpmaster lays out one strap parallel (lengthwise) to the load. He then lays out two more straps parallel to each other and perpendicular to the load. He ensures that the strap fasteners for these straps are on the same side of the load and that straps are at least 16 inches from each other, and are centered on the load.

a. Center the load to be rigged on the straps. Route the strap that is running parallel (lengthwise) to the load over the top of the load and through the two D-rings. Center the two D-rings on top of the load. Route the strap through the strap fastener and tighten it down. Ensure that the strap fastener does not rest on the top of the load. Roll the excess hand over hand towards the load and secure it with a non-slip knot above the strap fastener, but not on top of the load. Route the straps that are running perpendicular to load through the D-rings on the top of the load, from the inside toward the outside so that the D-rings are pointing towards each other. Route the straps through the strap fasteners, and tighten them down. Roll the excess hand over hand towards the load and secure it with a non-slip knot above the strap fastener, but not on top of the load.

b. Once all straps are tightened and secured, the free running ends should not rest on top of the load. The bundle should have one smooth side for ease in ejecting it from the aircraft.

14-7. FOUR-STRAP BUNDLE

The jumpmaster lays out two straps parallel (lengthwise) to the load and centered, ensuring that both strap fasteners are on the same side of the load. He then lays out two straps parallel to each other and perpendicular to the load, ensuring that both strap fasteners are on the same side of the load and that all straps are centered on the load.

a. Route the two straps that are running lengthwise on the load over the top of the load and through the two D-rings (one D-ring per strap). Center the D-rings on top of the load. Route the straps through the appropriate strap fasteners and tighten them down. Roll the excess webbing hand over hand towards the load and secure it with a non-slip knot above the strap fastener. Route the straps that are running perpendicular to the load from inside to outside through the appropriate D-rings (ensuring that both D-rings point in the same direction). Then, route the straps through the appropriate strap fasteners and tighten them down. Roll the excess webbing hand over hand towards the load and secure it with a non-slip knot above the strap fastener.

b. When tying off the free running ends of the straps, tie them in a non-slip knot above the strap fastener; ensure that the excess webbing does not rest on top of the load. The load should have one smooth side for ease in ejecting it from the aircraft.

Section III. A-21 CARGO BAG

The A-21 cargo bag consists of the following components.

- Canvas cover: Cotton duck material, 97 inches by 115 inches, with eight strap keepers.
- Sling assembly with scuff pad: One 188-inch main strap, two 144-inch side straps, scuff pad 30 inches by 48 inches, and four lifting handles.
- Quick-release assembly: Quick-release device with safety clip, three quick-release straps, and one fixed quick-release strap.
- Two-ring straps: The ring strap has one 9-inch strap that has a friction adapter, and one 7-inch strap with a D-ring.

14-8. CHARACTERISTICS

Container components weigh 18 pounds with a maximum weight of 500 pounds (not including the parachute). (The minimum weight depends on the parachute.) Dimensions are a maximum 30 inches wide by 48 inches deep by 66 inches high, or 69 inches high for the Stinger missile (see FM 10-550). Dimensions include the cargo parachute.

NOTE: See FM 10-500-3 for further information on rigging containers.

14-9. METHOD OF RIGGING

The jumpmaster spreads the canvas cover on a level surface with all strap keepers facing up. He positions the sling assembly webbing straps down on the canvas cover and threads the straps through the keepers. The sling and canvas cover are turned over as a unit so the sling is beneath the cover. The parachutist centers the load on the canvas cover, using cushioning material, as needed. He wraps the load in the canvas cover, side flap first, and folds all excess material under.

a. Attach the two-ring straps to the 188-inch main strap, keeping the D-ring-to-D-ring contact and ensuring they are centered. Attach the four quick-release straps to the 144-inch side straps. Ensure that the rotating disk is facing up when the quick-release assembly is placed on top of the load (thick-lip portion of the friction bar facing out).

b. Thread the fixed, quick-release strap with the quick-release assembly attached through the nearest steel rod ring. Thread the remaining quick-release straps through the nearest steel rod rings. Insert the lugs into the quick-release assembly.

c. Tighten the quick-release straps and the two-ring straps; roll all excess webbing. Ensure that it is tied off below the friction adapter with a surgeon's knot and locking knot and that the quick-release device is centered on the bundle.

Section IV. CARGO PARACHUTE RIGGING ON A-SERIES CONTAINERS

After the A-series containers are rigged, the jumpmaster inspects the cargo parachutes and attaches them to the load.

14-10. INSPECTION

The cargo parachute is placed on the center of the bundle and is inspected for—

- Four tie-down straps.
- Two risers complete (clevis, clevis pin, safety wire).
- Static line complete with drogue device (clevis, clevis pin, safety wire). The drogue device must be attached to the breakcord attaching loop, unless a breakaway static line is used.

14-11. ATTACHMENT

The jumpmaster ensures that the risers go directly to the attaching point (D-ring), tie-downs are attached (tied to side straps), static line is free to deploy, and risers are not routed around or under any part of the bundle.

NOTE: The cargo parachute should be attached with the side of the pack where the risers come out, collocated to the rough side of the bundle.

PART FOUR
Aircraft Used in Airborne Operations

CHAPTER 15
AIRCRAFT AND JUMP ALTITUDES

This chapter contains general aircraft descriptions, jumpmaster procedures, and aircraft preparations. Critical elements of airborne operations are the aircraft and drop altitudes. These aircraft are service tested and approved for troop jumping. Minimum jump altitudes and considerations that apply to basic airborne jumps, tactical jumps, and combat jumps are discussed.

15-1. TYPES OF AIRCRAFT

Commonly used types of aircraft for airborne operations are:

a. **High-Performance Aircraft**. The C-5, MC/C-130, KC-130, C-17, and C-141 are high-performance fixed-wing aircraft used for paratroop drops. (Nonstandard jump aircraft are found in Chapter 19.)

b. **Utility Rotary-Wing Aircraft**. The UH-1 series, UH-60, CH-47, CH-53 (USMC), and CH-46 (USMC) are the most commonly used rotary-wing aircraft for troop airdrop.

15-2. JUMP ALTITUDES

The minimum criteria discussed below include a 125-foot aircraft altimeter error and a 100-foot canopy control requirement for the MC1-series.

a. **Minimum Jump Altitudes**. The minimum jump altitudes for **all** aircraft are as follows.

(1) ***Peacetime Tactical Training***.

- Basic airborne training—1,250 feet AGL.
- Tactical training—1,000 feet AGL; however, 800 feet AGL drop altitude may be employed after a mature Command Decision Risk Assessment has been completed.
- Aircraft with a drop speed of 90 knots or less—1,500 feet AGL.

(2) ***Combat (Wartime).*** (Restricted to aircraft with a drop speed of 90 knots or higher.)

(a) *Rotary-Wing Aircraft with a Drop Speed of 90 Knots or Greater.*

- T-10-series parachute—600 feet AGL.
- MC1-series parachute—600 feet AGL.
- Minimum airspeed—90 knots.

(b) *Fixed-Wing Aircraft with a Drop Speed of 125 Knots or Higher.*

- T-10-series parachute—435 feet AGL.
- MC1-series parachute—475 feet AGL.
- Minimum airspeed—125 knots.
- Reserve parachute—optional.

b. **Prejump Training**. Sustained prejump training is modified to ensure that the individual parachutist accomplishes the following tasks:

(1) Upon receiving the opening shock of the main, immediately lowers individual equipment.

(2) Assumes the landing attitude (T-10-series) or canopy control (MC1-series).

c. **Data Summarizing Parachute Reliability**. Table 15-1 summarizes parachute reliability data at varying altitudes with personnel parachutes (mains only—not reserves).

	T-10-Series PARACHUTE		MC1-Series PARACHUTE	
Exit altitude in feet AGL	**Percentage of canopies that would be open**	**Percentage of parachutists who would have 100 feet or more to prepare to land**	**Percentage of canopies that would be open and would provide canopy control (no twists).**	**Percentage of parachutists who would have canopy control plus 100 feet or more to land**
200	76.11	4.00	13.00	NONE
300	99.92	76.11	90.66	13.00
400	99.98	99.92	99.98	90.66
500	99.98	99.98	99.98	99.98
600	99.98	99.98	99.98	99.98
700	99.98	99.98	99.98	99.98
800	99.98	99.98	99.98	99.98

Table 15-1. Parachute reliability data for varying altitudes.

d. **Jump Altitude Risk Assessment Decision-Making**. When making a training parachute jump altitude risk assessment decision, the airborne commander should consider soldier experience levels, soldier fatigue at jump time, whether the jump altitude is critical to mission success, and the data from Table 15-2 below. Risk reduction measures include alternate methods of entry for personal equipment (CDS, door bundle, and so on) and action-in-the-aircraft rehearsals, with particular attention devoted to jumper exit procedures.

Planned Altitude (Feet AGL)	**Possible Aircraft Altimeter Error** (Feet)	**Actual Jump Altitude Feet** (AGL)	**Total Time Available to Activate Reserve** (Seconds)	**Time Available Minus 4000-Count** (Seconds)
1,250	+125	1,375	9.4	5.4
	0	1,250	8.8	4.8
	-125	1,125	8.2	4.2
1,000	+125	1,125	8.2	4.2
	0	1.000	7.6	3.6
	-125	875	6.9	2.9
800	+125	925	7.1	3.1
	0	800	6.4	2.4
	-125	675	5.6	1.6

Table 15-2. Time available to activate reserve parachute.

15-3. HIGH-ELEVATION JUMPING

The term high-elevation jumping refers to airdrop operations that begin at normal altitude above ground level (that is, 800 feet AGL) but where the DZ is 5,000 to 10,000 feet above mean sea level, such as in mountainous terrain. Commanders must consider that lower air density or higher altitude will increase the canopies' rate of descent.

a. **Parachutes**. Standard troop-type parachutes are suitable for the airdrop of personnel onto DZs with ground elevations up to 10,000 feet. Current jump procedures are valid.

b. **Injuries**. Combat or training exercises onto DZ elevations of 5,000 to 10,000 feet that place safety secondary to tactical considerations can produce injury rates of up to four times those expected for similar DZ operations near sea level. Injury rates can be reduced by intensive instruction, training, and practice to include the following.

(1) Exit body position and PLF upon ground impact.

(2) Jumper awareness of increased opening shocks and faster rates of descent.

(3) The ability to control the parachute during sudden wind shifts and changes in wind velocity.

(4) Wearing equipment correctly and ensuring serviceability.

CHAPTER 16

HIGH-PERFORMANCE TRANSPORT AIRCRAFT

The C-5, MC/C-130, C-17, and C-141 are US Air Force high-performance aircraft for paratroop drops. These high-performance transport aircraft, provided by the USAF, are configured to meet the needs of the unit mission request. These configurations are termed tactical airdrop personnel (TAP) for the C-130 and the C-141 in USAF publications. The initial JA/ATT mission request from the airborne unit determines aircraft configuration. The USMC version of this aircraft is the KC-130; characteristics and parachute procedures are similar.

- *Tactical mass airdrop (both jump doors)—full seating configuration.*
- *In-flight rigging—tactical mass airdrop (both doors)—full seating configuration; comfort pallet with/without litters.*
- *Other load considerations—combination of air-land and airdrop mission; single door—reduced seating configurations.*
- *Over-the-ramp—combination equipment and personnel.*

Section I. C-130 HERCULES

The C-130 (Figure 16-1) is a medium-range, high-wing transport aircraft powered by four turboprop engines. Parachutists may be dropped using either the two jump doors or the ramp. (See notes regarding over-the-ramp operations at paragraph 16-3.)

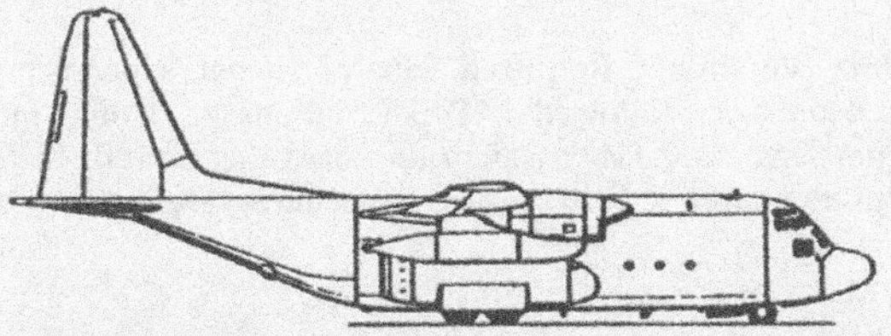

CONFIGURATION	PARACHUTIST LOAD
TAP 1 - MASS DROP	64
TAP 2 - IN-FLIGHT RIGGING	44
TAP 3 - HALO RAMP OPERATIONS AND WEDGE	42
TAP 4 - ARCTIC OPERATIONS	40

Figure 16-1. C-130 Hercules.

16-1. SEATING CONFIGURATION

The seating configuration for the C-130 Hercules is as follows:

a. **Peacetime Training (TAP 1)**. A total of 64 parachutists can be seated in two sticks of 32. Numbers 1, 2, and 3 are seated outboard aft of the wheel well; 4 through 23 are seated on the inboard seats; and numbers 24 through 32 are seated outboard forward on the wheel well (Figure 16-2). Jumpers are normally loaded over the aft end loading ramp.

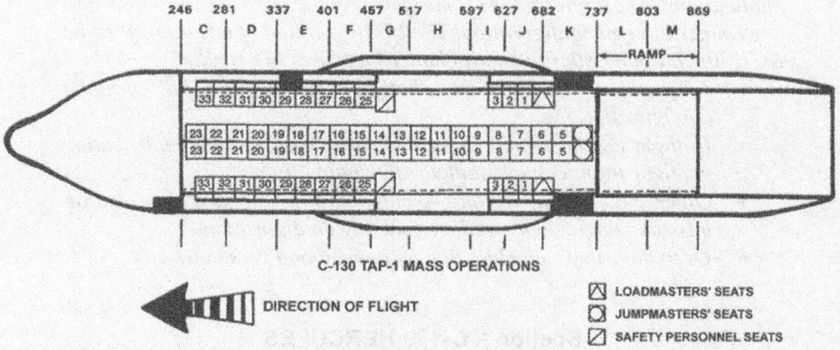

1. 68 TROOP SEATS - SEAT BELTS ON 24-INCH CENTERS - (66 TROOP SEATS - 2 LOADMASTER SEATS).
2. NORMALLY SEATS #10, LEFT AND RIGHT, ARE NOT RIGGED. SEATS #11, 12, AND 13, LEFT AND RIGHT, ARE RIGGED ON 20-INCH SPACING VERSUS 24-INCH. THIS PROVIDES 4 SEATS AFT WHEEL WELL BOTH SIDES.

Figure 16-2. C-130 seating configuration.

b. **Supervisory Personnel Required**. Six personnel supervise parachutists and ensure safety measures are followed (ADEPT options 1, 2, and mass exit. (USAF operations may not have an AJM or safety on board the aircraft. In these cases, prior coordination with the aircrew will ensure procedures are known by all personnel involved.)

- One JM.
- One AJM.
- Two safety personnel.
- Two airdrop certified loadmasters.

NOTE: USMC and USN utilize one door, the other door remains closed and unmanned, therefore only 1 JM and 1 safety are used. These units do not fall under AR 350-2.

c. **Jump Commands**. Jump commands are given in the following sequence:

GET READY.
OUTBOARD PERSONNEL, STAND UP.
INBOARD PERSONNEL, STAND UP.

HOOK UP.
CHECK STATIC LINES.
CHECK EQUIPMENT.
SOUND OFF FOR EQUIPMENT CHECK.
STAND BY.
GO.

16-2. IN-FLIGHT RIGGING PROCEDURES

In-flight rigging procedures include the following:

a. **Personnel**. These procedures provide in-flight rigging for the airdrop of 44 personnel (with equipment) including one JM and four AJMs. Three AJMs are designated from the parachutists onboard to assist in rigging. Two nonjumping safety personnel are also required. For USAF operations, a safety is not used and either the JM or loadmaster fulfills the duty after coordination.

b. **Briefing**. All parachutists must be briefed and rehearsed on their actions before executing this type mission. The preferred method for in-flight rigging is **buddy rigging**. This allows faster rigging and reduces parachutists' movement in the aircraft. The other method, **station rigging**, is seldom used. (See paragraphs 16-2e and f.)

c. **Aircraft Configuration**. The aircraft is TAP 2 configured to seat a total of 48 personnel to include 2 loadmasters, 2 nonjumping safeties, and 44 parachutists (Figure 16-3).

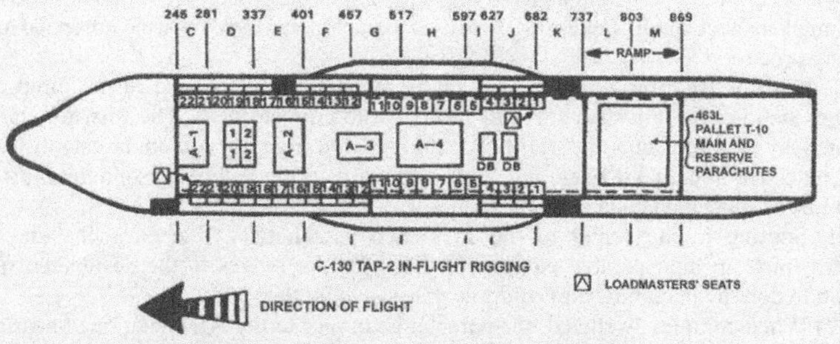

1. 48 TROOP SEATS-SEAT BELTS ON 20-INCH CENTERS-46 TROOP SEATS OFFERED. THIS CONFIGURATION IS FOR IN-FLIGHT RIGGING OF PARACHUTES ONLY (LONG-RANGE MISSIONS).

2. A-1: STORAGE AREA FOR ROLLER CONVEYORS, TOOL BOXES, AND CREW BAGGAGE.

A-2: 22 EACH MAIN AND RESERVE PARACHUTES IN KIT BAGS TO BE RELOCATED FROM CARGO RAMP TO AREA A-2 PRIOR TO RIGGING.

A-3: M1950 WEAPONS CASE STORAGE FOR WEAPONS BELONGING TO TROOPS THAT OCCUPY WHEEL WELL SEATS.

A-4: WEAPONS IN EQUIPMENT CONTAINERS.

DB: DOOR BUNDLES STORAGE AREA.

Figure 16-3. C-130 configuration for in-flight rigging (48 seats).

d. **Storage of Equipment**. Storing equipment involves the following:

(1) Forty-four parachutes and reserves, in kit bags, are palletized (covered with a cargo net or tie-down devices) on the ramp.

(2) Door bundles are placed in the center aisle just forward of the ramp.

(3) Special items of equipment are placed in the center aisle just forward of the door bundles.

(4) M1950 weapons cases should be placed behind individual seats. Weapons cases for personnel occupying wheelwell seats are placed on the floor at Station 477.

(5) Individual equipment should be placed under seats or, if too large, may be placed in the center aisle forward of the special items of equipment.

(6) All equipment placed in the center aisle must be secured.

e. **Buddy Rigging**. Buddy rigging begins 2 hours and 20 minutes before drop time. Rigging must be completed by the 20-minute warning.

(1) The main and reserve parachutes, with kit bags, are passed forward until each man has a parachute. Once everyone has his parachute, buddy rigging begins under the supervision of the JMs, designated JM qualified parachutists, and safety personnel. Each parachutist must know who his rigging partner is.

(2) Once a parachutist is rigged, he sits down and waits to be inspected by one of the JMs. The safety personnel serve as roving correction inspectors. The attachment of the static line snap hook to the top carrying handle of the reserve signifies an inspected parachutist.

(3) The JM supervises the entire rigging operation and assists, as needed.

(4) Once all the parachutists have been inspected, the JMs don equipment, and safety personnel inspect them. Other JM qualified parachutists may be used to speed up the rigging process.

f. **Station Rigging**. One rigging station is established forward of the jump doors; another station is established forward in the cargo compartment. The forward station is manned by a safety and the AJM. As the forward rigging station is established, 22 parachutes (in aviator kit bags) are passed forward to the safety personnel. One JM kit bag must be present at each station.

(1) Starting in the center of the stick, two parachutists (one on each side of the aircraft) pick up their combat equipment. Each jumper moves to the designated rigging station to don the parachute and equipment, assisted by the safety.

(2) When completely rigged, the parachutist moves to the AJM (stationed nearby) for inspection. The AJM attaches the static line snap hook to the top carrying handle of the reserve parachute when he reaches that portion of the inspection sequence. Only the JM or AJM removes the static line snap hook from the reserve before the command HOOK UP, since this indicates that the jumper has received JMPI.

(3) After JMPI, the jumper returns to his proper seat (or stick position). To ensure minimum time loss, the next parachutist is waiting to be rigged by safety personnel.

16-3. OVER-THE-RAMP OPERATIONS

These procedures provide for over-the-ramp airdrop of 40 personnel including one JM, one AJM, and two nonjumping safety personnel (Figure 16-4, page 16-6). For USAF operations, a safety is not used and either the JM or loadmaster fulfills the duty after coordination.

NOTES:

1. IAW message Headquarters, USAF AMC, DTG 021431Z MAY 1990, subject: C-130 Tailgating; over-the-ramp C-130 airdrop operations are restricted "to combination airdrops, tests requiring over-the-ramp operations to satisfy a specific objective, jumpers required to jump with snow skis, support for special operations forces training, MC-130, pathfinder and SOLL II operations." SOF were defined as "USA Special Forces (Green Berets), Ranger units, and Ranger Training Brigade, Navy Seals, USAF Combat Control Teams, Pararescue Teams, and those assigned to special tactics units." Message further stated, "These forces will continue to train for over-the-ramp operations which is their normal method for deployment."
2. IAW message Headquarters, USAF AMC, DTG 111345Z JUL 1991, subject: Authorization of Arctic Equipped Parachutists to Tailgate; authorization was granted "to allow arctic equipped parachutists with large field pack and/or skis to tailgate from C-130 aircraft." (Large field pack mentioned in message DTG 111345Z JUL 1991 is the FPLIF.)
3. IAW letter CG, Marine Corps Combat Development Command, C42 over 5600, dated 22 AUG 95, USMC Force Reconnaissance and Air Naval Gunfire Liaison Company (ANGLICO) personnel are authorized to conduct C-130 over-the-ramp parachute operations if required by their mission. These units train for over-the-ramp parachute operations as a normal method of employment.
4. Based on Airborne materiel and training proponency issues, the 1st Battalion (Abn), 507th Infantry, 11th Infantry Regiment, United States Army Infantry School is authorized to plan and conduct over-the-ramp operations from the C-130 aircraft.

a. **Equipment Drop**. Over-the-ramp operations can include the combination of an equipment drop followed by parachutists. Equipment is defined as heavy equipment rigged for airdrop, or equipment packaged and contained in the A-7A or A-21 containers. Parachutists may be dropped over the ramp without an equipment drop (maximum of 20 parachutists for each pass).

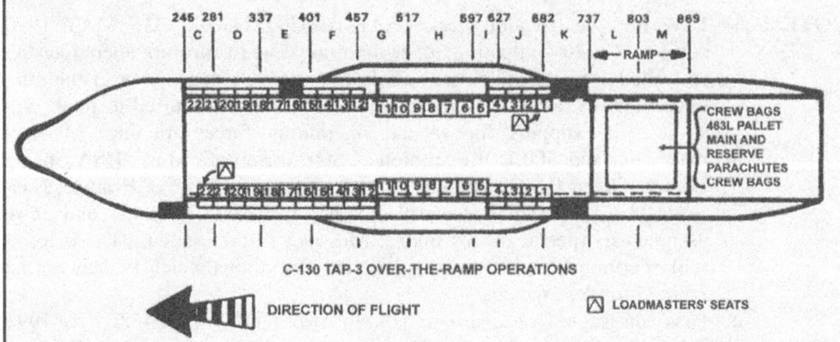

1. 44 TROOP-SEATS - SEAT BELTS ON 20-INCH CENTERS-42 SEATS OFFERED. THIS CONFIGURATION IS FOR TROOP DOOR HALO, TAILGATING, AND WEDGE.
2. EVERYTHING DOWN THE CENTER WILL BE REMOVED. PALLET ON THE RAMP WILL BE THE WEDGE (AS APPLICABLE).
3. 27 EACH MAIN AND RESERVE PARACHUTISTS IN KIT BAGS TO BE RELOCATED FROM AFT PALLET PRIOR TO RIGGING IN-FLIGHT.

Figure 16-4. C-130 configuration for over-the-ramp operations.

b. **Aircraft Operation**. The anchor line cables (only two are used—one on each side) are rigged from the forward outboard anchor line cable attachments to the aft inboard anchor line cable attachments. The anchor line cable stop (a small clevis, padded and taped) must be installed on the anchor line cable 20 inches aft of the center anchor line cable support bracket. The center anchor line cable support brackets at the door are disconnected and secured at the top of the fuselage. The static line retriever cables are tied or taped to the sides of the fuselage aft of the doors to ensure that they remain secured.

c. **Jump Commands**. Jump commands are given in the following sequence:

GET READY.
STAND UP.
HOOK UP.
CHECK STATIC LINES.
CHECK EQUIPMENT.
SOUND OFF FOR EQUIPMENT CHECK.
STAND BY. (Parachutists maintain a reverse bight in the static line.)
GO. (On the command GO, movement onto the ramp is a normal walking or shuffle pace. Exits are made at an angle of about 30 degrees toward the side of the aircraft and are not vigorous. Parachutists place their hands on the ends of their reserve parachute prior to exit.)

d. **Jump Procedures**. If the JM and AJM elect to jump, they will be the number 1 parachutist of each stick. Therefore, the safeties control the flow of the parachutists.

(1) An oral 10-minute warning is given, and the JM begins the jump commands. Prior to the 10-minute warning, the JMs hook up to the anchor line cable, hand the static line to the safety, and announce, SAFETY, CONTROL MY STATIC LINE. The JM then issues the jump commands.

(2) Jumpers hook up to the appropriate anchor line cable with the open gate of the static line snap hook facing toward the skin of the aircraft. They maintain a reverse bight (Figure 16-5).

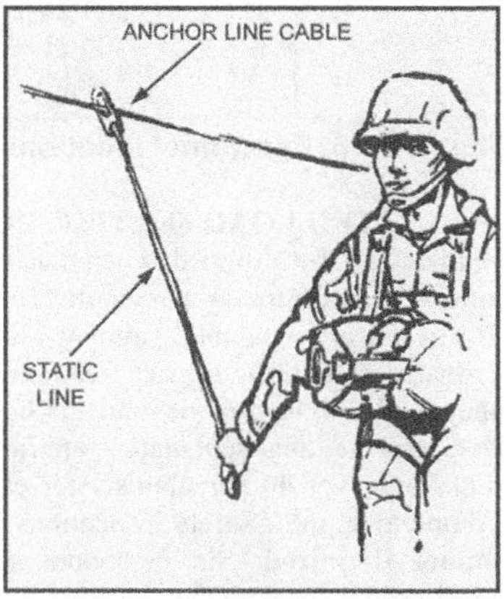

Figure 16-5. Static line grasped with reverse bight.

(3) Door check is not required. If using GMRS, the JM must spot the ground marking from the left side of the ramp.

(4) After giving the command STAND BY, the JM moves to the center edge of the ramp and exits on green light. The safety positions himself immediately behind the hinged portion of the ramp and controls the flow of parachutists.

(5) Each parachutist walks off toward the center of the ramp at an angle away from the anchor line cable (Figure 16-6, page 16-8).

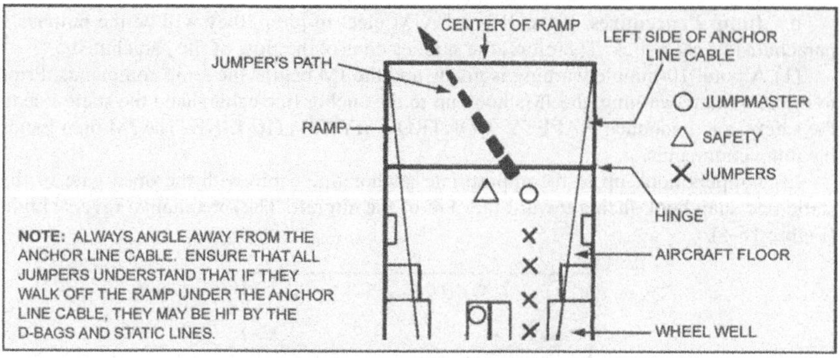

Figure 16-6. Personnel locations.

16-4. COMBAT CONCENTRATED LOAD SEATING CONFIGURATION

A maximum of 80 parachutists can be dropped when these procedures are used for wartime emergency operations. Concentrated parachutist loading is used when not enough C-130 and C-141B aircraft are available. Combat procedures reduce individual space inside the aircraft; crowded conditions restrict freedom of movement and present potential hazards if not supervised. Flight time beyond 1.5 hours may intensify adverse effects on parachutists. Data and seating schematics are included under appropriate aircraft titles. When the sticks are over 40 parachutists for each pass, additional safety personnel may be used to complete required safety inspections within the time warnings.

a. **Supervisory Personnel Required**. Six personnel supervise parachutists and ensure safety measures are followed.

- One JM.
- One AJM.
- Two (jumping) safety personnel.
- Two loadmasters.

b. **Loading Procedures and Seating Arrangement**. The aircraft is configured to provide 19 outboard seats and 21 inboard seats with seat belts installed on 20-inch centers (Figure 16-7).

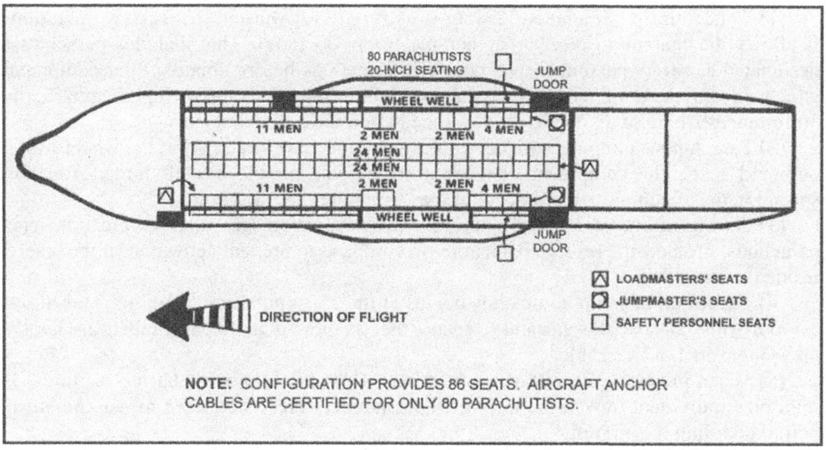

Figure 16-7. C-130 combat concentrated load.

(1) Jumpers are loaded through the aft end loading ramp and seated in two sticks of 40 jumpers each. Before entering the aircraft, all jumpers release their adjustable leg straps so they can lift their equipment over the seats. These safety straps are retied when the jumper is seated with seat belt in place.

(2) Each jumper must be assisted in seating by the AJM or safety personnel. They ensure that jumpers select the correct seat belt and that it is properly fastened.

NOTE: Parachutists sitting in outboard wheel well seats must place their knees forward or aft in the space of the raised seat. The area is not spacious enough to allow inboard or outboard parachutists to face each other.

c. **Jump Procedures**. The jump procedures for the C-130 remain the same with the following exceptions:

(1) The 30-minute time warning replaces the 20-minute time warning, and the 15-minute time warning replaces the 10-minute time warning.

(2) On the command STAND UP, inboard parachutists stand up and then stand on their seats, supporting themselves by grasping the center stanchion seat support rail. They remain in this position until outboard parachutists stand up and raise and secure their seats in the up position. On the command GET DOWN, inboard parachutists step off their seats, then raise and secure them in the up position. Standard jump commands are then resumed.

(3) On the command HOOK UP, no more than 20 parachutists hook up and exit on any single anchor line cable.

(4) When the command GO is given, outboard personnel exit in numerical sequence, followed by inboard personnel.

d. **Safety Procedures**. The standard safety procedures for the C-130 apply, with the following changes:

(1) Concentrated parachutist loading is used only for wartime emergency operations. It allows the maximum number of combat troops to jump. This includes parachutists designated as safety personnel who conduct inspections before donning their equipment, which allows them to move freely about the aircraft. This inspection begins at the 30-minute warning to allow the safeties time to don their parachutes.

(2) Parachutists jumping with equipment containers are seated aft of the wheel well in outboard seats; the equipment containers are stowed on the aircraft ramp. Attaching special items of equipment will occur at the 30-minute time warning.

(3) During movement in the aircraft and execution of jump commands, each parachutist protects the reserve parachute rip cord grip to prevent activation of the reserve inside the aircraft.

(4) Execution of jump commands begins at the 15-minute time warning. This allows for more time to stow all seats and to ensure that no more than 20 parachutists are hooked on a single anchor line cable.

(5) When preparing for airborne operations, JMs should provide latrine facilities. To minimize individual movement during flight, jumpers are encouraged to use the latrine before boarding the aircraft.

16-5. C-130 JUMPMASTER CHECKLIST

The JM inspects the aircraft for the following:

a. **Seats.**

- Enough seats are available for troop load.
- All seats have safety belts.
- Seats are not torn.
- No seat projections are present.
- Seat legs are locked into floor.
- Diagonal leg brace is attached in each set section (if required).
- Seats along the wheel well are removed (if required).

b. **Floor.**

- Nonskid covering is in good condition.
- Floor is clean and safe to walk on.
- All cargo compartment roller conveyors and dual rail system floor sections are removed when jumping the door.
- Loose equipment is lashed and does not interfere with movement or comfort.

c. **Jump Platform.**

- Nonskid surface is present.
- No cracks or bends are present.
- Hinge plate slots are engaged by tie-down studs and screw (with washer) so that the hinge fittings do not slide off the tie-down studs. When the screw and washer are missing, the hinge plate must be safety wired with four turns of stainless steel wire to the tie-down ring bracket.
- Two spring-down lock catches engage the edge of the door.
- Clutch mechanism is present on the spring-down lock catches.

d. **Jump Doors**.

- No excessive grease is on the door tracks.
- No sharp edges or protrusions are on door frames.
- Pip pin in top, forward edge of door frame is present.
- Auxiliary hydraulic ramp pump handle is secured (after takeoff).
- Doors open and close easily. (Doors are operated in flight by aircrew members.)

e. **Air Deflectors**.

- No sharp edges are on trailing edge.
- Deflectors operate electrically.
- No items or trash are stored in wells.

f. **Jump Caution Lights**. (Check all seven sets for operation.)

- Set 1—crew entrance door.
- Sets 2 and 3—top leading edges of right and left doors.
- Sets 4 and 5—trailing edges of right and left doors, waist high.
- Sets 6 and 7—right and left anchor cable aft supports.

g. **Anchor Cable System**.

(1) ***Forward Support Beam***.

- Four U-bolts, with self-locking nuts or nuts with cotter pins, are attached.
- Anchor cables are attached to first and second U-bolts right and left of center line for personnel jump.
- Forward latch assembly is in the locked position and secured with locking pins.

(2) ***Anchor Cable.***

- No breaks (within acceptable tolerances of TO 1C-130A-9), frays, or kinks exist.
- Cable is clean and free of rust.

(3) ***Anchor Cable Intermediate Center Support.***

- Cables run through slots after ramp is closed.
- Quick-release retaining pins are present

(4) ***Anchor Cable Aft Support.***

- Aft latch assembly is closed.
- U-bolts, nuts, and safety pins are present.
- Support anchor bolts, nuts, and safety pins are present.

(5) ***Static Line Retrievers.***

- Motor is operational.
- Retriever cables are not broken (within acceptable tolerances of TO 1C-130A-9), frayed, kinked, dirty, or rusty.
- Spool clamp and shackle are attached forward of intermediate cable support and are tied to support with two turns of 1/4-inch cotton webbing.
- Retriever cables are secured with two turns of 1/4-inch cotton webbing to litter brackets at station 627. If cable clips are installed on the wheel well, the tie at station 627 is not used. (Retriever cable must be at least 4 inches above the anchor line cable.)

h. **Emergency Equipment**.
- First aid kits (4) are present.
- Fire extinguishers (3) are present.
- CGU1-B cargo tie-down straps (2) (for retrieval of towed parachutists) are present.
- Alarm system is operational.
- Emergency exits are operational and accessible.
- Sufficient emergency parachutes are available.

i. **Miscellaneous**.
- Lighting system is operational.
- Airsickness bags are available.
- Comfort facilities are available.
- JM kit bag (extra equipment) is onboard.

Section II. C-141B STARLIFTER

The C-141B (Figure 16-8) is a swept-wing, long-range transport powered by four turbofan jet engines and equipped for in-air refueling. The aircraft can be configured in different peacetime troop-carrying modes for airborne operations. Parachutists jump from the two aft doors of the aircraft.

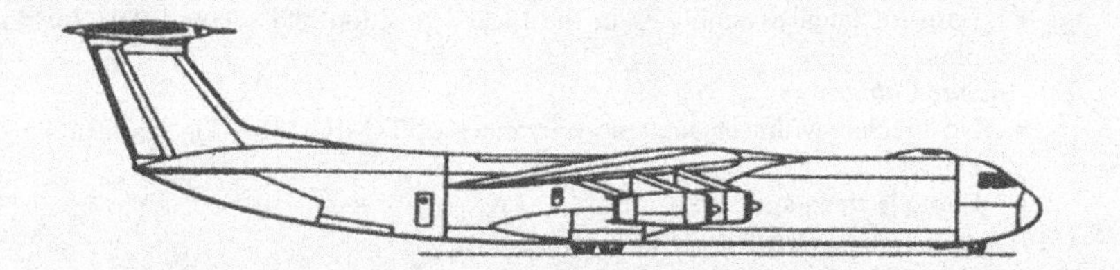

TACTICAL TRAINING	
CONFIGURATION	**PARACHUTIST LOAD**
TAP 1 - WITHOUT COMFORT PALLET	152
TAP 2 - IN-FLIGHT RIGGING WITH COMFORT PALLET	130
COMBAT OPERATIONS	
CONFIGURATION	**PARACHUTIST LOAD**
CONCENTRATED LOADING	180
NOTE: THE C-141B IS NOT USED FOR PERSONNEL JUMPING (STATIC LINE) OVER THE RAMP.	

Figure 16-8. C-141B Starlifter.

16-6. SEATING CONFIGURATION WITHOUT COMFORT PALLET

The seating configuration for the C-141B is as follows:

a. **Peacetime Training (TAP 1)**. In this configuration, the C-141B seats 156 personnel: 152 parachutists, 2 static safety personnel, and 2 airdrop qualified loadmasters (Figure 16-9). The maximum number of personnel for each anchor line cable is 45.

b. **Supervisory Personnel Required**. Six personnel supervise parachutists and ensure safety measures are followed.

- One JM.
- One AJM.
- Two static safety personnel.
- Two airdrop qualified loadmasters.

NOTE: USMC and USN utilize one door, the other door remains closed and unmanned. Therefore, only one JM and one safety are used. These units do not fall under AR 350-2.

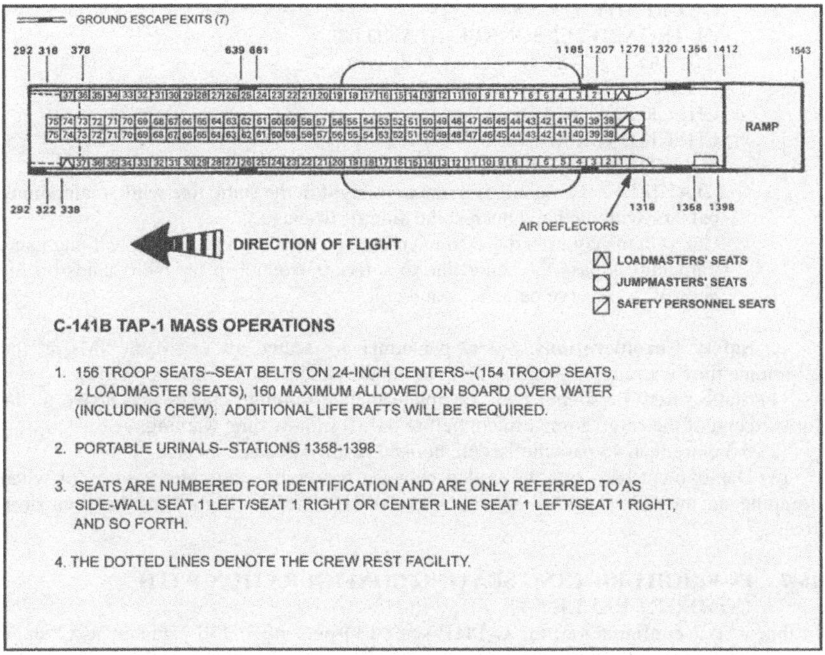

Figure 16-9. C-141B seating configuration without comfort pallet.

c. **Time Warnings.** Time warnings include the following:

(1) The 20-minute time warning may be increased to 30 minutes to provide enough time for static safety personnel to complete safety checks and to rig equipment containers.

(2) The 10-minute time warning may be increased to 15 minutes to allow time for parachutists to release and stow troop seats and for static safety personnel to complete safety checks before the 1-minute time warning. Prior to the 10-minute time warning, the JMs hook up to the inboard anchor line cable (or anchor line cable if only one is available), hand the static line to the safety, and announce, SAFETY, CONTROL MY STATIC LINE. The JM then issues the jump commands. If the aircraft is configured with only one anchor line cable, the JMs hook up to the one cable.

d. **Jump Commands.** The loadmaster gives the JM an oral 10-minute time warning. Emphasis is on using a public address system to give jump commands, since parachutists in the forward end of the cargo compartment may not be able to see JM hand-and-arm signals.

GET READY.
OUTBOARD PERSONNEL, STAND UP.
INBOARD PERSONNEL, STAND UP.
HOOK UP.
CHECK STATIC LINES.
CHECK EQUIPMENT.
SOUND OFF FOR EQUIPMENT CHECK.
STAND BY. (Parachutists maintain a bight in the static line while maintaining balance with the hand nearest the aircraft fuselage.)
GO. (On the command GO, movement into the door is a normal walking pace. Parachutists pass the static line to safety personnel, place their hands on the ends of the reserve parachute, and exit.)

e. **Safety Considerations.** Safety personnel are seated aft to aid the JMs, at the 20-minute time warning, in positioning door bundles and performing other duties.

(1) Static safety personnel must complete their 20-minute checks and arrive at the forward end of the cargo compartment before the 10-minute time warning.

(2) No more than 45 parachutists are hooked to any one anchor cable.

(3) Deployment bags may be trailed between passes on single drop zones, or when dropping on multiple drop zones with less than 10 minutes flight time between drop zones.

16-7. IN-FLIGHT RIGGING SEATING CONFIGURATION WITH COMFORT PALLET

In this TAP 2 configuration, the C-141B seats 134 personnel: 130 parachutists, 2 static safety personnel, and 2 loadmasters. Floor space is provided forward, midway, and aft for stowage of parachute assemblies. Seats are on a 24-inch center (Figure 16-10).

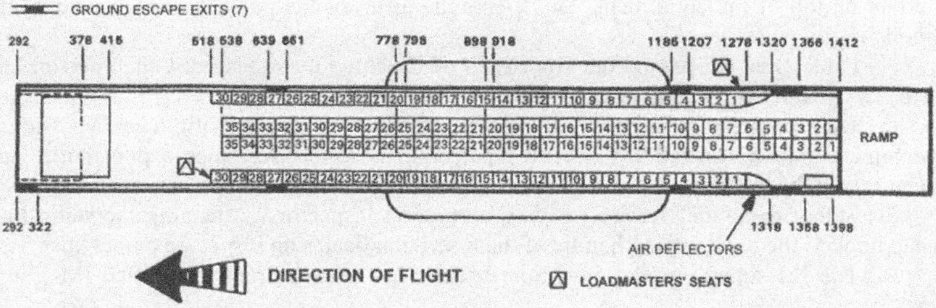

C-141B TAP-2 IN-FLIGHT RIG WITH COMFORT PALLET

1. 130 TROOP SEATS--SEAT BELTS ON 24-INCH CENTER (128 TROOP SEATS, 2 LOADMASTER SEATS, 160 MAXIMUM ALLOWED ON BOARD OVER WATER [INCLUDING CREW]). ADDITIONAL LIFE RAFTS WILL BE REQUIRED.

2. SEATS ARE NUMBERED FOR IDENTIFICATION ONLY AND ARE REFERRED TO AS SIDE-WALL SEAT 1 LEFT/SEAT 1 RIGHT OR CENTER LINE SEAT 1 LEFT/SEAT 1 RIGHT.

3. THE DOTTED LINES DENOTE THE CREW REST FACILITY.

4. FOR IN-FLIGHT RIGGING, DELETE CENTER LINE SEATS FORWARD OF FS 538, BETWEEN FS 778-898 AND AFT OF FS 1318. LEFT AND RIGHT OUTBOARD SEATS WILL BE RAISED FORWARD OF FS 518 AND BETWEEN FS 798-918. 102 TROOP SEATS--SEAT BELTS ON 24-INCH CENTERS--100 TROOP SEATS OFFERED (39 SEATS IN FORWARD SECTION AND 61 SEATS IN AFT SECTION).

5. IF REQUIRED, SEAT BELTS MAY BE INSTALLED ON 20-INCH CENTERS FOR IN-FLIGHT RIGGING ONLY. DELETE CENTER LINE SEATS FORWARD OF FS 538, BETWEEN FS 778-898 AND AFT OF FS 1318. LEFT AND RIGHT OUTBOARD SEATS WILL BE RAISED FORWARD OF FS 518 AND BETWEEN 798-918. 130 TROOP SEATS--SEAT BELTS ON 20-INCH CENTERS--128 SEATS OFFERED (51 SEATS IN FORWARD SECTION AND 77 SEATS IN AFT SECTION).

Figure 16-10. C-141B seating configuration with comfort pallet.

a. **Supervisory Personnel Required**. Fifteen personnel supervise parachutists and ensure safety measures are followed.

- One JM.
- Six AJM.
- Six safety personnel (two static).
- Two loadmasters.

b. **Storage of Equipment**. Parachute assemblies (133) are placed in aviator kit bags and stowed and secured in three locations on the cargo floor. There are 130 parachutists seated in two sticks of 65 each on the left and right.

(1) Door bundles are stowed on the cargo ramp.

(2) Individual equipment, web gear, ballistic helmets, and ALICE packs are stowed under the troop seats; weapon containers are stowed behind the seats; and individual equipment containers are stowed in the aisles or with the parachutes.

(3) One JM kit bag is stowed at each of the six rigging stations (two forward, two midway, and two aft).

c. **Buddy Rigging**. The JM initiates in-flight buddy rigging 2 hours before the 20- or 30-minute time warning. Parachutists unzip the seats, place their equipment on the seats, and sit on top of the equipment. This clears the aisle for the parachutists to stand while buddy rigging.

(1) Parachutes are passed out from each of the three locations until each person has one. Then buddy rigging begins.

(2) Each parachutist must know exactly who he will rig up with. The JMs, safety personnel, and designated JM qualified parachutists supervise. Once a parachutist has been rigged, he sits down and waits to be inspected by one of the JMs.

(3) Safety personnel serve as roving correction inspectors. Attaching the static line snap hook to the top carrying handle of the reserve indicates an inspected parachutist.

(4) The JM supervises the operation. If there are other currently qualified JMs, they may be used.

(5) Once all the parachutists have been inspected, the JMs rig up, and safety personnel inspect them.

d. **Station Rigging**. Station rigging involves the following:

(1) Six AJMs and six safety personnel (two static and four jumping safety personnel) are required to man the six rigging stations.

(2) Parachutists are divided into three segments and assigned to rigging stations.

(3) The JM initiates in-flight rigging 2 hours before the 20- or 30-minute time warning.

(4) Before initiation of in-flight rigging, jumpers are instructed to unzip the troop seats, place their equipment on the seats, and sit on it. Placing equipment on the seats clears the aisles for jumpers moving to and from the rigging stations.

(5) AJMs and safety personnel at each rigging station stow their seats to provide more rigging space.

(6) Starting with two parachutists from the end of the divided sticks, each parachutist moves to a designated rigging station and is rigged by safety personnel. When completely rigged, the parachutist moves to the AJM (stationed nearby) for inspection. The AJM attaches the static line snap hook to the top carrying handle of the reserve parachute when he reaches that portion of the inspection sequence.

(7) Only the JM or AJM removes the static line snap hook from the reserve (before the command HOOK UP) since this indicates that the parachutist has received the JM's inspection.

16-8. COMBAT CONCENTRATED LOAD SEATING CONFIGURATION

In this configuration, the C-141B seats 182 personnel: 180 parachutists and 2 loadmasters (Figure 16-11). Personnel are seated on the side-facing seats on 20-inch centers. Parachutists (180) are seated in two sticks of 90 each on the left and right sides. If 45 parachutists cannot be seated on the outboard seats, the overflow may be seated inboard aft on available center seats. The JM must ensure that only 45 parachutists hook up to the outboard/inboard anchor cables.

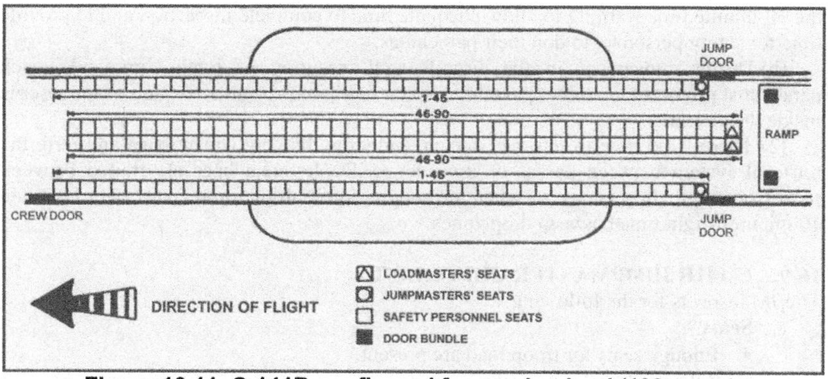

Figure 16-11. C-141B configured for combat load (182 seats).

NOTE: Concentrated parachutist loading is used in wartime emergency operations when not enough C-130 and C-141B aircraft are available. These procedures reduce individual space inside the aircraft. Crowded conditions restrict freedom of movement and present a potential safety hazard if jumpers are not supervised. Flight time beyond 1.5 hours may intensify adverse effects on parachutists. Data and seating schematics are included under appropriate aircraft titles. Additional safety personnel may be used (if the sticks have over 40 personnel for each pass) to complete the required safety inspection within the time warnings.

a. **Supervisory Personnel Required**. Six personnel supervise parachutists and ensure safety measures are followed.

- One JM.
- One AJM.
- Two (jumping) safety personnel.
- Two airdrop qualified loadmasters.

b. **Jump Procedures**. Jump procedures involve the following:

(1) Jump procedures for the C-141B remain the same with the following exceptions: the 30-minute warning replaces the 20-minute time warning, and the 15-minute time warning replaces the 10-minute time warning.

(2) Jump commands are echoed by jumpers since parachutists in the forward end of the cargo compartment cannot see the JM's arm-and-hand signals.

c. **Safety Considerations**. At the 30-minute time warning, safety personnel are seated aft to aid the JMs in positioning door bundles and in performing other duties. The JMs ensure that no more than 45 parachutists are hooked to any one anchor line cable.

(1) The standard safety procedures for the C-141B apply with the following changes:

(a) Concentrated parachutist loading is used only for wartime emergency operations and allows the maximum number of combat troops to jump. This includes parachutists designated as safety personnel who conduct safety inspections before donning their equipment, which allows them to move freely about the aircraft. This inspection begins at the 30-minute time warning to allow adequate time to complete inspection and to provide time for safety personnel to don their parachutes.

(b) During movement in the aircraft and execution of jump commands, each parachutist protects the reserve parachute rip cord grip to prevent activation of the reserve inside the aircraft.

(2) There is no time to retrieve, detach, and stow the deployment bags and rerig the retrieval system between passes or drop zones. Deployment bags are trailed between passes on single drop zones, or when airdropping on multiple drop zones with less than 10 minutes flight time between drop zones.

16-9. C-141B JUMPMASTER CHECKLIST

The JM inspects for the following:

a. **Seats**.
- Enough seats for troop load are present.
- All seats have safety belts.
- Seat backs are secured.
- Seats are not torn.
- No projections exist through seats.
- Legs are locked into floor.

b. **Floor**.
- Nonskid covering is in good condition.
- Floor is clean and safe to walk on.
- Roller conveyors are stowed.
- Loose equipment is secured in the cargo ramp area and does not interfere with troops.

c. **Jump Platforms**.
- Nonskid surface is present.
- No cracks or bends exist.
- Studs are locked in seat track receptacles.
- Tie-down fitting is locked.
- All bolts and nuts are present.
- Platforms swing in and out easily.
- The two spring-down lock catches engage the edge of the door.

d. **Jump Doors**.
- No excessive grease on the door tracks exists.
- No sharp or protruding edges exist on door frames or on loading strut door (aft and down from jump door).
- Doors open and close easily.
- Door lever catches are operational.

e. **Air Deflectors**.
- No sharp edges present.
- Deflectors operate electrically.

f. **Jump Caution Lights**. (Check seven sets for operation.)
- Set 1—crew entrance door.
- Sets 2 and 3—top leading edges of right and left doors.

- Sets 4 and 5—trailing edges of right and left doors, waist high.
- Sets 6 and 7—right and left anchor cable aft supports.

g. **Anchor Cable System.**

(1) ***Forward Support Beam.***

- Bolts and nuts are present and tight.
- Anchor cables are attached to first and third anchor points right and left of enter line for personnel jumps.
- Cable, bolts, nuts, and safety wire are present.
- Turnbuckle is secured with safety wire.

(2) ***Anchor Cable.***

- No breaks (within acceptable tolerances of TO 1C-141B-9), frays, or kinks exist.
- Cable is clean and free of rust.
- Cable swage is present.

(3) ***Anchor Cable Intermediate Support.***

- Cables run through slots after ramp is closed.
- Quick-release retaining pip pins are present.

(4) ***Static Line Retrievers.***

- Motor is operational.
- Retriever spools are secured forward of intermediate cable support and tied to supports with one turn of double 1/4-inch cotton webbing.
- Retriever cables are not broken (within acceptable tolerances of TO 1C-141B-1), frayed, or kinked.
- Retriever cables are secured in spring clips.
- Retriever bars are available (one for each door).

NOTE: Retriever cable must be at least 4 inches above the anchor line cable.

h. **Emergency Equipment.**

- Public address system is operational.
- First aid kits (4) are present.
- Fire extinguishers (3) are present.
- Alarm system is operational.
- Emergency exits are operational and accessible.
- Sufficient emergency parachutes are available.

i. **Miscellaneous.**

- Lighting system is operational.
- Airsickness bags are available.
- Comfort facilities are present.
- JM kit bag (extra equipment) is onboard.

Section III. C-5 A/B/C GALAXY

The C-5 A/B/C Galaxy is a swept-wing, long-range transport powered by four turbofan jet engines and equipped for in-flight refueling. The aircraft can be configured in different peacetime troop-carrying modes for airborne operations. Parachutists jump the two aft doors of the aircraft.

16-10. SEATING CONFIGURATION WITHOUT COMFORT PALLET

The configuration for the C-5 A/B/C for peacetime training is TAP-1. This configuration offers seats for 73 parachutists in the troop compartment and provisions for personnel airdrop as follows: 71 jumpers and 2 static safety personnel. The maximum number of personnel for each anchor line cable is 40.

16-11. IN-FLIGHT RIGGING SEATING CONFIGURATION WITH COMFORT PALLET

This configuration offers a partial palletized seat kit (eight seat pallets and one comfort pallet) for 78 jumpers in the cargo compartment and provisions for personnel airdrop. It does not provide ditching exits or sufficient ground emergency exits. Therefore, this configuration should be carefully evaluated when considered for use. This configuration should be used only during over-land missions.

16-12. JOINT PREFLIGHT INSPECTION

The purpose of the joint preflight inspection (JPI) is to verify the readiness of the aircraft for the conduct of aircraft missions and to take actions necessary to achieve this readiness.

a. **Supervisory Personnel Required**. To ensure command and control, a JM, AJM, and two safety personnel are used for C-5 A/B/C airdrops.

b. **Jumpmaster/Aircraft Commander/Loadmaster Coordination**. It is the responsibility of the JM to inform the aircraft commander and loadmaster of the exact time sequence of prejump procedures. Following preliminary orientation and before loading personnel onboard the aircraft, the JM and loadmaster make a joint inspection of the aircraft.

c. **Exterior Inspection**. An external inspection of the aircraft is made to detect hazards to the airdrop of personnel. Particular attention is directed to the rear of the aft paratroop doors. Any protruding objects and sharp edges must be removed or padded and taped.

d. **Interior Inspection**. This inspection is conducted to detect and correct any interior safety hazards.

(1) Any sharp edge or protrusion is securely taped and padded, as required.

(2) All equipment in the cargo compartment is securely stowed and lashed.

(3) The floor is clean and free of lubricants; no obstructions are on the walkway or along the paratroop exit route (outboard area between the safety fence and fuselage).

(4) Anchor line cables are installed and under correct tension. Maximum deflection of the cable at midpoint (FS 1465) must produce a minimum cable height of 73 inches above the cargo compartment floor.

(5) A seat and seat belt are available in the troop compartment for each parachutist.

(6) The retrieval system is installed in the aircraft with the winch cable retained in clips and free of the anchor line cable. The retriever spool is secured to the aft support frame with one turn of double 80-pound, 1/4-inch cotton webbing. A phenolic (plastic) block is installed on the leading edge of the paratroop doors at the retriever bar level.

(7) Troop compartment lavatory units are installed and operational.

(8) Air deflector systems are installed.

(9) Jump platforms are installed.

(10) Jump signal lights are operational.

(11) Cargo compartment and troop compartment lights are operational.

(12) The public address intercomm system is operational.

NOTE: To reduce confusion in loading, the safety fence can be numbered at this time.

e. **Jumpmaster Preload Inspection of Parachutists**. The JM/AJM inspects each parachutist, parachute, and parachutist's equipment prior to loading the aircraft. The JM/AJM inspects—

(1) Parachutist's helmet.

(2) Parachutist's ID tags and ID card.

(3) Rigging of parachutist's equipment:

- ALICE pack with H-harness, AIRPAC, and PIE/R2, or CWIE containers.
- Dragon missile jump pack, Stinger missile jump pack, or AT4 jump pack.
- M1950 weapons case.
- Lowering lines.

f. **Parachute Stowage**. Personnel stow parachutes in aviator kit bags, secure all equipment, and move to the aircraft.

g. **Aircraft Loading Sequence**. Loading is determined by the USAF loadmaster.

h. **Aircraft Loading Through Forward Ramp**. When loading through the forward ramp, parachutists enter in normal sequence (1 through 36 right side, 1 through 37 left side). When loading through the aft right door, parachutists enter in reverse sequence (36 through 1 right side, 37 through 1 left side).

16-13. PERSONNEL AND EQUIPMENT CONFIGURATION

Personnel seating configuration and equipment stowage for the C-5B Galaxy are as follows:

a. **Seating Configuration**. Parachutists are assembled into two sticks: 37 parachutists for the left door and 36 parachutists for the right door. Each parachutist is assigned a number in the stick. This number is the parachutist's rigging station and seat number. Each parachutist is issued a main and reserve parachute. Each parachutist must inspect his own parachute for safety wires and for fitting of the parachute harness.

b. **Equipment Configuration**. Personnel face inboard, cover on the nylon net assembly equipment stowage sections, and place their equipment on the floor of the cargo compartment. Equipment is stowed on the safety fence under the parachutist's assigned stick number. To facilitate equipment stowage on the safety fence, the two top retaining straps of the nylon net assembly must be fully extended. The individual parachutist's equipment (load-bearing equipment, weapons, and main and reserve parachutes) is stowed under the retaining straps.

16-14. MOVEMENT TO THE TROOP COMPARTMENT

On jumpmaster order, parachutists begin movement to the troop compartment as follows:

a. **Left Stick**. The left stick faces aft, moves in stick order, and ascends the aft stair/ladder to the troop compartment. Upon reaching the troop compartment, the stick leader faces the forward end of the aircraft and moves to the forward row of seats on the left side. The left stick is seated row by row. Seats are occupied in the following forward to aft order: outboard, center, and aisle.

b. **Right Stick**. As the left stick begins to ascend the aft stair/ladder, the right stick faces aft, moves past the end section of the safety fence, crosses the airdrop system guide rails/rollers to the left side of the cargo compartment, and moves to the base of the stair/ladder. Parachutists must exercise caution to avoid tripping over the rails and rollers of the airdrop system. The right stick ascends the stair/ladder into the troop compartment and is seated on the right side, row by row, forward to aft, outboard to inboard in the same manner as the left stick.

NOTE: The troop compartment contains 75 seats: 73 seats are for parachutists and the remaining 2 are reserved for USAF loadmasters.

16-15. LOADMASTER BRIEFING

As soon as all parachutists are seated, the loadmaster uses the public address/intercomm system to brief the parachutists on aircraft safety, emergency procedures, and comfort facilities.

16-16. MOVEMENT TO THE CARGO COMPARTMENT FOR IN-FLIGHT RIGGING PROCEDURES

Movement to the cargo compartment for in-flight rigging proceeds as follows:

a. **Movement of Left and Right Sticks**.

(1) The loadmaster alerts the jumpmaster 1 hour 20 minutes prior to drop time. At that time, the JM moves the two sticks to the cargo compartment, right stick followed by left stick, both sticks in reverse stick order.

(2) Each stick descends the stair/ladder to the floor of the cargo compartment and moves to the equipment stowed on the safety fence. Routes used are the same as those used in moving to the troop compartment. The right stick must exercise care in crossing the airdrop system rails and rollers.

NOTE: The 1-hour 20-minute time warning can be modified to a shorter time depending on the situation and number of parachutists.

> **CAUTION**
>
> Each parachutist must descend to the cargo compartment facing the troop ladder with both hands on the handrail, using caution at each step.

b. **In-Flight Rigging Procedures**.

(1) Individuals face inboard and cover on their stowed equipment.

(2) Using the buddy system, half of the parachutists in each stick don their parachutes and equipment. After the first group have completed donning their equipment, the second half don their parachutes and equipment.

(3) JMPI begins as soon as rigging is complete. The JM and AJM inspect each parachutist and correct any errors. As the JM/AJM completes inspection of each parachutist, the rigged parachutists are seated on the walkway. Safeties assist the JM/AJM during JMPI.

> **CAUTION**
> When on, lights located under the walkway can get hot enough to melt nylon.

16-17. JUMP COMMANDS

The following eight jump commands are used:

GET READY.
STAND UP.
HOOK UP.
CHECK STATIC LINES.
CHECK EQUIPMENT.
SOUND OFF FOR EQUIPMENT CHECK.
STAND BY.
GO.

16-18. JUMP PROCEDURES

The jump procedures for the C-5 A/B/C are as follows:

a. **Jumpmaster and Assistant Jumpmasters**. The JM and AJM hook up to the anchor line cable and stand adjacent to the rear of their respective aft personnel door. The JM and AJM face their sticks to give jump commands. The JM initiates the jump commands, which are then relayed by the AJM.

b. **Parachutists**. All parachutists are hooked up to the paratroop anchor line cable. Personnel must close up tightly, reserve to backpack, to the aft end of the aircraft to permit sufficient room for all parachutists to hook up to the anchor line cable. The number 1 parachutist in each stick does not move aft of the air deflector while it is positioned inside the aircraft. When parachutists are rigged with combat equipment, it may become necessary to stagger the parachutists to have enough room for all parachutists to hook up.

c. **Commands**. The following commands are used:

(1) ***SOUND OFF FOR EQUIPMENT CHECK.*** At this command, the two safety personnel move from forward to aft, checking each parachutist's static line. The safety personnel report to the JM or AJM at the completion of the check, hook up, and then position themselves by the aft personnel door to control the static lines of exiting

parachutists. The two safety personnel must be JM qualified and may jump at the end of the stick. The JM may jump from either aft personnel door. He should jump from the door from which he can best observe checkpoints and the drop zone.

(2) ***STAND BY.*** At this command, the number 1 parachutist moves toward the door and assumes a normal number 2 man position.

(3) ***GO.*** At this command, parachutists' movement into the door will be at a normal walking pace, without shuffling. Exits from the troop doors are not vigorous and are made at an angle of about 70 to 80 degrees to the tail of the aircraft. Each parachutist passes his static line to the safety, places both hands on the ends of his reserve parachute, and exits the aircraft.

16-19. TIME WARNINGS

The following actions occur at each time warning:

a. **1-Hour 20-Minute Time Warning**. Parachutists are moved from troop compartment to cargo compartment under the supervision of the JM, AJM, and safeties. In-flight rigging and JMPI begins.

b. **20-Minute Time Warning**. The JM makes necessary checks of personnel and equipment. Special items of equipment are attached to the parachutists, and door bundles are moved to the vicinity of the personnel doors.

c. **6-Minute Time Warning**. JMPI of all parachutists is complete. The loadmaster gives the JM a verbal and visual 6-minute warning, and the JM begins his jump commands.

d. **Slow-Down Warning**. Approximately 3 minutes from drop time, the jump commands are completed. Personnel doors are opened, air deflectors deployed, and jump platforms extended and locked.

e. **1-Minute Time Warning**. Troops are alerted. The JM and AJM make safety checks from personnel doors. The AJM informs the JM that the AJM's side is clear and that it is safe to jump.

f. **10-Second Time Advisory**. The loadmaster gives the JM a visual 10-second time advisory. At this time, the JM gives the command STAND BY, and the first parachutist assumes a normal number 2 position. The remainder of the stick moves aft to close up the stick.

g. **Green Light, GO**. At the green light signal, the JM taps out the first man. If T-10-series type parachutes are issued, the AJM taps out the first man in the opposite door.

16-20. SAFETY PRECAUTIONS

Safety precautions for the C-5 A/B/C are as follows:

a. **Platforms, Air Deflectors, Aft Cargo Door**. The JM ensures that jump platforms are extended and locked and that the air deflectors are fully deployed. The aircraft will **not** be jumped unless these conditions are met. These conditions are mandatory for each aft personnel door that is to be used.

b. **Door Bundles**. When personnel are to follow door bundles, the door bundle static line is outfitted with three drogue devices, a breakaway static line, or a T-10-series parachute.

c. **Movement into the Door**. Parachutists must exercise caution as they move to the door to avoid becoming entangled with the static lines of preceding parachutists. This precautionary action may slow movement into and out of the door.

d. **MC1-Series Type Parachutes**. Only single door exits will be executed when MC1-series type parachutes are used. However, both doors must be open to conduct outside air safety checks.

e. **Ramp Exits**. Static line parachutists are not authorized to make aft ramp exits of the C-5 A/B/C aircraft.

16-21. C-5 A/B/C JUMPMASTER CHECKLIST

The jumpmaster inspects the following at the departure airfield:

a. **Seats**.
- Enough seats are available for troop load.
- All seats have safety belts.

b. **Safety Fence**.
- Safety fence is installed.
- Sufficient equipment restraint nets are available for troop load.

c. **Floor**.
- Nonskid surface covering is in good condition.
- Floor is clean and safe to walk on.
- Roller conveyors are stored.
- Loose equipment is secured in the cargo ramp area and does not interfere with troops.

d. **Jump Platforms**.
- Nonskid surface covering is present and in good condition.
- No cracks or bends are present.
- Studs are locked in seat track receptacles.
- Tie-down fitting is locked.
- All bolts and nuts are present.
- Platforms swing in and out easily.

e. **Jump Doors**.
- No excessive grease is on door tracks.
- No sharp or protruding edges are on door frames or on loading strut door (aft and down from jump door).
- Doors open and close easily.
- Door level catches operate.
- Phenolic (plastic) block is installed on leading edge at retriever bar height.

f. **Air Deflectors**.
- No sharp edges are present.
- Deflectors operate electrically.

g. **Jump Lights**. (Check five sets for operation.)
- Set 1—crew entrance door.
- Sets 2 and 3—leading edges, top of right and left doors.
- Sets 4 and 5—trailing edges, right and left doors, waist high.

h. **Anchor Line Cable System**.

(1) ***Forward Support Beam.***

- Bolts and nuts are present and tight.
- Cable, bolts, nuts, and safety wire are present.
- Turnbuckle is secured with safety wire or tied with 1/4-inch cotton webbing.

(2) ***Anchor Line Cable.***

- Cable has no breaks.
- Cable has no frays.
- Cable has no kinks.
- Cable is clean and free of rust.
- Cable swage is present.
- Cable is at least 73 inches above floor.

(3) ***Static Line Retrievers.***

- Motor is operational.
- Retriever spools are secured and tied to supports with one turn of double 1/4-inch cotton webbing.
- Retriever cables are not broken, frayed, or kinked.
- Retriever cables are secured in spring clips.
- Retriever bar is available.

NOTE: Retriever cable must be in position, not less than 4 inches above the anchor line cable.

i. **Emergency Equipment**.

- Public address system is operational (upper deck only).
- First aid kits (4) are present.
- Fire extinguishers (3) are present.
- Alarm system is operational.
- Emergency exits are operational and accessible.
- Sufficient emergency parachutes are available.

j. **Miscellaneous**.

- Lighting system is operational.
- Airsickness bags are available.
- Comfort facilities are available.
- JM kit (extra equipment) is onboard.

k. **Aircraft Slowdown Warning** (about 3 minutes).

- Doors are opened and locked in place.
- Air deflectors are extended.
- Jump platforms are locked in place.

l. **Loadmaster/Jumpmaster Safety Checks**.

(1) ***Equipment.***

- Door bundles employ 15-foot-long static lines with three drogue devices, breakaway static lines, or T-10-series cargo parachute.

(2) ***Aircraft.***

- Jump platform is secure and will sustain parachutist's weight.

NOTE: The C-5 A/B/C aircraft cannot be used for aft ramp static line jumping. HALO personnel may conduct aft ramp jumps from this aircraft. Due to the number of jumpers that can be hooked up to one anchor line cable, the C-5 A/B/C aircraft have not been certified for combat concentrated loads. Any deviation to the listed configurations requires authorization from HQ, AMC/DOT/DOV.

Section IV. C-17A GLOBEMASTER III

The C-17A (Globemaster III) is a swept-wing, long-range transport aircraft powered by four turbofan engines capable of airlifting large payloads over intercontinental ranges without refueling. An in-flight refueling capability increases the deployment range. The cargo compartment is designed to permit safe and efficient operation with one loadmaster for any mission. The aircraft accommodates outsize or oversize cargo, tactical vehicles, container loads (which can be configured to conduct a combination of airland, personnel and equipment airdrop), and MEDEVAC operations. The C-17A can be configured with or without a comfort pallet. The jump platforms and air deflectors are an integral part of the aircraft fuselage. A dedicated antenna system enables use of organic Army tactical satellite (TACSAT) communication systems en route.

16-22. SEATING CONFIGURATION

The seating configuration for the C-17A (Globemaster III) is as follows:

a. 102 parachutists, includes seating for JM, AJM, and static safety personnel.

b. Two static safety personnel seated in crew compartment.

c. One loadmaster seated in forward loadmaster station.

d. Two door bundles (one per door) placed on the aircraft floor aft of the inboard seats.

16-23. SUPERVISORY PERSONNEL REQUIRED

The following personnel are required for all C-17A personnel airdrop operations:

- One JM.
- One AJM.
- Two JM qualified and current static safety personnel.
- One loadmaster.

16-24. TIME WARNINGS

Time warnings include the following:

a. **2-Hour 20-Minute Time Warning** (used only on in-flight rigging missions).

b. **20-Minute Time Warning**. At the 20-minute time warning, all in-flight rigging is complete. All jumpers are alert with ballistic helmets fastened. Door bundles are positioned in the vicinity of the jump doors, hooked up to the outboard anchor line cable and inspected. Special items of equipment are attached to their respective jumpers and inspected.

c. **10-Minute Time Warning**. The aircrew turns on the red jump caution lights, then notifies the JM and all that the inspections are complete. The JMs hook up to the inboard anchor line cables and begin issuing jump commands. If both doors are used on one pass, then both JMs issue the commands at the same time.

d. **6-Minute Time Warning and Aircraft Slowdown**. The aircrew completes their 6-minute slowdown check: aircraft deck is set between 6 and 7 degrees, aircraft slows to a drop speed of 130 knots (plus or minus 3 knots indicated airspeed), air deflectors are deployed, and troop doors opened. The loadmaster takes a position between both jump doors and gives control of the doors to the JMs. The JMs perform troop door safety checks, outside air safety checks, and spot for checkpoints en route to the drop zone.

e. **1-Minute Time Warning**. The JM issues the time warning to the parachutists with his lead hand and then spots for checkpoints.

f. **30-Second Time Warning**. At the 30-second reference point, the JMs make a final outside safety check, return inside the aircraft, gain eye-to-eye contact with each other, and give each other a "thumbs up" indicating that all conditions outside the aircraft are safe for the jump. The amber jump caution light is turned on. The JMs face the jumpers and issue the eighth jump command, STAND BY.

g. **Green Light**. The green light is the final time warning on an Air Force aircraft. When it comes on, the JM issues the ninth jump command, GO. The JM's first jumper exits the aircraft. The AJM taps his first jumper out a half-second after the JM's number 1 jumper exits.

16-25. JUMP COMMANDS

Jump commands are given in the following sequence:

a. **GET READY**. Jumpers undo their seatbelts.

b. **OUTBOARD PERSONNEL, STAND UP**. All outboard jumpers stand, place their seatbelts in their seats, and push in the seats to retract them to the raised position. (This allows the jumpers to use the handhold straps on the cargo wall and assists in keeping the seatbelts off the floor, reducing tripping hazards.)

c. **INBOARD PERSONNEL, STAND UP**. Jumpers stand up. (The inboard seats do not fold.) Safety moves forward in the aircraft and inspects each jumper's quick release in the waistband, quick-release snap on the M1950 weapons case, ejector snap of the hook-pile tape lowering line, and the adjustable leg straps of the HSPR.

d. **HOOK UP**. During mass tactical operations, a maximum of 27 jumpers on the outboard anchor line cables and 24 jumpers on the inboard anchor line cables are allowed.

e. **CHECK STATIC LINES**. The last two jumpers turn toward the skin of the aircraft. The second to the last jumper checks the last jumper's static line. Each jumper checks the static line of the jumper to his front.

f. **CHECK EQUIPMENT**. The JMs immediately check their own equipment.

g. **SOUND OFF FOR EQUIPMENT CHECK**. At this command, both safeties move from the forward portion of the aircraft to the aft end. They check the jumpers' static lines and tell the jumpers to gain eye-to-eye contact when handing their static lines to the safety.

h. **STAND BY**. The amber light activates 10 seconds before the green light activates, and both number 1 jumpers assume a position in the door but not on the jump platform. Each JM takes a position facing his door to control the flow of the jumpers to the platform. The safeties stand directly under the aft end of the anchor line cable support bracket facing the jumpers. They control the static lines by vigorously pushing them back and to the upper trailing corner of the door and not allowing slack to fall down into the door.

i. **GO**. During mass exits, on the command GO, the JM exits his number 1 jumper. The AJM taps out his number 1 jumper a half second after the JM's number 1 jumper exits the aircraft. Movement to the door is a normal walking pace without shuffling. The jumper keeps his arms straight and elbows locked while passing his static line to the safety; he places both hands on the ends of the reserve parachute and exits the aircraft. Exits are made at a 90-degree angle off the jump platform. The AJM follows his last jumper out the aircraft. The JM waits until the AJM clears the platform, then exits the aircraft. The safeties immediately conduct a towed jumper inspection and turn the door over to the loadmaster with a "thumbs up." (If using ADEPT options 1 or 2, use the appropriate exit procedures.)

16-26. DOOR CHECK PROCEDURES

After the JM receives the announcement "All okay, jumpmaster," from the jumpers, he regains control of his static line from the safety and takes the number 1 jumper position. Once the aircrew has completed their slowdown checks, the loadmaster transfers control of the door by giving the JM a "thumbs up" signal and moves to the center of the cargo floor area where he can observe both doors. It is important for the safeties to control the JM's static line with two hands and also to keep an eye on the loadmaster and the jumpers. The JM is now ready to perform his door safety check. He transfers control of his static line back to the safety. The JM then grasps the troop door guide tracks with his trail hand and conducts the door safety check procedures.

a. **Troop Door Up-Lock**. With the lead hand, the JM grasps the troop door lifting bar, raises the door up, and then pulls it back down to the door up-lock. A visual inspection of the door up-lock verifies that it is in the locked position.

b. **Trail Edge of the Troop Door**. With the lead hand, the JM starts at the top of the trail edge of the troop door and traces down to the inside edge, then back up to the door clearance faring guard cover, then back down the trail edge of the door clearance faring. At this point, the JM ensures the troop door clearance faring is fully retracted and locked in the stowed position. He traces back up to the top of the door. Then he secures a handhold with his lead hand by grasping the troop door guide or the handhold provided on the lead edge of the door

CAUTION

The JM must be aware that the reserve parachute ripcord grip can contact the trail edge of the door and could cause inadvertent reserve parachute deployment.

c. **Air Deflector**. To check the air deflector, the JM leans toward the trail edge of the door and performs a visual inspection of the air deflector, ensuring it is fully deployed.

d. **Outside Air Safety Check**. The JM places either foot on the center of the platform with his trail foot a comfortable distance behind the lead foot. He leans out by bending forward at the waist and locking both elbows. He maintains a firm grasp with both hands at all times and performs his initial 360-degree outside air safety check:

- Direction of flight.
- Overhead.
- Rear.
- Straight down.
- Straight to the front.
- Back toward the direction of flight.

The JM continues his duties at subsequent time warnings.

WARNING

At no time during airborne operations should the cargo ramp door be open when the troop doors are open. If this situation exists, all jumpers, including JMs and safety personnel must be seated and strapped in, guarding the ripcord grip of their reserve parachutes. If Army personnel must work around the open troop doors, caution must be taken to prevent personnel from being swept out of the aircraft. Reserve activation with troop doors and cargo door open can cause serious injury or death.

16-27. DOOR BUNDLE PROCEDURES AND EJECTION

Bundles may be pushed out the door. Use the same procedures as for the C-130 or C-141 aircrafts.

16-28. SAFETY PRECAUTIONS

In accordance with a memorandum from Headquarters, US Army Test and Evaluation Command (Subject: Safety of Personnel Formation Airdrop Operations from the C-17, Date: 8 March 1997), personnel parachutes MUST be equipped with a 20-foot static line.

a. **Ramp Jumps**. Static line personnel airdrops from the cargo ramp are prohibited. The cargo ramp must be closed during static line jumps from the paratroop doors.

b. **Aircraft Configuration**.

(1) The aircraft deck angle is set between 6 and 7 degrees during static line personnel airdrops with an air speed of 130 knots, plus or minus 3 knots indicated airspeed.

(2) The total aircraft gross weight should not exceed 360,000 pounds at the time static line personnel airdrops are initiated.

(3) Troop air deflector is deployed.

c. **Bundle Static Line**. All door bundles are equipped with a 20-foot static line, with one static line drogue device attached.

d. **Static Safety Personnel**.

(1) Static safety personnel are seated in the crew compartment, but they move to the troop compartment after takeoff.

(2) Static safety personnel must complete their 20-minute checks and arrive at the forward end of the cargo compartment before the 10-minute time warning.

(3) The safeties stand directly under the aft end of the anchor line cable support bracket facing the jumpers. They control the static lines by vigorously pushing them back and to the upper trailing corner of the door and not allowing slack to fall down into the door.

e. **Jumper Exit Technique**. All jumpers must be trained to exit the aircraft with a strong effort in leaving the platform at a 90-degree angle straight out the door as close to the lead edge of the door as possible.

f. **Aircraft Inspection.** Before the flight, the JM or safety inspects the exterior aft of the troop doors for any protrusions, sharp edges, or snag hazards. He closes the cargo ramp during this inspection to allow for examination of the actuator door bulb seal and jack pads.

g. **Paratroop Retrieval System**. Static line personnel airdrops are restricted to C-17A aircraft with installed paratroop retrieval system (Canadian retrieval systems).

h. **Anchor Line Cable**. The JMs ensure that no more than 27 jumpers are hooked up to the outboard anchor line cables and no more than 24 jumpers are hooked up to the inboard anchor line cables during mass tactical operations.

i. **Formation**. For personnel airdrop operations from separate aircraft, the formation separation time between aircrafts is 2.5 to 3 minutes or 26,000 feet to avoid any aircraft wake vortex interference with jumpers.

16-29. IN-FLIGHT RIGGING PROCEDURES

The C-17A seats 102 jumpers, 2 static safeties, and 1 loadmaster. Ramp space is provided for storage of parachute assemblies on pallets and door bundles.

a. **Supervisory Personnel Required.** The following are required for in-flight rigging:

- One JM (supervises in-flight rigging).
- Eight AJMs.
- Two static safety personnel.
- One loadmaster.

b. **Equipment Stowage**. Parachute assemblies (mains and reserves, 106 each) are placed in kit bags and stowed and secured in two pallets on the ramp. Jumpers are seated in two sticks each on the left and right sides of the aircraft. All ALICE packs and weapons containers are placed in the center aisles and are strapped down by the loadmaster. All DMJPs, SMJPs, AT-4s, and door bundles are strapped down to the cargo floor section aft of the inboard seats or on the ramp. The jumper keeps his ballistic helmet, protective mask, and LBE under the seat.

c. **Parachute Issue**. All ALICE packs and weapons containers are retrieved from the center aisles and placed on the jumper's lap. Starting with the left outboard stick, each jumper stands up, places his ALICE pack and weapon on his seat, walks to the pallets on the ramp, draws an aviator kit bag, continues in a counterclockwise motion between the two sticks of jumpers seated on the right side of the aircraft, and returns to his original seat. The left inboard stick repeats the same procedure following the outboard jumpers. After the left side is complete, the right outboard stick jumpers stand up, place their ALICE packs and weapons on their seats, and draw aviator kit bags in the same manner except that they return in a clockwise motion moving through the left door center aisles. Right inboard jumpers follow the outboard jumpers.

d. **Buddy Rigging**. The JM initiates in-flight buddy rigging two hours before the 20-minute time warning.

(1) Once the parachutes are issued, buddy rigging begins. The JM, safety personnel, and designated AJMs supervise. Once a parachutist has been rigged, he sits down and waits to be inspected by one of the JMs.

(2) The JM personnel serve as roving inspectors. Attaching the static line snap hook to the top carrying handle of the reserve designates inspected jumpers.

(3) The JM supervises the operation and ensures all personnel are inspected. If there are other current and qualified JMs onboard the aircraft, they may be used.

(4) Once all the jumpers have been inspected, the JMs don their equipment, and the safety personnel inspect them.

16-30. JUMPMASTER AIRCRAFT INSPECTION

The JM inspects the aircraft as follows:

a. **Exterior of Aircraft Fuselage**.
- Inspect aft of doors for any protrusions, sharp edges, or snag hazards.
- Inspect ramp area with the ramp in the closed position. Special attention must be made to the actuator door bulb seal on both sides of the exterior of the cargo ramp. Ensure that the rubber seal is in place and that there are no sharp edges or snag hazards.

b. **Seats**.
- There should be enough seats for troop load (102, normal load; 108, maximum), and all seats should be serviceable.
- All seats should have safety belts.
- Seats are secured to the floor or sidewall of the aircraft.

c. **Floor**.
- Nonskid covering is in good condition, clean, and safe to walk on.
- Roller conveyors are stowed.
- Loose equipment is secured in the cargo area.
- Deck tie-down ring covers (4) in place and secured just before each door.

d. **Jump Platforms**.
- Nonskid surface present.
- Platform lights operate for night jumps.

e. **Jump Doors**.
- No excessive grease on the door tracks.
- No sharp or protruding edges on door frames.
- Doors open and close without excessive force.
- Door lever catches operate in the up-lock position.

f. **Troop Door Clearance Faring**.
- No sharp edges, retracts properly, and can be locked into place.

g. **Air Deflector**.
- No sharp edges.
- Deploys properly to 40 degrees, as indicated by the interior angle gauge.

h. **Jump Caution Lights**.
- Check for proper operation: red, amber (10 seconds), to green.

i. **Anchor Line Cable System**.
- Properly installed with no breaks, kinks, or frays. Ensure the cables are clean and free of rust.
- Cables run through slots.
- Quick release retaining pip pins are present and installed.
- Paratroop retrieval system (Canadian retrieval system) attachment point on aft anchor line cable stanchion present.

j. **Static Line Retrievers**.
- Paratroop retrieval system assembly present and complete.
- Motor is operational.
- Retriever cables are not broken, frayed, or kinked.
- Retriever cables are secured in spring clips.

k. **Emergency Equipment**.
- Public address system is operational.
- First aid kits and fire extinguishers present.
- Alarm system is operational.
- Emergency exits are operational and accessible.
- Sufficient emergency parachutes are available.

l. **Miscellaneous**.
- Lighting system is operational, especially red lights for night operations.
- Air sickness bags and hearing protection are available.
- Comfort facilities, water, urinals, and so forth, are available for in-flight rigging.
- Complete JM kit bag is on the aircraft.

16-31. TOWED JUMPER PROCEDURES

When a jumper becomes towed, follow these procedures:

a. **Supervisory Personnel Actions**. When a jumper becomes towed, the JM stops the stick and notifies the loadmaster. The loadmaster notifies the pilot, and requests that the drop altitude and speed be maintained. The JM identifies how the jumper is being towed. If the jumper is being towed by anything other than his static line, the JM attempts to free the jumper. If the jumper is being towed by the static line, recovery procedures are initiated. The JM observes the jumper and recommends whether to retrieve or cut the jumper free. The recommendation is relayed by the loadmaster to the pilot. The pilot makes the decision to retrieve or cut the jumper free from the aircraft. If directed by the pilot, the USAF loadmaster cuts the jumper free from the aircraft.

b. **Jumper Actions**. If a towed jumper is conscious, he maintains a good tight body position with both hands on the ends of the reserve, right hand protecting the rip cord grip, and he is prepared to activate the reserve if cut free from the aircraft.

c. **Priority for Action**.

(1) ***First Priority (Jumper is Towed by the Static Line):*** Retrieve the jumper using the paratroop retrieval system.

(2) ***Second Priority:*** Cut the jumper free.

(3) ***Third Priority (Least Preferred):*** Land with the jumper outside the aircraft. When using this method, if the jumper cannot be completely retrieved into the aircraft, he should be retrieved as close as possible to the door and secured. The aircraft lands at the closest available runway.

d. **AJM and Static Safety Actions**. If the jumper is to be retrieved, the AJM and safeties move the remaining jumpers toward the front of the aircraft. All personnel stay clear of the door and line of travel of the static line retriever cable. When the jumper has been retrieved to the door, the JM and the safety gain physical control of the jumper. At no time should the JM or the safety grasp the jumper's reserve parachute. The retrieved jumper is moved all the way forward in the aircraft, and he does not exit the aircraft during the airdrop operation. Instruct the jumper not to touch any of his equipment or de-rig in any way. If emergency medical treatment is needed, cancel the airdrop operation during training. During combat operations, the mission may be continued.

e. **Emergency Jettisoning**. To jettison the HSPR in an emergency, the parachutist lowers the ALICE pack, lowers the M1950 weapons case, and then pulls out on the yellow safety lanyard (attached to the ejector snap on the HPT lowering line), which allows the ALICE pack to fall free.

CHAPTER 17

ROTARY-WING AIRCRAFT

Rotary-wing aircraft can be used for airdrop operations when special missions are conducted to deploy small-unit forces. The aviation unit supporting the airdrop is responsible for preparing the aircraft for equipment and personnel drops to include seat and door removal (if required) and installation or rearrangement of seat belts. The installation of the field-expedient anchor line system is the JM's responsibility. Aircraft preparation is usually accomplished jointly by the crew chief and JM.

Section I. SAFETY CONSIDERATIONS

Although safety considerations for each aircraft are discussed, the requirements below apply to all Army aircraft (unless otherwise indicated).

17-1. GROUND TRAINING

Unit commanders require all personnel to participate in ground training immediately before the jump. The parachutists are shown the correct movement procedures inside the aircraft and the exit procedures. Parachutists are required to practice and demonstrate these procedures to JM satisfaction before the jump. Different techniques are involved in jumping from rotary-wing aircraft; failure to conduct ground training may result in a serious jump accident.

17-2. MOVEMENT IN AIRCRAFT

The pilot is briefed to expect rapid shifts in the aircraft's center of gravity during stand up, hook up, and exit of parachutists.

17-3. RESERVE PARACHUTE

Crowded conditions inside the cargo compartment could cause accidental activation of a reserve parachute, creating an extremely hazardous situation. During movement, the rip cord grip of the reserve parachute is protected by placing the right hand and forearm over the front of the reserve. This method allows the parachutist to control the pilot chute and canopy in case of accidental activation.

17-4. SPACE LIMITATIONS

The total number of parachutists and air delivery containers must conform to the weight and space limitations of the specific aircraft involved.

17-5. 6-SECOND COUNT

Due to the slow forward speed of helicopters and the downward rotor wash, the time interval between exit and full deployment of T-10-series or MC1-series parachutes requires about 100 feet more altitude. Due to the longer opening time, the parachutist extends the normal 4000-count to a 6000-count.

17-6. STATIC LINES AND DEPLOYMENT BAGS

Static lines and deployment bags are retrieved by the JM or crew chief immediately after the last parachutist is clear. The static lines and deployment bags are secured as soon as they are retrieved inside the aircraft. If the door on the aircraft can be closed, the static lines can be removed from the anchor cable or attaching point; otherwise, the static lines are not detached until the aircraft is on the ground.

17-7. CROWDED CONDITIONS

Crowded conditions inside these aircraft dictate that caution be used to prevent entanglement or misrouting of static lines during the parachutist's exit. Each parachutist is cautioned to watch the static line of the preceding parachutist and to observe all the static lines trailing from the lower aft corner of the cargo or personnel door. This precaution ensures that succeeding parachutists do not jump until the parachute of the preceding parachutist has deployed, and that the deployment bag has trailed to the rear of the aircraft.

17-8. CONTAINER LOADS

If container loads are to be airdropped from bomb shackles (wing load), helicopter door, bomb bay, or the doors of utility airplanes, they must be rigged using parachutes equipped with breakaway static lines. A-7A, A-21, and CDS container loads using breakaway static lines may be airdropped from the ramp or rear end (tailgate) of cargo and transport-type aircraft without jumpers following. CRRC loads may be dropped followed by jumpers with the 15-foot extraction parachute packed in a main parachute deployment bag IAW FM 10-542. Container loads with breakaway static lines are not rigged for airdrop from the troop door (side door) of cargo and transport-type aircraft. IAW message DTG 311313ZOCT00, for paratroop door drops, loads that will be followed immediately by parachutists will be rigged with cargo parachutes having nonbreakaway static lines. Each static line must have a drogue attached to it as outlined in the parachute's technical manual. When using the T-10-series cargo parachute or the 68-inch pilot parachute for the paratroop door load, the deployment bag serves as the drogue. Parachutists are **not** dropped at the same time as bundles that are rigged for release from bomb shackles.

17-9. HOOKUP PROCEDURES

When using rotary-wing and small fixed-wing aircraft for airborne operations, parachutists might use different hookup procedures from the standard hookup procedures used in USAF large fixed-wing troop carrier aircraft. This difference is due to the location of the anchor cables. Also, the JM may hook up the individual jumper. Unless otherwise specified in the hookup procedures for specific aircraft, the rule is to hook the open portion of the snap hook to the front of the aircraft with all static line snap hooks facing the same direction. This permits rapid, visual inspection before the jump and easy removal of the static lines after the jump.

17-10. TOWED PARACHUTIST PROCEDURES

In the event of a towed jumper on a rotary-wing aircraft, the jumpmaster will prevent any other jumpers from exiting and will notify the pilot. The parachutist stays in a tight body position and protects the rip cord grip. The jumpmaster will ensure the jumper is securely attached to the aircraft and will not break free during descent. If the jumper is not securely attached, the jumpmaster will attempt to shake or cut him free. If the jumper is attached, the aircraft will slowly descend to the DZ and come to a hover, and the jumper will be freed from the aircraft.

Section II. UH-1H IROQUOIS/UH-1N HUEY

The Army's UH-1H is powered by a single gas turbine engine; the USMC UH-1N has two gas turbine engines. Up to 8 combat-equipped parachutists can jump from the UH-1H/UH1-N, consistent with weight limitations (Figure 17-1). The JM is a static JM.

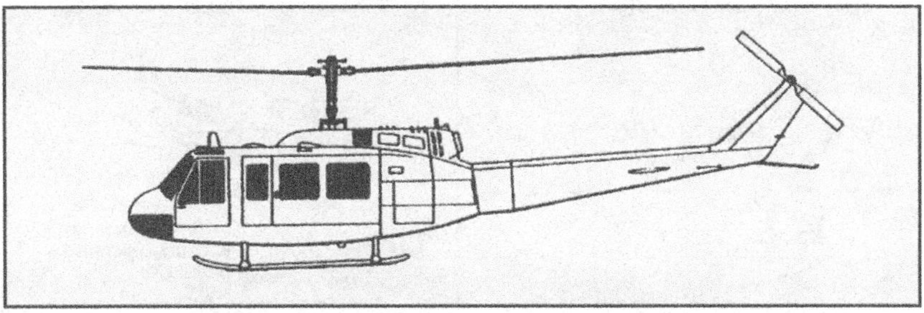

Figure 17-1. UH-1H Iroquois/UH-1N Huey.

17-11. PREPARATION AND INSPECTION

Prepare and inspect the UH-1H as follows:

a. **Preparation**. The following steps prepare the UH-1H for jumping:

(1) Both cargo compartment doors are locked in the open position. If the doors cannot be locked, they are removed.

(2) All troop seats are removed except one seat on each side (located to the rear of the pilot and copilot seats). These two seats are installed so they are facing to the rear of the aircraft. If the parachutists are equipped with combat equipment and eight parachutists are to jump, all seats in the cargo compartment are removed.

(3) The door and frame are inspected to ensure there are no sharp edges that could cut or fray static lines. If these are determined to be hazardous, corrective action is taken before the helicopter is jumped.

(4) Under field conditions, the door and frame can be padded and taped to preclude a mission abort. Otherwise, the aircraft is returned to maintenance for correction of the deficiency.

(5) Safety belts are attached to the tie-down rings on each side of the compartment for floor-seated parachutists.

(6) The door gunner/crew chief foot-operated radio switch may be unscrewed (by hand) before jumping. The exposed radio switch wires are taped to prevent an electrical short. If the switch is not removed, it is padded and taped. The ground-handling wheel-mount brackets on both landing skids are padded with cellulose wadding and taped

(Figure 17-2, page 17-4). Some aviation units have fabricated special covers that may be used to cover the wheel-mount brackets.

b. **Anchor Line Systems**. Two anchor line systems are available with the UH-1H aircraft for airdrop of personnel. They are the **standard overhead system** and the **expedient system** (Figure 17-3, page 17-4). The expedient system (modified STABO strap or Type XXVI nylon webbing anchor line cable assembly) consists of a nylon A-7A strap, four D-rings, and four connector snaps (TM 10-1670-298-20&P authorized the fabrication of the modified STABO strap or Type XXVI nylon webbing strap from Type XIII nylon webbing). Nylon modified STABO straps or Type XXVI nylon webbing must be used with cotton buffers on the D-rings and connector snaps.

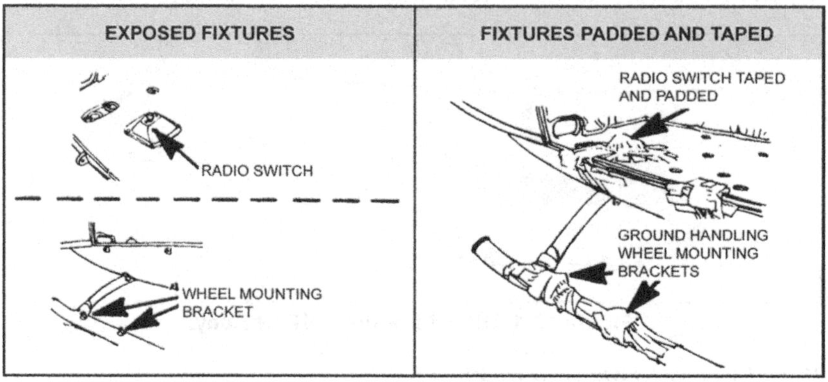

Figure 17-2. UH-1 exposed fixtures padded.

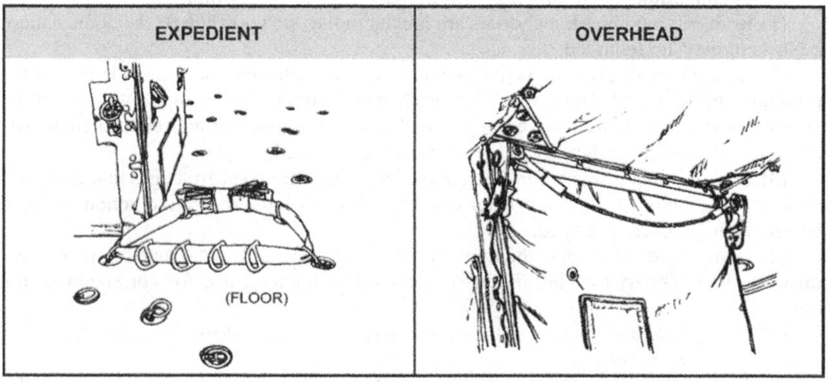

Figure 17-3. UH-1 anchor line system.

c. **Anchor Line Assembly Installation**. An anchor line assembly is installed on each side of the aircraft. It can be installed quickly by means of four tie-down rings located on the floor on the right and left sides of the aircraft compartment. The modified STABO strap or Type XXVI nylon webbing strap is threaded through the D-rings, which are used for attachment of the static line snap hooks (Figure 17-4).

(1) ***Left Door.*** For the left door, one connector snap on the modified STABO strap or Type XXVI nylon webbing strap is attached to the tie-down ring number G2. The strap is connected to the tie-down ring number F4. Four D-rings are on the strap with the round part of the rings facing outboard (of aircraft). The strap is then connected to tie-down ring number K3 and tie-down ring number J4. The free end of the strap is secured to the strap fastener, and any excess between tie-down rings number J4 and number G2 is taped. All connector snaps must be safety wired.

(2) ***Right Door.*** The same procedures apply to the right door as the left except that the A-7A strap is attached to tie-down ring number G1, then to F2. Four D-rings are on the strap with the round part of the rings facing outboard (of aircraft). The free end of the strap is secured to tie-down rings number K2 and number J3, and the strap fastener is secured. Excess strap between tie-down rings number J3 and number G1 is taped.

d. **Inspection**. Before enplaning, the JM and pilot, or pilot's representative, jointly inspect the aircraft to determine the following:

(1) All protruding objects near the cargo compartment doors are removed or taped.

(2) The lower right and left aft edges of both the cargo compartment doors are padded and taped.

(3) The anchor line cable or field-expedient anchor line system is secure, serviceable, and properly installed.

(4) A safety belt is available for each parachutist.

(5) A headset is available for the jumpmaster to effect coordination among the jumpmaster, the pilot, and the ground.

17-12. LOADING TECHNIQUES AND SEATING CONFIGURATION

During loading, jumpers should not approach directly from the front or sides but at a 45-degree angle to the nose of the aircraft. Jumpers 1 through 4 enter the cargo compartment through the right door, are hooked up by the JM in numerical order, and seat themselves (Figures 17-4 and 17-5). Jumpers 5 through 8 enter the cargo compartment through the left door, are hooked up by the JM in numerical order, and seat themselves.

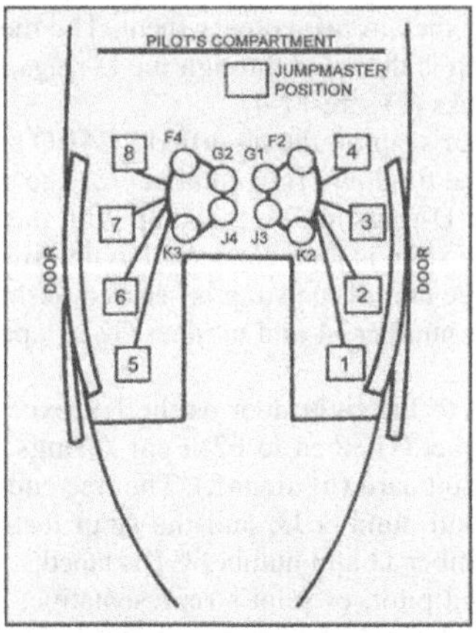

Figure 17-4. UH-1 seating configuration, expedient anchor line system.

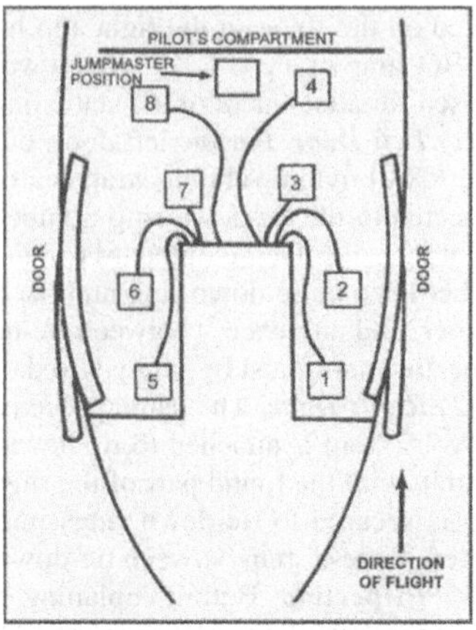

Figure 17-5. UH-1 seating configuration, overhead anchor line system.

a. The JM ensures excess static line is stowed as he hooks up each parachutist.

b. The open portion of the static line snap hook faces the front of the aircraft.

c. For flights less than 25 minutes long, jumpers may sit in the door with their feet outside the cargo compartment.

17-13. JUMP COMMANDS

The JM issues the following commands:

a. **GET READY**. This command is given 4 minutes or less from drop time, and the aircraft is level for final approach. All seat belts are unlatched and moved to the rear of the parachutists. The JM visually inspects each safety belt to ensure that it is clear of the parachutist and the equipment.

b. **CHECK STATIC LINES**. The JM rises and checks the routing of static lines from the point of attachment to the pack tray to ensure they are properly routed and hooked up.

c. **CHECK EQUIPMENT**. All parachutists check their equipment.

d. **SOUND OFF FOR EQUIPMENT CHECK**. On this command, number 1 parachutist orally indicates "Okay" to the JM. The remaining parachutists follow in order.

e. **SIT IN THE DOOR**. This command is given by the JM 30 seconds from drop time. Numbers 1 and 2 swing their legs to the right and take sitting positions in the door with feet together outside the cargo compartment. (Numbers 3 and 4 extend their legs outside and move to sitting positions.) They place both hands, palms down, on the floor alongside their thighs, turn their heads toward the JM, and wait. Numbers 5 and 6 swing their legs to the left, take sitting positions in the left door, and follow the same procedure as numbers 1 and 2. (Numbers 7 and 8 extend their legs outside and move to sitting positions.) This command is omitted if the parachutists are already sitting in the door on a short flight.

f. **STAND BY**. This command is given 8 to 10 seconds before the command GO.

g. **GO**. At this command, the following occurs:

(1) ***Personnel.*** The JM controls the jumper's exit and ensures a 1-second interval between jumpers by giving each jumper the oral command GO after the preceding jumper has exited and cleared the aircraft. The jumpers exit in numerical sequence.

(2) ***Air Delivery Containers.*** When an air delivery container is being released from the cargo hook, the pilot releases the container and informs the JM when the load has cleared the aircraft. The jumpers exit as explained above.

17-14. ARCTIC OPERATIONS

If the helicopter has skis, the ski attaching bolts and the sharp edges of the skis are padded and taped on the outboard side of the landing skids aft of the leading edge of the cargo door. Due to the bulk and weight of arctic clothing, individual equipment is not worn. The equipment is dropped either as an internal or external load.

17-15. SAFETY PRECAUTIONS

Safety precautions on the UH-1H are as follows:

a. **Parachutists**. During movement inside the aircraft, the parachutist protects the rip cord grip. Crowded conditions inside the cargo compartment and the open doors on both sides of the fuselage pose a hazardous situation regarding accidental activation of the reserve parachute.

b. **Jumpmaster**. The JM ensures all parachutists remain secured by their safety belts until the command GET READY is given. The JM prevents (or corrects) excessive static line from flopping about the aircraft. (The JM does not jump from this aircraft.) The JM wears a safety harness.

c. **Equipment**. Equipment prescribed in Chapter 12 can be worn by parachutists when jumping this aircraft.

(1) Standard air delivery containers rigged with G-14 cargo parachutes can be delivered from the cargo hook, using the breakaway static line (FM 10-500-2). The snap hooks of the static lines are hooked to the anchor line system before lift-off. Door bundles reduce the number of parachutists that can be carried, depending on the size and number of bundles.

(2) When CWIEs are jumped (two per load maximum), they are attached to number 3 and number 7 parachutists (one for each door). The DMJP may **not** be jumped from aircraft that require parachutists to sit on the floor.

d. **Aircraft**. The indicated airspeed of the aircraft during jumps is not less than 50 knots or more than 70 knots. The minimum drop altitude is 1,500 feet AGL. After the last parachutist has cleared the aircraft, the static lines are retrieved inside the aircraft and secured in an aviator kit bag or secured by a safety belt to the aircraft floor. The static line snap hooks **are not removed** from the anchor line cable until the aircraft lands.

Section III. UH-60A BLACK HAWK

The UH-60A is a twin-turbine, medium-speed, single-main-rotor helicopter (Figure 17-6). Eight combat-equipped parachutists can jump from this aircraft.

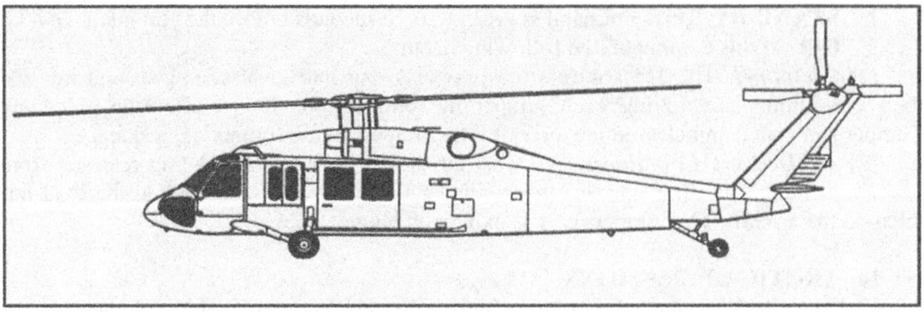

Figure 17-6. UH-60A Black Hawk.

NOTE: On missions requiring a window gunner, the maximum number of parachutists is reduced to six. Static line parachute operations require a static JM.

17-16. PREPARATION AND INSPECTION

The JM prepares and inspects the UH-60A as follows:

a. **Preparation**. To prepare the UH-60A for jumping, adhere to the following procedure (Figure 17-7):

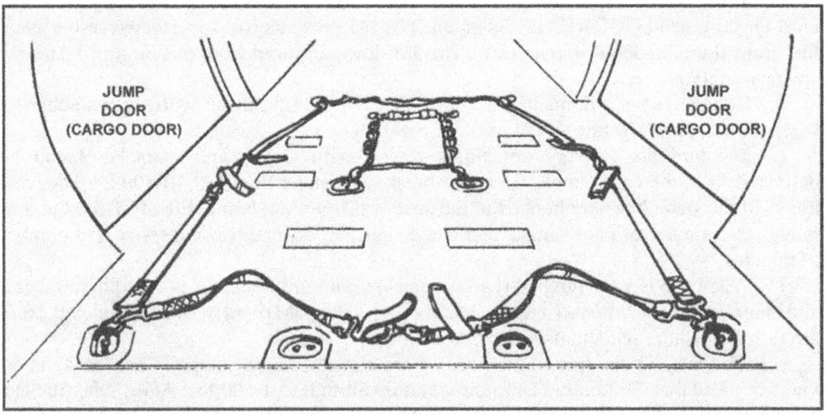

Figure 17-7. UH-60A compartment prepared for jumping.

(1) Lock both cargo doors in the open position.

NOTE: For arctic or other cold-weather operations, or during flights of long duration, the aircraft doors may be closed and locked. Doors cannot be opened during flight. The aircraft must either land or hover near the ground to open the doors. This procedure requires coordination between the supporting aviation and airborne units for the jump.

(2) Remove seat belts in the cargo compartment (except as required by aircraft crew).

(3) Tape cargo floor troop seat and tie-down fitting wells in front of the cargo doors.

(4) Tape sharp edges and tie-down fitting wells on the cargo floor and door jambs that could cut or fray static lines or snag parachutists' equipment.

(5) Tape the weather stripping on cargo doors below the door catch (Figure 17-8).

(6) Tape the entire trail edge of the door.

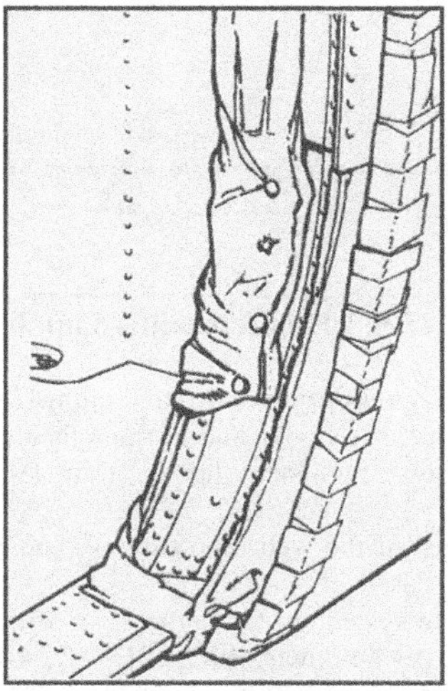

Figure 17-8. UH-60A door edge padded and taped.

NOTE: Tape must not interfere with closing, locking, unlocking, or opening cargo doors in flight. If the weather stripping below the cargo door catch is missing, pad the door edge with felt and tape in place. Padding must not preclude closing the cargo doors.

b. **Modified Anchor Line System**. Install a floor-mounted anchor line system (Figure 17-9, page 17-10) using a modified STABO extraction system anchoring strap assembly made from type XXVI nylon webbing (NSN 1670-00-999-3544, TM 10-1670-262-12&P and TM 10-1670-251-12&P). To modify the STABO, remove two of the connector snaps (leaving four) and add two D-rings (NSN 1670-00-360-0466). The cotton buffers may be locally manufactured.

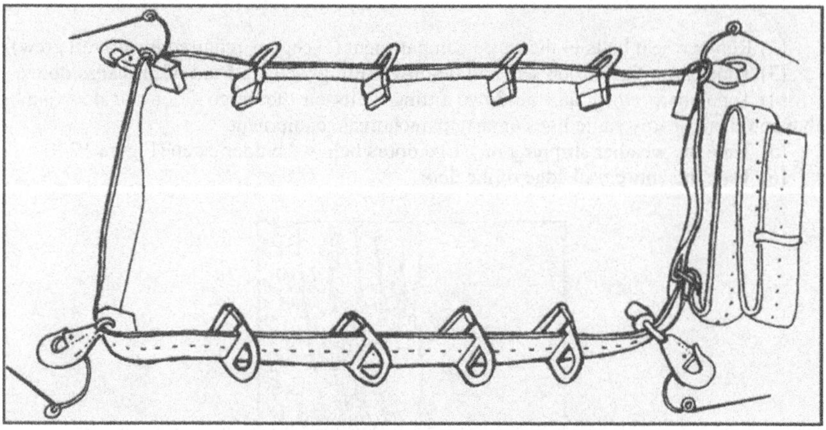

Figure 17-9. UH-60A modified anchor line.

(1) Install four snap hooks with safety wires and eight D-rings with cotton buffers on the anchor web loop, with the snap hooks and D-rings facing out in the following order: one snap hook, four D-rings; two snap hooks, four D-rings; and one snap hook (Figure 17-10).

(2) Insert about 30 inches of the web loop running end into the quick-fit adapter to secure the loop.

(3) Center the anchor line system on the cargo floor with the quick-fit adapter to the rear. Attach the snap hooks to tie-down fittings 3B, 3C, 4B, and 4C. Insert the safety wires and tape the snap hooks.

(4) Center the quick-fit adapter between tie-down fittings 4B and 4C, and tighten the web loop by pulling on the loop running end. Secure the web loop running end with an overhand knot. Fold and tape excess webbing to the web loop.

c. **Safety Belt Installation**. Install three floor-mounted safety belts.

(1) Attach a standard safety belt to tie-down fittings 5A and 5C for the JM. (This is necessary only if a seat has not been left for the JM.)

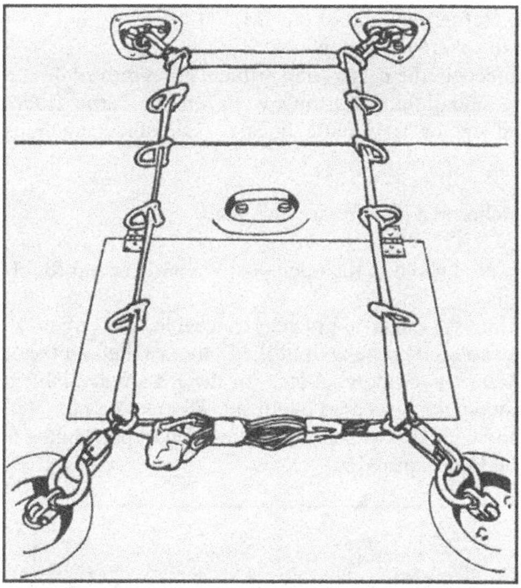

Figure 17-10. UH-60A modified anchor line secured to floor.

NOTE: The UH-60A cargo compartment configuration and floor tie-down fitting pattern preclude use of standard (individual) safety belts. Therefore, parachutists are restrained in groups of two and three, using modified safety belts.

(2) Attach an 86-inch-long (extended) safety belt to forward tie-down fittings 1A and 1D.

(3) Attach a 112-inch-long (extended) safety belt to tie-down fittings 1A and 5A, left door.

(4) Attach a 112-inch-long (extended) safety belt to tie-down fittings 1D and 5C, right door.

(5) Ensure that a serviceable safety harness is available for the JM (and the crew chief, when required). The JM's safety line is attached to tie-down fitting 5B. The crew chief's safety line is attached to tie-down fitting 1A or 1D, as required. If safety harnesses are not available, a backpack-type parachute may be used.

WARNING

Movement in the cargo compartment must be minimized to preclude inadvertent parachute activation.

d. **Inspection**. Before enplaning, the JM and pilot, or pilot's representative, jointly inspect the aircraft to ensure the following.

(1) All loose objects in the cargo compartment are removed or secured forward.

(2) Sharp edges and tie-down fitting wells on the cargo floor and doorjambs (or anything that could cut or fray static lines or snag the parachutists' equipment) are padded and taped.

NOTE: Door catches and handles are not taped.

(3) Cargo doors are locked in the open position and cleared for closing, depending on mission requirements.

(4) The anchor line system is complete, serviceable, and properly installed.

(5) Three serviceable safety belts (modified) are installed on the cargo floor.

(6) A headset/helmet and intercom jack for the JM are available and operational, and the intercom extension cord is secured overhead (Figure 17-11).

(7) Safety harnesses and backpack-type emergency parachutes are available for the JM and the crew chief, as required.

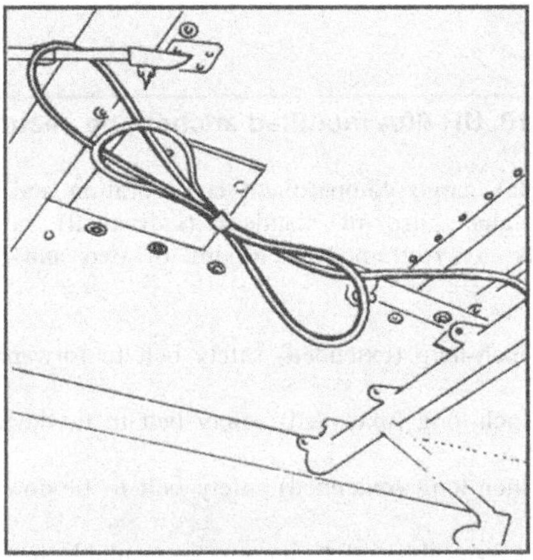

Figure 17-11. UH-60A JM's intercomm stowed overhead.

17-17. LOADING TECHNIQUES AND SEATING CONFIGURATION

Personnel are organized into a stick of eight parachutists. They approach the aircraft from the left or right side at a 90-degree angle in reverse order: numbers 8, 7, 6, 5, 4, 3, 2, 1 (Figure 17-12).

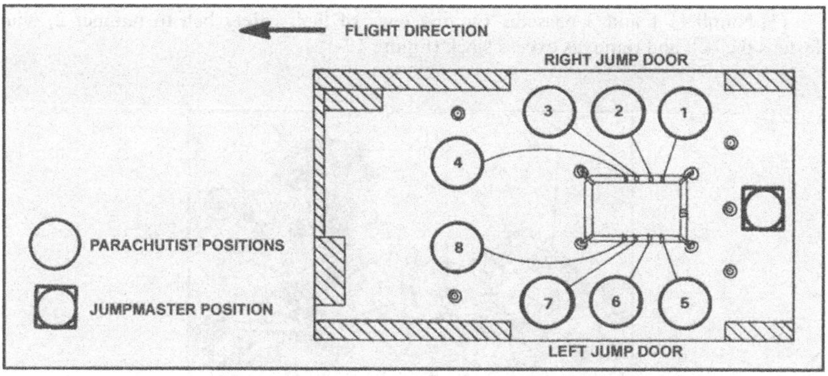

Figure 17-12. UH-60A seating and static line routing.

> **WARNING**
> **Do not approach the aircraft directly from the front because the lowest arc of the turning rotor blades occurs at that point.**

a. Number 8, followed by numbers 7, 6, and 5, enter the left door on command from the static JM. Numbers 4, 3, 2, and 1 enter the right door on command from the static JM. They are seated and hooked up by the JM in reverse numerical sequence, beginning with parachutist number 8, as they enter the aircraft. The open portion of static line snap hooks face the front of the aircraft.

b. Numbers 4 and 8 hold their static lines with a reverse bight: number 4 with the right hand and number 8 with the left hand. The static lines of the remaining parachutists, seated in the left and right doors, are routed directly behind them and down to the anchor line.

c. The JM ensures that any excess static line is stowed in the pack tray retainer band and that numbers 4 and 8 have correctly routed their static lines with the proper reverse bight.

NOTE: To preclude binding during exit, excess static lines of numbers 1, 2, 3, 5, 6, and 7 are stowed through the static line slack retainer on the parachutist's backpack.

d. When the JM commands FASTEN SAFETY BELTS, parachutists do the following:

(1) Numbers 4 and 8 pass the running ends of their safety belt to the center, fasten the belt, and remove excess slack (Figure 17-13, page 17-14).

(2) Numbers 5 and 7 pass the running ends of their safety belt to number 6, who fastens the belt and removes excess slack (Figure 17-14, page 17-15).

(3) Numbers 1 and 3 pass the running ends of their safety belt to number 2, who fastens the belt and removes excess slack (Figure 17-15).

Figure 17-13. Numbers 4 and 8 with static line bight and safety belt secured.

e. The JM inspects all safety belts to ensure that they are securely fastened and properly fitted. He is seated aft with his safety belt fastened for lift-off and landing. (One seat should have been left in place for the JM.)

f. For airdrop operations requiring the crew chief and window gunner (seat installed), the number of combat-equipped parachutists is reduced to six. The seating configuration is modified—positions 4 and 8 are deleted, and positions 5, 6, and 7 are renumbered 4, 5, and 6.

Figure 17-14. Parachutists (left door) with safety belt secured.

Figure 17-15. Parachutists (right door) with safety belt secured.

17-18. JUMP PROCEDURES

If the cargo doors are to be closed en route to the drop zone, the JM briefs numbers 3 and 7 on door opening procedures before loading. At 6 minutes before the drop, the pilot either lands or brings the aircraft to a hover (near the ground) and notifies the JM to open the cargo doors. The JM directs numbers 3 and 7 to open them. He ensures that the cargo doors are opened and locked. A 4-minute, a 30-second, and an 8- to 10-second warning are relayed to the JM by the pilot through the intercom system.

17-19. JUMP COMMANDS

The JM issues the following commands:

a. **GET READY**. This command is given at the 4-minute warning to alert the parachutists. All safety belts are removed at 1,000 feet AGL.

NOTE: Safety belts are removed when directed by the JM. Numbers 2, 6, and 8 release them. The running ends are stowed forward and aft to clear the static lines and the exit path.

b. **CHECK STATIC LINES**. The JM checks the routing of all static lines (from pack trays to anchor cable) to ensure they are correctly routed and hooked up. He ensures excess static line is stowed through the slack retainer on the backpacks of numbers 1, 2, 3, 5, 6, and 7, and that numbers 4 and 8 have the prescribed reverse bight in their static lines.

c. **CHECK EQUIPMENT**. All parachutists check their equipment.

d. **SOUND OFF FOR EQUIPMENT CHECK**. Number 1 indicates orally (and with a hand signal) to the JM the status of his equipment, followed by the remaining parachutists in numerical order.

e. **SIT IN THE DOOR**. This command is given by the JM at the 30-second warning. Numbers 1, 2, 3, 5, 6, and 7 assume door positions (Figure 17-16) with feet together outside the cargo compartment. Numbers 4 and 8 remain in place, ensuring that their feet are clear of their static lines. (This command is omitted if the parachutists are already sitting in the door.)

f. **STAND BY**. This command is given at the 8- to 10-second warning. The JM ensures that all parachutists hear and understand this command, particularly number 1, who places both hands, palms down, on the cargo floor alongside his thighs and awaits the next command. Numbers 2, 3, 5, 6, and 7 place both hands, palms down, on the cargo floor and await the next command; numbers 4 and 8 remain in place.

g. **GO**. The JM gives this command by an oral GO and a sharp tap on the rear of the parachutist's helmet. Each parachutist is tapped out. The jump sequence is in numerical order, 1 through 8. As soon as number 3 clears the door, number 4 moves into the door and assumes the door position before being tapped out. The static JM assumes control of number 4 jumper's static line as the parachutist begins moving to the door. Numbers 5, 6, 7, and 8 repeat the sequence in the left door.

Figure 17-16. Numbers 1, 2, 3, 5, 6, and 7 exit positions.

17-20. SAFETY PRECAUTIONS

Safety precautions on the UH-60A are as follows:

a. **Parachutists**. If CWIEs will be jumped, they may be attached to numbers 1 or 5 parachutists, or both. No more than two CWIEs, one for each door, can be jumped. DMJPs may not be jumped from aircraft that require parachutists to sit on the floor. (Procedures for towed parachutists are in paragraph 17-10.) Crowded conditions inside the cargo compartment make accidental activation of the reserve parachute more likely. During movement inside the aircraft, parachutists must protect the rip cord grip.

b. **JM**. The static JM wears a safety harness that is attached to the aft cargo floor tie-down fitting (5B). Backpack-type emergency parachutes may be used if a safety harness is not available. The JM is equipped with a headset or flight helmet that allows direct communications with the aircraft crew. The static JM immediately notifies the pilot of a towed parachutist.

c. **Equipment**. Parachutists can wear combat equipment when jumping this aircraft. Without detaching the static lines, the JM retrieves static lines and D-bags, places them inside an aviator kit bag, and secures the kit bag until the aircraft has landed. The static line snap hooks are then removed from the anchor line attaching points. The UH-60A is **not** used for static line parachute operations with the cargo doors removed. The static line anchor line cable is **never** rigged to the cargo door or overhead tie-down rappelling rings, since trailing D-bags might foul the main rotor system (due to the high position in which the bags would trail).

d. **Aircraft**. The indicated airdrop speed of the aircraft should not be less than 65 knots or more than 75 knots. The minimum jump altitude is 1,500 feet AGL.

NOTE: The pilot must maintain level flight and airdrop speed during D-bag retrieval to preclude D-bag entanglement with the cargo doors.

17-21. SAFETY BELT MODIFICATION

The UH-60A cargo compartment floor configuration does not provide a specific design of tie-down fittings for restraining personnel seated on the cargo floor. The safety belts used for restraining personnel are part of the troop seat assembly and are removed when conducting parachute operations.

a. The three modified C-3A troop-type safety belts, using the cargo floor tie-down fittings, restrain parachutists in groups of two and three by a single safety belt (Figure 17-17).

b. Two safety belts, 112 inches long and adjustable to 86 inches, restrain parachutists numbers 1 through 3 and 5 through 7, who are seated in the left and right cargo doors.

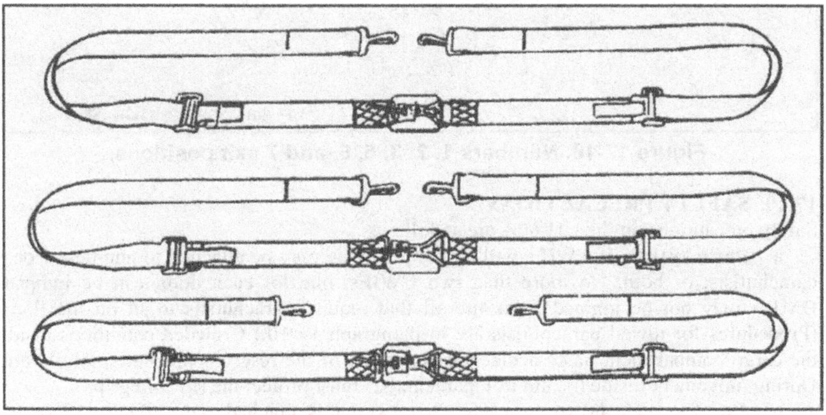

Figure 17-17. Modified C-3A troop safety belts.

c. One safety belt, 86 inches long and adjustable to 60 inches, restrains numbers 4 and 8 seated in the cargo compartment. Belt modifications are as follows:

(1) Place three standard C-3A troop-type safety belts (NSN 1670-00-447-9504) on a flat surface with hardware facing up.

(2) Remove the 8-inch lengths of webbing located between the end snap hooks and the quick-fit adapters of each belt.

(3) Cut two 32-inch and four 46-inch lengths of number 3 nylon webbing and heat-sear the ends.

(4) Reassemble one belt using the two 32-inch lengths of webbing.

(5) Thread the running ends of the webbing up through the bar of the snap hooks and quick-fit adapters. Make a 5-inch fold-back and tack in place.

(6) Sew a 4-inch, 4-point, WW stitch formation on each fold-back using number 3 nylon thread and a medium-duty machine (TM 10-1670-298-20&P).

(7) Reassemble the other two belts as indicated, using the 4-inch lengths of webbing.

Section IV. CH-47 CHINOOK

The CH-47 is a tandem-rotor, medium-transport helicopter. Twenty-eight combat-equipped parachutists can jump from this aircraft (Figure 17-18). The jumpmaster may be a jumping JM or a static JM.

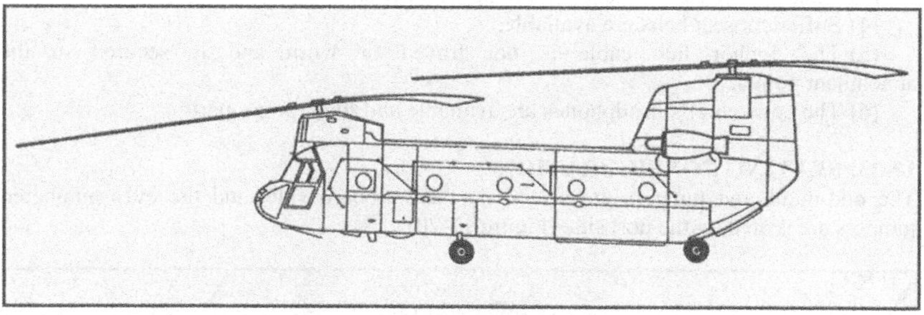

Figure 17-18. CH-47 Chinook.

17-22. PREPARATION AND INSPECTION

The JM prepares and inspects the CH-47 as follows:

a. **Preparation.** The following steps prepare the CH-47 for jumping.

(1) Install safety belts for each parachutist and extend all the way out to ensure positive hookup while seated.

(2) Secure the permanently installed anchor line cable to the attachment points on the starboard side of the aircraft (Figure 17-19).

(3) Incline the ramp for personnel parachute drops during flight.

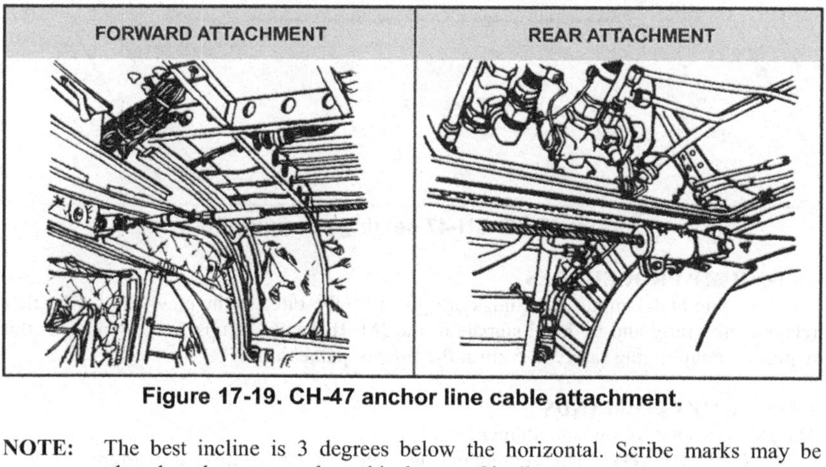

Figure 17-19. CH-47 anchor line cable attachment.

NOTE: The best incline is 3 degrees below the horizontal. Scribe marks may be placed on the ramp to show this degree of incline.

b. **Inspection**. Before enplaning, the JM and the pilot, or pilot's representative, jointly inspect the aircraft to determine the following:

(1) Troop seats can be easily lifted and secured before jumping.

(2) The ramp is clean and free of oil and water.

(3) Seats are securely fastened in the down position.

(4) Sufficient seat belts are available.

(5) The anchor line cable is not frayed or worn and is secured to the attachment points.

(6) The crew chief's headphones are available and function properly.

17-23. SEATING CONFIGURATION

The odd-numbered jumpers are seated on the starboard side, and the even-numbered jumpers are seated on the port side (Figure 17-20).

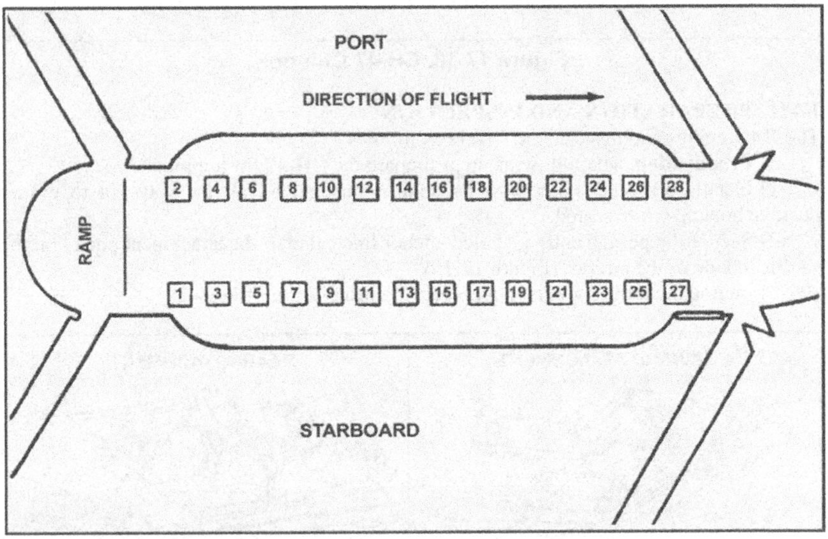

Figure 17-20. CH-47 seating configuration.

17-24. JUMP PROCEDURES

The 6-minute and 1-minute warnings are given by the pilot to the crew chief, who then relays them orally and by hand signals to the JM. If the JM jumps, he is number 1; this requires a nonjumping safety to control the flow of parachutists.

17-25. JUMP COMMANDS

The JM issues the following commands:

a. **GET READY**. This command is given at the 6-minute warning to alert parachutists.

b. **PORT SIDE PERSONNEL, STAND UP**. Parachutists seated on the port side of the aircraft stand up and secure their seats in the up position.

c. **STARBOARD SIDE PERSONNEL, STAND UP**. Parachutists seated on the starboard side of the aircraft stand up and secure their seats in the up position.

d. **HOOK UP**. On this command, odd-numbered personnel hook up, followed by the even-numbered personnel, who hook up (the open portion of the snap hook facing starboard) between the odd-numbered personnel to form one continuous stick of 28 parachutists.

NOTE: After hooking up, each parachutist controls the static line in a reverse bight at waist-level (left hand).

e. **CHECK STATIC LINES**. The JM or safety checks the routing of all static lines.

f. **CHECK EQUIPMENT**. All parachutists check their equipment.

g. **SOUND OFF FOR EQUIPMENT CHECK**. Beginning with number 28, the jumpers pass the status of their equipment toward the aft end of the aircraft. Number 1 orally (and with a hand signal) indicates to the JM the status of his and all other jumpers' equipment by stating, "All OK, jumpmaster."

h. **STAND BY**. The command is given 8 to 10 seconds before the command GO. Number 1 assumes a standing position at the ramp hinge (near center) of the aircraft. The remaining personnel close up interval behind the first parachutist.

i. **GO**. Number 1 walks off the port side rear corner of the ramp. The remaining parachutists follow at 1-second intervals.

NOTE: The JM or safety controls the flow from his position on the port side near the ramp hinge. Less than a 1-second interval between parachutists may result in entanglement of parachutists and static lines.

17-26. SAFETY PRECAUTIONS

Safety precautions on the CH-47 are as follows:

a. **Parachutists**. Parachutists ensure that seats are secured in the up position with seat legs rotated inside the seats. When following internal drop loads, parachutists exit between the ramp roller conveyor sections, staying as close to the port side section as possible. The parachutists jumping after external load drops, who are forward of the open floor hatch (used to check a load drop), remain clear of the opening until the load leaves the aircraft and the hatch is closed by the crew chief.

b. **Jumpmaster**. The JM or safety personnel ensure that parachutists are hooked up consecutively, 1 through 28 (Figure 17-21, page 17-22). If the JM does not jump, he wears a safety harness or back emergency parachute. He checks each parachutist after they hook up and controls the flow of parachutists. When an external load is delivered, the JM ensures the external load is clear and the aircraft has accelerated to a safe airdrop speed before dropping cargo bundles from inside the aircraft or before permitting parachutists to exit.

c. **Safety Personnel**. If the JM jumps, one nonjumping safety is required; the safety wears an emergency parachute.

d. **Equipment**. When cargo bundles are delivered, JMs use 15-foot breakaway static lines with cargo parachutes. The ramp roller conveyor section is installed on the starboard side of the ramp and is used to help eject the bundles from the cargo ramp; numbers 1 and 2 push the bundles out.

e. **Aircraft**. Aircraft safety requires that the speed during jumps is not less than 80 knots or more than 110 knots, with 90 knots being optimum speed. No special preparation is required if the aircraft has skis. Minimum jump altitude is 1,500 feet AGL when jumping below 90 knots. Minimum jump altitude 90 knots and above is 1,250 feet AGL. After the last parachutist has cleared the aircraft, the static lines are retrieved (using the static line retriever) inside the aircraft and secured in an aviator kit bag. The ramp must not be lowered until all jumpers have hooked up to the anchor line cable.

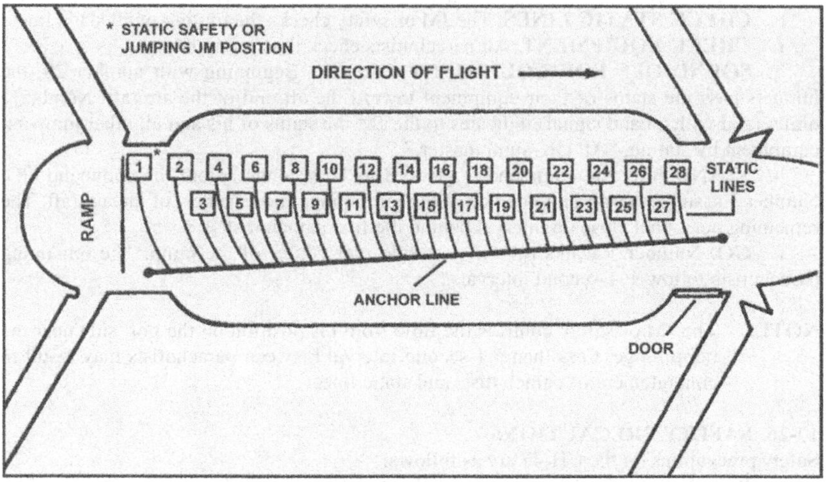

Figure 17-21. CH-47 static line routing.

CHAPTER 18
OTHER SERVICE AIRCRAFT

Other service rotary-wing aircraft can be used for parachuting operations. In addition to the procedures described here, Chapter 17, Section I, also applies.

Section I. CH-53 SEA STALLION (USMC)

The CH-53 is a twin-engine, single-rotor, medium-transport helicopter. Twenty combat-equipped parachutists, using the ramp, can jump from this aircraft (Figure 18-1).

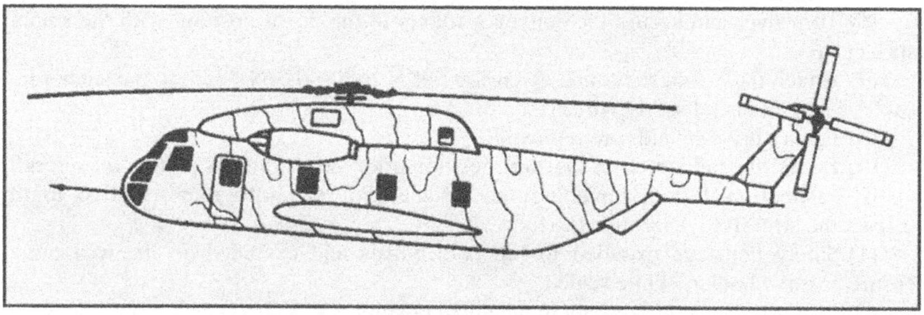

Figure 18-1. CH-53 Sea Stallion (USMC).

18-1. PREPARATION AND INSPECTION

The jumpmaster prepares and inspects the CH-53 as follows:

a. **Preparation.** The following procedures prepare the CH-53 for jumping:

(1) Install the anchor line cable on the port side of the floor (Figure 18-2, page 18-2) using the appropriate 5,000-pound tie-down fittings for the length of cable used (forward attachment point) and station number 522 (rear attachment point).

(2) Use one 1/4-inch or 3/8-inch steel cable (3/8-inch preferred), of an acceptable length, with four 1/4-inch or 3/8-inch cable clamps, lock washers, nuts, and bolts. The minimum installed length of the anchor line cable will be 8 feet between forward and rear attaching points (four to five jumpers). Ensure the installed anchor line cable is long enough so that a jumper's static line extends from the jumper to the anchor line cable without touching other jumpers in front of him in the stick.

(3) Route the anchor line cable directly through the forward and rear attaching point deck rings or use a steel stubai 85 snap link (with locking gate) between the anchor line cable and the aircraft attaching points. Ensure the locking gates on the snap links are facing up, locking to the aft of the aircraft, and taped. Ensure all tape is removed from the anchor line cable components and that the clamps are tight prior to every operation.

CAUTION

Do not use a clevis assembly (G-13) to attach the anchor line cable to the aircraft. If the cable is loose, the clevis may rotate sideways and drastically reduce the strength of the anchor line cable.

(4) Place a 4- by 4- by 6-inch wooden block between the anchor line cable and floor and attach a clevis assembly (G-13) to the tie-down ring at station number 502 for use as a static line snap hook stop.

(5) Remove excess slack from the anchor line.

(6) Turn over and secure the conveyor rollers in the down position with the smooth surface up.

(7) Attach the 5-foot static line extension (NSN 1670-00-368-4225) to the anchor line cable; insert and bend safety wires.

(8) Ensure drive-on aids are removed.

(9) Ensure the tail skid is in the "up" position prior to the jumpers exiting the aircraft.

b. **Inspection**. Before enplaning, the JM and pilot, or pilot's representative, jointly inspect the aircraft to determine the following:

(1) Safety belts are installed for all parachutists and extended all the way out to ensure positive hookup while seated.

(2) Seats are fastened securely in the down position.

(3) Seats are lifted and secured before jumping.

(4) The anchor line cable is not worn or frayed and is secured to the attachment points in the prescribed manner.

(5) The ramp and deck are clean and free of oil and water.

(6) All protruding objects near the ramp are removed or taped.

(7) The crew chief's headphones are available and function properly.

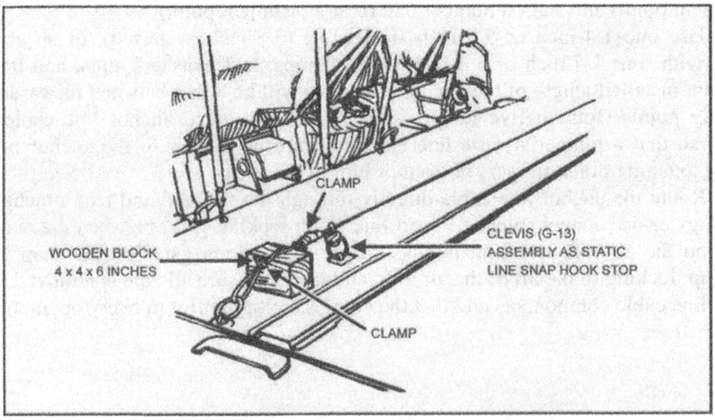

Figure 18-2. CH-53 anchor line installation.

18-2. LOADING TECHNIQUES AND SEATING CONFIGURATION
Parachutists enter the aircraft over the ramp with the static line over the right shoulder. Odd-numbered parachutists are seated on the port side, and even-numbered parachutists are seated on the starboard side (Figure 18-3).

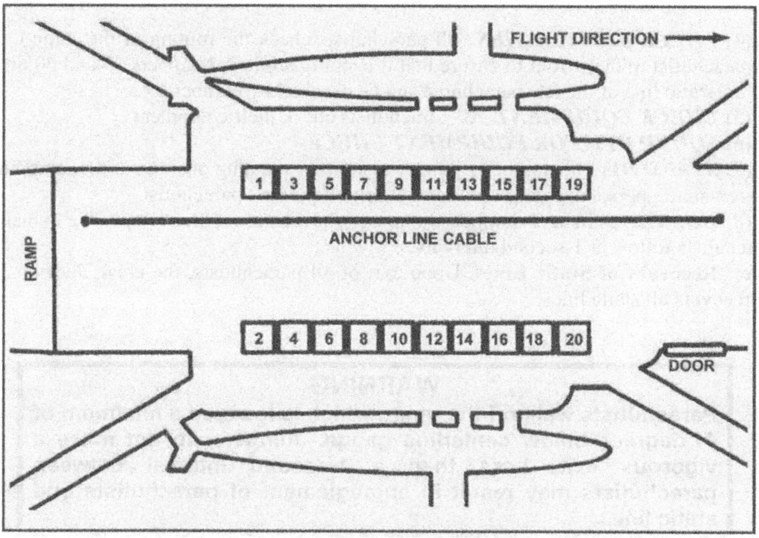

Figure 18-3. CH-53 seating configuration.

18-3. JUMP COMMANDS AND PROCEDURES
Jump commands and procedures for the CH-53 are as follows:

a. **Time Warnings**. The 6-minute and 1-minute warnings are given by the pilot to the crew chief, who in turn relays them orally and by hand signals to the JM.

b. **Jump Commands**.

(1) ***GET READY.*** All parachutists remove seat belts.

(2) ***STAND UP.*** All parachutists stand up and secure their seats in the up position.

(3) ***HOOK UP.*** The 5-foot static line extensions are attached to the anchor line cable by the snap hooks; insert and bend safety wires. The JM or AJM passes the static line extension to each jumper starting with the last jumper and working forward. When the command HOOK UP is given, the jumpers ensure the cotton sleeve is on the static line extension, attach their static line snap hooks, insert and bend safety wires, and hold the static line snap hook exposed for inspection by the JM. Once inspected, place the sleeve over the snap hook and take a reverse bite at waist level.

WARNING
The JM must ensure the jumper has a proper reverse bite. If an improper reverse bite is taken, the static line could become misrouted under the arm and cause a towed jumper or severe injury.

(4) ***CHECK STATIC LINES.*** All parachutists check the routing of the static line of the parachutist to their front to ensure that it is not misrouted. Numbers 19 and 20 turn so that the static line of the last parachutist can be checked by number 19.

(5) ***CHECK EQUIPMENT.*** All parachutists check their equipment.

(6) ***SOUND OFF FOR EQUIPMENT CHECK.***

(7) ***STAND BY.*** The number 1 jumper assumes a standing position at the ramp hinge. The remaining personnel close up intervals behind the first parachutist.

(8) ***GO.*** The number 1 jumper walks off the center of the ramp. The remaining parachutists follow at 1-second intervals.

c. **Recovery of Static Lines**. Upon exit of all parachutists, the crew chief or static JM recovers all static lines.

WARNING
Parachutists walk off the ramp, which is lowered a minimum of 11 degrees below centerline gauge. Jumpers do *not* make a vigorous exit. Less than a 1-second interval between parachutists may result in entanglement of parachutists and static lines.

18-4. SAFETY PRECAUTIONS

No more than two parachutists should jump with CWIE in one pass over the DZ. These parachutists should be numbers 1 and 2 in the stick. The static JM or safety personnel ensure that parachutists are hooked up consecutively (1 through 20), and that all seats are secured out of the way of the jumpers. The speed of the aircraft during jumps is 90 to 110 knots. The minimum drop altitude for aircraft with a 90- to 110-knot drop speed is 1,250 feet AGL. Either the JM or AJM must remain with the aircraft to handle the D-bags and towed jumpers. The JM spots the aircraft using WSVC, VIRS, or GMRS. The JM may spot from the crew chief door or the ramp as long as a safety is controlling the jumpers. If spotting from the ramp and jumping, the JM leads the stick and the AJM stays with the aircraft.

Section II. CH-46 SEA KNIGHT (USMC)

The CH-46 is a tandem-rotor, medium-transport helicopter. Twelve combat-equipped parachutists can jump from either the personnel door or the ramp (Figure 18-4).

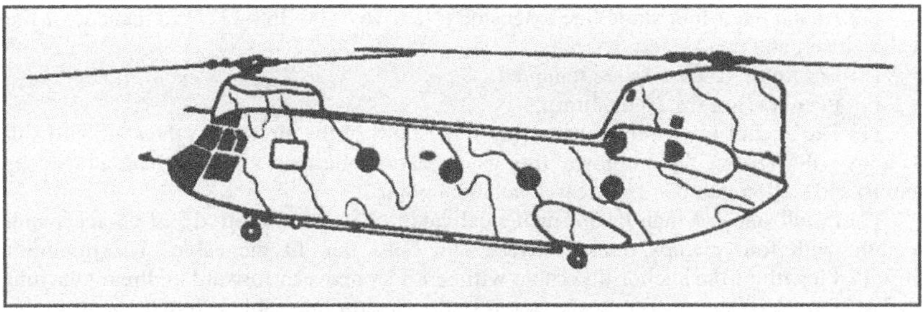

Figure 18-4. CH-46 Sea Knight.

18-5. PREPARATION AND INSPECTION

The jumpmaster prepares and inspects the CH-46 as follows:

a. **Preparation for a Ramp Jump**. The following procedure prepares the CH-46 for jumping:

(1) Install the anchor line cable on the port side of the floor using the appropriate 3,000-pound tie-down fittings for the length of the cable used (forward attachment point) and station number 410 (rear attachment point) (Figure 18-5, page 18-7).

(2) Use one 1/4-inch or 3/8-inch steel cable (3/8-inch preferred) of an acceptable length with 1/4-inch or 3/8-inch cable clamps and nuts. The minimum installed length of the anchor line cable will be 8 feet between forward and rear attaching points (four to five jumpers). Ensure the installed anchor line cable is long enough so that the jumper's static line goes from the jumper to the anchor line cable without touching other jumpers in front of him in the stick.

(3) Route the anchor line cable directly through the forward and rear attaching point deck rings or use a steel stubai 85 snap link (with locking gate) between the anchor line cable and the aircraft attaching points. Ensure the locking gates on the snap links are facing up, locking to the aft of the aircraft, and taped. Ensure all tape is removed from the anchor line cable components and that the clamps are tight prior to every operation.

CAUTION

Do not use a clevis assembly (G-13) to attach the anchor line cable to the aircraft. If the cable is loose, the G-13 clevis may rotate sideways and drastically reduce the strength of the anchor line cable.

(4) Place two 4-s by 6-inch wooden blocks between the anchor line cable and floor and attach one clevis assembly (G-13) to the tie-down ring at station number 390 for use as a static line snap hook stop.

(5) Remove excess slack from the anchor line.

(6) Turn the conveyor rollers over and secure them in the down position with the smooth surface up.

(7) Attach the 5-foot static line extension (NSN 1670-00-368-4225) to the anchor line cable; insert and bend safety wires.

(8) Ensure drive-on aids are removed.

b. **Preparation for Door Jump.**

(1) The anchor line cable is installed on the floor of the aircraft on the starboard side (Figure 18-5) using the tie-down fittings at station number 170 (forward attachment point) and station number 410 (rear attachment point).

(2) Install one 1/4-inch or 3/8-inch steel cable (3/8-inch preferred), of an acceptable length, with four clamps, lock washers, and bolts that fit the cable. The minimum installed length of the anchor line cable will be 8 feet between forward and rear attaching points (four to five jumpers). Ensure the installed anchor line cable is long enough so that the jumper's static line goes from the jumper to the anchor line cable without touching other jumpers in front of him in the stick.

(3) Route the anchor line cable directly through the forward and rear attaching point deck rings or use a steel stubai 85 snap link (with locking gate) between the anchor line cable and the aircraft attaching points. Ensure the locking gates on the snap links are facing up, locking to the aft of the aircraft, and taped. Ensure all tape is removed from the anchor line cable components and that the clamps are tight prior to every operation.

CAUTION

Do not use a clevis assembly (G-13) to attach the anchor line cable to the aircraft. If the cable is loose, the G-13 clevis may rotate sideways and drastically reduce the strength of the anchor line cable.

(4) Place two 4- by 6-inch wooden blocks between the anchor line cable and floor and attach a clevis assembly (G-13) to the tie-down ring at station number 190 for use as a static line snap hook stop.

(5) Remove excess slack from the anchor line.

(6) Turn the conveyor rollers over and secure them in the down position with the smooth surface up.

(7) Attach the 5-foot static line extension (NSN 1670-00-368-4225) to the anchor line cable; insert and bend safety wires.

(8) Ensure drive-on aids are removed.

(9) Remove the personnel door and install a plywood cover over the aft side of the door frame and secure it with a metal plate. Pad and tape the bottom and aft edges of the door frame (Figure 18-6, page 18-8).

c. **Inspection**. Before enplaning, the JM and pilot, or pilot's representative, jointly inspect the aircraft to determine the following:

(1) Seat belts are installed for all parachutists and extended all the way out to ensure positive hookup while seated.

(2) Seats are fastened securely in the down position.

(3) Seats can be lifted and secured before jumping.

(4) The anchor line cable is not worn or frayed and is secured to the attachment points in the prescribed manner.

(5) The wooden cover is installed and secured to the aft edge of the door frame when jumping the personnel door.

(6) The bottom edges of the door frame are properly padded and taped.

(7) The ramp and deck are clean and free of oil or water.

(8) All protruding objects near the ramp and personnel door are removed or taped.

(9) The crew chief's headphones are available and function properly.

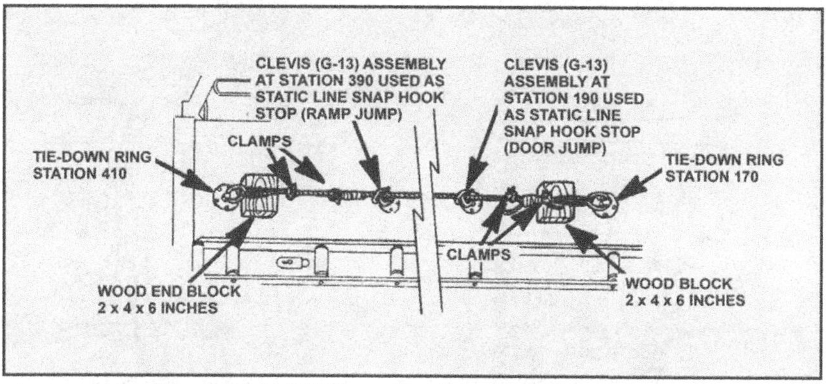

Figure 18-5. CH-46 anchor line cable installation.

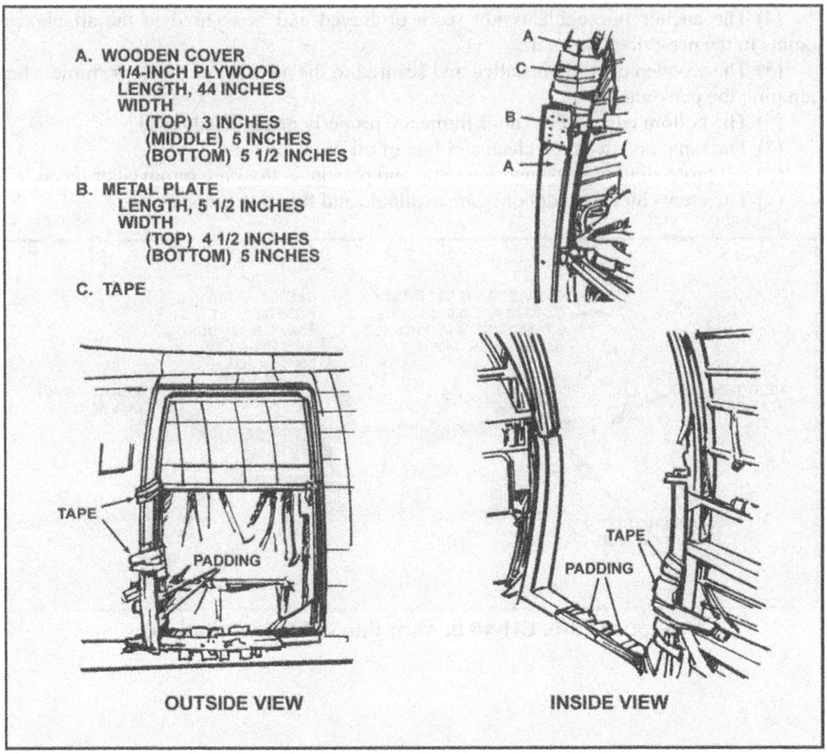

Figure 18-6. CH-46 door frame preparation.

18-6. LOADING TECHNIQUES AND SEATING CONFIGURATION

There are two ways a parachutist can jump from the aircraft: the ramp or the door. The parachutists enter the aircraft with the static line over the right shoulder; they are seated on the port and starboard side.

18-7. JUMP COMMANDS AND PROCEDURES

The 6-minute and 1-minute warnings, whether jumping the door or ramp, are given by the pilot to the crew chief, who in turn relays them orally and by hand signals to the static JM or AJM.

a. **Jump Commands for Door Jump.**

(1) ***GET READY.*** All parachutists in the first stick unfasten their seat belts.

(2) ***STAND UP.*** Parachutists stand up and move to the starboard side (Figure 18-7, page 18-10).

(3) ***HOOK UP.*** The 5-foot static line extensions are attached to the anchor line cable by the snap hooks; insert and bend safety wires. The JM or AJM passes the static line extension to each jumper starting with the last jumper and working forward. When the

command HOOK UP is given, the jumpers ensure the cotton sleeve is on the static line extension, attach their static line snap hooks, insert and bend safety wires, and hold the static line snap hook exposed for inspection by the JM. Once inspected, place the sleeve over the snap hook and take a reverse bite at waist level.

WARNING

The JM must ensure the jumper has a proper reverse bite. If an improper reverse bite is taken, the static line could become misrouted under the arm and cause a towed jumper or severe injury.

(4) ***CHECK STATIC LINES.*** The parachutists take a reverse bight (at waist level) of about 8 inches in the static line with the right hand, keep the arm close to the side, and check the static line of the parachutist to the front.

(5) ***CHECK EQUIPMENT.*** All parachutists check their equipment.

(6) ***SOUND OFF FOR EQUIPMENT CHECK.***

(7) ***STAND IN THE DOOR.*** Number 1 parachutist moves to the door and assumes the door position. He must crouch low to allow at least a 2-inch clearance between his helmet and the top of the door. The other parachutists close up behind number 1 at normal intervals.

(8) ***GO.*** Number 1 exits by jumping straight out the door and assuming the proper body position. The succeeding parachutists move up, make a 90-degree turn at the door, take up a correct door position, and exit in the same manner as number 1, maintaining a 1-second interval. Upon exit of all parachutists, the crew chief or static JM recovers all static lines.

WARNING

Parachutists must not spring upward when jumping this helicopter.

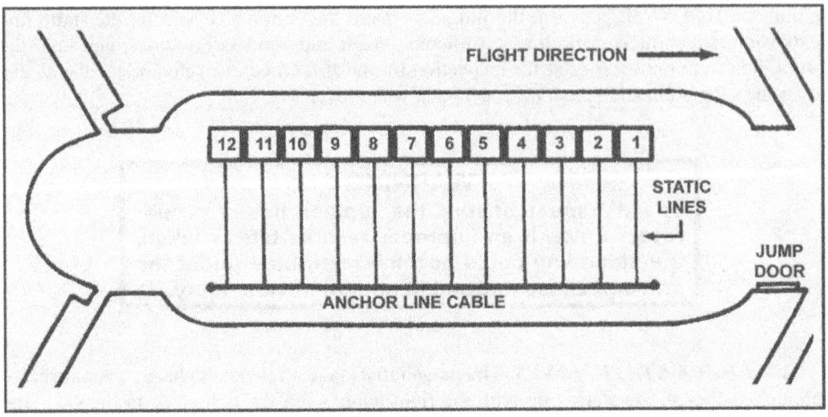

Figure 18-7. CH-46 door jump.

b. **Jump Commands for Ramp Jump.**

(1) ***GET READY.*** All parachutists in the first stick unfasten their seat belts.

(2) ***STAND UP.*** Parachutists stand up and move to the port side (Figure 18-8).

(3) ***HOOK UP.*** When the command HOOK UP is given, the jumpers ensure the cotton sleeve is on the static line extension, attach their static line snap hooks, insert and bend safety wires, and hold the static line snap hook exposed for inspection by the JM. Once inspected, place the sleeve over the snap hook and take a reverse bite at waist level.

(4) ***CHECK STATIC LINES.*** The parachutists take a reverse bight at waist level of about 8 inches in the static line with the right hand, keep the arm close to the side, and check the static line of the parachutist to the front.

(5) ***CHECK EQUIPMENT.*** All parachutists check their equipment.

(6) ***SOUND OFF FOR EQUIPMENT CHECK.***

(7) ***STAND BY.*** Number 1 assumes a standing position at the ramp hinge. The remaining parachutists close up the interval.

(8) ***GO.*** Number 1 walks off the starboard rear of the ramp and assumes a normal body position. The remaining parachutists follow and exit in the same manner as number 1, maintaining a 1-second interval between parachutists. Upon exit of all parachutists, the crew chief or static JM recovers all static lines.

WARNING

Parachutists walk off the ramp (which is lowered to 11 degrees below centerline gage) in a crouched position to avoid hitting their heads on the upper ramp door. They do *not* make a vigorous exit. Less than a 1-second interval between parachutists may result in entanglement of parachutists and static lines. Upon exit of all parachutists, the crew chief or static JM recovers all static lines.

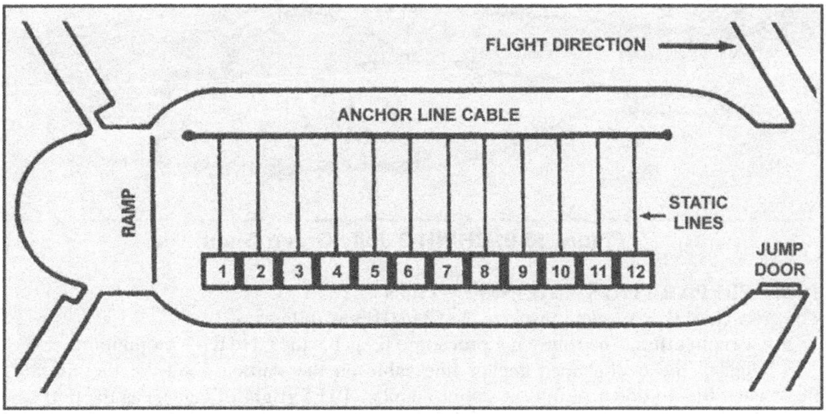

Figure 18-8. CH-46 ramp jump.

18-8. SAFETY PRECAUTIONS

On a single pass over the DZ, only the number 1 parachutist jumps with a CWIE from either the ramp or the door. The static JM or AJM ensures that parachutists are hooked up consecutively (1 through 12). The JM ensures that seats along the starboard side are secured in the up position when parachutists are jumping from the personnel door, or that seats along the port side are secured in the up position when they are jumping from the ramp. The speed of the aircraft is 80 to 90 knots when jumping. The minimum drop altitude is 1,250 feet AGL when jumping at 90 knots and 1,500 feet AGL when jumping at 80 knots. Either the JM or AJM must remain with the aircraft to handle the D-bags and towed jumpers. The JM spots the aircraft using WSVC, CIRS, or GMRS. The JM may spot from the crew chief door or the ramp as long as a safety is controlling the jumpers. If spotting from the ramp and jumping, the JM leads the stick and the AJM stays with the aircraft.

Section III. CH/HH-3 JOLLY GREEN GIANT (USAF)

The CH/HH-3 is a twin-engine, single-rotor, medium-transport helicopter. Fifteen combat-equipped parachutists can jump from this aircraft (Figure 18-9, page 18-12).

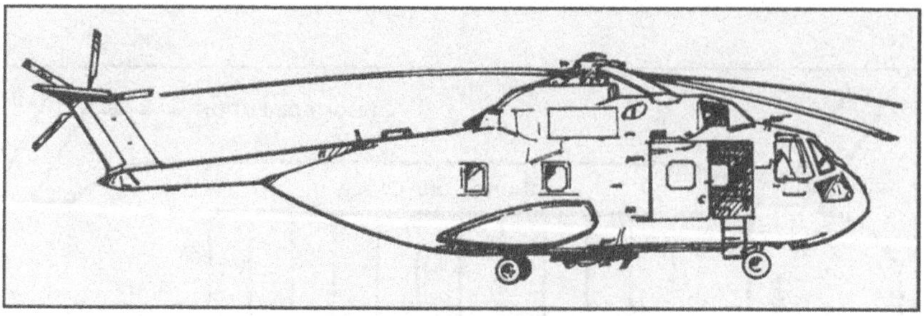

Figure 18-9. CH/HH-3 Jolly Green Giant.

18-9. PREPARATION AND INSPECTION

The crew chief prepares and inspects the CH/HH-3 as follows:

a. **Preparation**. The following procedure prepares the CH/HH-3 for jumping:

(1) Install the oval-shaped anchor line cable on the starboard side of the aircraft's floor using the tie-down fittings at station number 193.5 (right of center) as the port side forward attachment point, station number 212.5 as the starboard side forward attachment, station number 256.5 as an intermediate starboard side attachment point, and station number 276.5 as the starboard side rear and port side rear attachment points (Figure 18-10).

(2) The anchor line cable is constructed of 1/4-inch diameter, 6,400-pound test steel cable. Thread the anchor line cable through four static line snap hooks, where the static line is normally attached. These static line snap hooks connect the anchor line cable to the tie-down fittings. Complete the oval by overlapping both ends of the steel cable, then by securing the overlap with four cable clamps spaced intermittently between the swaged cable ends.

(3) Manufactured cables have the date of initial manufacture and weight testing capacity (2,500 pounds) permanently marked on the starboard side forward static line snap hook. Inspect cables each time the anchor line cable is installed for jumping. Remove from service cables showing excessive wear, corrosion, or more than three broken strands per inch. Weight test cables to a 2,500-pound capacity within each 12-month calendar period. (For example, cables weight tested on 1 January will be due weight testing by 31 January the following year.) Document annual weight testing on a DD Form 1574, which is attached to the cable.

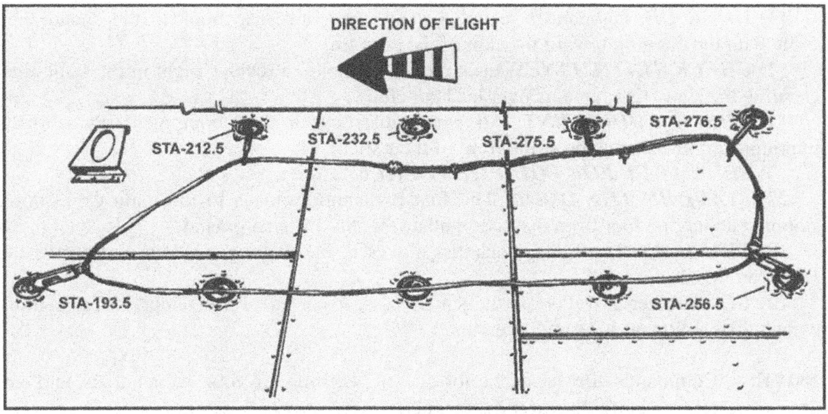

Figure 18-10. CH/HH-3 anchor line cable.

b. **Inspection**. Before enplaning, the JM and pilot, or pilot's representative, jointly inspect the aircraft to determine the following:

(1) The main cabin door is secured to the rear and taped, including the door handle and latch.

(2) Any external cargo slings are removed before conducting jump operations.

(3) All protruding objects near the doors are removed or taped. The penetrator is removed before conducting jump operations.

(4) Safety belts are installed for all parachutists and are extended completely to ensure positive hookup while seated.

(5) Seats are fastened securely in the down position with backs loose (except the four seats raised for anchor line cable installation).

(6) The anchor line cable is tight and free of frays.

(7) The deck is clean and free of oil and water.

(8) The flight engineer's headphones and the JM intercom cord are available and function properly.

18-10. LOADING TECHNIQUES AND SEATING CONFIGURATION

Parachutists enter the aircraft through the starboard side cabin door with their static line over their right shoulder. They enter the aircraft in reverse stick order with numbers 1 through 8 seated on the port side and numbers 9 through 15 seated on the starboard side.

18-11. JUMP COMMANDS AND PROCEDURES

Jump commands and procedures on the CH/HH-3 are as follows:

a. **Time Warnings**. The 6-minute and 1-minute warnings are given by the pilot to the flight engineer, who in turn relays them orally and by hand signals to the JM. Due to the limited space available to hook up, only four parachutists are airdropped each pass.

b. **Jump Commands**. The commands on the CH/HH-3 are—

(1) ***GET READY.*** The first four parachutists remove seat belts.

(2) ***STAND UP.*** Parachutists stand up and move to the anchor line cable.

(3) ***HOOK UP.*** Parachutists connect their static line snap hook to the anchor line cable with the opening toward the skin of the aircraft.

(4) ***CHECK STATIC LINES.*** Each parachutist takes a reverse bight in the static line, ensuring the static line remains over his bent elbow.

(5) ***CHECK EQUIPMENT.*** All parachutists check their equipment. Parachutists jumping equipment must be at the front of their stick.

(6) ***SOUND OFF FOR EQUIPMENT CHECK.***

(7) ***STAND IN THE DOOR.*** The first parachutist moves to the main cabin door, stopping about one foot from the door, and awaits the JM's commands.

(8) ***STAND BY.*** The first parachutist moves to the main cabin door and awaits the JM's commands.

(9) ***GO.*** Number 1 walks off the starboard rear corner of the ramp. The remaining parachutists follow at 1-second intervals.

NOTE: Commands are repeated for the next group of four parachutists and are repeated until the aircraft is empty.

WARNING

Parachutists walk out the door 90 degrees to the aircraft in a crouched position to avoid hitting their heads on the upper door frame. They do not make a vigorous exit.

c. **Exits.** Parachutists step out the same as exiting the tailgate of a fixed-wing aircraft, maintaining about a 1-second interval between parachutists. Less than a 1-second interval may result in entanglement of parachutists and static lines. Upon exit of all parachutists of each pass, the JM or flight engineer recovers all deployment bags and static lines.

18-12. SAFETY PRECAUTIONS

Approaching or loading the aircraft is performed only after visual clearance by the pilot or flight engineer. Before clearing any parachutists to jump, the JM confirms that the main gear is in the up position. The speed of the aircraft during all jump operations is between 70 knots and 90 knots indicated air speed.

CHAPTER 19

NONSTANDARD AIRCRAFT USED DURING AIRBORNE OPERATIONS

This chapter contains aircraft descriptions, JM procedures, and aircraft preparation techniques for nonstandard rotary-wing and fixed-wing aircraft. The aviation supporting unit prepares the aircraft for equipment and personnel drops to include seat and door removal and installation or rearrangement of seat belts. The installation of a field-expedient anchor line cable is the jumpmaster's responsibility. Aircraft preparation is usually accomplished jointly by the loadmaster/crew chief and JM. These aircraft are service tested and approved for personnel airdrop operations.

Section I. MODIFICATIONS TO JUMP COMMANDS AND JUMPERS' MOVEMENT IN NONSTANDARD AIRCRAFT

On some nonstandard aircraft, jumpers are required to shuffle in the aircraft and assume a stand-in-the-door position. The standard jump commands are modified by substituting the command STAND IN THE DOOR for STAND BY. The parachutists execute the shuffle and the stand-in-the-door position in the following manner.

19-1. SHUFFLE

The shuffle is a method of moving to the jump door without losing balance or tripping. To perform the shuffle—

a. The jumper's outboard arm is extended down and out to assist in maintaining balance and to assume the door position. The other hand grasps the static line in the correct bight for the aircraft.

b. Both feet are slightly spread, directly beneath the body, and staggered 6 to 8 inches. The jumper faces the rear of the aircraft and places his foot nearest the side of the aircraft forward; this foot is the shuffle foot. The foot nearest the center of the aircraft is the trail foot.

c. The jumper moves by stepping forward with his shuffle foot 6 to 8 inches and then with his trail foot. He keeps both feet staggered in the same relative heel-and-toe position throughout the shuffle.

19-2. STAND IN THE DOOR

At the command STAND IN THE DOOR—

a. The jumpers shuffle toward the jump door, and the number 1 jumper hands his static line to the safety.

b. The number 1 jumper ensures his arm is not entangled with the static line and shuffles into the door so that the toe of his outboard foot is extended about 2 inches over the edge of the doorsill or jump platform and his trail foot is about 6 inches to the rear.

c. The number 1 jumper keeps his feet shoulder-width apart. His weight is equally distributed on both feet, knees are bent, upper body is straight, head and eyes are to the front, and hands are outside the aircraft, with fingers extended and joined. He is in a position of "coiled alertness" and is ready to exit the aircraft without further weight adjustment. His hands are not grasping the aircraft but are used to maintain balance.

d. The number 2 jumper is in the shuffle position, roughly even with the leading edge of the jump door and facing the rear of the aircraft.

e. Follow-on jumpers close up behind the preceding jumper, using the shuffle position to maintain balance.

19-3. GO

At the command GO, the number 1 jumper makes a vigorous up-and-out exit, 6 inches up and 36 inches out, depending on aircraft requirements. The exit action is gained from the legs alone, and the hands are used only to guide the jumper. On some aircraft, the vigorous exit is essential to avoid hitting the aircraft or coming in contact with another jumper beneath or behind the aircraft. The number 2 jumper and all following jumpers—

a. Shuffle toward the jump door.

b. Hand the static line to the safety and ensure the arm is not entangled with the static line.

c. Shuffle into the door, take up a proper door position, and exit the aircraft without command, maintaining a 1-second interval between jumpers.

Section II. C-7A CARIBOU

The C-7A is a high-wing transport powered by two piston engines (Figure 19-1). A total of 24 parachutists may be dropped using the ramp or doors. The ramp is normally used for dropping parachutists.

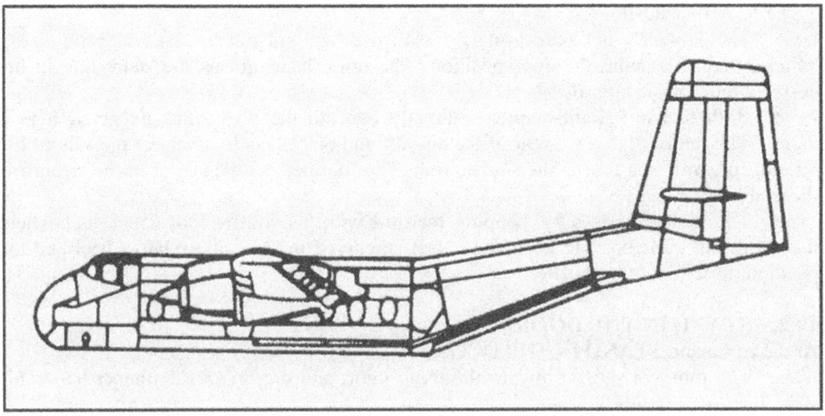

Figure 19-1. C-7A Caribou.

19-4. SEATING CONFIGURATION

Twenty-four parachutists are seated in two 12-parachutist sticks. Parachutists are loaded over the loading ramp or through the doors. The odd-numbered personnel are seated on the starboard side, and even-numbered personnel are seated on the port side.

19-5. SUPERVISORY PERSONNEL REQUIRED

Three personnel supervise safety procedures: one JM who performs standard aircraft check procedures, one safety NCO, and one loadmaster.

19-6. ANCHOR LINE CABLE ASSEMBLIES

There are two anchor line cable assemblies in the C-7A.

a. The anchor line cable for ramp jumps runs from the reinforced anchor line attachment plate on the forward bulkhead to the anchor line connector near the right side of the aft starboard door.

b. The anchor line cable for door jumps runs from the reinforced anchor line attachment plate down the center of the cargo compartment. It is permanently installed.

19-7. JUMP COMMANDS

Jump commands for the C-7A are as follows:

a. **GET READY**. Jumpers respond in the same manner as for other fixed-wing aircraft.

b. **PORT SIDE PERSONNEL, STAND UP**. Jumpers on the left side of the aircraft stand up, raise and secure their seats, and face the ramp of the aircraft.

c. **STARBOARD SIDE PERSONNEL, STAND UP**. Jumpers on the right side of the aircraft stand up, raise and secure their seats, and face the ramp of the aircraft.

d. **HOOK UP**. Even-numbered jumpers hook up between the odd-numbered jumpers to form a continuous stick of jumpers. The jumpers detach the static line snap hook from the top carrying handle of the reserve parachute and hook up to the anchor line cable with the open portion of the snap hook facing outboard, ensuring that the snap hook locks properly. The safety wire is inserted in the hole, pointing toward the rear of the aircraft, and folded down. The static line is controlled by each parachutist in a reverse bight at waist level in the left hand.

e. **CHECK STATIC LINES, CHECK EQUIPMENT, and SOUND OFF FOR EQUIPMENT CHECK**. These are executed in the same manner as with other fixed-wing aircraft.

f. **STAND BY/STAND IN THE DOOR**.

(1) ***STAND BY.*** The number 1 jumper, upon receiving the command STAND BY, assumes a standing position near the starboard side of the ramp hinge. The number 2 jumper stands on the port side of the aircraft slightly to the right of jumper number 1. The remaining personnel close up the interval behind the number 1 jumper.

(2) ***STAND IN THE DOOR.*** The proper door position is taken by the parachutists in both doors, with the appropriate foot resting on the elevated doorsill. There is no jump platform.

g. **GO**. Personnel exit the aircraft at 1-second intervals.

19-8. RAMP JUMPING

The number 1 jumper, upon receiving the command GO, walks off the port side rear corner of the ramp. The remaining jumpers follow at a 1-second interval. After the command GO, each jumper visually checks his body position for correctness and begins the 4000-count. (See Figure 19-2, page 19-4, for C-7A configuration for ramp jumping.)

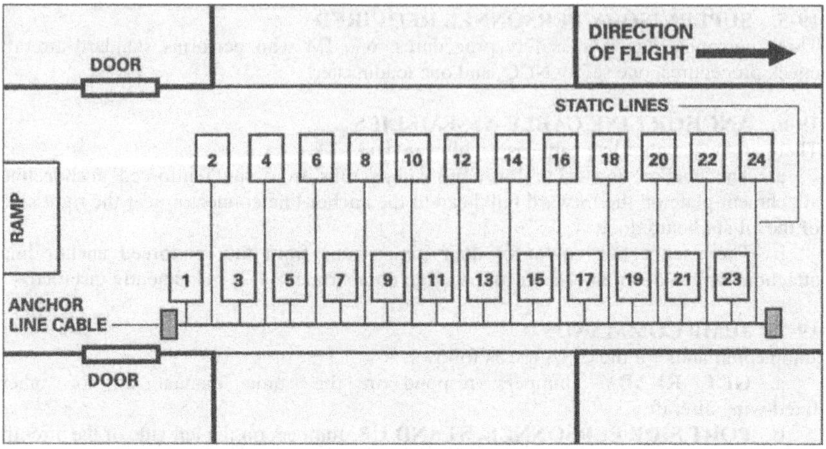

Figure 19-2. C-7A configuration for jumping from the ramp.

19-9. DOOR JUMPING

When the troop doors are used, simultaneous exits must not be made. Number 1 jumper exits the starboard door and number 2 exits the port door 1 second after number 1. The remaining parachutists alternate in numerical order at 1-second intervals. (See Figure 19-3 for C-7A configuration for door jumping.)

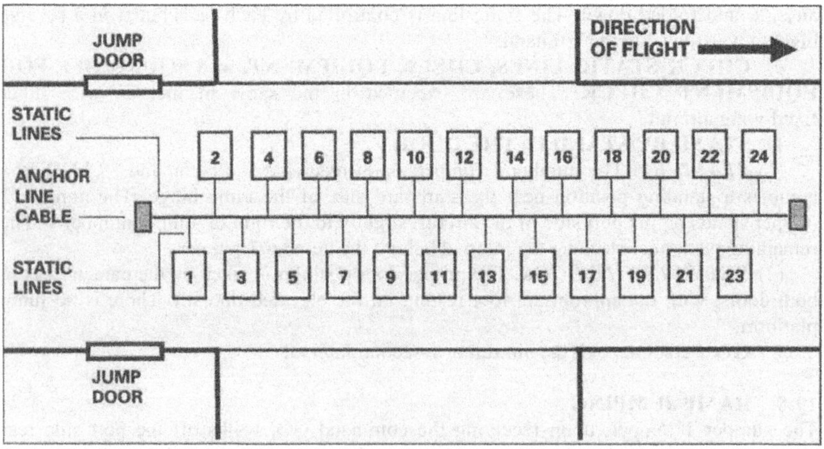

Figure 19-3. C-7A configuration for jumping from the doors.

19-10. SAFETY PRECAUTIONS

Safety precautions for the C-7A are as follows:

a. **Parachutists.**

(1) Parachutists ensure that all seats are secured in the up position when they stand to hook up. During extreme air turbulence, parachutists take a short bight on the static line and use the center anchor line to steady themselves.

(2) All parachutists remain off the ramp while it is being lowered to the 15-degree incline for aft end jumping. Parachutists walk down the ramp with feet spread wide to prevent striking the side of the aircraft. Upon exit from the aircraft, the parachutist brings the feet and knees together to form a tight body position. When following heavy equipment loads, parachutists exit between the roller conveyers of the aerial unloading kit.

b. **Jumpmaster.**

(1) The JM or safety ensures personnel are hooked up in an alternating manner to the same anchor line cable and form one continuous stick of jumpers.

(2) For door jumping, the JM or safety taps out the jumpers alternately to preclude a simultaneous exit from both sides of the aircraft.

(3) Normally, a safety is required on the aircraft, but if no safety personnel are in the aircraft, the JM jumps last. He must hook up to the center anchor line cable and exercise caution to control his own static line and ensure it does not become fouled.

NOTE: The left troop door may be removed before the operation to allow the JM to look for the DZ. If worn, the restraint harness is attached to the centerline anchor cable as a safety measure.

c. **Equipment.**

(1) When adjustable individual weapon cases are jumped from the doors, they must be reduced to 36 inches in length.

(2) When accompanying supplies and equipment are dropped from the doors, the bundles must be standard air delivery containers no larger than 40 by 24 by 36 inches.

(3) When ramp bundles are dropped, either the 15-foot static line with drogue or the breakaway static line may be used. When door bundles are dropped, the 15-foot static lines with drogues are used with cargo parachutes. The ramp roller conveyor section for the air unloading kit is installed on the port side of the ramp and is used to assist in ejecting the bundles from the ramp. Parachutists number 1 and number 2 push the bundles.

d. **Aircraft.**

(1) The speed of the aircraft during the jump will not be less than 90 nor more than 120 knots.

(2) When jumping from the doors, the crew chief must remove the doors and tape the rear portion of the door frames prior to takeoff.

(3) The rear tie-down ring, located beneath the tail section, should be removed prior to jumping. The ring can be unscrewed with a breaker bar or similar device.

Section III. C-23B/B+ SHERPA

The C-23B/B+ Sherpa is a twin-engine, nonpressurized, turboprop aircraft (Figure 19-4). It has a cruise speed of 180 knots with a range of 800 nautical miles. The aircraft can drop 12 combat-equipped parachutists or 16 Hollywood parachutists in the airdrop configuration. This is a base planning figure; actual troop capacity may vary due to aircraft limitations based on weight, density, altitude, and fuel loads. Troops may be loaded over the ramp or through the port side door.

This section outlines the procedures regarding the conduct of airborne operations by USASOC units from the C-23B/B+ Sherpa. The primary jump door for personnel and cargo is the rear ramp door. Only over-the-ramp procedures are authorized for static line personnel parachute operations. Only military free fall operations are authorized from the port side door. **Use of the port side door is *no longer authorized* for static line personnel jump operations.** Static line personnel drops using the port side jump door are authorized ***only for emergencies***.

Figure 19-4. C-23B/B+ Sherpa.

19-11. DROP PROCEDURES

The primary method for determining the exit point for jumpers for static line operations is using wind drift indicators (WDI). GMRS and VIRS can also be used based on the situation and the mission. The aircraft is also capable of a GPS release if the pilot is given the release point coordinates. A thorough briefing between the aircrew and all key personnel is mandatory before any operations involving the C-23B/B+. Standard drop altitude and speed is 1,500 feet AGL at 105 knots. Military free fall operations can be conducted up to 17,500 feet MSL.

19-12. SEATING CONFIGURATION

Parachutists are seated in two sticks along the port and starboard side of the aircraft (Figure 19-5). Numbers 1 through 8 are seated on the port side and numbers 9 through 16 are seated on the starboard side.

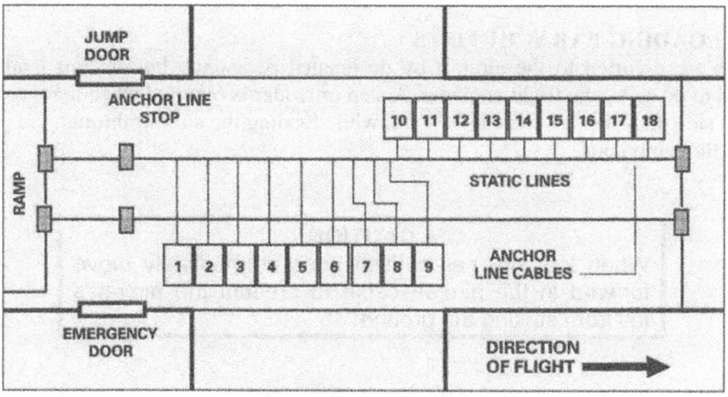

Figure 19-5. Seating configuration.

19-13. ANCHOR LINE CABLE ASSEMBLIES

There are two anchor line assemblies located overhead and running the length of the cabin down the center. The cables run from the reinforced anchor line attachment plate on the forward bulkhead to the anchor line connector at the center of the ramp hinge. Only the starboard side anchor line cable is used for personnel para-drop operations. Either cable can be used for cargo para-drop operations.

19-14. STATIC LINE RETRIEVAL SYSTEM

The static line retrieval system is a 5,000-pound winch located forward in the cabin on the floor and against the bulkhead. The retrieval cable runs up from the winch and along the port side anchor line cable and is attached to the starboard side anchor line cable forward of the anchor line stop. The retrieval system is operated from the rear of the cabin by the flight engineer. It is only operated in case of a towed jumper.

19-15. SUPERVISORY PERSONNEL REQUIRED

The supervisory personnel required for the C-23B/B+ Sherpa are as follows:

a. **Jumpmaster.** The jumpmaster will lead the stick out when jumping. The aircraft may be jumped using a nonjumping/static jumpmaster.

b. **Safety.** The safety will be nonjumping. Either one or two safeties may be used when jumping this aircraft.

c. **Flight Engineer.** The flight engineer is responsible for all operations in the cabin.

19-16. PREPARATION AND INSPECTION

The jumpmaster and the flight engineer jointly inspect the aircraft. The JM should follow basic aircraft inspection criteria as outlined in the JM handbook. At a minimum, inspection should include:

- Seat configuration and seatbelts (correct number and location).
- Static line retrieval cable and winch (attached correctly and secured).
- Jump lights (may have to wait until the aircraft is powered up).

19-17. LOADING PARACHUTISTS

Jumpers are escorted to the aircraft by designated personnel, but are not loaded until directed to do so by the flight engineer. A step or ladder is required when loading through the port side door. Rucksacks can be worn while loading the aircraft through the port side door or the ramp door.

CAUTION

When loading, parachutists must immediately move forward in the aircraft cabin to prevent the aircraft's tail from striking the ground.

a. **Cold Loading.** Cold loading is when the aircraft is shut down and the engines are not turning. Jumpers may be loaded over the rear ramp or the port side door. The C-23B/B+ has a double-hinged ramp that operates differently from conventional ramps on other aircraft. During ground operations, the ramp door can be lowered to its lowest position (resting on the ground) and equipment and jumpers can easily be loaded. The ramp can also be opened to the half-lowered position. The port side door with steps may also be used during cold loading.

b. **Hot Loading.** Hot loading is when the engines are turning. During multilift operation, the aircraft may be hot loaded in order to expedite the airborne operation. Jumpers may be loaded by either the ramp door or through the port side door. The ramp cannot be lowered to the ground during hot-load operations. After loading the aircraft, the jumpers must immediately take their seats and fasten their seatbelts. The safety ensures the jumpers are secured and signals the JM when completed. The JM then signals the flight engineer that he is ready for takeoff.

WARNING

Hot loading aircraft is dangerous. Special control measures should be implemented to ensure jumpers and ground personnel remain clear of the propellers.

19-18. JUMP COMMANDS AND TIME WARNINGS

Jump commands and times warnings are as follows:

a. **Jump Commands.**

- GET READY!
- PORT SIDE PERSONNEL, STAND UP!
- STARBOARD SIDE PERSONNEL, STAND UP! (if required).
- HOOK UP! (gates toward starboard skin).
- CHECK STATIC LINES!

- CHECK EQUIPMENT!
- SOUND OFF FOR EQUIPMENT CHECK!
- STAND BY! (30 seconds).
- FOLLOW ME! (jumping JM).
- GO! (static JM).

b. **Time Warnings.**

(1) ***20-Minute Warning.***

(a) *Flight Engineer.* Gives JM verbal and visual time warning.

(b) *Jumpmaster.* Gives time warning to jumpers.

(c) *Safety.* If jumping rucksacks, releases the cargo tie-down straps and facilitates the attachment of the jumpers' rucksacks.

(d) *Jumpers.* If jumping rucksacks, stand up; secure seatbelts and fold seats to the upright position and secure; attach rucksacks for jumping.

(2) ***10-Minute Warning.***

(a) *Flight Engineer.* Gives JM verbal and visual time warning.

(b) *Jumpmaster.* Hooks up his static line to the **starboard side** anchor cable. Hands static line to the safety. Begins jump commands.

(c) *Safety.* Takes static line from JM.

(d) *Jumpers.* All rigging completed.

(3) ***6-Minute Warning.***

(a) *Flight Engineer.* Gives JM verbal and visual time warning. After aircraft slow-down, the port side door and the rear ramp door are opened. When the rear ramp door and port side door are open and secure, the JM is signaled that the door and ramp are ready. Red light comes on.

(b) *Jumpmaster.* Continues giving jump commands. Upon completion of jump commands, moves to the port side door to begin spotting procedures, if required.

(c) *Safety.*

- Procedure with one safety: After the command of CHECK EQUIPMENT, hands the JM's static line to the JM and begins inspecting the jumpers from forward to the rear. After completing the inspection, he again secures the JM's static line. After the JM completes the jump commands, moves to the port side door with the JM and controls the JM's static line.
- Procedure with two safeties: Rear safety continues to maintain control of JM's static line while forward safety conducts standard jumper safety checks.

(d) *Jumpers.* Stand up and secure the seats. Upon the command of HOOK UP, jumpers hook up to the **starboard side** anchor line cable, gates facing starboard fuselage. Follow jump commands from JM. Take **standard bight** on static line.

(4) ***1-Minute Warning*** (all procedures).

(a) *Flight Engineer.* Gives verbal and visual time warning to the JM.

(b) *Jumpmaster.* Continues to monitor flight path from the port side door and identifies the DZ. Announces, "ONE MINUTE" to the jumpers.

(c) *Safety.* Controls JM's static line.

(d) *Jumpers.* Keep eyes on JM.

(5) ***30 Seconds*** (JM release procedures using GMRS or WDI).

(a) *Flight Engineer.* Ensures green light comes on.

(b) *Jumpmaster.* Checks green light. Tracks panels/exit point. Gives command of STAND BY.

(c) *Safety.* Controls JM's static line.

(d) *Jumpers.* Keep eyes on JM.

(6) ***30 Seconds*** (pilot release procedures using GPS/VIRS).

(a) *Flight Engineer.* Moves to starboard side of aircraft.

(b) *Jumpmaster.* Gives the command "STAND BY," turns toward the open ramp, takes control of his static line from the safety, and keeps eye on jump lights.

(c) *Safety.* Gives JM his static line and moves to starboard side of aircraft next to flight engineer.

(7) ***10 Seconds*** (JM release procedures).

(a) *Flight Engineer.* Ensures he is clear of the ramp.

(b) *Jumpmaster.* Keeps the panels in sight. Stands up and turns to face the ramp. Takes the static line from the safety. Waits until panels are 90 degrees from aircraft.

(c) *Safety.* Gives static line to the JM and moves to the starboard side of the aircraft.

(d) *Jumpers.* Keep eyes on JM.

(8) ***Exit*** (JM release procedures using GMRS/WDI).

(a) *Flight Engineer.* Ensures he is clear of the ramp.

(b) *Jumpmaster.* Identifies exit point, gives the command "FOLLOW ME," and walks straight out the ramp.

(c) *Safety.* Controls the jumper interval (one second). Retrieves static lines.

(d) *Jumpers.* Exit the ramp straight at one-second intervals.

(9) ***Exit*** (pilot release procedures using GPS/VIRS).

(a) *Flight Engineer.* Ensures he is clear of the ramp.

(b) *Jumpmaster.* When green light is illuminated, gives the command "FOLLOW ME" and walks out the ramp along the port side.

(c) *Safety.* Controls the jumper interval (one second). Retrieves static lines.

(d) *Jumpers.* Exit the ramp straight along the port side at one-second intervals.

CAUTIONS

Low overhead clearance may require jumpers to duck their heads while exiting the ramp door.

Jumpers must walk STRAIGHT out the ramp and along the port side fuselage and NOT at a 45-degree angle towards the center of the ramp.

Safety must remain clear of the ramp door and against the starboard side fuselage while jumpers exit.

19-19. CARGO OPERATIONS

The C-23B/B+ Sherpa is capable of both low level and high altitude cargo delivery operations. Bundle weight on the ramp should not exceed 500 pounds.

a. **Cargo Airdrops Without Personnel.** The following are procedures for cargo airdrops without personnel.

(1) The pilot sets up the approach, airspeed, and altitude. He commands the flight engineer to "STAND BY."

(2) The flight engineer ensures the bundle static line is hooked up and moves the cargo to the edge of the ramp.

(3) The pilot gives countdown to the flight engineer—"5, 4, 3, 2, 1, Now!"

(4) The flight engineer releases the cargo over the ramp and retrieves the static line/clevis.

b. **Cargo Airdrops with Personnel.** The following are procedures for cargo airdrops with personnel.

(1) The JM and safety coordinate and rehearse cargo release procedures with the flight engineer before the mission.

(2) Cargo may be released before or after jumpers exit.

(3) The safety retrieves static line/clevis after the drop.

NOTE: Based on the mission, the JM selects either the breakaway or nonbreakaway 15-foot static line.

CAUTIONS

No more than four jumpers are authorized aft of the port side jump door prior to exit.

Excessive weight load and cargo shift in the ramp door area prior to exit should be avoided.

19-20. MILITARY FREE-FALL OPERATIONS

All MFF operations are conducted IAW FM 31-19.

Section IV. C-27A (AERITALIA G-222)

The C-27A is a pressurized, medium transport aircraft developed from the Aeritalia G-222. It is a twin-engine, high-wing-mount, tailgate-equipped aircraft that is similar to a downsized C-130. The C-27A can carry 34 fully equipped combat troops, 28 static line parachutists, 34 military free fall (MFF) parachutists, or 16 MFF parachutists on oxygen. It can airdrop up to six CDS bundles. Typical internal loads are two HMMWVs or three full-sized 463L pallets that are turned sideways. Static line parachutists may be dropped using either of the two jump doors but may not use the ramp. MFF personnel may use both jump doors or the ramp.

19-21. SEATING CONFIGURATION

The seating configuration for the C-27A is as follows:

a. **Stick Configuration**. Parachutists are assembled into two sticks of jumpers. Jumper number 28 (the JM) is seated on the port side of the aircraft forward of the jump door. Forward of him is number 13, then numbers 1 through 12. Jumper number 27 (the AJM) is seated on the starboard side of the aircraft just forward of the jump door. Forward of him is jumper number 26; then jumpers 14 through 25. The safeties sit on each side to the rear of the jump doors.

b. **Anchor Line Cables**. There are two anchor line cable assemblies in the C-27A. The anchor line cables are from the attachment point on the forward bulkhead, through the anchor line support bracket just behind both doors, then to the side of the aircraft over the tailgate. Each parachutist is issued main and reserve parachutes. Each parachutist is responsible for inspecting his parachute for safety wires and for fitting of the parachute harness.

19-22. SUPERVISORY PERSONNEL REQUIRED

To ensure command and control when jumping one jump door, one jumpmaster, one nonjumping safety, and an airdrop-certified USAF loadmaster are required. These personnel requirements double when using both troop doors.

19-23. JUMP COMMANDS

Jump commands for the C-27A are as follows:

a. **Jump Commands**. The following nine jump commands are used whether the doors or the ramp are jumped.

(1) ***GET READY.*** Jumpers respond in the same manner as for other fixed-wing aircraft.

(2) ***PORT SIDE PERSONNEL, STAND UP.*** Jumpers on the left side of the aircraft stand up, raise and secure their seats, and face the ramp of the aircraft.

(3) ***STARBOARD SIDE PERSONNEL, STAND UP.*** Jumpers on the right side of the aircraft stand up, raise and secure their seats, and face the ramp of the aircraft.

(4) ***HOOK UP.*** The jumpers detach the static line snap hook from the top carrying handle of the reserve parachute and hook up to the anchor line cable with the open portion of the snap hook facing inboard, ensuring that the snap hook locks properly. The safety wire is inserted in the hole, pointing toward the rear of the aircraft, and folded down.

(5) ***CHECK STATIC LINES, CHECK EQUIPMENT, and SOUND OFF FOR EQUIPMENT CHECK.*** These commands are executed in the same manner as with other fixed-wing aircraft.

(6) ***STAND IN THE DOOR.*** A proper door position is taken by the parachutist.

(7) ***GO.*** Personnel exit the aircraft at 1-second intervals.

NOTE: Port side personnel exit first. After all port side personnel except the JM have cleared the aircraft, the starboard side personnel (except the AJM) exit the aircraft. JMs "clear to the rear" of the aircraft, the AJM exits, and the JM follows. All jumpers exit using the stand-in-the-door type door exit.

b. **Modification of Jump Commands**. At the 10-minute warning, the JM and AJM send jumpers number 13 and number 26 to the forward end of the aircraft to take their correct place in the stick. The two seats forward of both jump doors are folded upright and secured. The two safeties fold their seats upright and secure them.

19-24. SAFETY PRECAUTIONS

Safety precautions for the C-27A are as follows:

a. **Jumpmasters**. The JMs inspect the door platforms after the doors are opened. The JMs hook up to the cables on their side of the aircraft. They control and observe the personnel as they exit. JMs exit last.

b. **Safeties**. The safeties assist the loadmaster in installation of the door platforms if they are to be installed in-flight. They ensure personnel hook up **consecutively** and that jumpers number 13 and 26 are in the correct position. The safeties control the static lines as the jumpers approach the door to exit. They assist the loadmasters when retrieving the deployment bags.

c. **Equipment**. Standard combat equipment can be jumped out of the doors, which are 36 by 75 inches. Standard door bundles (that is, A-7A/A-21) can be dropped out of the doors. The 15-foot static line with drogue is used. Troops may follow.

d. **Aircraft**. The drop speed of the aircraft is 125 knots. Both doors **cannot** be jumped at the same time. The ramp of this aircraft cannot be used for static line ramp exits.

19-25. OVER-THE-RAMP OPERATIONS

Considerations for over-the-ramp operations follow.

a. **Static Line Operations**. The erratic behavior of the deployment bags poses a serious safety hazard; the C-27A **cannot** be used for static line over-the-ramp operations.

b. **Equipment Drop**. Door bundles can be pushed off the ramp. The rollers can be installed on the ramp to aid in handling larger bundles.

c. **Military Free Fall**. MFF exits can be made over the ramp when both doors are closed. The C-27A will hold 34 MFF jumpers; however, it is recommended the number of MFF jumpers be limited to 16 when oxygen consoles are installed, due to overcrowding in the aircraft. The using unit must provide two console positions for the loadmasters to use during MFF jumps above 10,000 feet. It is very difficult for the JM to spot the release point from the aircraft during ramp exits. Therefore, the JM should not wear an ALICE pack for this type operation, and the unit should use a nonjumping JM.

19-26. JOINT PREFLIGHT INSPECTION

The C-27A is inspected as follows:

a. **JM/Aircraft Commander/Loadmaster Coordination**. The JM is responsible for informing the aircraft commander and loadmaster of the exact time sequence of prejump procedures. Following preliminary orientation, and before loading personnel onboard the aircraft, the JM and loadmaster make a joint inspection of the aircraft. The purpose of the joint inspection is to verify the readiness of the aircraft for the conduct of the mission and to take actions necessary to achieve this readiness.

b. **Exterior Inspection**. An external inspection of the aircraft is made to detect hazards to the airdrop of personnel. Particular attention is directed to those areas to the rear of the aft paratroop doors. Any protruding objects and sharp edges are removed, or padded and taped.

c. **Interior Inspection**. An interior inspection checks for the following:

(1) Any sharp edge or protrusion is securely taped and padded, as required.

(2) All equipment in the cargo compartment is securely stowed and lashed.

(3) The floor is clean and free of lubricants; no obstructions are on the walkway or along the paratroop exit route (outboard area between the safety fence and fuselage).

(4) Anchor line cables are installed.

(5) A seat and seat belt are available in the troop compartment for each parachutist.

(6) The retrieval system is installed in the aircraft with the winch cable retained in clips and free of the anchor line cable.

(7) Windscreen systems are available/installed.

(8) Jump platforms are available/installed.

(9) Jump caution lights are operational.

(10) Troop compartment lights are operational.

d. **Jumpmaster Preload Inspection of Parachutists**. The JM/AJM inspects each parachutist, parachute, and parachutist's equipment prior to loading the aircraft.

19-27. LOADMASTER BRIEFING

As soon as all parachutists are seated, the loadmaster briefs them on aircraft safety, emergency procedures, and comfort facilities.

19-28. TIME WARNINGS

Time warnings are as follows:

a. **20-Minute Warning**. JMs check personnel and equipment. Missile jump packs are attached to the parachutists, their HSPR leg straps are secured, and door bundles are moved near the personnel doors.

b. **10-Minute Warning**. Final onboard JMPI of all parachutists is complete. A verbal and visual 10-minute warning is given to the JMs by the loadmaster, and the JMs begin jump commands.

c. **Slow-Down Warning**. About 3 minutes from drop time, the jump commands are completed. Personnel doors are opened, and jump platforms are extended and locked.

d. **1-Minute Warning**. JMs alert troops and make safety checks from personnel doors. The AJM informs the JM that his side is clear and that it is safe to jump.

e. **10-Second Warning**. The loadmaster gives the JMs a visual 10-second warning. At this time the JM gives the command STAND IN THE DOOR, and the number 1 jumper assumes a proper door position. The remainder of the stick shuffles aft to close up the stick.

f. **Green Light, GO**. At the green light, the JM taps out the first man. Port side personnel exit first. After all port side personnel have cleared the aircraft, the starboard side personnel exit the aircraft. JMs "clear to the rear" of the aircraft, the AJM signals to the JM that all jumpers are clear of the aircraft, the AJM exits, and then the JM exits. All jumpers exit using the stand-in-the-door exit.

19-29. ADDITIONAL SAFETY PRECAUTIONS

Further safety precautions for the C-27A follow.

a. **Platforms, Air Deflectors, Aft Cargo Door**. The JMs must ensure that jump platforms and windscreens are available. This equipment is mandatory for each aft personnel door that is to be used.

b. **Door Bundles**. When personnel follow door bundles, the door bundle static line will be outfitted with a drogue.

c. **Movement into the Door**. Parachutists exercise caution as they move to the door to avoid becoming entangled with the static lines of preceding parachutists. This precautionary action may slow movement into and out of the door.

19-30. C-27A JUMPMASTER CHECKLIST

The jumpmaster follows this checklist.

a. **Seats**.

- Adequate seats for troop load are onboard.
- All seats have safety belts.
- Seat backs are secure.
- Seats are serviceable.
- There are no projections through seats; pairs of seats forward of each troop door have a strap attached to secure them in the upright position.

b. **Floor**.

- Nonskid surface covering is in good condition.
- Floor is clean and safe to walk on.
- Roller conveyors are stored.
- Loose equipment is secured in the cargo ramp area and does not interfere with troops.
- Equipment tie-down rings are depressed into their recesses.

c. **Jump Platforms**.

- Nonskid surface covering is present and in good condition.
- There are no cracks or bends.
- Studs are locked in seat track receptacles.
- Tie-down fitting is locked.
- All bolts and nuts are present.
- Platforms swing in and out easily.

d. **Jump Doors**.

(1) ***Ground Check.***

- There are no sharp or protruding edges on door frames.
- Doors open and close easily.

(2) ***Prior to Exit.***

- The platforms are locked into the two "keyholes" on the floor and slid to the rear of the aircraft. The large portion of the keyhole slot should be visible.
- The platform locking lever on the leading edge of the door should be in its locked position. The lug this lever controls should be engaged to the door frame.
- The platform locking lever should be taped in place to help prevent any jumpers from inadvertently unlocking it.
- The flange on the trailing side of the platform must overlap the inside of the door frame approximately 1/2 inch.

e. **Jump Lights** (five total).

- Rear at the forward left door.

- Rear of both troop doors.
- High above and to the rear of the ramp on both sides.

f. **Static Line Anchor Cable System**.
- Forward end of cable is firmly secured to bracket on bulkhead with three threads showing on turnbuckle.
- Rear of cable has a pin in it with tape.

g. **Anchor Cable**.
- Cable has no breaks.
- Cable has no frays.
- Cable has no kinks.
- Cable is clean and free of rust.
- Static line stop is present.
- Support bracket at the trailing edge of the door is locked in place to support the cables.

h. **Static Line Retrievers**.
- Motor is operational.
- Cable is secured to ceiling with one turn of double 1/4-inch cotton webbing.
- The Y attachment is in place so one retrieval cable can pull in two groups of deployment bags. (Y cable is used if both jump doors are used.)
- Retriever cables are not broken, frayed, or kinked. They are clean and free of rust.
- Phenolic block/anchor cable spool is installed on both anchor cables with the pull ring on the forward end.
- Retriever equipment is available.

i. **Emergency Equipment**.
- First aid kit is onboard (one).
- Fire extinguishers are onboard (two).
- Alarm system is operational.
- Emergency exits are operational and accessible.
- Sufficient emergency parachutes are available.

j. **Miscellaneous**.
- Day lighting system is operational.
- Night lighting system is operational.
- Airsickness bags are available.
- JM kit (extra equipment) is onboard.
- Earplugs are available.
- Heavy tape is available to secure the platform and windscreen locking lever.
- If jump platforms and windscreen are not installed in the doors, they must be secured to the upper ramp.
- All equipment and crew baggage is secured to the floor.
- During the jump briefing, the jumpers are warned to avoid striking or grabbing the door platform or windscreen locking lever on the leading edge of the door.

WARNING

It is a serious hazard to the exiting parachutist if the windscreen locking lever swings into the open door.

k. **Tailgate Drops (MFF and Bundles Only).**
- Ensure that the loadmaster installs the stops on both sides of the tailgate so that it will be level with the aircraft floor when open.
- Disengage the support bracket near the door for bundle drops that use the retrieval system to pull in the static lines.
- Secure the retrieval cable against the anchor line in several places with breakaway ties starting at the rear of the cable and ending at the tailgate hinge. This will prevent the tailgate from cutting the retrieval cable during operation.

l. **Aircraft Slow-Down Warning at 3 Minutes.**
- Doors are opened and locked in place.
- Air deflectors are extended.
- Jump platforms are locked in place.

m. **Loadmaster/JM Safety Checks.**
- Door bundles employ approximately 15-foot-long static lines with three drogue parachutes.
- Jump platform is secure and will sustain parachutist's weight.

Section V. C-46 COMMANDO/C-47 SKYTRAIN

The C-46 and the C-47 are twin-engine, short-range transport aircraft. Because of the similarity in characteristics and procedures, both aircraft are discussed in this section.

19-31. SEATING CONFIGURATIONS

Seating configurations for the C-46 and C-47 are as follows:

a. **C-46**. A total of 27 parachutists can be jumped from the C-46 using both aft troop doors. This aircraft can accommodate two sticks: a 14-man stick sits on the starboard side, and a 13-man stick sits on the port side.

b. **C-47**. A total of 24 parachutists can be jumped from the C-47 using the aft troop door. This aircraft can accommodate two sticks: a 12-man stick sits on the starboard side and a 12-man stick sits on the port side.

19-32. JUMP PROCEDURES

Jump procedures for the C-46 and C-47 are as follows:

a. **C-46 Jump Commands**.

(1) ***GET READY.*** Jumpers respond in the same manner as for other fixed-wing aircraft.

(2) ***STAND UP.*** The parachutists near the doors stand and steady themselves by firmly grasping a structural member of the cargo compartment wall.

(3) ***HOOK UP.*** The open portion of the static line snap hook is away from the parachutist and toward the floor of the aircraft when he hooks up. The elbow of the arm holding the static line is kept close to the body. The static line is controlled by each parachutist in a reverse bight at waist level in the left hand. When the anchor line cable is installed along the top of the aircraft, the static line is controlled in a standard bight held at shoulder level.

(4) ***CHECK STATIC LINES.*** Each jumper checks the man in front to ensure the static line snap hook cover on the 5-foot static line extension covers the snap hook.

(5) ***CHECK EQUIPMENT.*** Each jumper checks his equipment and checks the man in front to ensure that the man's elbow is close to his body and the static line extension hangs below and behind the arm.

(6) ***SOUND OFF FOR EQUIPMENT CHECK.*** This command is executed in the same manner as with other fixed-wing aircraft.

(7) ***STAND IN THE DOOR.*** A proper door position is taken by the parachutists in both doors.

(8) ***GO.*** Personnel exit the aircraft at 1-second intervals.

b. **C-47 Jump Commands.**

(1) ***GET READY.*** Jumpers respond in the same manner as for other fixed-wing aircraft.

(2) ***PORT SIDE PERSONNEL, STAND UP.*** The odd-numbered jumpers seated on the left side of the aircraft stand up, secure their seats in the down position, and face the rear of the aircraft.

(3) ***STARBOARD SIDE PERSONNEL, STAND UP.*** The even-numbered jumpers seated on the right side of the aircraft stand up, secure their seats in the down position, and face the rear of the aircraft.

(4) ***HOOK UP.*** The even-numbered jumpers hook up between the odd-numbered jumpers to form a staggered stick of jumpers. The static line is controlled by each parachutist in a reverse bight at waist level in the left hand.

(5) ***CHECK STATIC LINES.*** Each jumper checks the man in front to ensure the static line snap hook cover on the 5-foot static line extension covers the snap hook.

(6) ***CHECK EQUIPMENT.*** Each jumper checks the man in front to ensure that the man's elbow is close to his body and the static line extension hangs below and behind the arm.

(7) ***SOUND OFF FOR EQUIPMENT CHECK.*** Jumpers respond to this command in the same manner as with other fixed-wing aircraft.

(8) ***STAND IN THE DOOR.*** A proper door position is taken by the parachutists in both doors.

(9) ***GO.*** Personnel exit the aircraft at 1-second intervals.

19-33. SAFETY PRECAUTIONS

Safety precautions for both aircraft are as follows:

a. Parachutists ensure all seats are in the **down** position when they stand to hook up.

b. Upon exit from the aircraft, parachutists bring their feet and knees together to form a tight body position.

19-34. SAFETY PERSONNEL AND JUMPMASTER RESPONSIBILITIES

Responsibilities for the safety and jumpmaster follow.

a. **C-46.**

(1) The JM or safety ensures personnel hook up in an alternating and consecutive fashion.

(2) One safety is required when jumping the C-46. After the safety has checked each parachutist, the safety then moves aft of the door and physically controls the JM's static line.

(3) Due to the movement of the safeties and length of the aircraft, it is advisable to use a USAF BA-18 emergency parachute when using one safety for each door. When using the safety harness, two safeties are required for each door. Safety number 1 is positioned forward to inspect jumpers. Safety number 2 is positioned aft to assist the JM.

NOTE: Although the C-46 has two aft troop jump doors, normally only the **left** aft troop door is used for jumping. When both doors are used, one safety for each door is required. The JM (or safety) taps out parachutists alternately at one-second intervals to preclude simultaneous exits from both sides of the aircraft.

(4) Normally, the JM exits first when jumping this aircraft. When the JM gives the command, STAND IN THE DOOR, he moves to the door and assumes a proper door position. On the command GO, the JM looks at the number 2 man, gives the command GO, and then exits the aircraft. Once the JM exits, the safety controls the static lines and the parachutists' flow out of the aircraft.

b. **C-47.**

(1) The JM or safety ensures personnel hook up consecutively.

(2) One safety is required when jumping the C-47. After the safety has checked each parachutist, the safety then moves aft of the door and physically controls the JM's static line.

(3) Due to the movement of the safeties and length of the aircraft, it is advisable to use a USAF BA-18 emergency parachute when using one safety. When using the safety harness, two safeties are required. Safety number 1 is positioned forward to inspect jumpers. Safety number 2 is positioned aft to assist the JM.

(4) The JM must exit first when jumping this aircraft. When the JM gives the command STAND IN THE DOOR, he moves to the door and assumes a proper door position. On the command GO, the JM looks at the number 2 man, gives the command GO, and then exits the aircraft. Once the JM exits, the safety controls the static lines and the parachutists' flow out of the aircraft.

c. **Parachute Fouling.** To prevent fouling of the T-10-series or MC1-series type parachute on the aircraft when jumping the C-46 or C-47, a 5-foot static line extension must be used. This extension has a snap hook at one end and a nondetachable connector link at the other end. The connector link on the extension is attached to the snap hook of the parachute static line. The snap hook on the parachute static line is safetied with a short piece of wire covered with a canvas duck sleeve and taped in place. The extension is stowed by using rubber retainer bands, continuing to stow the extension on the pack body. When the stow is completed, the static line and extension should have four stows on the right and three stows on the left.

NOTE: When jumping the C-46 utilizing the MC1-series, parachutists use only the left aft door.

d. **Anchor Line Cables.**

(1) The C-46 has two anchor line cables. Both anchor line cables must be detached from their floor fittings and anchored to the floor in the rear section of the cargo compartment. A wooden block is used to support the cable at the anchor point.

(2) The C-47 has one permanently installed anchor line cable that must be secured to the overhead attachment points provided in the center of the aircraft.

- The aft edge of the jump door is lined with a smooth metal tubular stripping, or it is padded and taped.
- The flooring of the jump door is made smooth by the insertion of an additional plywood section to butt against the tubing and existing flooring.
- If aft troop exit doors are installed, they are opened at the 20-minute warning.
- A slowdown from cruise airspeed (when applicable) is initiated in enough time to allow drop-speed cruise two minutes prior to drop time.
- Personnel and cargo drops are normally made from 105 to 125 knots.
- When possible during personnel drops, propeller RPM should be reduced to lessen the blast effect on the jumpers.

Section VI. DC-3 (CONTRACT AIRCRAFT/CIVILIAN SKYTRAIN)

The DC-3 is the civilian version of the C-47 Skytrain. Like the C-47, the DC-3 is a twin-engine, short-range transport aircraft.

19-35. SEATING CONFIGURATION

The seating configuration on the DC-3 is as follows:

a. A total of 24 parachutists can be jumped from the DC-3 using the aft troop door. This aircraft can accommodate two sticks: a 12-man stick sits on the starboard side, and a 12-man stick sits on the port side.

b. Contract DC-3 aircraft are not rigged with paratroop seats or with individual seat belts. Jumpers are required to sit on the floor. They are restrained by one safety belt over the entire stick of jumpers. USAF 10,000-pound tie-down straps or C-3A (NSN 1670-00-447-9504) modified safety belts must be supplied by the using unit.

19-36. JUMP COMMANDS AND PROCEDURES

Jump commands and procedures for the DC-3 follow.

a. **GET READY**. Jumpers respond in the same manner as for other fixed-wing aircraft.

b. **PORT SIDE PERSONNEL, STAND UP**. The odd-numbered jumpers seated on the left side of the aircraft stand up.

c. **STARBOARD SIDE PERSONNEL, STAND UP**. The even-numbered jumpers seated on the right side of the aircraft stand up.

d. **HOOK UP**. The open portion of the static line snap hook is toward the port side of the aircraft. The elbow of the arm holding the static line is kept close to the body. The static line is controlled by each parachutist in a reverse bight at waist level in the left hand. The odd-numbered jumpers hook up first, then the even-numbered jumpers hook up between the odd-numbered jumpers to form a staggered stick of jumpers.

e. **CHECK STATIC LINES**. Each jumper checks the man in front to ensure the static line snap hook cover on the 5-foot static line extension covers the snap hook.

f. **CHECK EQUIPMENT**. Each jumper checks the man in front to ensure that the man's elbow is close to his body and the static line extension hangs below and behind the arm.

g. **SOUND OFF FOR EQUIPMENT CHECK**. Jumpers respond in the same manner as with other fixed-wing aircraft.

h. **STAND IN THE DOOR**. A proper door position is taken by the parachutists.

i. **GO**. Personnel exit the aircraft at 1-second intervals.

19-37. SAFETY PRECAUTIONS

Safety precautions for the DC-3 are as follows:

a. **Parachutists**.

(1) Jumpers hook up using a reverse bight, with the elbow of the arm holding the static line kept close to the body.

(2) Upon exit from the aircraft, jumpers bring the feet and knees together to form a tight body position.

b. **Safety Personnel Duties**.

(1) When using DC-3 aircraft, the safety also performs duties as a loadmaster. Prior to takeoff, the safety ensures all jumpers are secured and prepared for takeoff.

(2) After he has checked each parachutist, the safety moves to the aft end of the aircraft, aft of the door, and physically controls the JM's static line.

(3) The safety maintains communications with the pilots through the ICS located in the aft of the aircraft, and relays all information to the JM.

(4) Due to the movement of the safeties and length of the aircraft, it is advisable to use an USAF BA-18 emergency parachute when using one safety. When using the safety harness, two safeties are required. Safety number 1 is positioned forward to inspect jumpers. Safety number 2 is positioned aft to assist the JM.

c. **Jumpmaster**. The JM must exit first when jumping this aircraft. When the JM gives the command, STAND IN THE DOOR, he moves to the door and assumes a proper door position. On the command GO, the JM looks at the number 2 man and gives the command GO, and then he exits the aircraft. Once the JM exits, the safety controls the static lines, the parachutists' flow out of the aircraft, and the jumper interval.

d. **Anchor Line Cables**. The permanently installed anchor line cable must be secured to the overhead attachment points provided in the center of the aircraft. On the C-47, the aft anchor point for the cable is located at the aft right side of station number 542. When the anchor line cable anchor point on the DC-3 is at station number 542, all procedures relevant to the C-47 (Section V) apply.

e. **Jump Door**.

(1) The aft DC-3 cargo door must be removed and rigged for jumping prior to takeoff.

(2) The aft edge of the cargo/jump door is rigged with a smooth metal tubular stripping or padded and taped.

(3) The two aft cargo door hinges, door hasp, and the door knob are padded and taped.

Section VII. C-212 (CASA 212)

The C-212 is a twin-engine, high-wing, multipurpose light transport designed for operations involving short, rough airfields (Figure 19-6). The aircraft can transport 24 personnel in a troop lift mode and 15 combat-equipped parachutists in the airdrop mode using the port door. Troops are loaded over the ramp.

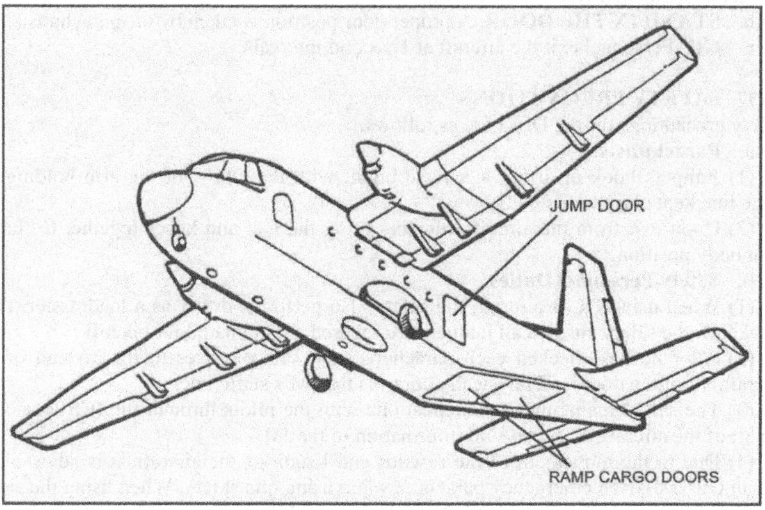

Figure 19-6. C-212 (Casa 212).

19-38. SEATING CONFIGURATION

Fifteen parachutists are seated in two sticks of jumpers (Figure 19-7). The odd-numbered personnel (eight) are seated on the starboard side and the even-numbered personnel (seven) are seated on the port side.

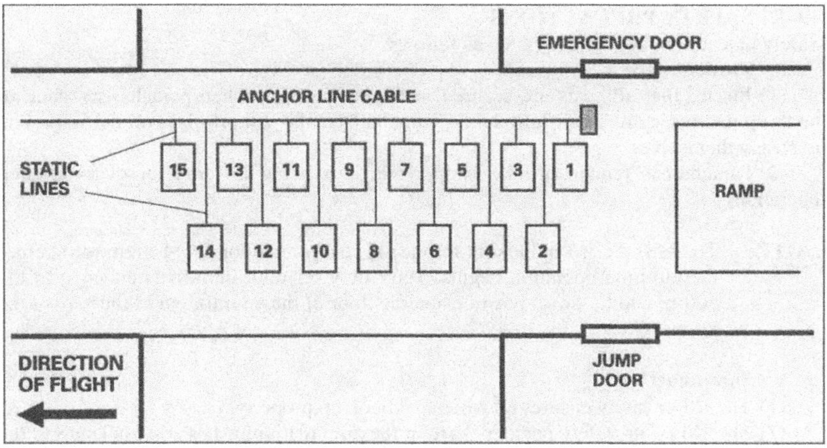

Figure 19-7. C-212 seating configuration.

19-39. ANCHOR LINE CABLE ASSEMBLY

There is one anchor line cable assembly in the C-212. It runs from the reinforced anchor line attachment plate on the forward bulkhead to the anchor line connector near the right side of the aft starboard emergency door.

19-40. SUPERVISORY PERSONNEL REQUIRED

The following personnel are required for airdrop operations from the C-212: one jumpmaster to perform standard aircraft check procedures, one safety, and one loadmaster/crew chief.

19-41. JUMP COMMANDS

The following jump commands are used with the C-212 aircraft.

a. **GET READY**.

b. **STARBOARD SIDE PERSONNEL, STAND UP**.

c. **PORT SIDE PERSONNEL, STAND UP**.

d. **HOOK UP**. On this command, the odd-numbered personnel hook up between the even-numbered personnel to form a continuous stick of parachutists, hooking the open portion of the snap hook facing inboard over the left shoulder. All parachutists take up a reverse bight.

e. **CHECK STATIC LINES, CHECK EQUIPMENT, and SOUND OFF FOR EQUIPMENT CHECK**. These commands are executed in the same manner as with other fixed-wing aircraft.

f. **STAND IN THE DOOR (Door)/STAND BY (Ramp)**. A proper exit position is taken by the parachutist.

g. **GO**. Personnel exit the aircraft at one-second intervals.

19-42. SAFETY PRECAUTIONS

Safety precautions for the C-212 are as follows:

a. **Parachutists.**

(1) Ensure that all seats are secured in the **up** position when parachutists stand to hook up. During extreme air turbulence, parachutists take a short bight on the static line to steady themselves.

(2) Parachutists remain off the ramp while it is being lowered for over-the-ramp operations.

NOTE: To assist the JM in looking for the DZ, the troop door may be removed before the airborne operation begins. The safety restraint harness is attached to the 500-pound tie-down positions on the floor of the aircraft, out of the way of the jumpers.

b. **Jumpmaster.**

(1) The JM or safety ensures all personnel hook up properly.

(2) The JM (if no safety personnel are in the aircraft) jumps last. He hooks up to the anchor line cable, ensuring his static line does not become fouled.

NOTE: On aircraft that do not have a positive communication system, the following safety measure is recommended: one ring on the alarm bell signals the JM to look at the jump light or communicate with the cockpit.

c. **Equipment.**

(1) When adjustable individual weapons cases are jumped from the door, they must be reduced to 36 inches in length.

(2) When accompanying supplies and equipment are dropped from the door, the bundles must be standard air delivery containers no larger than 40 by 24 by 36 inches.

(3) When ramp bundles are dropped, either the 15-foot static line with drogue or the breakaway static line may be used. When door bundles are dropped, the 15-foot static line with drogue is used with cargo parachutes.

(4) When ramp bundles are dropped, troops may follow out the troop door. The JM and safety or loadmaster push the bundles out.

d. **Aircraft.**

(1) Aircraft speed during the jump is 90 to 110 knots.

(2) When parachutists are jumping from the troop door, the door may be opened or removed and set into the door recess provided on the ramp.

(3) When conducting bundle operations from the ramp, the JM must close the door. The door may be opened or removed before the ramp is lowered.

(4) The Omega antenna, located beneath the tail section, must be removed prior to ramp bundle operations.

19-43. TOWED PARACHUTIST PROCEDURES

Procedures outlined for other fixed-wing aircraft will be followed for the C-212.

19-44. AIRCRAFT CONFIGURATION FOR RAMP STATIC LINE PERSONNEL AIRDROP

The loadmaster configures the aircraft. The JM should verify the configuration. Static line ramp parachute operations are authorized only when the retrieval system is operational.

a. The aircraft is configured for a static line personnel airdrop. One of each of the following items of equipment is needed:

(1) Hand winch.
(2) Static line deflector block.
(3) Retrieval bar.
(4) Retrieval strap.
(5) Extended interphone cord.
(6) 2,500-pound tie-down strap.
(7) 5,000-pound tie-down strap.
(8) One 3-foot length of 1-inch tubular nylon.
(9) Cloth-backed adhesive tape.
(10) Anchor cable.
(11) Two restraint harnesses.

b. Install and preflight-inspect the following equipment:

(1) Attach hand winch to right tie-down row in zone 1 and check for security.
(2) Inspect cable for broken wires or kinks and check for operation.
(3) Ensure static line deflector block is attached to the right side of the ramp. Cover the bolt head with tape.
(4) Inspect retrieval base on board and attaching brackets.
(5) Install and check extended interphone cord for operation.
(6) Fit and adjust restraint harnesses.
(7) Ensure that the 3-foot length of 1-inch tubular nylon and the 5,000-pound tie-down strap are secured and available for immediate use.

19-45. C-212 JUMPMASTER CHECKLIST

The jumpmaster follows this checklist for the C-212:

a. **Seats**.
- Adequate seats are available for troop load.
- All seats have safety belts.
- Seat backs are secure.
- Seats are serviceable.
- There are no projections through seats.

b. **Floor**.
- Nonskid surface covering is in good condition.
- Floor is clean and safe to walk on.
- Loose equipment is secured and does not interfere with troops.

c. **Jump Door**.
- There are no sharp or protruding edges on door frame.
- Door opens and closes easily.
- Door sits in ramp recess properly.

d. **Jump Lights**. Check sets for operation.

- Set 1—above port aft jump door.
- Set 2—above starboard aft emergency door. Check alarm bell; it is the signal for exiting.

e. **Static Line Anchor Cable System**.

(1) *Forward support beam.*

- Bolts, nuts, and safety wire are present.
- Anchor cable is attached to centerline anchor point.
- Cable bolt, locking bolt, nut, and safety wire are present.
- Check anchor line tension indicator—red line indicator should not be seen.

(2) *Anchor cable.*

- Cable has no breaks, frays, or kinks.
- Cable is clean and free of rust.
- Swage is present.

(3) *Anchor line cable aft support.*

- Cable, locking bolt, nut, and safety wire are present.

f. **Emergency Equipment**.

- First aid kits are onboard (2).
- Fire extinguishers are onboard (2).
- Alarm system is operational.
- Sufficient emergency parachutes are available.

g. **Miscellaneous**.

- Lighting system is operational.
- Airsickness bags are available.
- JM kit (extra equipment) is onboard.
- Earplugs are available.

NOTE: Loose equipment and jump door (removed) are lashed to the cargo ramp or to the rear of the forward bulkhead.

PART FIVE
Drop Zones

CHAPTER 20
PROCEDURES ON THE DROP ZONE

A drop zone is any designated area where personnel and equipment may be delivered by means of parachute or free drop. The DZ is located where it can best support the ground tactical plan; it is selected by the ground unit commander. For tactical training, the USAF assault zone availability report should be checked for an approved DZ within the tactical area. If the selected DZ is not on the AZAR, a tactical assessment must be conducted.

Section I. DROP ZONE SELECTION AND METHODS

The GUC uses the tactical analysis to select an area that can best support his mission. This section discusses several technical selection factors that must be considered.

20-1. AIR DROP AIR SPEED

Table 20-1 provides recommended drop speeds for various aircraft.

TYPE OF AIRCRAFT	DROP SPEED
UH-1	50 to 70 knots—optimum 70 knots
UH-60	65 to 75 knots—optimum 70 knots
CH-46 (USMC)	80 to 90 knots
CH-47	80 to 110 knots—optimum 90 knots
CH-53 (USMC)	90 to 110 knots
CH/HH3 (USAF)	70 to 90 knots
C-5/130/141/17/KC-130	130 to 135 knots (personnel)
C-5/130/141/17/KC-130	130 to 150 knots—optimum 130 knots for all loads (door bundles, CDS, and heavy equipment)

Table 20-1. Aircraft drop speeds.

20-2. AIRCRAFT DROP ALTITUDES

See Table 20-2 for aircraft drop altitudes.

AIRCRAFT	DAY (AGL) (feet)	NIGHT (AGL) (feet)
Rotary Wing		
(All Services, Rotary Wing))		
Personnel < 90 KTs	1,500	1,500
> 90 KTs	1,500	1,500
Bundles	300	300
Fixed Wing (Troop Carrier)*		
Personnel		
Basic Airborne Training	1,250	1,250
*Training***	1,000	1,000
Tactical	800	800
Door Bundles	300	500
Heavy Equipment	1,100	1,100

* AWADS/SKE – Drop altitude is 500 feet above highest obstacle that falls within 3 miles either side of DZ run-in. CDS using G-13 or any other parachute, minimum drop altitude is 400 feet AGL. CDS using G-14 parachute can drop from 300 feet AGL.

** The training drop altitude of 1,000 feet AGL may be waived to 800 feet AGL by completing a mature risk assessment decision cycle.

Table 20-2. Aircraft drop altitudes.

20-3. TYPE OF LOAD

Type of load includes personnel (one-second drop interval between jumpers), CDS, CRRC, door bundles, heavy equipment (three-second exit interval), or heavy equipment followed by personnel (three seconds per CDS/CRRC/heavy equipment and one second for each personnel).

CAUTION

There must be a three-second interval between equipment drops and the exit of parachutists to avoid possible jumper entanglement. The DZSO or DZSTL must follow the procedures for heavy drop operations, but observe the jumpers as they exit the aircraft.

a. **Obstacles**. To ensure that the airdrop is safe and that equipment and personnel can be recovered or employed to accomplish the mission, the drop zone and adjacent areas should be free of obstacles. Examples of obstacles are—

- Trees 35 feet or higher impeding recovery of personnel or equipment.
- Water 4 feet deep within 1,000 meters from any edge of the DZ.

- High tension wire that is carrying active current of 50 volts or greater. (Should be turned off prior to drop.)
- Any other conditions that may injure parachutists or damage equipment (inactive electric wires, barbed wire fences, swamps, ditches, gullies, and so forth).

NOTE: See Appendix D for DZ risk assessment.

b. **Air Approach and Departure Routes**. Air routes to and from the DZ should not conflict with other air operations or restrictive terrain, or with manmade objects (television or radio towers).

20-4. METHODS OF DELIVERY

Different drop methods are described as follows:

a. **High Velocity**. A drogue chute stabilizes and keeps equipment upright but does not slow the descent (for example, a 12-foot, high-velocity [HV] parachute on a door bundle).

b. **Low Velocity**. A parachute slows the rate of descent for a soft landing (for example, personnel and cargo parachutes).

c. **Free Drop**. This is cargo that has no device to stabilize or slow the rate of descent (for example, durable items such as clothing bundles).

20-5. ACCESS TO AREA

The unit must have access to and from the DZ to recover equipment or conduct troop movement. DZs with no roads leading to them or next to a river with no bridges are examples of impeded access to areas.

20-6. SIZE

The following information provides minimum peacetime sizes when using fixed-wing aircraft and must be adhered to unless a waiver is issued. During contingency or wartime missions, DZ sizes may also be waived. However, size requirements remain a joint responsibility of the COMALF and the airborne commander.

NOTE: To convert yards to meters, multiply yards times .9144 (yards x .9144). To convert meters to yards, divide meters by .9144 (meters ÷ .9144).

a. **Personnel from Military Rotary-Wing and Small Fixed-Wing Aircraft**. The distance required for personnel is determined by the use of the formula D=RT.

b. **GMRS, VIRS, or WSVC DZ**. Minimum size is 275 meters by 275 meters.

c. **Personnel from Fixed-Wing C-130 and Larger Aircraft Using a CARP DZ**. The ground space required is 550 meters by 550 meters for one parachutist from a single aircraft (add 70 meters to the length for each additional parachutist).

(1) For drop altitudes above 1,000 feet AGL, add 28 meters (14 meters each side) to the width and 28 meters (trail edge) for each additional 100 feet.

(2) From official sunset to sunrise, add 90 meters to the width (45 meters each side) and length (both ends) for visual drops.

(3) For visual formation, add 90 meters (45 meters each side) to the width.

(4) For AWADS and SKE, add 370 meters (185 meters each side) to the width.

d. **Heavy Equipment Drops from Fixed-Wing Aircraft Using a CARP DZ**. The ground space required is 550 meters by 915 meters for one platform from a single aircraft.

(1) For a C-130, add 370 meters to the length (trail edge) for each additional platform.

(2) For a C-141, add 460 meters to the length (trail edge) for each additional platform.

(3) For drop altitudes above 1,100 feet AGL, add 28 meters to the width (14 meters each side) and to the length (trail edge).

(4) From official sunset to sunrise, add 90 meters to the width (45 meters each side) and to the length (45 meters each end) for visual drops.

(5) For visual formation drops, add 90 meters to the width (45 meters each side).

(6) For AWADS and SKE, add 370 meters to the width (185 meters each side).

e. **Containerized Delivery System (CDS) Drops for the C-130 Using a CARP DZ**. The ground space required is 370 meters by 370 meters for one container from a single aircraft (182.5 meters each side).

f. **Containerized Delivery System Drops for the C-141 Using a CARP DZ**. The ground space required is 410 meters by 540 meters for one container from a single aircraft.

(1) For altitudes above 600 feet, add 37 meters for each additional 100 feet to the width (18.5 meters each side) and to the length (trail edge).

NOTE: Altitudes above 1,000 feet are not recommended.

(2) From official sunset to sunrise, add 90 meters to the width (45 meters each side) and to the length (45 meters each end) for visual drops.

(3) For visual formation drops, add 90 meters to the width (45 meters each side).

(4) For AWADS and SKE, add 370 meters to the width (185 meters each side).

NOTE: The size of the CDS drop zone depends on drop altitude, number of bundles, formation, and type of aircraft. (Refer to AFI 13-217.)

Section II. AIRDROP RELEASE METHODS AND PERSONNEL

The number and type of aircraft that air-delivers personnel and equipment, using one of the four methods, usually dictate the type and composition of the ground support party, which can be tailored for a mission.

NOTE: The unit mission request for aircraft specifies the type of drop method to be used, such as CARP, GMRS, VIRS, or WSVC, and the composition of the ground support party (STT and DZSO; DZST and or DZSO).

20-7. METHODS

To ensure accurate delivery on the DZ, JMs use four different methods. Each method uses various input from the ground and air in the calculation formula.

a. **Computed Air Release Point (CARP)**. The CARP is the most often used method in aerial delivery for conventional airborne operations. The CARP is computed by the aircrew (navigator) and determines the release point from the air.

b. **Ground Marking Release System (GMRS)**. The GMRS is the method used mostly by special operations forces (SOF). The GMRS is computed by the DZSTL and determines the release point from the ground.

c. **Verbally Initiated Release System (VIRS)**. The VIRS is one of the two methods used by services (Army and USMC) having rotary-wing and fixed-wing aircraft for small DZs, for dropping a specified number of personnel. VIRS is computed by the DZSTL; the release point is indicated by an oral command to the aircraft.

d. **Wind Streamer Vector Count (WSVC)**. The WSVC is one of two methods used by services having rotary-wing and fixed-wing aircraft for DZs for dropping a specified number of personnel. The release point is JM-directed and is the only method not requiring markings on the DZ.

20-8. ORGANIZATION

To become operational, drop zones require key personnel to be located on the DZ for controlling, marking, medical evacuating, wind readings, and malfunctions.

a. **USAF STT and DZSO**. STT and DZSO personnel are normally used in joint airborne operations of more than four troop carrier aircraft. Reference MOA 87 airdrop operations without STT.

b. **DZST**. The DZST consists of trained military personnel. It is normally used in small joint airborne operations involving four or fewer troop carrier aircraft. CARP, GMRS, JSJR, WSVC, or VIRS is used.

c. **DZSO**. Acting alone, without STT support, the DZSO operates the drop zone, with a small number of aircraft dropping a limited number of personnel. The airdrop release method used is CARP or GMRS, or, if rotary-wing and small fixed-wing aircraft are employed, VIRS or WSVC.

20-9. DROP ZONE SAFETY OFFICER DUTIES

The USAF STT and Army DZSO have specific duties, which are discussed as follows:

a. When the USAF STT is supporting an airborne operation, the DZSO is the airborne commander's direct representative on the drop zone. He is responsible for the safe operation of the DZ. No personnel or equipment is dropped if the DZSO is not physically on the DZ.

NOTE: The prerequisites to perform the duties of the DZSO are outlined in Chapter 7.

(1) ***Special Duties***. The duties of the DZSO are—

(a) Coordinating with the USAF STT.

(b) Ensuring the drop zone is fully operational 1 hour before drop time.

(c) Opening the drop zone through range control and closing it when accountability of personnel, air items, and equipment is completed.

(d) Before the drop, conducting ground or aerial reconnaissance of the DZ for obstacles or safety hazards.

(e) Collocating with USAF STT and taking initial wind readings 1 hour before the scheduled drop time.

(f) Establishing communications with the DACO NLT 1 hour before drop time.

(g) Conducting continuous surface wind readings NLT 12 minutes before the scheduled drop. Giving the CLEAR TO DROP or NO DROP to the STT (to relay to aircraft) 2 minutes before the scheduled drop.

(h) Monitoring surface winds from the parachutists' point of impact and at the highest point of elevation on the drop zone. A NO DROP situation exists when surface winds exceed the maximum allowable limits within 10 minutes of the actual drop. (See paragraph 22-5.)

(i) Controlling all ground and air medical evacuations. Priority for airspace must be given to medical evacuations. This is particularly important when rescue or medical aircraft are involved, since they may be delayed if follow-on jumps continue.

(j) Ensuring that any water obstacle is covered by a boat detail. A water obstacle is water more than 4 feet deep and 40 feet wide that is within 1,000 meters from any edge of the surveyed DZ. Water that is deeper than 4 feet but less than 40 feet wide is not an obstacle that requires boats, but does require approved flotation devices for the jumpers.

(k) Submitting postmission reports (for example, MAC Form 168 or incident/accident forms) to the appropriate agency.

NOTE: The CLEAR TO DROP or NO DROP that is relayed at 2 minutes does not indicate the final wind reading. If surface winds increase beyond authorized limits, a NO DROP can be relayed at any time thereafter. If readings exceed the limits, the DZSO must reestablish a 10-minute window.

(2) ***Support Requirements.*** These apply to multiple aircraft formations (fixed-wing aircraft), personnel, and equipment, or to single aircraft operations on drop zones more than 2,100 meters in length or with 20 seconds exit time ("green light").

(a) The DZSO ensures the ground support team is in place on the DZ one hour before the drop. The support team includes:

- Assistant DZSO (not required for USN or USAF operations).
- Four medical personnel (with two FLA); USMC/USN/USAF require one qualified support person.
- Malfunction officer (with camera).
- Parachute recovery detail (with saw and tree-climbing equipment).
- Parachute turn-in detail (with vehicles).
- Radios—one for the DZSO and one for the assistant DZSO (minimum).
- Anemometers—Services should only use approved anemometers to measure surface winds during all personnel and cargo parachute operations. The approved anemometers are the DIC, DIC3, TurboMeter, and AN/PMQ-3A. The AN/ML433A/PM and the anemometers that use floating balls or small floating lightweight aluminum devices in a tube are not authorized for use during personnel or cargo airdrop operations. The DIC, DIC3, and

TurboMeter cannot be calibrated; they must be given an expedient check just before use.

— Ensure fresh batteries are installed in the anemometer.

— Check the anemometer in a no-wind condition such as in a vehicle cab or a building. Turn on the anemometer and, if any reading other than zero registers, the anemometer is not fit for use and must be discarded.

— Use a three-anemometer check by comparing the reading on three anemometers in identical conditions. Discard the one anemometer that doesn't read the same as the other two.

— The TurboMeter must be held within 20 degrees of wind line with the wind entering the rear of the meter to ensure accurate readings.

— Calibration requirements for the AN/PMQ-3A will be conducted IAW appropriate TMs. Other anemometers not tested and recommended for use should be employed only after a command-initiated risk assessment is completed. Regardless of the method or device used to measure DZ winds, the airborne commander is responsible for ensuring winds on the DZ do not exceed 13 knots during static line personnel airdrops.

- Compasses—two (one each for the DZSO and the assistant DZSO).
- Smoke grenades.
- Vehicles.
- Road guards.
- Military police (to control vehicles and spectators or appropriate crowd control).
- Boat detail.

NOTE: These requirements may be supplemented based on the type of drop, size of operation, number of aircraft, number of parachutists, and geographical location of the DZ.

(b) For single aircraft (no more than a 20-second exit time or no more than 2,100 meters of usable DZ), the composition of a partial control group is as follows:

- One DZSO (an assistant DZSO is not required).
- Wind reading from a single location on the DZ.
- Two medics (with FLA); USMC/USN require one qualified Navy corpsman.
- Malfunction officer (with camera, binoculars, and night vision goggles for night operations).
- Parachute recovery detail.
- Radio.
- Anemometers—Services should only use approved anemometers to measure surface winds during all personnel and cargo parachute operations. The approved anemometers are the DIC, DIC3, TurboMeter, and AN/PMQ-3A. The AN/ML433A/PM and the anemometers that use floating balls or small floating lightweight aluminum devices in a tube are not authorized for use during personnel or cargo airdrop operations. The DIC, DIC3, and

TurboMeter cannot be calibrated; they must be given an expedient check just before use.

— Ensure fresh batteries are installed in the anemometer.

— Check the anemometer in a no-wind condition such as in a vehicle cab or a building. Turn on the anemometer and, if any reading other than zero registers, the anemometer is not fit for use and must be discarded.

— Use a three-anemometer check by comparing the reading on three anemometers in identical conditions. Discard the one anemometer that doesn't read the same as the other two.

— The TurboMeter must be held within 20 degrees of wind line with the wind entering the rear of the meter to ensure accurate readings.

— Calibration requirements for the AN/PMQ-3A will be conducted IAW appropriate TMs. Other anemometers not tested and recommended for use should be employed only after a command-initiated risk assessment is completed. Regardless of the method or device used to measure DZ winds, the airborne commander is responsible for ensuring winds on the DZ do not exceed 13 knots during static line personnel airdrops.

- Compass.
- Smoke grenades.
- Boat detail and road guards.

b. The DZSO has operational responsibility for the drop zone. In addition to the DZSO's duties for drop zones, the DZSO must also—

(1) Be positioned at the point of impact 15 minutes before drop time. The assistant DZSO is at the highest point of the drop zone or at the opposite end. For combination airdrop operations, the DZSO/DZSTL must follow the procedures for heavy drop operations, but observe the jumpers as they exit the aircraft.

(2) Relay a ground weather decision and CLEAR TO DROP or NO DROP signal to the lead aircraft 2 minutes before the drop for each pass.

(3) During night drops, ensure all lights that are on or next to the drop zone and are not a part of the drop zone marking system are turned off 5 minutes before drop time and remain off during the drop (except those lights that mark obstacles).

(4) Contact the pilot of the aircraft immediately after the drop and ask if any personnel or equipment did not drop. He relays this information to the airborne commander on the drop zone.

c. When advised of the coordination appointment by the airborne commander or his representative, the DZSO is furnished the following information:

- Number of lifts.
- Type of aircraft.
- Drop zone requirements.
- Unit SOP.
- Station time.
- Drop time.
- Number of personnel for each pass.

20-10. DROP ZONE SUPPORT TEAM AND DROP ZONE SUPPORT TEAM LEADER DUTIES

In operations in which the STT is not present, the DZSTL has overall responsibility for the conduct of operations on the drop zone. He represents both the airborne and airlift commanders. The DZSTL assumes all the responsibilities normally associated with the USAF STT and DZSO. Refer to DZSO MOA between USAF and Army/USMC/Navy dated 6/26/87.

a. DZSTs have the primary mission of supporting wartime CDS airdrops for battalion-size units and below and peacetime airdrops of personnel, CDS, and heavy equipment for one- to four-aircraft operations. With some exceptions, these primary mission airdrops are limited to day and night visual conditions.

b. DZSTs also maintain the secondary mission of supporting other types of airdrops. The secondary missions may include wartime force projection and sustainment of personnel, equipment, and CDS; peacetime airdrops under AWADS and IMC conditions; and VMC formation drops with four or more aircraft.

c. The DZST consists of one or more individuals. More members may be required, depending on the complexity of the mission. The senior member of the DZST functions as the DZSTL. For combination airdrop operations, the DZSO/DZSTL must follow the procedures for heavy drop operations, but observe the jumpers as they exit the aircraft.

d. If an individual assumes the duties of both the DZSO and the DZSTL, he also is responsible for the following:

(1) Conducting premission coordination.

(2) Evaluating the DZ for suitability and safe operating conditions.

(3) Ensuring all DZ markings are properly displayed.

(4) Operating all visual acquisition aids.

(5) Ensuring NO DROP signals are relayed to the aircraft.

e. Once the DZSTL has been notified and assigned a mission, he must conduct accurate premission coordination. The recommended DZST crew mission briefing checklist (Figure 20-1) reflects the minimum essential information that must be addressed and confirmed by the DZSTL. Normally, peacetime drops should employ every acquisition aid and safety device available, including air-to-ground radio communications, PIBAL mean effective wind measurement, air traffic control light gun, and smoke or flares. During contingency or wartime operations, limited airdrop support equipment is available; therefore, it is important for premission coordination and briefings to be comprehensive with respect to visual signals (drop cancellation, postponement, and authentication procedures). The coordination must be timely to ensure the DZST has enough time for planning and for moving to and establishing the drop zone.

20-11. BRIEFING CHECKLIST

Figure 20-1 shows a recommended DZST/aircrew mission briefing checklist.

DZ name/location and JA/ATT mission sequence number verified. ____________

TOT(s) block time (NO DROP procedures, for example, race track). ____________

Valid DZ survey (date) verified. ____________

Type drop (HE, PE, CDS). ____________

Type release (VIRS, CARP, GMRS, AWADS, WSVC, visual). ____________

Type parachutes. ____________

Ground quick disconnects. ____________

Number of jumpers/bundles. ____________

Number and type of aircraft. ____________

DZ information. ____________

<u>Markings/signals:</u>

- Panels/lights. ____________
- Block letter identification. ____________
- Smoke, flares. ____________
- Emergency NO DROP procedures. ____________
- Mission cancellation indication. ____________

<u>DZ support capabilities:</u>

- Radios available/frequencies. ____________
- Visual acquisition aids available. ____________
- NAVAIDS available. ____________
- MEW equipment. ____________

Airspace coordination verified. ____________

Aircraft (mission) commander's name, unit of assignment, telephone number. ____________

DZSTL name, rank, unit of assignment, telephone number. ____________

Drop score/incident/accident reporting procedure. ____________

Figure 20-1. DZST/aircrew mission briefing checklist.

20-12. EQUIPMENT

The DZSTL should maintain an inventory of basic equipment to support an airdrop mission:

- VS-17 panels.
- Smoke (red, yellow, green).
- White steady lights, white omni-directional.
- Amber rotating beacon.
- Air traffic control gun, SE-11 light gun, or 4-cell MAGLITE flashlight.

NOTE: ATC gun requires a special power source and plug to function properly.

- Signal mirror.
- Binoculars.
- Night vision goggles for night operations.
- Anemometers—Services should only use approved anemometers to measure surface winds during all personnel and cargo parachute operations. The approved anemometers are the DIC, DIC3, TurboMeter, and AN/PMQ-3A. The AN/ML433A/PM and the anemometers that use floating balls or small floating lightweight aluminum devices in a tube are not authorized for use during personnel or cargo airdrop operations. The DIC, DIC3, and TurboMeter cannot be calibrated; they must be given an expedient check just before use.
  - — Ensure fresh batteries are installed in the anemometer.
  - — Check the anemometer in a no-wind condition such as in a vehicle cab or a building. Turn on the anemometer and, if any reading other than zero registers, the anemometer is not fit for use and must be discarded.
  - — Use a three-anemometer check by comparing the reading on three anemometers in identical conditions. Discard the one anemometer that doesn't read the same as the other two.
  - — The TurboMeter must be held within 20 degrees of wind line with the wind entering the rear of the meter to ensure accurate readings.
  - — Calibration requirements for the AN/PMQ-3A will be conducted IAW appropriate TMs. Other anemometers not tested and recommended for use should be employed only after a command-initiated risk assessment is completed. Regardless of the method or device used to measure DZ winds, the airborne commander is responsible for ensuring winds on the DZ do not exceed 13 knots during static line personnel airdrops.
- Compass.
- Signal flares.
- PIBAL system with helium source.

NOTE: Other items of equipment/signals may be required by premission coordination and mission complexity.

CHAPTER 21

DROP ZONE COMPUTATIONS AND FORMULAS

Once the composition of the ground party and the selection of a drop zone have been established, several technical aspects must be considered and planned for marking the DZ. These technical aspects are critical because of the data that must be used (ground, winds, drift-distance formula, forward throw, release point).

21-1. DROP ZONE FORMULAS FOR GMRS AND VIRS

The following procedures for using the distance and time formulas apply to GMRS and VIRS.

a. **Distance Formula (D = RT)**. Compute DZ length for a specific mission by using the D = RT formula. (D is the required length of the DZ in meters; R is the ground speed of the aircraft in meters per second; and T is the time required for the aircraft to release its cargo.) To use this formula, some conversions and mathematics are required.

(1) ***Airspeed Conversion to Ground Speed***. To find the aircraft ground speed, convert aircraft airspeed (expressed in knots) to ground speed (meters per second). Do this by multiplying knots times .51 (knots x .51) (1 knot equals .51 meter per second). The following table from Chapter 20 is repeated to assist the estimation of aircraft airspeeds.

TYPE OF AIRCRAFT	DROP SPEED
UH-1	50 to 70 knots—optimum 70 knots
UH-60	65 to 75 knots—optimum 70 knots
CH-46 (USMC)	80 to 90 knots—optimum 90 knots
CH-47	80 to 110 knots—optimum 90 knots
CH-53 (USMC)	90 to 110 knots
CH/HH3 (USAF)	70 to 90 knots
C-5/130/141/17/KC-130	130 to 135 knots (personnel)
C-5/130/141//17KC-130	130 to 150 knots—optimum 130 knots for all loads (door bundles, CDS, and heavy equipment)

Table 21-1. Aircraft drop speeds.

(2) ***Time Over DZ Requirement***. To determine the time over the DZ that is needed to release a parachutist or equipment, use the following factors:

(a) Allow 1 second for each parachutist to exit the aircraft; do **not** include the first parachutist (10 parachutists require 9 seconds). (Mathematically, this is represented as 10 x 1 - 1.)

(b) Allow 3 seconds per bundle to exit the aircraft; do **not** include the first bundle (3 bundles would require 6 seconds). (Mathematically, this is represented as 3 x 3 - 3.)

(c) Personnel jumping T-10-series parachutes may exit both doors simultaneously. The door with the most parachutists is used to calculate the time required.

EXAMPLE: D = RT

What length DZ would 8 jumpers require when jumping from an aircraft flying at a drop speed of 90 knots?

Step 1: Solve for R (answer is expressed in meters per second): airspeed x .51 (90 knots x .51) = 45.90 meters per second.

Step 2: Solve for T (answer is expressed in seconds): number of jumpers x 1 - the first jumper (8 x 1 - 1) = 7 seconds.

Step 3: Solve for D (answer is expressed in meters): 45.90 meters per second x 7 seconds = 321.30 meters. Always round **up** to the nearest whole number. Therefore, D = 322 meters, the required DZ length.

b. **Time Formula (T = D/R).** Solving this formula provides the seconds required to exit the jumpers over the DZ (time = meters divided by meters per second). If a DZ less than the required length must be used, compute the flight time over the DZ to determine how much of the load can be released in one pass. Use the T = D/R formula: T is the time the aircraft is over the DZ in seconds, D is the length of the DZ in meters, and R is the ground speed (rate) of the aircraft in meters per second.

(1) ***Airspeed Conversion.*** Convert the aircraft's airspeed (expressed in knots) to its ground speed (expressed in meters per second) as in the D = RT formula (knots x .51). Round **up** the answer to the next whole number.

(2) ***Determination of T.*** Divide the ground speed conversion number into D (the DZ length); this determines T. Any fractional answer is rounded **down** to the next whole number.

EXAMPLE: T = D/R

How many parachutists from a CH-47 (drop speed of 90 knots) can land on a 750-meter DZ each pass?

T = Number of parachutists.

D = DZ length is 750 meters (given).

R = Airspeed is 46 meters per second (90 knots x .51 = 45.9; round up to 46).

Solution: T = D/R (D ÷ R).

D/R = 750 meters divided by 46 meters per second = 16.3 seconds.

T = 16 seconds (round **down**).

16 seconds over DZ x 1 parachutist per second + 1 parachutist (the first parachutist exiting the aircraft does not affect the number of seconds spent over the DZ) = 17 parachutists. Thus, 17 parachutists per pass can land on the 750-meter DZ.

21-2. WIND DRIFT

Two means of determining wind drift are the WSVC method and the D = KAV formula.

a. **Wind Streamer Vector Count**. The WSVC method (Figure 21-1) is used when the release point is determined from the air. It is normally jumpmaster-executed and does not require markings to be placed on the DZ.

(1) ***Streamer Drop.*** On the first aircraft pass over the desired point of impact, a streamer is dropped from the aircraft. The aircraft then turns to allow the JM to keep the streamer in sight. The pilot adjusts his route so that the flight path is over the streamer on the ground and the desired impact point (in a straight line).

(2) ***Count.*** As the aircraft passes over the streamer, the JM begins a count, stopping the count directly over the impact point. He immediately begins a new count. When that count equals the first count, the aircraft is over the release point for the first parachutist.

(3) ***Aircraft Flight Adjustment.*** The pilot then maneuvers the aircraft to fly along the axis of the DZ and over the release point. Slight adjustments may be made by observing the parachutists as they land on the DZ.

NOTE: This method should not be used for tactical employment, since the aircraft is required to make multiple passes over the DZ.

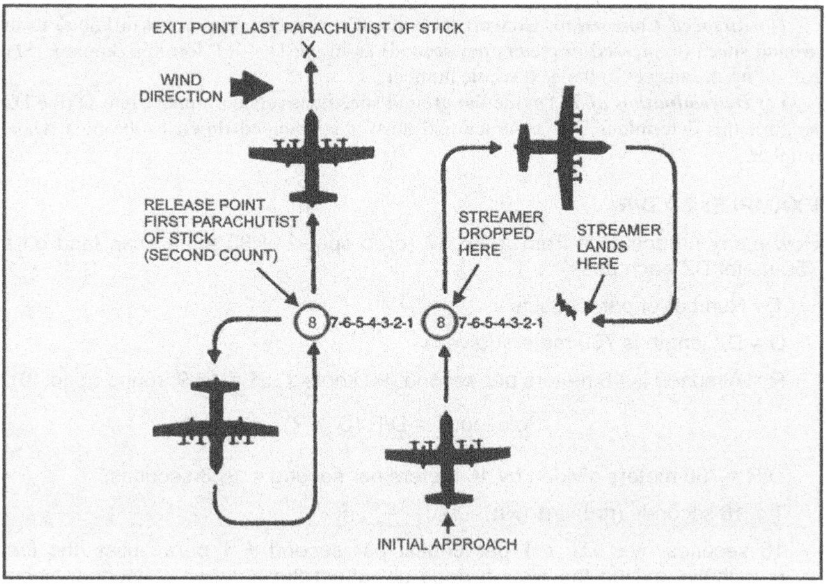

Figure 21-1. Determination of the release point by WSVC.

b. **D = KAV Formula**. This is another method for determining the effects of wind on a parachute: D = drift of parachute (in meters) from a given altitude; K = constant that represents the typical drift characteristic for a type of parachute. These constants are—

- 1.5 for cargo parachutes and heavy equipment (HD or HE).
- 3.0 for personnel parachutes.
- 2.4 for tactical training bundles.

A = drop altitude (expressed in hundreds of feet AGL); and V = velocity of wind. The mean effective wind (MEW) should be used for personnel. MEW is the preferred method for all other types of loads; however, the surface wind may be used during low-risk operations.

NOTE: If aircraft must be shut down for a long period, a wind drift indicator should be thrown at the last release point to ensure the release point is still valid.

EXAMPLE:

An aircraft is dropping cargo from 500 feet AGL with a surface wind of 10 knots. What is the calculated parachute drift? (The parachute drift is calculated using the D = KAV formula.)

D = Wind-induced drift in meters.

K = Wind drift constant for type of parachute.

A = Drop altitude expressed in hundreds of feet (500 feet would be expressed as 5).

V = Velocity of wind in knots (either MEW or surface wind measurement).

Step 1: K = 1.5 (cargo parachute or HE constant).

Step 2: A = 5 (500 feet).

Step 3: V = 10 (10 knots).

Step 4: D = 1.5 x 5 x 10 = 75.0, or 75 meters of drift. (Any fractional answer is rounded **up** to the nearest whole number.)

21-3. WIND VELOCITY

Two options are available for determining wind velocity.

a. **Mean Effective Wind**. The most effective option is the use of MEW. This reading is a constant wind speed average from drop altitude to the ground. The PIBAL system determines the MEW. This system should be used when possible; it is more reliable than the other option, which measures surface wind velocity only.

b. **Surface Wind Measurement**. Either the AN/PMQ-3A or commercial anemometers authorized by USAIS messages DTG 101000Z MAR 94, subject: Use of Anemometers During Airdrop Operations, and DTG 211200Z OCT 94, subject: Use of Turbometer During Static Line Airdrop Operations, are recommended for use. Other anemometers not recommended for use should be employed only after a command-initiated risk assessment is completed. Regardless of the method or device used to measure DZ winds, the airborne commander is responsible for ensuring winds on the DZ do not exceed 13 knots during static line personnel airdrops.

(1) ***Equipment.*** The equipment needed to compute the MEW by the PIBAL method is as follows:

- Helium source.
- Pilot balloons (10 or 30 grams).
- Clinometers or other devices for measuring from 0 to 90 degrees.
- Balloon measuring tape (to measure balloon circumference) (10-gram: 57 inches day, 74 inches night; 30-gram: 78 inches day, 94 inches night).
- PIBAL lighting units (Type 5) for night use (liquid-activated lights).
- Compass.
- Conversion charts (10- and 30-gram) (Tables 21-2 and 21-3).
- Watch with second hand.

(2) ***Procedure.*** The procedure for measuring MEW using the PIBAL is as follows:

(a) Fill the 10-gram or 30-gram balloon with helium to the required size.

(b) Check the conversion chart for drift time to drop altitude (Table 21-2, page 21-6, and Table 21-3, page 21-7).

(c) Release the balloon and begin timing.

(d) Keep the balloon in sight.

(e) Once the required time has elapsed, determine the azimuth to the balloon with the compass and read the degrees from the drift scale.

(f) Refer to the conversion chart and read down the angle column to the number closest to the angle on the scale.

(g) Read across the top of the chart (altitude in feet) to the drop altitude in use. Read down this column until the two lines (6 and 7) intersect.

(h) Where the two lines intersect is the MEW at drop altitude, in knots. The direction of the MEW is the back azimuth of the compass reading that was taken at the same time as the angle measurement.

(i) The MEW becomes the variable V in the D = KAV formula to determine the amount of drift in meters.

NOTE: A parachute's K-factor is based on the parachute's flight characteristics, not on its mode of use. IAW Chapter 2 of FM 10-500-3/TO 13C7-1-11, the K-factor for the T-10-series parachute used in the cargo mode is the same (3.0) as for personnel drops using the T-10-series parachute.

WIND SPEED IN KNOTS
10-GRAM HELIUM BALLOON
Inflate balloon to 57" circumference for day and 74" circumference for night.

ELEVATION ANGLE	DROP ALTITUDE IN FEET												ASCENSION TABLE	
	500	750	1000	1250	1500	1750	2000	2500	3000	3500	4000	4500	TIME	ALT (FT)
70	02	02	01	01	01	01	01	01	01	01	01	01		
60	03	02	02	02	02	02	02	02	02	02	02	02		
55	03	03	03	03	03	03	03	03	03	03	03	03		
50	04	04	03	03	03	03	03	03	03	03	03	03	0:10	80
45	05	04	04	04	04	04	04	04	04	04	04	04	0:20	170
40	06	05	05	05	05	05	05	04	04	04	04	04	0:30	250
35	07	06	06	06	06	05	05	05	05	05	05	05	0:40	330
30	08	07	07	07	07	07	07	07	06	06	06	06	0:50	400
25	10	09	09	09	08	08	08	08	08	08	08	08	1:02	500
24	11	10	09	09	09	09	08	08	08	08	08	08	1:10	540
23	11	10	10	09	09	09	09	08	08	08	08	08	1:20	610
22	12	11	10	10	10	10	09	09	09	09	09	09	1:23	670
21	12	11	11	10	10	10	10	10	10	10	10	10	1:43	750
20	13	12	11	11	11	11	11	10	10	10	10	10	1:50	790
19	14	13	12	12	11	11	11	11	11	11	11	11	2:25	1000
18	15	13	13	12	12	12	12	12	11	11	11	11	2:44	1100
17	16	14	13	13	13	13	12	12	12	12	12	12	3:05	1250
16	17	15	14	14	14	13	13	13	13	13	13	13	3:49	1500
15	18	16	15	15	14	14	14	14	14	14	14	14	4:30	1750
14	19	17	16	16	16	15	15	15	15	15	15	15	5:11	2000
13	21	19	18	17	17	17	17	16	16	16	16	16	6:34	2500
12	22	20	19	19	18	18	18	18	17	17	17	17	7:58	3000
11	24	22	21	21	20	20	20	19	19	19	19	19	9:22	3500
10	27	25	23	23	22	22	22	21	21	21	21	21	10:44	4000
09	30	27	26	26	25	24	24	24	23	23	23	23	12:08	4500

Table 21-2. The 10-gram PIBAL chart.

21-4. FORWARD THROW

Forward throw is the effect that inertia has on a falling object. When an object leaves an aircraft, it is traveling at a speed equal to the speed of the aircraft. The parachutist (or bundle) continues to move in the direction of flight until the dynamics of parachuting takes effect.

a. **Forward Throw for Rotary-Wing Aircraft**. To determine the amount of forward throw for rotary-wing aircraft, divide the drop speed of the aircraft in half. This yields the forward throw in meters. (For example, an aircraft flying at 70 knots would have a forward throw of 35 meters.)

b. **Forward Throw for Fixed-Wing Aircraft**. To determine the forward throw for fixed-wing aircraft, the following distances apply (Table 21-4).

WIND SPEED IN KNOTS
30-GRAM HELIUM BALLOON
Inflate balloon to 78" circumference for day and 94" circumference for night.
DROP ALTITUDE IN FEET

ELEVATION ANGLE	500	750	1000	1250	1500	1750	2000	2500	3000	3500	4000	4500
80	01	01	01	01	01	01	01	01	01	01	01	01
70	03	03	03	02	02	02	02	02	02	02	02	02
60	04	04	04	04	04	04	04	04	04	04	04	04
55	05	05	05	05	05	05	05	05	05	05	04	04
50	06	06	06	06	06	06	06	06	05	05	05	05
45	07	07	07	07	07	07	07	07	07	06	06	06
40	09	08	08	08	08	08	08	08	08	08	08	08
35	10	10	10	10	10	10	10	10	09	09	09	09
30	12	12	12	12	12	12	12	11	11	11	11	11
25	15	15	15	15	15	15	14	14	14	14	14	14
24	16	16	15	15	15	15	15	15	15	15	15	15
23	17	17	16	16	15	15	15	15	15	15	15	15
22	18	18	17	17	17	17	17	16	16	16	16	16
21	19	19	18	18	18	17	17	17	17	17	17	17
20	20	20	19	19	19	19	18	18	18	18	18	17
19	21	20	20	20	20	20	19	19	19	19	19	18
18	22	22	21	21	21	21	20	20	20	20	20	20
17	23	23	23	22	22	22	22	22	21	21	21	21
16	25	25	24	24	24	24	23	23	23	23	22	22
15	27	27	26	26	25	25	25	25	24	24	24	24
14	29	19	18	17	17	17	17	16	16	16	16	25
13	31	20	19	19	18	18	18	18	17	17	17	27

ASCENSION TABLE

TIME	ALT (FT)
0:10	120
0:20	240
0:30	360
0:42	500
0:50	400
1:02	600
1:10	830
1:17	1000
1:46	1250
2:10	1500
2:34	1750
2:56	2000
3:43	2500
4:31	3000
5:21	3500
6:09	4000
7:00	4500

Table 21-3. The 30-gram PIBAL chart.

	C-130	C-141	C-5
PERSONNEL/DOOR BUNDLES	229 METERS/ 250 YARDS	229 METERS/ 250 YARDS	229 METERS/ 250 YARDS
HEAVY EQUIPMENT	458 METERS/ 500 YARDS	668 METERS/ 730 YARDS	668 METERS/ 730 YARDS
CONTAINER DELIVERY SYSTEMS	503 METERS/ 550 YARDS	686 METERS/ 750 YARDS	686 METERS/ 750 YARDS
TACTICAL TRAINING BUNDLE	147 METERS/ 160 YARDS	147 METERS/ 160 YARDS	147 METERS/ 160 YARDS
NOTE: To convert yards to meters, multiply yards by .9144. To convert meters to yards, divide meters by .9144.			

Table 21-4. Fixed-wing forward throw data.

21-5. DROP HEADINGS, POINT OF IMPACT, WIND DRIFT COMPENSATION, AND FORWARD THROW COMPENSATION

For CARP operations, the navigator onboard the aircraft determines when the load is to be released from the aircraft (when the green light is turned on). For GMRS and VIRS operations, ground personnel determine the release point (Figure 21-2).

a. **Drop Heading**. Drop heading on all DZs depends on three factors—the long axis, prevailing winds, and obstacles on approach and departure ends. The DZSO or DZSTL uses all three when the situation permits; however, the long axis is the primary concern. With a GMRS, WSVC, or CARP DZ, drop heading can be obtained from USAF Form 3823, DZ Survey (formerly MAC Form 339). A circular/random approach DZ does not have a set drop heading. The mission commander notifies the aircrew and the DZ commander of the drop heading to be used NLT 24 hours in advance of the airdrop operation.

NOTE: On some DZs, predetermined drop headings must be used.

b. **Point of Impact**. The location selected where the first bundle or parachutist should land is known as the PI. The PI should be located along the DZ centerline. However, due to the tactical situation, the PI may need to be located near a wood line. The DZSO or DZSTL uses a buffer zone of 100 meters on one end of the DZ for safety reasons. PI location for GMRS or VIRS is 100 meters in from the leading edge centerline for personnel. CARP PI is designated on USAF Form 3823, DZ Survey (formerly MAC Form 339).

c. **Wind Drift Compensation**. To compensate for wind drift, the DZSO or DZSTL moves from the desired PI into the wind the number of meters calculated using the D = KAV formula. (For example, if drift equals 350 meters from the PI, he faces into the wind and walks 350 meters in a straight line.)

d. **Forward Throw Compensation**. To compensate for aircraft forward throw, the DZSO or DZSTL faces the back azimuth of the drop heading and walks the appropriate forward-throw distance to the release point.

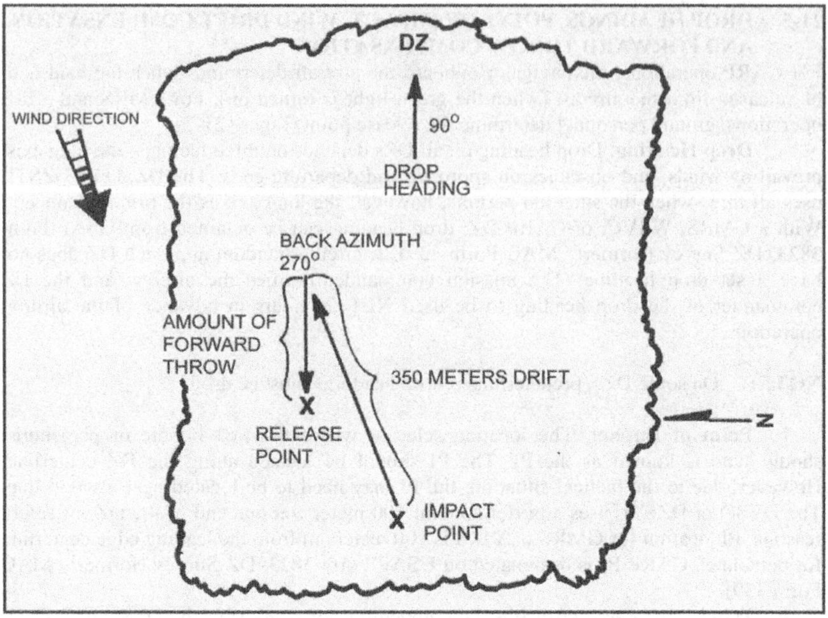

Figure 21-2. RP location for VIRS and GMRS.

CHAPTER 22

ESTABLISHMENT AND OPERATION OF A DROP ZONE

Five methods may be used to establish or operate a drop zone. Four of these require markings to be placed on the DZ: CARP, GMRS, and VIRS. The WSVC method requires no markings on the DZ.

22-1. COMPUTED AIR RELEASE POINT

CARP is used only by fixed-wing aircraft in conjunction with a CCT or qualified DZSTL.

a. **CARP Points of Impact** (Figures 22-1 and 22-2, page 22-2). The PIs for CARP operations are as follows:

(1) ***Personnel.*** For personnel, drops at the PI are 300 yards (day) or 350 yards (night) from the leading edge.

(2) ***CDS.*** For CDS bundles from a C-130, drops at the PI are 200 yards (day) or 250 yards (night) from the leading edge. For CDS drops from a C-141, the PI is 225 yards (day) and 275 yards (night).

(3) ***HE.*** For heavy equipment, drops at the PI are 500 yards (day) or 550 yards (night) from the leading edge.

NOTE: On surveyed DZs, the PI for a particular type load is predetermined. Its surveyed location can be found on AF Form 3823 or MAC Form 339. (Use of MAC Form 339 is authorized until supplies are exhausted. USAF is converting all MAC Forms 339 to AF Form 3823 when a DZ comes due for recertification.)

b. **No-Drop Communication to Aircraft**. No-drop conditions are relayed to the aircraft in the following ways: red smoke, red flares, forming the code letter into two parallel bars perpendicular to flight, or the absence of a planned signal. Forming the code letter into an X indicates mission cancellation.

NOTE: The type of marking used for no-drop conditions is coordinated in the premission briefing.

c. **Control Center**. Control center locations (location of DZSTL) are as follows:

(1) ***Personnel Drops.*** Personal drops are normally located at the PI.

(2) ***CDS Drops.*** CDS is located 150 yards to the 6 o'clock position of the PI.

(3) ***HE, Free Drops, High Velocity, AWADS.*** Heavy equipment, free drops, high velocity, and AWADS with a ceiling of less than 600 feet are all off the DZ. For this type of operation, the DZSTL uses the best vantage point off the DZ to observe the airdrop.

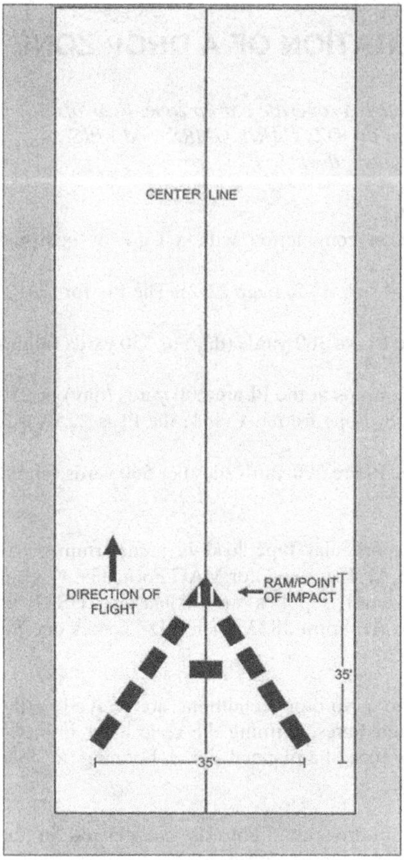

Figure 22-1. Day CARP drop zone markings.

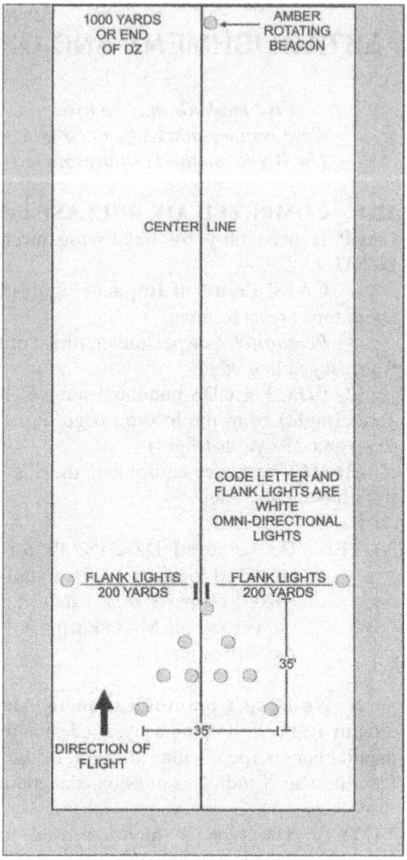

Figure 22-2. Night CARP drop zone markings.

22-2. DROP ZONE MARKINGS

A marked DZ has a PI or release point marked with a precoordinated visual or electronic signal. Standard DZ markings consist of raised angle markers (RAM), VS-17 marker panels, visible lighting systems, and light beacons. Virtually any type of lighting or visual marking system is acceptable if all participating units are briefed and concur. Night markings or visual acquisition aids may include a light gun, flares, fire pots, railroad fuses, flashlights, chem lights, and infrared (IR) lighting systems. Electronic NAVAID markings (ZM, SST-181, GAR-I, tactical aid to navigation (TACAN), and so forth) may be used for either day or night operations and placed as directed by mission requirements.

a. During day operations, the PI will be marked with a RAM or block letter. If authentication is required, a block letter will be used instead of the RAM. Authorized letters for PI markings are A, C, J, R, and S. The block letters H and O are authorized for random approach DZs. The block letters should be aligned with the surveyed DZ axis or with the aircraft line-of-flight, if different from the survey. The minimum size for block letters is 11 meters (35 feet) by 11 meters and consists of at least nine marker panels.

b. During night operations, the PI will be marked with a block letter and flanker lights. The apex of the block letter will be located on the PI. Flanker lights will be white and located 250 meters left and right abeam the PI. The minimum size is 11 meters by 11 meters and consists of at least nine white lights with a recommended minimum output rating of 15 candela. A trailing edge beacon will be used during actual personnel airdrops. When used, the amber trailing edge beacon will be placed along the surveyed DZ centerline. 1,000 meters from the PI, or at the DZ trailing edge, whichever is closer to the PI. During premission coordination for personnel drops, aircrews will identify their trailing edge beacon requirements to STS or DZC. For all airdrops, the DZ identification must be coordinated and briefed to the ground party and aircrews.

c. When mission requirements dictate and aircrews are qualified and equipped, IR lights may be substituted for overt lights using the DZ marking patterns specified in paragraph 22-2b.

22-3. GROUND MARKING RELEASE SYSTEM

The GMRS uses markings known as the four-panel inverted L, six-panel T, or seven-panel H. The T or H pattern is recommended for C-141/C-5 airdrops due to aircraft's side angle vision limitations (Figure 22-3, page 22-4).

a. **Inverted L Marking**. When the drop aircraft is 100 meters directly to the right of the corner (A) panel, the drop is executed.

b. **Marking Placement for Inverted L**. Markings (four panels) are placed as follows (Figure 22-3, page 22-4):

(1) From the RP, move 100 meters to the left (90 degrees) of drop heading for the location of the corner (A) panel. Emplace a VS-17G panel with the long axis of the panel parallel with the drop heading. Elevate the panel at a 45-degree angle toward the approaching aircraft. This aids the aircrew and the JM in visual identification of the DZ.

(2) From the corner (A) panel, move in the same direction as above for 50 meters for the location of the alignment (B) panel. Emplace this panel as described above.

(3) From the alignment (B) panel, move 150 meters in the same direction as above for the location of the flanker (C) panel. Emplace this panel as described above.

(4) From the corner (A) panel, move 50 meters on a back azimuth of the drop heading for the location of the approach (D) panel. Emplace this panel the same as described above.

(5) At night, replace all panels with a white omni-directional light. Lights may be shielded on three sides or placed in pits.

(6) During day operations, smoke may be displayed at the RP. During night operations, a white air traffic control light may be used to mark the RP.

(7) NO DROP may be signaled to the aircraft by red smoke, red flares, scrambled panels, or the absence of a planned signal.

c. **Mask Clearance.** Since the aircraft is required to fly along the markings on the DZ, these markings must be visible to the aircrew. The markings are placed where obstacles will not mask the pilot's line of sight. As a guide, a mask clearance ratio of 1:15 is used, that is, one unit of vertical clearance for every 15 units of horizontal clearance. For example, if a DZ marker must be positioned near a terrain mask, such as the edge of a forest that is on the DZ track, and the trees are 10 meters high (33 feet), the markings would require 150 meters (492 feet) of horizontal clearance from the trees (Figure 22-4, page 22-6). This applies to static line jumps only.

d. **Code Letters.** If any portion of the inverted L falls within a 15 to 1 (15:1) mask clearance ratio of obstacles on the approach end of the DZ, a code letter (H, E, A, T) or far panel is required on the departure end of the DZ for CDS or bundle drop and should be coordinated during the DZST/aircrew mission briefing. This far marking is on line with the corner (A) panel to allow the aircrew to begin alignment on the release point until the inverted L comes into view. If a code letter is used, it can be used to distinguish the DZ from other DZs in the area.

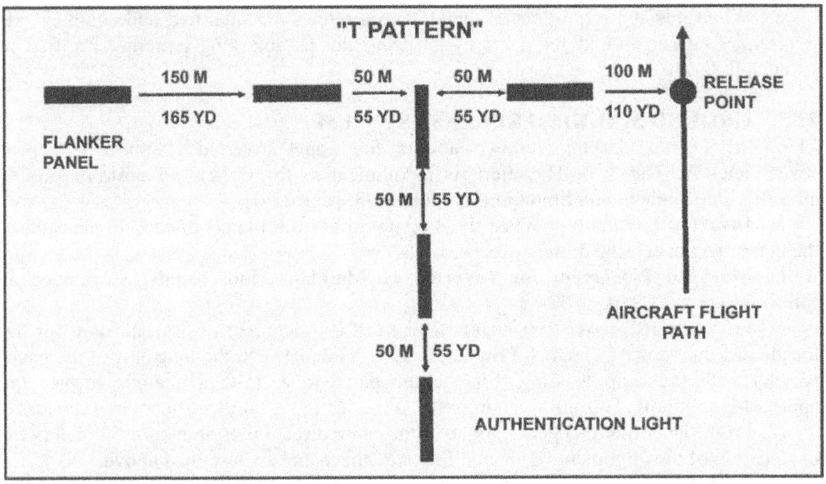

Figure 22-3. GMRS panel emplacement.

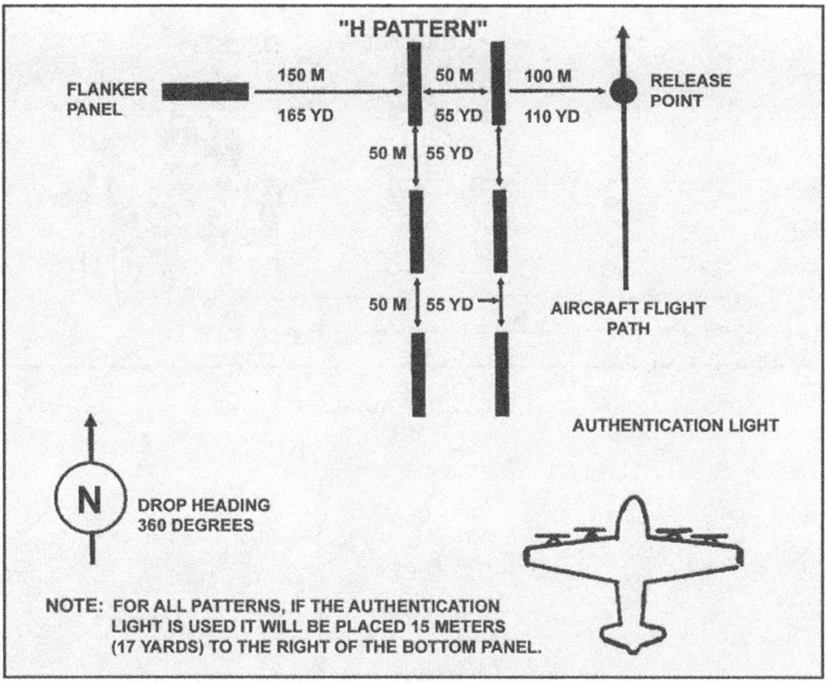

Figure 22-3. GMRS panel emplacement (continued).

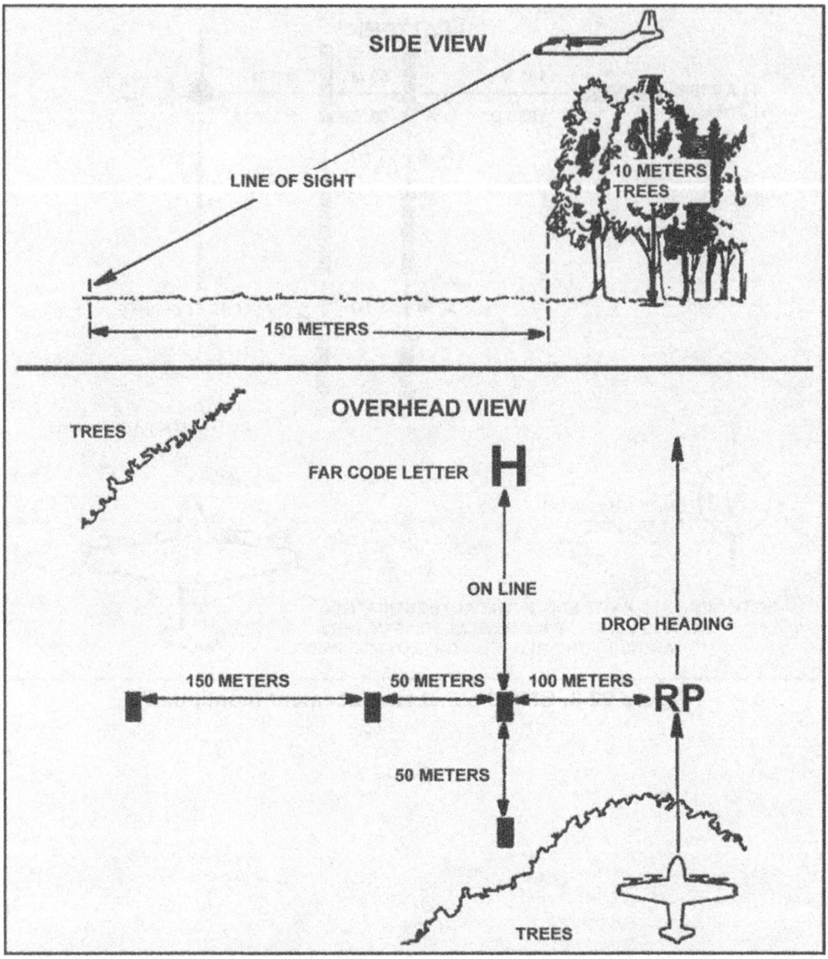

Figure 22-4. The 15:1 mask clearance ratio.

22-4. VERBALLY INITIATED RELEASE SYSTEM FOR ROTARY-WING AND FIXED-WING AIRCRAFT

VIRS is used to execute a drop over the RP by GTA verbal command. This method allows the conduct of the operation with a minimum amount of prior DZ information and coordination. The aircraft flies the given direction until the DZSTL sees the aircraft. A code letter (H,E,A,T) marks the RP. Once the crew identifies the DZ, the radio operator directs the aircraft over the drop heading RP. When the aircraft is directly over the RP, the command EXECUTE, EXECUTE, EXECUTE initiates the drop (Figure 22-5).

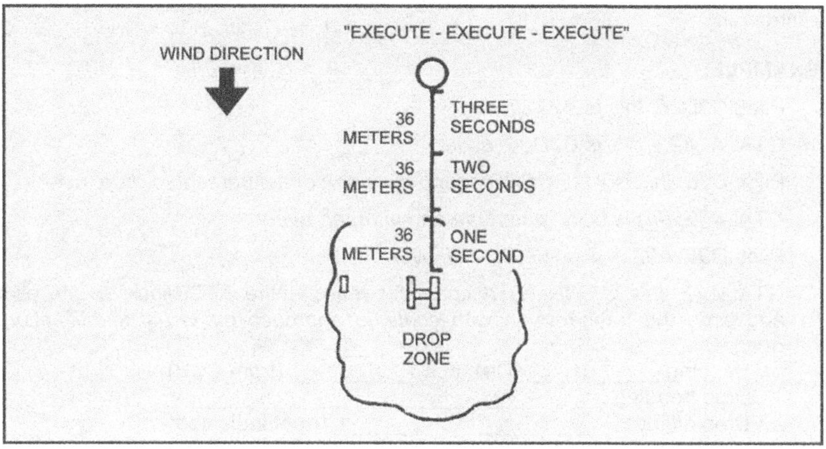

Figure 22-5. Army VIRS offset.

a. **Day DZ Markings.**

(1) ***Determination of Release Point.*** The DZSTL then places the code letter on the drop heading, with the base panel of the letter at the release point. The code letter is formed by VS-17G panels placed together. Each letter is two panels high and one panel wide.

(2) ***Placement of Flank Panel.*** The flank panel is placed parallel to the code letter and aligned with the base panel. It is placed 200 meters to the left of the code letter (or at the edge of the DZ, whichever is closer).

(3) ***Placement of Far Panel.*** The far panel is placed 500 meters from the base panel (or at the edge of the DZ, whichever is closer) and on line with the drop heading.

(4) ***Panel Construction.*** Both the far and flank panels consist of a single VS-17G panel. These panels may also be elevated at a 45-degree angle to improve visibility.

b. **Night DZ Markings.**

(1) ***Use of Lights.*** The procedures for establishing the DZ are the same for night operations except that white light is used for the code letter and far and flank markings.

(2) ***Code Letter Construction.*** Each code letter is four lights high and three lights wide. There is a distance of 5 meters between each light in the code letter. The far and flank lights are signal lights. Also, a white-and-red lens ATC (SE-11) light should be located at the RP.

(3) ***Security.*** Lights may be shielded on three sides or placed in pits to prevent enemy ground observation.

22-5. GUIDANCE PROCEDURES

During Army DZ operations, the GTA is responsible for guiding the jump aircraft to the DZ, over the DZ on the proper drop heading, and at the proper altitude and drop speed. He ensures the parachutists exit the aircraft at the proper release point. Once the parachutists have exited the aircraft, the GTA must then clear the aircraft from the control zone.

EXAMPLE:

Pilot: *C3D36, this is A2A22, over.*

GTA: *A2A22, this is C3D36, over.*

Pilot: *D36, this is A22, CCP inbound for a personnel parachute drop, over.*

GTA: *A22, this is D36, state type and number, over.*

Pilot: *D36, A22 is a single UH-1H, over.*

GTA: *A22, this is D36,* (GTA controller reads entire ATC block* to the pilot and ends the transmission with *continue approach for visual identification, over*).

Heading ___________ Distance _________ (from CCP)
Drop heading ___________________________________
Drop altitude ______________________ (feet indicated)
Drop speed ____________________________________
Number jumpers/bundles that can be accepted _______

Pilot: *Wilco.*

Upon sighting aircraft, the GTA tells the pilot:

GTA: *A22, this is D36, I am at your 11 o'clock, 500 meters, signal out, can you identify, over.*

Pilot: *D36, A22 identifies orange panel, over.*

GTA: *A22, D36 has visual contact, turn to drop heading, over.*

Pilot: *D36, A22 turning drop heading, over.*

GTA: *A22, this is D36, steer left/right, over.*

Pilot: *D36, A22 roger.*

GTA: *A22, this is D36, on course, over.*

Pilot: *D36, A22 roger.*

When aircraft is 8 to 10 seconds out from release point:

GTA: *A22, this is D36, with six jumpers, stand by, over.*

Pilot: *D36, this is A22, standing by, over.*

When aircraft is directly over release point:

GTA: *A22, this is D36, with six jumpers, execute, execute, execute.* (GTA must say EXECUTE or NO DROP at least three times, or until first load exits.)

At completion of operation, the GTA tells the pilot:

GTA: *A22, this is D36, I observe six jumpers away and clear, state intention and report when clear of my control zone,* (issue any advisories), *over.* (GTA must place aircraft into a closed traffic pattern with a reporting point if more than one pass is required.)

22-6. ACCEPTABLE WIND LIMITATIONS

Maximum allowable surface wind for static line parachute personnel airdrops is 13 knots (17 knots for WDZ). The maximum surface wind speed for static line heavy equipment airdrops is 17 knots with ground quick disconnects, 13 knots without ground quick disconnects, and 20 knots for CDS using G-13/14 parachutes. There is no wind speed restriction when HV parachutes are used with door bundles and CDS. There is no altitude wind limitation. Winds on the DZ are measured using the AN/PMQ-3A anemometer, or commercial anemometers authorized by USAIS messages DTG 101000Z MAR 94, subject: Use of Anemometers During Airdrop Operations, and DTG 211200Z OCT 94, subject: Use of Turbometer During Static Line Airdrop Operations)—two (one each for the DZSO and the assistant DZSO). Other anemometers not recommended for use should be employed only after a command-initiated risk assessment is completed. Regardless of the method or device used to measure DZ winds, the airborne commander is responsible for ensuring winds on the DZ do not exceed 13 knots during static line personnel airdrops.

22-7. THE 10-MINUTE WINDOW

On multiple aircraft operations or single aircraft operations using more than 2,100 meters of DZ, the surface wind is measured from the control center and the highest point of elevation on the DZ or the trail edge of the DZ. For single operations using less than 2,100 meters of DZ, the wind is measured from only one location, normally the control center. Beginning 12 minutes before TOT, the DZSO begins a constant monitoring of the surface wind using an anemometer.

a. **Surface Wind Exceeds Limits**. If the surface wind exceeds allowable wind limits, the aircraft is notified of a NO DROP, and a new 10-minute window is established. If the wind remains within limits during this new window, the drop takes place as planned. If the winds exceed allowable limits during the new window, NO DROP is relayed to the pilot and the entire procedure starts again.

b. **No-Drop Signal**. A NO DROP signal may be relayed to the aircraft by radio, red smoke, red flares, scrambled panels, or another planned signal.

22-8. POSTMISSION REQUIREMENTS

Immediately following the operation, several reports must be forwarded to higher headquarters.

a. **Required Reports**. Most of these reports are self-explanatory and require little time to complete. MAC Form 168 is used to record strike report information. All services will complete parachute malfunction and incident reports IAW AR 59-4/MCO 13480.1B/ OPNAVINST 4630.24C/AFJ 13-210(I). All services may also be required to report malfunctions and incidents on the loss of equipment and or injury to personnel IAW service regulations, ground mishap reporting MCO P5102.1 and OPNAVISNT 5102.1C.

- DZSO report.

- Malfunction report.
- MAC Form 168, Airdrop/Airland/Extraction Zone Control Log (Figure 22-6, page 22-11).
- Incident reporting format (Figure 22-7, page 22-12).

NOTE: MAC Form 168 will become AMC Form168 in the next printing.

b. **MAC Form 168 Completion**. Complete the MAC Form 168 as follows:
(1) DATE box—date of airdrop.
(2) LOCATION box—name of DZ.
(3) SST AND UNIT box—SST name and unit.
(4) DZ/LZ/EZ CONTROL OFFICER AND UNIT box—DZSTL name and unit.
(5) DROP ZONE SAFETY OFFICER AND UNIT box—enter names.
(6) LINE NO column—mission sequence number of each aircraft.

NOTE: Every aircraft has a mission sequence number (entered under LINE NO column). Subsequent passes by that same aircraft will all be scored on separate lines, in the order that they occur, immediately below the line for the first pass.

(7) TYPE ACFT column—type of aircraft.
(8) UNIT column—unit of aircraft.
(9) CALL SIGN column—call sign of pilot.
(10) TYPE MSN column—type of mission; refer to LEGEND for abbreviations.
(11) ETA column—estimated time of arrival, estimated TOT, S3 air brief.
(12) ETA column—actual arrival time of every pass.
(13) STRIKE RPRT columns—
- YDS column—distance first jumper/container lands from PI in yards; if within 25 yards, it is scored a PI.
- CLOCK column—using direction of flight as 12 o'clock and its back azimuth as 6 o'clock, estimated direction from PI to first jumper/bundle.

(14) SURF WIND column—surface wind; direction in degrees and velocity in knots.
(15) SCORE METHOD column—refer to LEGEND.
(16) MEAN EFFECTIVE WIND columns—
- TIME column—time taken.
- ALT column—what altitude taken to (should be drop altitude).
- DIR & VEL column—wind direction in degrees and velocity in knots.

c. **MAC Form 168 Routing**. The DZSTL forwards the completed MAC Form 168 to his air operations officer, who in turn submits it through the chain of command to the USAF representative.

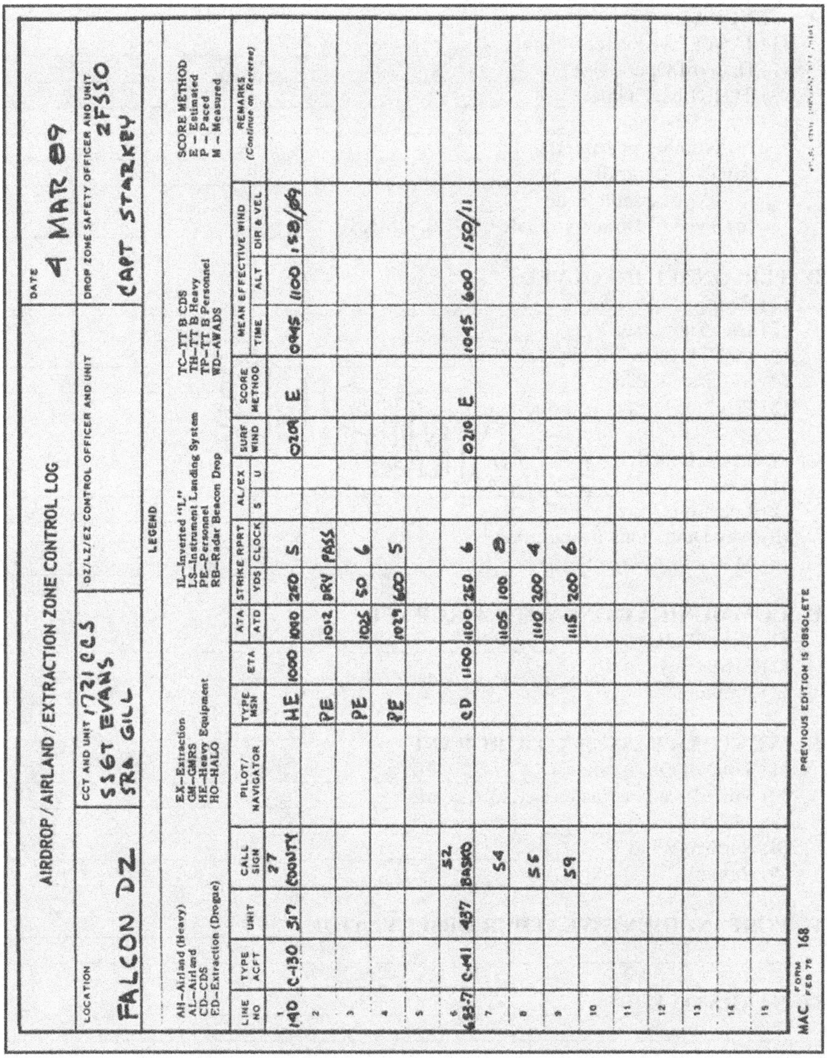

AIRDROP / AIRLAND / EXTRACTION ZONE CONTROL LOG

DATE: 4 MAR 89

LOCATION: FALCON DZ

CCT AND UNIT: 1721 CCS, SSGT EVANS, SRA GILL

DZ/LZ/EZ CONTROL OFFICER AND UNIT:

DROP ZONE SAFETY OFFICER AND UNIT: 2FSSO, CAPT STARKEY

LEGEND

AH–Airland (Heavy)
AL–Airland
CD–CDS
ED–Extraction (Drogue)

EX–Extraction
GM–GMRS
HE–Heavy Equipment
HO–HALO

IL–Inverted "L"
LS–Instrument Landing System
PE–Personnel
RB–Radar Beacon Drop

TC–TT B CDS
TH–TT B Heavy
TP–TT B Personnel
WD–AWADS

SCORE METHOD
E – Estimated
P – Paced
M – Measured

LINE NO	TYPE ACFT	UNIT	CALL SIGN	PILOT/ NAVIGATOR	TYPE MSN	ETA	ATA	STRIKE RPRT		AL/EX		SURF WIND	SCORE METHOD	MEAN EFFECTIVE WIND			REMARKS (Continue on Reverse)
							ATD	YDS	CLOCK	S	U			TIME	ALT	DIR & VEL	
1 140	C-130	317	27 COUNTY		HE	1000	1010	250	5			0209	E	0945	1100	158/09	
2					PE		1012	DRY	PASS								
3					PE		1025	50	6								
4					PE		1029	600	5								
5																	
6 483-7	C-141	437	52 BASKO		CD	1100	1100	250	6			0210	E	1045	600	150/11	
7			54				1105	100	8								
8			56				1110	200	4								
9			59				1115	200	6								
10																	
11																	
12																	
13																	
14																	
15																	

MAC FORM FEB 78 168 PREVIOUS EDITION IS OBSOLETE

Figure 22-6. Example of completed MAC Form 168.

A. GENERAL
- (1) JA/ATT Sequence Number ______
- (2) Date (of Operation) ______
- (3) TOT (Local Time) ______
- (4) Type Mission ______
  - (a) Number of Aircraft ______
  - (b) Type Aircraft ______
  - (c) Type Assault Zone ______
  - (d) Type of Delivery (CARP, VIRS, GMRS) ______

B. PERSONNEL INVOLVED
- (1) Flying Unit ______
- (2) Unit Supported ______
- (3) DZSTL (Name/Rank/Unit) ______
- (4) Medics (In Place) ______
- (5) POC for Further Information ______

C. ASSAULT ZONE
- (1) Name/Type ______
- (2) Location ______
- (3) Any Deviations from Survey ______
- (4) Marked IAW the Survey ______

D. COMMUNICATIONS WITH AIRCRAFT
- (1) Type Radios ______
- (2) Frequency Used ______
- (3) Problems ______

E. WEATHER PASSED TO AIRCRAFT
- (1) Time of Observation ______
- (2) Time Weather was Passed to Aircraft ______
- (3) MEW ______
- (4) Surface Wind ______
- (5) Remarks ______

F. POST-INCIDENT WEATHER OBSERVATION ______

G. NARRATIVE ______

Figure 22-7. Example of suggested format for incident reporting.

22-9. SURVEYS

USAF DZs are surveyed by qualified CCT/DZST (for SOF only—JM qualified at minimum). However, CCT are not required to be JM qualified. All DZs will be surveyed or tactically assessed by qualified STT/DZSTL/JM/Pathfinder personnel prior to use. Procedures can be found in FM 3-21.38(FM 57-38) and AFI 13-217. All information concerning the DZ is placed on MAC Form 339, Drop Zone Survey (Figure 22-8, page 22-14 and Figure 22-9, page 22-15), or AF Form 3823, Drop Zone Survey (Figure 22-10, page 22-16 and Figure 22-11, page 22-17). These forms provide the user the essential information needed to operate the DZ. Section 4 of the forms states what type of missions may be conducted on the DZ.

NOTE: When supplies of MAC Form 339 are exhausted, it will be replaced with AF Form 3823.

a. **Contingency/Wartime Operations.** During contingency/wartime and major exercises, DZSTLs may be expected to tactically locate, inspect, and approve a potential DZ for follow-up airdrop of resupply or reinforcements.

b. **Tactical Assessment.** All services will conduct airdrops on approved DZs. The tactical assessment is an approved means to certify a DZ for airdrop on both fixed-wing and rotary-wing aircraft. All tactical DZ assessments that can be used for future operations and that meet the standard for USAF aircraft will be forwarded to the USAF for inclusion in the AZAR. This will be accomplished using the following checklist, AFI 13-217, and FM 57-38. The JM/DZSTL will ensure all DZ requirements are within standards for the type of personnel, parachutes, and equipment airdropped. Once a DZ has been tactically assessed, it must be approved using AF Form 3823 (Drop Zone Survey).

- DZ name or intended call sign.
- Topographical map series and sheet number.
- Recommended approach axis magnetic course.
- Point of impact location (eight-digit grid).
- Leading edge centerline coordinates (eight-digit grid).
- DZ size in meters or yards.
- Air traffic restrictions/hazards.
- Name of surveyor and unit assigned.
- Recommended approval/disapproval (cite reason for disapproval).
- Remarks (include a recommendation for airdrop option, CARP, GMRS, VIRS, or blind drop).
- All other requirements must be met for a nontactical drop zone.

NOTE: Airdrop operations on tactically assessed DZs are made only under the following conditions:

- During training events, using rotary-wing or fixed-wing aircraft.
- The airdrop is located within a military reservation or on US government leased property.
- The supported service accepts the responsibility for any damage that occurs as a result of airdrop activity.

- There is adequate time for safe, effective planning.
- All hazards and obstacles are identified.
- A detailed risk assessment is completed.
- The tactical assessment is documented on AF Form 3823 and is approved by the first O6 in the chain of command.

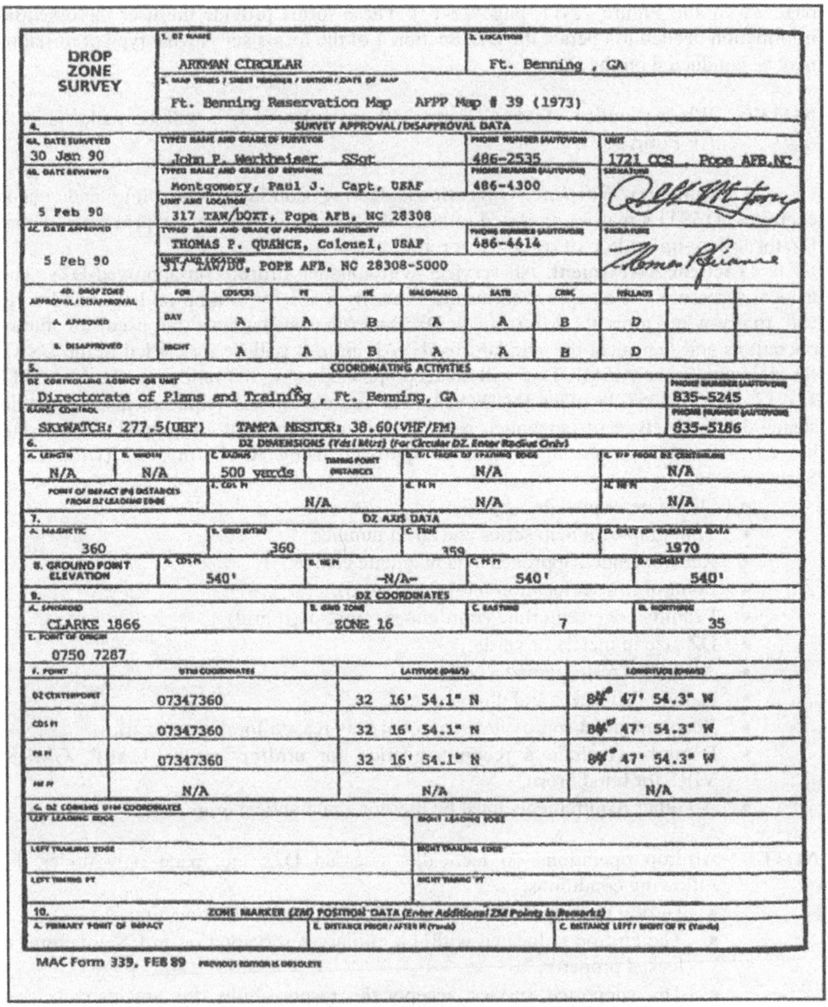

DROP ZONE SURVEY

1. DZ NAME: ARKMAN CIRCULAR
2. LOCATION: Ft. Benning , GA
3. MAP SERIES / SHEET NUMBER / EDITION / DATE OF MAP: Ft. Benning Reservation Map APPP Map # 39 (1973)

4. SURVEY APPROVAL/DISAPPROVAL DATA

4A. DATE SURVEYED: 30 Jan 90 — TYPED NAME AND GRADE OF SURVEYOR: John P. Warkheiser SSgt — PHONE NUMBER (AUTOVON): 486-2535 — UNIT: 1721 CCS , Pope AFB,NC

4B. DATE REVIEWED: 5 Feb 90 — TYPED NAME AND GRADE OF REVIEWER: Montgomery, Paul J. Capt, USAF — PHONE NUMBER (AUTOVON): 486-4300 — UNIT AND LOCATION: 317 TAW/DOXT, Pope AFB, NC 28308 — SIGNATURE

4C. DATE APPROVED: 5 Feb 90 — TYPED NAME AND GRADE OF APPROVING AUTHORITY: THOMAS P. QUANCE, Colonel, USAF — PHONE NUMBER (AUTOVON): 486-4414 — UNIT AND LOCATION: 317 TAW/DO, POPE AFB, NC 28308-5000 — SIGNATURE

4D. DROP ZONE APPROVAL / DISAPPROVAL (A. APPROVED, B. DISAPPROVED)

FOR	CDS/CRS	PE	HE	HALO/HAHO	SATB	CRRC	NIGHT		
DAY	A	A	B	A	A	B	D		
NIGHT	A	A	B	A	A	B	D		

5. COORDINATING ACTIVITIES

DZ CONTROLLING AGENCY OR UNIT: Directorate of Plans and Training , Ft. Benning, GA — PHONE NUMBER (AUTOVON): 835-5245

RANGE CONTROL: SKYWATCH: 277.5(UHF) TAMPA NESTOR: 38.60(VHF/FM) — PHONE NUMBER (AUTOVON): 835-5186

6. DZ DIMENSIONS (Yds/Mtrs) (For Circular DZ, Enter Radius Only)

A. LENGTH: N/A — B. WIDTH: N/A — C. RADIUS: 500 yards — TIMING POINT DISTANCES — D. T/C FROM DZ TRAILING EDGE: N/A — E. T/P FROM DZ CENTERLINE: N/A

POINT OF IMPACT (PI) DISTANCES FROM DZ LEADING EDGE — F. CDS PI: N/A — G. PE PI: N/A — H. HE PI: N/A

7. DZ AXIS DATA

A. MAGNETIC: 360 — B. GRID (UTM): 360 — C. TRUE: 359 — D. DATE OF VARIATION DATA: 1970

8. GROUND POINT ELEVATION — A. CDS PI: 540' — B. HE PI: -N/A- — C. PE PI: 540' — D. HIGHEST: 540'

9. DZ COORDINATES

A. SPHEROID: CLARKE 1866 — B. GRID ZONE: ZONE 16 — C. EASTING: 7 — D. NORTHING: 35

E. POINT OF ORIGIN: 0750 7287

F. POINT	UTM COORDINATES	LATITUDE (DMS)	LONGITUDE (DMS)
DZ CENTERPOINT	07347360	32 16' 54.1" N	84° 47' 54.3" W
CDS PI	07347360	32 16' 54.1" N	84° 47' 54.3" W
PE PI	07347360	32 16' 54.1" N	84° 47' 54.3" W
HE PI	N/A	N/A	N/A

G. DZ CORNERS UTM COORDINATES

LEFT LEADING EDGE	RIGHT LEADING EDGE
LEFT TRAILING EDGE	RIGHT TRAILING EDGE
LEFT TIMING PT	RIGHT TIMING PT

10. ZONE MARKER (ZM) POSITION DATA (Enter Additional ZM Points in Remarks)

A. PRIMARY POINT OF IMPACT	B. DISTANCE PRIOR / AFTER PI (Yards)	C. DISTANCE LEFT / RIGHT OF PI (Yards)

MAC Form 339, FEB 89 PREVIOUS EDITION IS OBSOLETE

Figure 22-8. Example of completed MAC Form 339 (front).

11. DZ DIAGRAM/12. REMARKS

DZ NAME

ARKMAN CIRCULAR DZ, GA

NOT TO SCALE

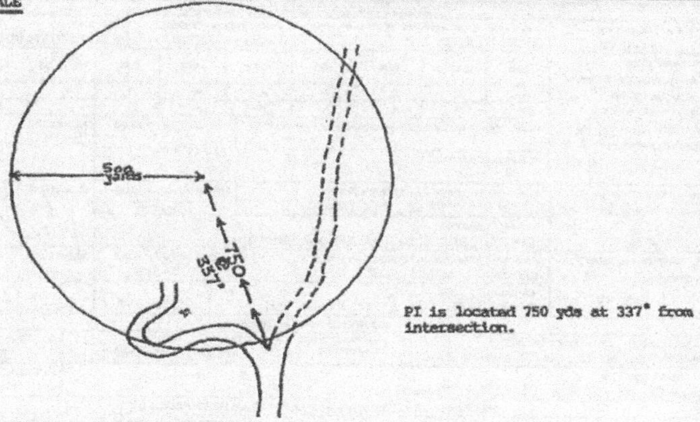

1. DZ is for single ship operations only.
2. User accepts responsibility for injury to personnel and damage of equipment.
3. DZ is approximately 30% trees from 15' to 20' tall.
4. For pathfinder use, run in heading and PI will be pre-briefed.
5. Aircraft must contact Range Control at least 10 minutes prior to entry into the restricted area (R3002).
6. Ground party must establish radio communications with Range Control 30 mintes prior to TOT and maintain contact for duration of mission.
7. Recommend all turn outs to the west to avoid over flying the town of Cusseta, GA located 026° at 2000m.
8. Cellular One Telephone tower located 026° at 2600m. Height of tower is 795'MSL.
9. Dirt roads located on the south and east side of DZ.
10. TV tower 2249' MSL located at LAT/LONG 32° 19' 25"N, 84° 46'46"W; UTM grid coordinates 091784.

13. PHOTOGRAPHY AVAILABLE

☐ YES ☒ NO

LOW LEVEL ROUTES

☐ NONE AVAILABLE

☐ ROUTE NAME/DESIGNATOR

Figure 22-9. Example of completed MAC Form 339 (back).

AIRBORNE UNIT ASSUMES RESPONSIBILITY FOR PERSONNEL INJURY AND EQUIPMENT DAMAGE ON DZ

DROP ZONE SURVEY

1. DZ NAME: LEDO I
2. LOCATION: FT. BENNING, GEORGIA
3. MAP SERIES/SHEET NUMBER/EDITION/DATE OF MAP: AFPP MAP #39 FT. BENNING RESERVATION 1973/TPC G-21D/ATLANTA SECTIONAL

4. SURVEY APPROVAL/DISAPPROVAL DATA

4A1. DATE SURVEYED	4A2. TYPED NAME AND GRADE OF SURVEYOR	4A3. PHONE NUMBER	4A4. UNIT
15 Aug 95	EDDIE D. KIDD, SSG	(DSN) 835-3218	HHC 1/507TH PIR

4B. DROP ZONE APPROVAL/DISAPPROVAL (A = APPROVED, D = DISAPPROVED)

FOR	CDS/CRS	PER	HE	MFF	SATB	CRRC	HSLLADS	HVCDS
DAY	A	A	D	A	A	D	A	A
NIGHT	A	A	D	A	A	D	A	A

4C. DATE APPROVED FOR GROUND OPERATIONS — NAME, GRADE AND SERVICE OF APPROVAL AUTHORITY: — PHONE NUMBER (DSN): — SIGNATURE: — UNIT AND LOCATION:

4D. DATE SAFETY OF FLIGHT REVIEW APPROVED: 31 Jan 95
NAME AND GRADE OF REVIEWING OFFICER: JAMES E. GOOD, JR., Capt, USAF
PHONE NUMBER (DSN): 731-3988
SIGNATURE: [signature]
UNIT AND LOCATION: 3134 OSS/OSTX, Little Rock AFB AR 72099-5043

4E. DATE OF MAJCOM APPROVAL: 17 Feb 95
NAME AND GRADE OF APPROVING AUTHORITY: RONALD L. MOREY, COL, USAF
PHONE NUMBER (DSN): DSN 576-5007
SIGNATURE: [signature]
UNIT AND LOCATION: HQ AMC/DOT, 402 Scott Drive Unit 3A1, Scott AFB IL

5. COORDINATING ACTIVITIES

A. DZ CONTROLLING AGENCY OR UNIT: 1/507th Para Inf. Reg. S-3, Ft. Benning GA
B. MEMORANDUM OF UNDERSTANDING/LAND USE AGREEMENT: YES ☐ NO ☒ ATTACHED ☐
C. PHONE NUMBER (DSN): 835-3012
D. RANGE CONTROL: Range Control FM 38.60 UHF 277.5 (Skywatch)
E. PHONE NUMBER (DSN): 835-2152

6. DZ DIMENSIONS *(YDS/METERS) (FOR CIRCULAR DZ, ENTER RADIUS ONLY)*

A. LENGTH	B. WIDTH	C. RADIUS	TIMING POINT DISTANCES	D. T/P FROM DZ LEADING EDGE	E. T/P FROM DZ CENTERLINE
925 yds	700 yds	N/A		N/A	N/A

POINT OF IMPACT DISTANCES FROM DZ LEADING EDGE	F. CDS PI	G. PE PI	H. HE PI
	350 yds	350 yds	N/A

7. DZ AXIS DATA *(OPTIONAL FOR CIRCULAR DZ)*

A. MAGNETIC	B. GRID (UTM)	C. TRUE	D. DATE OF VARIATION DATA
061 deg	058 deg	059 deg	1991

8. GROUND POINT ELEVATION	A. CDS PI	B. HE PI	C. PE PI	D. HIGHEST
	480'	N/A	480'	500'

9. DZ COORDINATES

A. SPHEROID	B. DATUM	C. GRID ZONE	D. EASTING	E. NORTHING
CLARKE 1866	1927 NA	16	7	35

F. GPS DERIVED COORDINATES: YES ☒ NO ☐
G. POINT OF ORIGIN: GL 1530 7913 Intersection of roads PI is 432m @ 224 deg

H. POINT	UTM COORDINATES	WGS84 LATITUDE (D-M.MM)	WGS84 LONGITUDE (D-M.MM)
DZ CENTERPOINT	GL 1509 7880	N 32d 19' 37.4"	W 84d 42' 54.1"
CDS PI	GL 1500 7874	N 32d 19' 35.7"	W 84d 42' 57.5"
PE PI	GL 1500 7874	N 32d 19' 35.7"	W 84d 42' 57.5"
HE PI	N/A	N/A	N/A

I. DZ CORNERS UTM COORDINATES

LEFT LEADING EDGE: GL 1456 7884	RIGHT LEADING EDGE: GL 1490 7830
LEFT TRAILING EDGE: GL 1527 7929	RIGHT TRAILING EDGE: GL 1562 7875
LEFT TIMING POINT: N/A	RIGHT TIMING POINT: N/A

AF FORM 3823, FEB 94 *(EF-V1) (PerFORM PRO)*

Figure 22-10. Example of completed AF Form 3823 (front).

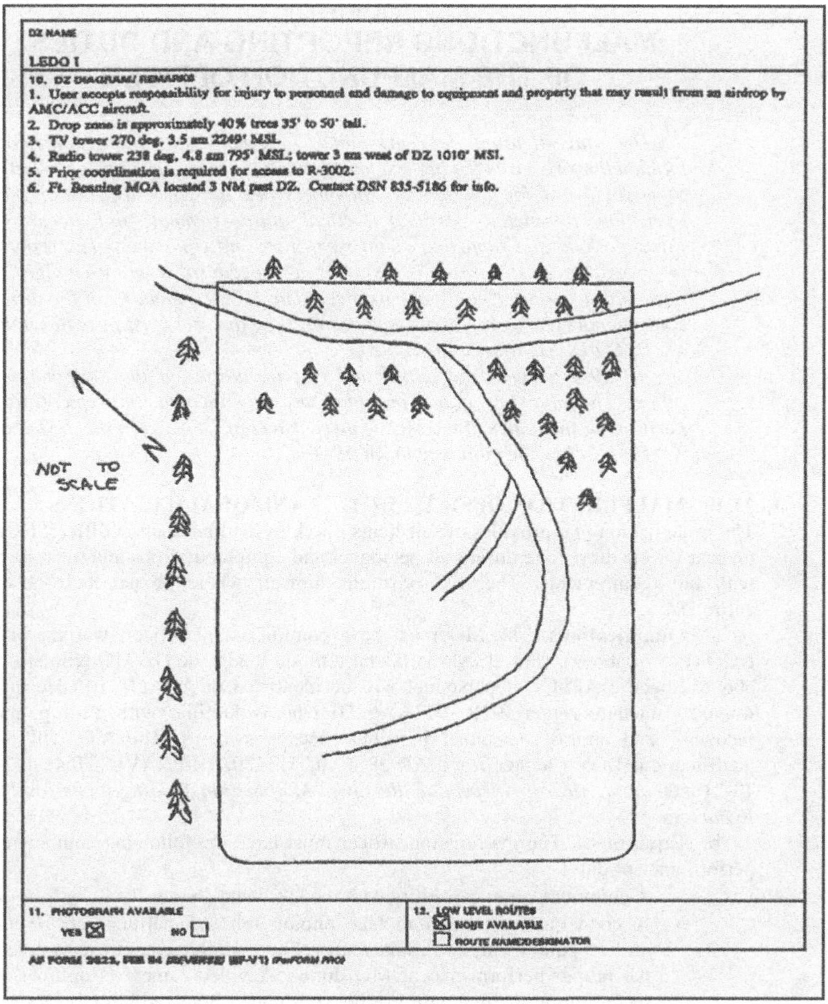

DZ NAME

LEDO I

10. DZ DIAGRAM/ REMARKS

1. User accepts responsibility for injury to personnel and damage to equipment and property that may result from an airdrop by AMC/ACC aircraft.
2. Drop zone is approximately 40% trees 35' to 50' tall.
3. TV tower 270 deg, 3.5 nm 2249' MSL
4. Radio tower 238 deg, 4.8 nm 795' MSL; tower 3 nm west of DZ 1010' MSL.
5. Prior coordination is required for access to R-3002.
6. Ft. Benning MOA located 3 NM past DZ. Contact DSN 835-5186 for info.

11. PHOTOGRAPH AVAILABLE

YES [X] NO [ ]

12. LOW LEVEL ROUTES

[X] NONE AVAILABLE

[ ] ROUTE NAME/DESIGNATOR

AF FORM 3823, FEB 94 (REVERSE) (EF-V1) (PerFORM PRO)

Figure 22-11. Example of completed AF Form 3823 (back).

CHAPTER 23

MALFUNCTIONS REPORTING AND DUTIES OF THE MALFUNCTION OFFICER

The investigation of personnel, parachutes, and equipment malfunctions receives the highest priority and is secondary in priority only to medical aid for the injured. It supersedes all other aspects of the operation to include ground tactical play. Prompt and accurate investigations and reporting could save lives and equipment. The report provides data to determine if a system or procedural training change is necessary to prevent future occurrences. The MO is a member of the drop zone support team. Any assistance required by the MO must pass through the DZSO/DZST, who controls the DZ.

AR 59-4 governs the duties and responsibilities of the malfunction officer. The following chapter provides only the minimum requirements for performing the duties of the MO. If there is a conflict between this FM and AR 59-4, follow the guidance in AR 59-4.

23-1. MALFUNCTION OFFICER DUTIES AND QUALIFICATIONS

The organization that provides the air items provides the malfunction officer. He must be present on the drop zone during all personnel and equipment drops and must be familiar with the requirements. The MO positions himself where he can best observe the entire DZ.

a. **Qualifications**. The MO must be a commissioned officer, warrant officer, or NCO (E-5 or above). (For USAF unilateral training loads, the DZ MO will be an E-4 or above. Specific AFSCs of personnel will be identified in AFI 11-410.) He must be a trained parachute rigger (92R, 921A, 92D) who is familiar with airdrop, parachute recovery, and aircraft personnel parachute escape systems. The MO will meet the qualifications IAW Chapter 7 and AR 59-4/MCO13480. 1B/PNAVINST 4630.24C/AFJ 13-210(I), *Joint Airdrop Inspection Records, Malfunction Investigations, and Activity Reporting*.

b. **Equipment**. The malfunction officer must have the following equipment during performance of duty:

- A communication capability with the DZ control party.
- A good quality camera to take photographs of malfunctions or incidents, airdrop equipment, and impact sites. Photographic equipment is essential for the proper performance of MO duties. A video camera is optimal. Pictures recording a malfunction or the equipment involved greatly assist in the investigation.
- The forms and clerical supplies necessary to tag equipment and initiate reports.
- Binoculars (day) or night vision devices (night).
- Transportation to move around the DZ.

c. **MO Duties**. The MO performs the following actions in case of a malfunction:

(1) ***Partial or Total Malfunction with No Serious Injuries.***

(a) Secure and guard the impact site. Conduct an on-site investigation to determine, if possible, the cause of the malfunction, and to ascertain whether a criminal or malicious act may have caused the malfunction. (Refer to AR 59-4, Appendixes B and C, for checklists.)

(b) Photograph the malfunction as it happens (if possible), the malfunctioned equipment, and the malfunction or impact site. Sketch the impact site in relation to the DZ.

(c) Collect, properly identify, tag, and maintain chain of custody for items of physical evidence involved in the malfunction (such as clothing, equipment, air items, and personal property).

(d) Conduct a detailed component-by-component examination of all equipment.

(e) Collect statements from any personnel present who witnessed the malfunction. Take statements from JM/AJM, safeties, previous jumper, follow-on jumper, ground observers, and aircraft personnel who are able to provide facts.

(f) Appropriately tag and bag the collected equipment and evacuate it to an appropriate area to conduct the TM-10-1670-series rigger type investigation.

(g) Ensure all air items and evidence are retained until the investigating authority releases them.

(h) Prepare and submit the appropriate reports IAW AR 59-4, Chapter 5.

(2) ***Partial or Total Malfunction with Serious Injury or Death.***

(a) Place the impact area off limits and post a guard to ensure the integrity of the scene and to limit access to authorized personnel only. Take immediate possession of the log record book (DA Form 3912), and limit access to the appointed investigating officer. In the event of a death, ensure the DZSTL notifies the military police immediately. Access to the site is limited to the malfunction officer, the SME assisting the MO, and responding CID and medical personnel only.

(b) Photograph the parachutist, the impact site, and any obvious defects in the equipment (to include damage done by the impact). Sketch the impact site in relation to the DZ and mark the impact location of the parachutist and his equipment.

(c) Collect **all** items of equipment involved and maintain chain of custody.

(d) Immediately initiate an investigation prior to the arrival of the CID and ensure the scene is not altered. If determination of tampering or intentional sabotage is made, the entire investigation is turned over to the CID. If there has been no tampering or sabotage, continue with the investigation.

(e) Request that medical personnel secure and preserve the parachutist's clothing and equipment that is removed from the DZ.

(f) Gather statements from previous jumper, follow-on jumper, JM/AJM, safeties, and from any other ground or aircraft personnel who may be able to provide facts.

(g) Record the name and unit of any personnel who observed the incident even if the information is not new to the investigation.

(h) Secure a copy of the jump manifest. Reconstruct the jump stick from personnel present if required.

(i) Conduct a detailed component-by-component examination of all equipment after the parachutist has been evacuated.

(j) Request the DZSO/DZSTL notify the DACO so the aircrew can inspect the aircraft for any defects or damage that may have contributed to or caused the malfunction. Request DACO ensure the parachute deployment bags from that aircraft be identified and segregated from those of other aircraft.

(k) Obtain the deployment bag serial number from the log record book or parachute pack sheet and secure it with the remainder of the parachute assembly until the investigation is completed.

(l) When removing the parachute and related air items to a suitable area to perform the TM-10-1670-series rigger type investigation, ensure the parachute is loosely rolled, leaving any turns, tangles, or twists in the suspension lines. Appropriately tag and bag the air items and maintain chain of custody. Release these to the investigating SME only.

(m)Ensure that the DZSO/DZSTL signs for any organizational and personal clothing, weapons, and equipment not collected as part of the investigation. This equipment will be returned to the jumper's unit, and the unit will preserve the equipment until the completion of the investigation.

23-2. MALFUNCTION OFFICER RESPONSIBILITIES DURING INVESTIGATIONS

The depth of any investigation varies according to the severity of the malfunction and resultant injuries. In cases apparently not involving death or serious injury, the MO conducts the on-sight investigation solely to determine the cause of the malfunction and actions required to prevent future occurrences.

a. **MO Follow-On Investigation**. In cases involving misconduct, serious incident or injury, or death, the MO conducts a follow-on investigation according to service directives and AR 59-4. His investigative notes, insights, reports, and physical evidence are available to these investigations. The MO must ensure that after the component-by-component examination is completed, the equipment involved in the malfunction is secured and accessible only to the appointed investigating SME.

b. **Confidentiality of Investigation**. During the investigation, the MO gathers items of information and evidence that are sensitive in nature. He ensures the information pertaining to the investigation is given only to authorized personnel on a need-to-know basis. The MO must exercise great care so that the rights of involved personnel are not compromised and the government is not placed in an unfavorable position.

23-3. REPORTING DATA

The MO discusses as much data as allowable with the DZSO/DZSTL prior to transmitting it to the control group. The MO and the DZSO/DZSTL normally discuss this feeder information immediately after the jump is complete.

NOTE: The MO uses DD Form 1748-2 to report all airdrop personnel malfunctions IAW AR 59-4.

PART SIX

Special Airborne Procedures

CHAPTER 24

ADVERSE WEATHER AERIAL DELIVERY SYSTEM

AWADS is a navigational system installed in some USAF C-130 aircraft. It enables the aircraft to fly to a DZ during reduced visibility, and provides flexibility to the airborne force commander in the accomplishment of all airborne missions. AWADS is effective in large, joint operations, tactical reinforcements, and specialized missions.

24-1. MULTIPLE MISSION SUPPORT

AWADS operations facilitate rapid and continuous aerial deployment or resupply in adverse weather or darkness. Units can execute a parachute assault without a pre-positioned AFCCT or an Army DZST. As a result, time (length) of the air formation is shortened, and the air corridor must be cleared only once by tactical air to heighten the element of surprise. AFCCTs are introduced with the assault elements and assist the GUC with additional CDS, airland, or HD missions. AWADS enables a commander to conduct a rapid vertical reinforcement during instrument meteorological conditions or visual meteorological conditions for units threatened by enemy penetration.

NOTE: A minimum ceiling of 200 feet and a minimum visibility of 1/2 mile for personnel and equipment are imposed for tactical training.

24-2. TRAINING AND PREPARATION

AWADS operations demand detailed planning, rehearsal, training, and coordination between USAF and Army units to be effective. Due to limited visibility in the air and during assembly on the ground, AWADS requires both technical training and psychological preparation of the parachutists. AWADS sustained prejump training requires modification to normal jump conditions.

24-3. MODIFIED JUMPMASTER DUTIES

The JM relies on the loadmaster to obtain and relay en route information. Under AWADS conditions, the JM still attempts to perform all of the required checks. He may not be able to observe safety hazards beyond the immediate area of the door.

24-4. MODIFIED PARACHUTIST ACTIONS

Parachutist actions during descent under normal conditions must be modified in AWADS training. Modifications are made when using the T-10-series and MC1-series parachutes.

a. **T-10-Series Points of Performance**.

(1) Check body position and count.

(2) Check canopy and gain canopy control. Any malfunction under these conditions requires activation of the reserve since the parachutist cannot effectively judge the rate of descent.

(3) Keep a sharp lookout during descent.

(4) Do not slip, except to avoid collisions, until after breaking through the clouds.

(5) Give way. Give lower canopies the right of way; higher canopies slip to avoid them.

(6) Recheck the canopy. Make this check after breaking through the clouds if the canopy could not be checked while in the clouds.

(7) Prepare to land. Do not release equipment until the ground can be seen and it is clear below.

(8) Land. Execute a proper PLF.

b. **MC1-Series Points of Performance**.

(1) Check body position and count.

(2) Check canopy and gain canopy control. If the TU modification is to the front, there is a complete inversion. Do **not** activate the reserve since the canopy can still be controlled. For any other malfunction, activate the reserve immediately.

(3) Identify steering toggles. Grasp each one and bring them both to chest level to reduce lateral movement and perform braking; this helps to eliminate midair collisions and extreme dispersion due to excessive drift.

(4) If either or both toggles are broken, steer the canopy by pulling a slip with the rear riser on the side of the intended turn. In the clouds, anyone with a broken toggle and not applying "brakes" automatically has a greater lateral drift than anyone else and must keep alert during descent.

(5) Keep a sharp lookout during descent. During reduced visibility prepare to take immediate evasive action.

(6) Give way. Give lower canopies the right of way. If parachutists see the possibility of converging at any altitude and from any direction, they immediately turn away from each other by pulling the toggle that is away from the other parachutist.

(7) Recheck canopy. Perform a thorough canopy inspection after breaking through the clouds. Release the toggles when the ground is in sight and prepare equipment for landing. Use the lowering lines.

(8) Prepare to land. Turn into the wind 75 to 150 feet above the ground. Obstacles spotted when coming out of low clouds normally require rapid preparation for the appropriate emergency landing.

(9) Land. Execute a proper PLF.

CHAPTER 25

DELIBERATE WATER DROP ZONE OPERATIONS

Units conducting water drop zone operations should use the following procedures as a guide to ensure the mission is conducted safely. Conduct all jumps under the following conditions:

- *Daylight; warm weather conditions.*
- *Surface winds do not exceed 17 knots.*
- *Drop control—visual marking system.*
- *Helicopter—flying under visual flight conditions.*
- *Wind conditions permit operation on the WDZ.*
- *Water should be more than 10 feet deep with no underwater obstacles that depth.*
- *Briefing held for WDZ operations by DZSO/DZSTL and WDZ control personnel.*
- *All parachutists are classified as swimmers or strong swimmers. (USMC S-1 or higher).*
- *All parachutists have completed drownproofing training within the preceding 12 months.*
- *All parachutists have been trained on activation procedures for the life preserver in use, to include manual inflation procedures.*
- *All parachutists have attended prejump training on deliberate water DZ procedures according to unit SOP.*
- *All combat equipment is float checked.*

NOTE: The parachute operation OIC briefs parachutists, aircrew, JM, and WDZ control party regarding WDZ operations.

WARNING

The MC1-1B should be used, if available. The porosity of the MC1-1C is extremely limited and may create a suffocation hazard if the jumper is trapped under the canopy.

25-1. PERSONNEL AND EQUIPMENT

The following personnel and equipment are required for deliberate WDZ operations:

a. **Personnel.**

(1) JM, AJM, safety personnel, as required for the type of aircraft used.

(2) DZSO/DZSTL, JM qualified and current IAW Chapter 7.

(3) Medic/corpsman with resuscitator.

(4) Safety vehicle driver.

(5) Boat commander/coxswain for each boat.

(6) Malfunction officer.

(7) Safety swimmers—minimum of one safety swimmer is required to be on board each recovery boat. The safety swimmer must have fins, facemask, knife, and an inflatable life preserver. The safety swimmers will be used to recover personnel and equipment, and assist parachutists as needed. The safety swimmer cannot be additionally assigned as the boat coxswain or corpsman. The safety swimmer should be a qualified parachutist and, as a minimum, will be qualified as a US Army Class 1 advanced survival swimmer (IAW TC 21-21), USMC CWS1, Navy 1st class, or lifeguard certified. The safety swimmer will attend prejump training for intentional water jumps with parachutists and wet silk training.

b. **Water Drop Zone Equipment**.

(1) Equipment for safety swimmers.

(2) Serviceable boats, motors, and fuel (RB-15 boats, 14-foot engineer assault boats, combat rubber raiding craft, or civilian equivalent boats with motors). If available, hard-body boats should be used.

(3) Equipment for recovery and command boats.

(4) Required panels and smoke.

(5) Required communications equipment for each boat to include spare batteries and one complete spare radio set for the operation. Boat-to-boat and boat-to-air communication should be checked.

(6) Floating, nonflammable container with suitable anchor for smoke grenades.

(7) First aid equipment to include resuscitator and backboard.

(8) Bailing cup per boat.

(9) Motor tie-down rope per boat.

(10) Sheath knife (boat commander and safety swimmer) per boat.

(11) Pliers per boat.

(12) One extra life preserver for emergencies per boat.

(13) Life jackets or life preservers for all personnel onboard each boat.

(14) Oars or paddles per boat.

(15) Boat hook per boat.

(17) Approved anemometers. The approved anemometers are the DIC, DIC3, TurboMeter, and AN/PMQ-3A. (The AN/ML433A/PM and the anemometers that use floating balls or small floating lightweight aluminum devices in a tube are not authorized for use during personnel or cargo airdrop operations. See Chapter 20, paragraph 20-9 for use and calibration information.)

c. **Parachutist Requirements.** Parachutists, at a minimum, must be classified as a US Army Class 3 basic survival swimmer, USMC CWS-1, or Navy 1st class swimmer before making a water parachute drop. Parachutists will be current static line jumpers before making water jumps. Wet silk training for intentional water jumps will be conducted at a minimum of once every six months by all parachutists and safety swimmers involved in water jumps.

a. An individual's first water jump must be performed during the day and without combat equipment. To be classified as a Class 1 advanced water survival swimmer, personnel will be tested and documented on an annual basis and must pass the requirements contained in TC 21-21.

b. Wet silk training is conducted at the unit level by certified personnel (Class 1 advanced water survival swimmer [IAW TC 21-21], USMC CWS1, Navy 1st class, or lifeguard-certified current JM) by putting an unserviceable parachute in a pool, in a controlled environment, with safety swimmers. The jumpers, one at a time, demonstrate the actions to take during a water landing, jump into the water, and swim under the canopy to experience what it is like to be under a parachute in the water. The jumper demonstrates his ability to follow a radial seam to get safely out from under the canopy, make an air pocket under the canopy to breathe from, and breathe from the apex of the canopy. If a jumper is trapped under the canopy during wet silk training, the safety swimmers pull the jumper out of the water. A minimum of two safety swimmers with mask, fins, snorkel, and dive knife and a medic/corpsman will be available during this training. A standby diver on SCUBA is also recommended.

25-2. ORGANIZATION AND EQUIPMENT OF DROP ZONE DETAIL

The organization and equipment of the drop zone detail include the following:

a. **DZSO Command Boat**. The command boat will be separate from the recovery boats. It will be used by the DZSO and medic/corpsman and will not be used to recover parachutists or equipment except in the case of an emergency. The command boat (boat number 1) includes the following personnel and equipment:

NOTE: All boats required for the operation will be on station with required personnel and the engines running before the release of parachutists.

(1) ***Personnel.***

- Drop zone safety officer.
- Boat commander/boat coxswain.
- Safety swimmer with equipment.
- Medic/corpsman with equipment.
- Malfunction officer.

(2) ***Equipment.***

- Marker panels.
- Medical aid kit.
- Backboard.
- Resuscitator.
- Boat hook.
- FM radio (complete) with extra battery for boat-to-boat, boat-to-air, and surface communications.
- Approved anemometers. The approved anemometers are the DIC, DIC3, TurboMeter, and AN/PMQ-3A. (The AN/ML433A/PM and the anemometers that use floating balls or small floating lightweight aluminum devices in a tube are not authorized for use during personnel or cargo airdrop operations. See Chapter 20, paragraph 20-9 for use and calibration information.)
- Bailing cup.
- Motor tie-down rope.
- Sheath knife (safety swimmer and boat commander).

- Pliers.
- Life jackets or life preservers for all personnel on board.
- One extra life preserver for emergencies.
- Oars or paddles per boat.

b. **Recovery Boats for Equipment Drops.** A minimum of one power driven boat is required for two equipment platforms dropped on the same pass. Equipment recovery boats are to be used in the recovery of equipment parachutes and platforms. Recovery boats assigned to recover jumpers doe not meet this requirement when jumpers and equipment are on the same pass. Equipment recovery boats must be large enough to recover cargo parachutes and platforms. The boat coxswain's only duty is to navigate the boat. The boat coxswain cannot act as the safety swimmer or corpsman.

(1) Use extreme caution when recovering CRRC and other equipment in the water.

(2) The jumpers are the recovery personnel. They will have assisted in the rigging of the CRRC and container equipment before the airdrop. This will help they know what straps to cut without losing the equipment or injuring other personnel.

(3) Ensure a safety tie is rigged between the platform and boat and is cut last after the recovery boat has secured the platform.

(4) Tow the platform back to shore with the recovery boat. Place sandbags on the back of the platform to raise the front to ease towing.

WARNING

Ensure that all personnel are clear of the platform and equipment and not fouled in any lines before cutting the equipment loose from the platform. Failure to do so may cause serious injury or death if the platform sinks.

b. **Recovery Boats for Personnel.** A minimum of one power driven recovery boat is required for every three parachutists being dropped on the same pass. Parachutist recovery boats must have an inflatable boat or ladder rigged along side if they have a freeboard of more than 3 feet and or the boats do not provide an easy platform for recovery of personnel. Boats assigned as personnel recovery platforms may only be used to assist in the recovery of airdropped equipment after all parachutists have been recovered. The boat coxswain's only duty is to navigate the boat. The boat coxswain cannot act as the safety swimmer or corpsman.

c. **Equipment and Personnel Requirements for Recovery Boats.** Recovery boats require the following personnel and equipment:

(1) ***Personnel (Personnel and Cargo Recovery Boats).***

- Boat commander/coxswain.
- Safety swimmer with equipment.

(2) ***Equipment (Personnel and Cargo Recovery Boats).***

- Boat hook.
- Bailing cup.
- Motor tie-down rope.

- Sheath knife (safety swimmer and boat commander).
- Pliers.
- Life jackets or life preservers for all personnel onboard.
- Extra approved life preserver for emergencies.
- Oars or paddles per boat.
- Radio (complete) with extra battery for boat-to-boat surface communications.

(3) ***Safety Swimmer Equipment.***

- Swim fins—one pair.
- Face mask.
- Snorkel.
- Sheath knife.
- Life vest.
- Wet suit—optional.

(4) ***Safety Vehicle Operator.***

- Driver (on standby).
- Radio (communicate with command boat).

NOTE: The JM briefs the parachutists, aircrew, and WDZ control party regarding WDZ operations.

(5) ***DZSO (Additional Duties).***

(a) Briefs boat crews, safety swimmers, medic, and safety vehicle driver on the following:

- Overall organization of WDZ.
- Number of lifts and personnel to be dropped.
- Drop altitude and aircraft heading.
- Surface winds.
- Water depth.
- Turnaround time between drops.
- Recovery procedures.
- Communications plan.
- Emergency recovery and evacuation plan.
- Drop and abort signals.
- Applicable special instructions.

(b) Maintains visual observation of all parachutists until safely recovered.

(6) ***Jumpmaster.***

(a) Performs JM duties.

(b) Helps recover personnel from the water in an emergency.

(c) Keeps parachutists under observation. Does not release the next pass until all parachutists have been recovered.

(7) ***Boat Commander/Coxswain.***

(a) Ensures all personnel and equipment are on board.

(b) Ensures that all equipment is operational.

(c) Ensures that the safety swimmer has been briefed and understands instructions.

(d) Maintains visual observations of parachutists from time of exit to safe recovery.

(e) Controls actions of safety swimmer.

(f) Ensures all personnel are trained and rehearsed in their duties.

(8) ***Safety Swimmer.***

(a) Maintains visual observation of parachutists from time of exit to safe recovery. Is alert at all times for parachutists in trouble and is prepared to enter the water to assist parachutists.

(b) Enters water (on order of boat commander) and assists parachutist recovery.

(9) ***Medic/Corpsman.***

(a) Ensures the resuscitator is complete and in operational condition.

(b) Ensures that safety personnel understand lifesaving techniques.

(c) Uses applicable medical equipment and provides required first-aid treatment and medical evacuation.

25-3. SAFE CONDITIONS

Units conducting WDZ operations should use the following to ensure the mission is conducted safely. Ensure all WDZ parachute jumps are conducted under the following conditions for all services.

- Water will be more than 10 feet deep.
- The WDZ will not be in or near the surf zone.
- Surface winds will not be in excess of 17 knots.
- Sea state is no more than 2. (IAW MCWP 2-15.3, Table 7-1, or Joint Pub 4-01.6, sea state 2 indicates that wave height will not exceed 1 2/3 feet (20 inches), as measured from a wave crest and the preceding trough.)
- Parachutists jumping in cold water (60 degrees or lower) will wear wetsuits or dry suits.
- All parachutists are classified as strong swimmers, USMC CWS-1, or Navy 1st class swimmers.
- All safety swimmers should be parachutists and will be classified as Class 1 advanced water survival swimmers (IAW TC 21-21), USMC CWS-1, Navy 1st class swimmer, or lifeguard certified.
- All parachutists and swimmers have completed wet silk training in the past six months.
- All parachutists and swimmers have been trained on activation procedures for the life preserver in their use to include manual inflation procedures.
- All floatation devices are approved, serviceable, and properly maintained.
- All parachutists have attended prejump training on deliberate water DZ procedures according to this chapter and unit SOP.
- All combat equipment is waterproofed and float checked.
- All jumpers and equipment have chem lights and strobes attached for night jumps.
- Night vision devices will be used on the WDZ for all night drops.
- Drop control—visual marking system: mirror, smoke, VS-17 panels.
- Release method: CARP or WSVC.
- Helicopter or fixed-wing—flying under visual flight conditions.
- Only one to three parachutist jumps for each recovery boat for each pass. (Boat-to-jumper ratio depends on unit and jumper experience with water jumps.

One boat to one jumper is strongly recommended for units that rarely conduct WDZ operations.)

- Sufficient boats are properly manned with the engines running during all live passes.
- Clear to drop is by radio or by all the boats driving in one big circle.
- Abort the pass is by radio or by all the boats being scrambled.

25-4. JUMP RECOVERY PROCEDURES

After parachutists enter the water, recovery boats (boat numbers 2 and 3 and any additional recovery boats) proceed to the location of designated parachutists and begin recovery operations.

a. **Boat Commander/Coxswain**. The boat commander/coxswain of each recovery boat identifies his assigned parachutists as soon as possible.

(1) The boat coxswain determines whether the parachutist's life preserver has been inflated. If the life preserver has not inflated, the recovery boat coxswain immediately proceeds to the impact point, taking care to stay out of the way of the other parachutists. Upon reaching the assigned parachutist, the boat coxswain treats the situation as a parachutist in distress and takes appropriate action.

(2) The DZSO command boat (boat number 1) is located so the DZSO can observe the landing of parachutists and reinforces recovery boats with additional safety swimmers and a resuscitator as required.

(3) The following procedure is used for recovery operations.

(a) When possible, recovery boats approach parachutists from alongside. This will help avoid getting the parachute stuck in the engine crop. Recover parachutes by the apex.

(b) If the parachutist experiences no difficulties after impact, he signals "All OK" by raising one arm straight up (without waving).

NOTE: **Any other signal** or **no signal** given by the parachutist is considered as a distress situation and immediate action is taken.

(c) The boat coxswain directs the boat alongside the parachutist; the safety swimmer extends the boat hook so that the parachutist can grasp it. If the parachutist is unable to grasp the hook, the safety swimmer secures a portion of the parachutist's equipment with the hook.

b. **Safety Swimmers**. On instruction from the boat commander/coxswain, the safety swimmer enters the water alongside the parachutist to assist the parachutist and to recover the parachute canopy and other equipment. If a parachutist has gone underwater, the following action is taken.

(1) Red smoke is displayed to indicate an emergency and a message is transmitted to the command boat by radio. All other activities cease. (A red smoke grenade may be activated and dropped in an ammunition can bolted to a wood plate, or a smoke machine may be used.)

(2) The safety swimmer dives to recover the parachutist and cuts him free of equipment as required.

(3) The DZSO moves his boat to the scene and the medic/corpsman prepares to use the resuscitator and backboard.

c. **Recovery Helicopter (if used).** The jump aircraft is alerted to assist in emergency recovery operations.

(1) When directed by the DZSO, the pilot stands-by the location to provide medical evacuation by air, if necessary.

(2) After normal recovery operations, parachutists and equipment are unloaded at the assembly area. All boats are repositioned for the resumption of jump operations.

25-5. WATER DROP ZONE PREJUMP TRAINING

Prejump training for water jumps should be conducted within 24 hours before the scheduled jump. This training includes demonstrations and practical exercises for all parachutists; use of the suspended harness is recommended. Prejump training includes, but is not limited to, the following:

a. Six points of performance for water jumps.
   - Check body position and count.
   - Check canopy and gain canopy control.
   - Prepare to land with a life preserver (B-7 or UDT, see below).
   - Keep a sharp lookout during descent and maneuver to indicated impact area.
   - Prepare to land by turning and facing into the wind, maintaining position until just before landing.
   - Prepare to make a PLF (if water is shallow or ground contact is made).

b. Emergencies in the air.

c. Emergencies in the water.
   - Ensure the life preserver is fully inflated.
   - If trapped under the canopy, the jumper finds a radial seam and follows it to the edge.
   - If trapped under the canopy, the jumper lifts his hand slowly and creates an air pocket to breath, then cuts a hole in the canopy material. (If the jumper lifts his hand too quickly, he will create a suction and the canopy will stick to his face and arm.)
   - Wet silk training will have been completed and documented within six months prior to the jump.

d. Recovery procedures.
   - Recover parachutes by pulling them in the boat by the apex.
   - Ensure the boat is kept free of gasoline and oil, which would ruin the parachutes.

e. Orientation at the WDZ for the WDZ crew.

f. Any special instructions.

25-6. PROCEDURES FOR DELIBERATE WATER LANDINGS WITH A LIFE PRESERVER

The procedures for deliberate water landings with a life preserver are as follows:

- Parachutists may wear combat equipment only after it has been waterproofed and float checked. To waterproof equipment, use bags and rigger tape inside equipment containers. Float check by submerging in a tank of water until saturated and ensure the equipment is neutral or positive buoyant (floats).

- Parachutists will wear service-approved life preservers.
- Parachutists will wear the ballistic helmet or a service-approved helmet during WDZ operations.
- Parachutists jumping in cold water (60 degrees or below) will wear wetsuits or dry suits and may wear a wetsuit/dry suit hood under the ballistic helmet.
- Parachutists may wear wetsuit booties, coral booties, or tennis shoes.
- Parachutists may jump with mask and fins. Fins to be worn on the feet will be taped or tucked under the left arm and should be dummy-corded to the jumper with 1/4-inch cotton webbing.
- All jumpers will wear a knife for emergencies.
- Jumpers may follow equipment, such as door bundles, CRRC, and RAMZ, on a single-ship operation. There will be three seconds between the cargo exit and the first jumper's exit to ensure the jumpers do not become entangled with the cargo.
- The cargo and jumpers are hooked to the same anchor line cable on the same side when jumping the ramp.

a. **Deliberate Water Landing with the B-7 Life Preserver.** The B-7 is worn under the parachute harness with the inflatable portions under the jumper's armpits (Figure 25-1, page 25-10). At altitude, the jumper releases all equipment tie-downs and lowers equipment, but does not jettison it.

(1) During the descent, the parachutist inflates the life preserver by discharging the attached CO2 cartridges. If necessary, the life preserver can be inflated by blowing air into the inflation valve hose.

(2) After entering the water, the parachutist activates both canopy release assemblies. He pulls the cable release safety clip out and away from his body (exposing the cable loops) and activates the canopy release assembly using one of the two methods (hand to shoulder or two-hand assist) as his feet touch the water.

(3) Since the B-7 life preserver will provide 500 pounds of positive buoyancy, the parachutist does not remove the harness and equipment.

(4) The parachutist swims upstream or upwind away from the parachute to avoid becoming entangled, and signals "All okay" to the recovery boat.

WARNING

Do not release the canopy release assemblies until the feet make contact with the water. Altitude is hard to judge over the water. If the jumper activates his canopy releases at even moderately high altitude (for example, 50 feet), serious injury or death may result.

CAUTION

The jumper must stay clear of the suspension lines to avoid entanglement. If the parachute and equipment are to heavy and hard to hold on to, the jumper lets go of the equipment to avoid being pulled under.

Figure 25-1. Landing with the B-7 life preserver.

b. **Deliberate Water Landing with UDT Life Preserver.** Parachutists wearing a UDT vest route the chest strap **under** the UDT vest so that the chest is not crushed if the UDT is inflated.

(1) Upon exiting the aircraft, the jumper checks the canopy and steers away from other jumpers. At altitude, the jumper releases all equipment tie-downs and lowers equipment, but does not jettison it.

(2) The jumper activates the quick-release on the waistband and unsnaps the left connector snap on the reserve parachute. He rotates the reserve to the right side of the parachute harness and seats himself well into the saddle.

(3) The jumper activates the ejector snap on the chest snap and inflates his life preserver with air by blowing into the inflation tube or activating the CO2 cartridge.

(4) Sitting back in the harness and using the right hand to hold onto the left main lift web, the jumper steers with his left hand. Just before entering the water, he places his hands on the ejector snaps on the leg straps.

(5) The jumper activates the ejector snaps on the leg straps, throws his arms up, and arches out of the harness when entering the water. He prepares to execute a PLF if the water is shallow.

(6) The jumper swims upstream or upwind away from the parachute to avoid becoming entangled, and signals "All okay" to the recovery boat. He recovers the parachute by grabbing the apex.

CAUTION

The jumper must stay clear of the suspension lines to avoid entanglement. If the parachute and equipment is too heavy to hold on to, the jumper legs go of the equipment to avoid being pulled under.

Figure 25-2. Landing with a UDT life preserver.

c. **Landing Procedures.** Upon landing, personnel must be prepared to perform the following procedures.

(1) If being dragged in the water, parachutists activate one (or both) canopy release assemblies after entering the water.

(2) Parachutists prevent fouling in the canopy or suspension lines during severe wind conditions.

(3) Parachutists signal "All okay" by extending one stationary arm overhead (DO NOT WAVE) or "Help" (any other signal or no signal) to the recovery boat, and stand by for pickup, remaining calm.

d. **OIC Responsibilities.** The parachute operation OIC ensures that—

- All personnel scheduled for a water jump are not less than Class 3 basic water survival swimmer qualified IAW TC 21-21 (Army), USMC CWS-1 (USMC), or Navy 1st class swimmers (USN).
- All jumpers and safety swimmers have completed wet silk training within the past six months.
- All equipment has been waterproofed and float checked.
- All jumpers are using approved serviceable floatation.
- All jumpers have completed the required prejump training.

CHAPTER 26

EXIT PROCEDURES

Two types of exit procedures can be used on fixed-wing aircraft: ADEPT exit and mass exit.

DANGER

ONLY ONE TYPE OF PARACHUTE (T-10- OR MC1-SERIES) MAY BE USED DURING ANY ONE GIVEN PASS OVER A DROP ZONE. THE MIXING OF PARACHUTES THAT HAVE DIFFERENT OPENING PERFORMANCE CHARACTERISTICS, SUCH AS NONSTEERABLE (T-10-SERIES) VERSUS STEERABLE (MC1-SERIES), CAN CAUSE HIGH-ALTITUDE INCIDENTS DURING ADEPT OPTION 2 OR MASS EXIT OPERATIONS.

26-1. ALTERNATE DOOR EXIT PROCEDURES FOR TRAINING (ADEPT) OPTIONS 1 AND 2

The ADEPT options are used when jumping the T-10- and MC1-series parachute. This training safety measure allows the maximum number of parachutists to exit the aircraft with a minimum risk of high-altitude entanglements.

a. **ADEPT Option 1**.

(1) During a single pass over the drop zone, only one stick of parachutists on one side of the aircraft jumps. After issuing the eighth jump command, the JM turns toward the paratroop door and regains control of his static line. He is positioned close enough to the door to control the flow of jumpers but far enough back not to impede the jumpers' movement to the door. The JM controls the number 1 jumper by grasping the jumper's saddle. When the jump caution lights turn green, the JM (on the active door for the pass) issues the ninth jump command, GO, to the first parachutist and taps him on the thigh. The JM controls the flow of parachutists (performing a visual inspection of each parachutist, his static line, and his equipment as he approaches the door) and observes the jump caution lights; the safety takes the static lines.

(2) Once the last parachutist exits the active door, the safety visually clears to the rear, gives the loadmaster a thumbs-up signal, and, with the assistance of the loadmaster and or static line retriever, pulls in the static lines and deployment bags.

(3) During the pass, the JM in the inactive door performs outside air safety checks and then observes the actions in the active door, since (under this option) he will not have any parachutists exiting his door.

(4) Subsequent passes alternate from door to door until all parachutists have exited. The AJM is the last parachutist on his side of the aircraft. The JM is the last parachutist onboard the aircraft; he exits from his door.

b. **ADEPT Option 2**.

(1) During a single pass over the drop zone, one stick of parachutists exit from the JM's door, followed by a stick of parachutists from the AJM's door. After issuing the

eighth jump command, the JM turns toward the paratroop door and regains control of his static line. He is positioned close enough to the door to control the flow of jumpers but far enough back not to impede the jumpers' movement to the door. The JM controls the number 1 jumper by grasping the jumper's saddle. When the jump caution lights turn green, the JM (on the primary door for the pass) issues the ninth jump command, GO, to the first parachutist and taps him on the thigh. The JM controls the flow of parachutists (performing a visual inspection of each parachutist, his static line, and his equipment as he approaches the door) and observes the jump caution lights; the safety takes the static lines.

(2) The AJM observes the actions in the JM's door. When he sees only three parachutists remaining in the JM's stick, he faces his parachutists and issues the eighth jump command. When the AJM sees the last parachutist clear the jump platform in the JM's door, he turns and rechecks his jump caution lights. (The JM is now observing the actions in the AJM's door.)

(3) If the jump caution lights are still green, the AJM verbally issues the ninth jump command, GO, to his first parachutist and taps him on the thigh. He controls the flow of parachutists (performing a visual inspection of each parachutist and his equipment as he approaches the door) and observes the jump caution lights; the safety takes the static lines.

(4) When the last parachutist exits from his side of the aircraft (last pass), the AJM gives his static line to the safety, checks the jump caution lights, and, if they are green, exits. The JM, seeing the AJM exit, passes his static line to the safety, checks the jump caution lights, and, if they are green, exits.

(5) Safety personnel visually clear to the rear of the aircraft and give the loadmaster a thumbs-up signal. They help the loadmaster recover static lines and deployment bags.

26-2. MASS EXITS

This exit procedure is used only when jumping the T-10-series parachute. During this type of exit, parachutists may exit from both doors at the same time. The JM gives the command GO, and the AJM then turns and gives his jumper the command GO to create a staggered effect.

CHAPTER 27

BUNDLE DELIVERY SYSTEM (WEDGE)

The bundle delivery system (referred to as a wedge because of its shape) is constructed with two Type II, 8-foot, modular airdrop platforms; a lightweight aluminum frame; four sections of skate wheel roller conveyors; and bundle release assemblies. The wedge weighs 930 pounds. It is designed to lock into the ramp restraint rails and to provide adequate tie-downs to secure various A-7A and A-21 configurations for flight. When positioned on the raised cargo ramp in flight, the wedge provides an inclined platform so that bundles can be released by rolling them out and off the ramp immediately before or after parachutists exit the doors.

27-1. APPLICATION

The wedge provides minimum dispersion between parachutists and equipment bundles on a single pass over the drop zone. The bundles can be released individually or in any group combination from the wedge when the loadmaster pulls the release pins from one or more release plates, depending on the number of bundles to be released at one time. Each release plate system employs a 1/8-inch-diameter steel release pin, a restraint release plate, and a 1-inch-wide tubular nylon restraint strap that secures each bundle, or each combination of bundles, to the wedge. A 1-inch-wide tubular nylon strap lanyard is attached to each steel release pin. Releasing the restraint strap allows the bundle(s) to roll off the aircraft ramp under the force of gravity.

27-2. RESTRICTIONS

The following restrictions apply when using the wedge:

a. **General Restrictions**. The wedge accommodates six A-7A or A-21 aerial delivery bundles, or a combination thereof, aboard the C-130 for cargo ramp airdrop. It limits parachutists to 20 for each door. (The anchor line cable stops are positioned about 26 inches forward of the center anchor line cable supports.)

NOTE: Wedge bundles can be dropped only on the first pass across the drop zone.

b. **Bundle Sizes**. Bundle sizes are limited to 27 inches long by 42 inches wide by 48 inches high, to include parachutes. The total rigged weight for each bundle must not exceed 538 pounds. Maximum allowable weight is 3,228 pounds for each six bundles. For example, maximum dimensions for mortar bundles are 27 inches long by 60 inches wide by 27 inches high (height dimension does not include the cargo parachute, paperboard honeycomb, and skid board). Mortar bundles must weigh between 320 pounds and 538 pounds (weights include skid board and cargo parachute). These dimensions allow up to three mortar bundles to be configured on the wedge, or mortar bundles can be mixed with smaller (27 inches by 42 inches) bundles to facilitate cross-loading. Two configurations are possible when mixing bundles: one mortar and four normal-size bundles (27 inches by 42 inches) or two mortar and two normal-size bundles.

c. **Aircraft**. During tactical training (visual or instrument flight regulations), bundle drops are restricted to the first three aircraft.

27-3. RIGGING PROCEDURES

Aerial delivery units usually rig the wedge; however, the system may be rigged in the airborne unit area. The areas of responsibility overlap; therefore, the JM must be familiar with the rigging procedures. However, the wedge must have a final inspection by a qualified rigger before the wedge leaves the rigging site.

27-4. BUNDLE DROP SEQUENCE

Bundles may be dropped from the wedge in the following sequence:

- A single bundle before and after the parachutists exit.
- A single unit of two to three bundles before or after the parachutists exit.
- Two units of two or three bundles for each drop with one unit dropped before and one unit after the parachutists exit.

The bundle sequence is determined before loading the wedge. If single bundles are to be dropped, a restraint strap and release lanyard are required for each bundle. If multiple bundles are to be dropped in units of two or three, a restraint strap and release lanyard are required for the aft bundle in each unit. Three exceptions are as follows:

a. **A-7 and A-21 Bundles**. The skid board size is 27 inches by 42 inches by 3/4 inch. Drill the skid board and center a 1-inch hole on the 42-inch side, 1 inch from the edge. Place two layers of honeycomb the size of the bundle on the skid board.

b. **Bundle Restraint Strap**. A restraint strap is formed by cutting a 12-foot length of 1-inch-wide tubular nylon webbing (requirement for maximum-size bundle). Form a 6-inch loop, with an overhand knot, 3 feet from one end of the strap. Lay the tie across the honeycomb with the loop next to the center hole in the skid. The 3-foot length of strap should be at the front of the skid (the front is the side with the 1-inch hole nearest the edge). Place the bundle on top of the honeycomb and rig the bundle (FM 10-500-3/ FMFM 7-47/TO 13C7-1-11, which supersedes FM 10-501/TO 13C7-1-11). Place the running ends of the restraint strap on top of the rigged bundle in preparation for loading on the wedge.

c. **Cargo Parachutes**. Use G-14 cargo parachutes, rigged with nonbreakaway static lines without drogue for A-7A or A-21 bundles dropped from the wedge. Attach the parachutes to the bundles after they are loaded aboard the wedge and restraint straps are secured.

27-5. INSPECTION

Before bundles are loaded, the wedge is inspected for completeness and serviceability and to ensure that the roller conveyors are properly positioned and secured to the platform. The bundle release plates are correctly installed as follows:

a. **Left-Side Release Plates**. Install left-side release plates (three each) between roller conveyors 1 and 2 at rollers 1, 2, 11, and 12, 21, 22, counting from the front to the rear of the wedge.

b. **Right-Side Release Plates**. Install right-side release plates between roller conveyors 3 and 4 at roller positions 1, 2, 11, and 12, 21, 22.

c. **Release Lanyard**. Install a 1-inch-wide tubular nylon webbing release lanyard with a 5 3/4-inch by 1/8-inch steel pin and ensure it is serviceable, marked, and secured to the correct release plate with a 20-inch length of Type III nylon cord. (The 20-inch cord forms a safety to prevent rebound of the lanyard and possible injury to personnel releasing the bundles.)

NOTE: If a lanyard is unserviceable, make a new one with a length of 1-inch tubular nylon webbing and a 5 3/4-inch by 1/8-inch steel pin. Using a bowline knot, attach the pin to one end of the webbing. Insert the pin in the appropriate release plate hole and extend the webbing forward to a point 30 inches from the forward edge of the wedge. Cut the webbing and form a 6-inch loop in the running end to provide a secure handhold for releasing the bundles.

27-6. LOADING, RIGGING, AND RESTRAINING BUNDLES TO WEDGE

Bundles are loaded, rigged, and secured to the wedge at the rigging site, and then transported to the aircraft for loading. The following procedure is for loading and rigging bundles to be dropped individually.

a. Load bundle number 3, left unit, with the loop of the restraint strap forward.

b. Position the loaded skid board with the 1-inch hole centered over the hole in the forward release plate. Pass the 6-inch loop of the restraint strap down through the hole in the board and release plate. Insert the release lanyard pin through the restraint strap loop and the hole in the release plate.

c. Form a loop in the forward running end of the restraint strap. Pass the forward running end of the restraint strap through the loop, cinch the strap tight around the bundle, and secure with two half hitches and an overhand knot.

d. Secure the release lanyard pin with one complete turn (attach cotton ticket number 8/7). Pass the cord down through the hole in the release plate and through the pin loop. Tie the running end with a square knot and a locking knot.

NOTE: Repeat the above procedure for each bundle in the following sequence: bundles 2 and 1, left side; bundles 3, 2, and 1, right side.

e. Position, attach, and secure G-14 cargo parachutes to the bundles. Using tie-down straps, secure the bundles to the wedge to prevent movement and possible damage to the release pin during transport and to the wedge when loading aboard the aircraft.

NOTE: If bundles are to be dropped in multiples of two or three, only the aft bundle requires a restraint strap and release lanyard. Bundles are loaded on the wedge in reverse order—left to right, front to rear—with the loop of the restraint loop strap forward.

27-7. JUMPMASTER PROCEDURES

JM procedures include managing anchor line cables and performing inspections before and after loading.

a. **Anchor Line Cables**. The JM ensures that—

(1) Anchor line cables are disconnected from the center anchor cable supports and secured to preclude obstruction to personnel.

(2) Stops are installed and taped on the inboard anchor line cable at fuselage station 749 for the bundle static lines, and about 26 inches forward of the center anchor cable support for parachutist static lines.

NOTE: Static line retrieval spools are forward of these stops. The stops and spools are adjusted so that the distance from the forward edge of the anchor cable supports to the forward edge of the spool is 26 inches.

b. **Inspection Before Loading**. The JM is responsible for inspecting, with a qualified parachute rigger, the following items on the wedge before loading. (He also inspects them with the loadmaster after loading.)

- Bundles A-7A or A-21 are present and are properly secured to plywood skid board (FM 10-500-3/TO 13C7-1-11).
- Quick-release assemblies are properly seated and safety clip is inserted (A-21 bundle).
- Correct size and layers of honeycomb are present.
- Bundles are properly secured to plywood skid board.
- A one-inch tubular nylon restraint strap is routed vertically around bundles, with a six-inch lanyard overhand loop at the release point.
- G-14 cargo parachute(s) is rigged for nonbreakaway and without drogue.
- Parachute is properly attached and secured to bundle(s).

c. **Inspection After Loading**. The JM is responsible for inspecting the wedge for correct positioning and locking into the aircraft restraint rails when the following actions are accomplished:

- Wedge is positioned so that aft restraint rail lock in platform number 6 indent is in place (counting from the rear of the platform).
- Cargo parachute static lines are connected to the inboard anchor line cables forward of the stops at station 749.
- Bundle release lanyards are correctly routed.
- Release pins are seated in the release plates.
- Safety tie is in place.
- Bundles are secured for flight.

27-8. BRIEFING RELEASE PROCEDURES

Briefing release procedures include the following:

a. Identify the jumpmaster, loadmaster, safety personnel, and assistants; brief each individual on his responsibilities.

b. Coordinate type signals to be used to conduct the drop—for example, thumbs-up for DROP, thumbs-down for NO DROP.

c. Coordinate bundle release and parachutist exit sequence to be used—for example, on the green light, the number 1 loadmaster releases the bundle(s) on his side of the aircraft. All parachutists will exit after the last bundle clears the aircraft ramp. After the last parachutist exits the aircraft, safety personnel give a thumbs-up, signaling the number 2 loadmaster to release the bundle(s) on his side of the aircraft.

d. Review procedures for a bundle hung up in the aircraft—for example, the loadmaster will signal NO DROP and notify the pilot; the pilot turns on the NO DROP red light. The drop is canceled if the problem cannot be immediately identified and safely corrected.

27-9. LOADMASTER AND JUMPMASTER DUTIES DURING FLIGHT

Airdrop sequence of bundles and personnel include:

a. **Loadmaster**. At the 3-minute slowdown, the loadmaster raises the cargo door and installs the anchor line cables in the center anchor line cable supports.

b. **Jumpmaster**. The JM hooks up, issues the 6-minute warning, and begins the jump commands.

NOTE: All commands are the same (whether using the wedge or not) unless all bundles are to be released before the exit of the parachutists. Then, the command STAND BY is given. The number 1 parachutist in each door assumes a position at a 10-degree angle to the forward edge of the doors and observes the JM.

(1) ***Bundles Precede Jumpers.*** On the green light, the loadmasters release the bundles. When the JM observes that the last bundle has cleared the aircraft ramp, he gives the command GO. All parachutists then exit the aircraft.

(2) ***Jumpers Precede Bundles.*** If all bundles are to be released after the parachutists exit, the JM gives the command STAND BY for the first two parachutists. On the green light, the command GO is given. When the parachutists have exited, the static safety personnel give a thumbs-up signal for the loadmaster to release the bundles.

(3) ***Bundles Released Before and After the Jumpers.*** On the green light, the number 1 loadmaster releases the required bundles. When the JM observes that the last bundle has cleared the aircraft ramp, he gives the command GO. All parachutists then exit, and the static safety personnel signal thumbs-up to the number 2 loadmaster to release the remaining bundles. Static lines are retrieved, and the cargo and troop doors are secured by the loadmaster.

CHAPTER 28
COMBAT AIRBORNE OPERATIONS

This chapter describes allowable combat techniques for airborne operations, which may deviate from those for peacetime operations. Commanders should conduct a detailed risk assessment before modifying normal peacetime techniques to which their soldiers have grown accustomed in training. Failure to do so can result in unnecessary loss of combat power and additional casualties that must be cared for on the drop zone. Commanders should ensure any changes are thoroughly briefed, understood, and rehearsed during troop leading procedures and pre-jump training. Within the constraints of this chapter, MACOMs should develop their own combat procedures based upon their mission, organizational structure, and level of training.

28-1. MODIFICATIONS TO PERSONNEL AND EQUIPMENT PROCEDURES

Personnel and equipment can be dropped in any sequence that best supports the ground tactical plan. Commanders may modify peacetime aircraft, serial, and formation interval rules based upon METT-T.

a. All aircraft are loaded to the maximum.

b. The static line snap hook safety wire is not required to be inserted in the snap hook.

c. The right connector snap of the reserve parachute does not require a safety wire.

d. The waistband, if not used, is rolled and secured with a piece of masking tape to keep it out of the way.

e. Jumpers should only carry minimum mission-essential equipment to make room for ammunition.

f. Commanders should determine, based upon METT-T, when jumpers should hook up. If faced with a well-developed enemy air defense threat, jumpers should be stood up prior to entering the threat zone.

g. The green light should go on to achieve the desired PI. There is no requirement for safety buffers on the lead or trail edge of the DZ, provided there are no obstacles. The red light should not go on unless there is a significant hazard (such as a large body of water) or the aircraft is scheduled to continue to another DZ.

h. Weapons should be rigged in the M1950 weapons case. (A magazine is loaded, weapon placed on SAFE, and no round is chambered.) Jumping weapons exposed does not significantly decrease the time required to place the weapon into action. More importantly, with jumpers and jumpmasters executing procedures that they have not rehearsed, jumpers are more likely to become entangled or towed. The M1950 weapons case is equipped with a quick release zipper. The jumper should not attempt to unzip the zipper but should grasp the tab thong and pull upward sharply (as if to finish zipping completely). When the slide fastener disengages, the zipper teeth will uncouple the length of the M1950.

i. Leaders and JMs should consider the heavy and nonstandard loads paratroopers carry. An excessively heavy load or protruding object may prevent a jumper's strong exit.

More importantly, the load may become an obstacle and may result in the failure of a stick to exit.

j. Jumpmasters should either lead or follow their sticks out on combat jumps. Commanders should load-plan their jumpmasters to come from the lead or trail edge units, thereby landing closer to their objectives. Jumpmasters may jump in the stick closest to their exit points. If safeties are designated to jump, they should come from the trail edge unit. They should rig as late as possible in order to conduct their required safety checks unencumbered. Safeties must continue to control static lines until all jumpers have exited the aircraft.

CAUTION

Do not abandon your training techniques just because you are in combat. Do a detailed risk assessment to ensure changes are necessary. Ensure changes are well briefed, understood by all, and rehearsed during pre-jump training.

k. Parachutists may be briefed to "pop and go," leaving all their air items where they land. If the DZ is under enemy attack and it is necessary to find cover and start fighting, they may drop their weapons cases and reserves, activate both canopy release assemblies, and move out, leaving their harnesses on to be discarded later. METT-T may require a follow-on airland. Jumpers should be briefed to "bag and go" if they land near a runway or taxiway. This means the jumper places his parachute in the kit bag and leaves it upside down no closer than 10 meters from the active runway or taxiway. METT-T determines whether units designate a "clean side" or "dirty side" of the active runway or taxiway.

l. Drop altitude is determined by coordination between the airborne commander and the air mission commander. The JTF commander makes the final jump altitude decision. When conducting risk assessment, commanders are advised of the altitude chart located in Chapter 26 and are reminded of the 135-foot altimeter error. If the decision is made to jump at 600 feet AGL or lower, pre-jump training should include the proper modifications (such as lowering equipment after the second point of performance).

m. Leaders should jump their own radios. Radio frequencies should be accessible, short whip antennas screwed in, and handsets plugged in. Leader radios are ready to go immediately after landing.

28-2. MOVEMENT FROM ASSEMBLY AREAS

The quicker assault objectives are secured, the better the odds for a successful mission. Units should proceed to the assault objective IAW unit SOP.

a. Determine minimum force required to seize and secure the assault objective. When that force has assembled (to include any "lost" troopers from other units), the senior leader of the group may begin the assault on the objective IAW the commander's intent.

b. Leaders should never bypass a soldier on the DZ. Every soldier "wandering around" ought to be either pointed in the right direction or integrated with that leader's unit.

> **CAUTION**
> If you train as you fight, you will fight as you train. Do not let heavy loads exceed the capabilities of your jumpers just because you are in combat.

28-3. LANDING PLAN

The development of the landing plan must be thorough. Assembly areas must be selected. A heavy drop plan with multiple HD impact points along the length and across the width of the DZ may be necessary. Offset HDs as required.

a. Vehicle drivers and equipment operators should land in the same sector of the DZ where the impact point for their equipment is located. Commanders must spread the HD impact points along the entire DZ; they do not saturate one small sector of the DZ, thus making a lucrative enemy target. They must consider the time and distance factors where load number 1 and load number 2 from the same aircraft will land. For example, delivering a howitzer and its prime mover from the same aircraft ensures they both land on the same DZ but physically separated. The crew should be cross-loaded accordingly.

b. GUCs adjust loads and HD impact points so howitzers and prime movers land at or near the same point. If the crew cannot locate their assigned prime mover right away, they should take the nearest one they can find and move out.

c. Crew members should not waste time looking for a particular bumper number on the DZ. If the vehicle belongs to the unit, recover it.

d. GUCs carefully plan CDS delivery. Rather than dropping the CDS in the center of the DZ, they must drop it along the edge of the DZ or on the unit assembly area. If a CDS is needed later, GUCs deliver it on or near the assault objective. Regardless of the decision as to where and when the CDS is needed, they must plan carefully so that backhauling is reduced.

28-4. HEAVY DROP LOADS

Minimum-essential HD loads should precede the personnel drop. Other important HD equipment should follow the assault personnel drop as soon as possible. Units cannot fight without heavy weapons, vehicles, and ammunition bundles. Peacetime safety rules require a 30-minute to one-hour separation between personnel drops and HD unless the HD is delivered first. These rules **do not** apply in combat. Door bundles or wedge loads can be dropped from any aircraft. Bundles should be exited on the green light with troops following immediately.

28-5. INJURED PERSONNEL

Mission completion is paramount in combat. Soldiers will want to provide aid to their buddies injured during the jump, but **speed is critical**, and **every fighter is needed to**

seize and secure the assault objective. Therefore, injured personnel are cared for as follows:

a. Medics must quickly join and move to the assault objectives with their assigned units. Medics should not stop on the DZ to care for injured parachutists (non-life-threatening casualties); injured jumpers will be cared for later.

b. Jumpers must be briefed and understand that they cannot stop on the DZ to help their buddies who were hurt during the jump and who received non-life-threatening injuries.

NOTE: The most effective way to aid an injured paratrooper on the DZ is to eliminate the enemy threat.

28-6. SUPPLIES

Units should obtain more critical items than are actually needed for mission accomplishment to compensate for those items that are damaged or destroyed during the airdrop.

a. Resupply and ammunition replenishment are the parachutist's most serious problem. Commanders should devise alternate methods of insertion for items such as LAWs, antitank mines, RAAWs, mortar rounds, or radio batteries. They should **not** rely on the individual paratrooper.

b. Keep door bundles rigid and light if you are exiting jumpers after them. Litters provide an excellent platform to build door bundles on. They provide rigidity and a means to move the load quickly while on the ground.

CAUTION

If you train as you fight, you will fight as you train. Do not let heavy loads exceed the capabilities of your jumpers just because you are in combat.

APPENDIX A

AIRBORNE REFRESHER TRAINING

Airborne refresher training is required for personnel who have not jumped within a 6-month period. The length of the refresher training depends on the proficiency of the parachutist. The minimum requirements are in Table A-1. Airborne refresher training will be instructed and documented by a qualified and current JM. Individual service components may modify these requirements depending on training aids and equipment availability.

PERIOD	HOURS	LESSON	TRAINING AIDS/EQUIPMENT
1	1	Sustained airborne training (pre-jump) to include PLFS (front, side, and rear) and methods of recovery.	Two-foot PLF platforms, sawdust pit, and mock door.
2	1	Fitting and wearing the T-10-series or MC1-series parachute and B-5/B-7 life preservers; rigging individual equipment in the HSPR; and packing individual weapons/equipment containers.	TOE and TDA equipment, ALICE pack and HSPR, T-10-series and MC1-series parachute assemblies, B-5/B-7 life preservers, and individual weapons/equipment containers.
3	1	Actions in the aircraft rehearsal, proper exit procedures, sequence of jump commands, all jump commands, first two points of performance.	Mock door structures and 5-foot static lines with snap hooks.
4	1	Control of canopy, turns, slips, entanglements, emergency landings, landing attitude, and activation of the reserve.	Suspended harness apparatus (if available) and swing landing trainer apparatus (if available).
5	1	Door exit procedure rehearsal, sequence of first three points of performance, releasing equipment containers, and activation of reserve.	34-foot tower (if available) and troop parachute harness.
6	1	Parachute jump.	Aircraft, parachutes, DZ, unit equipment.

Table A-1. Minimum requirements, airborne refresher training.

APPENDIX B

JUMPMASTER TRAINING COURSE

The Jumpmaster Training Course is the standard US Army course of instruction designed to qualify individuals as jumpmasters for conventional static line parachuting. It is the standard for all airborne unit activities. The course is approximately 2 weeks long and contains 97 hours of instruction.

SUBJECT	SCOPE	HOURS
In-processing	Administrative in-process of class.	1.0
Orientation and Administration	Orientation on the course requirements that the student must meet to include student handouts, questionnaires, the entire training schedule, and the grading system used to evaluate students.	1.0
Duties and Responsibilities of the Jumpmaster and Safety	Discussion of the duties and responsibilities of the jumpmaster and safety personnel from the time of notification until completion of the airborne operation.	2.0
Individual Equipment Containers	Discussion of the characteristics and nomenclature of individual equipment containers to include a demonstration on the correct rigging, attachment, and lowering procedures for the ALICE pack, M1950 weapons case, and AT4JP.	2.0
Army Aircraft Orientation	Familiarization with the preparation, inspection, and jump procedures for the UH-1H, UH-60A, and CH-47 helicopters.	1.0
Jumpmaster Personnel Inspection	Demonstration of the correct method of inspecting a parachutist and attached combat equipment. Remaining hours are spent on practical exercise using two-man buddy teams, where the student is required to conduct a personnel inspection and find and report major and minor rigging deficiencies that have been placed in the parachute assembly and attached equipment. Students are changed over often to ensure all receive the same amount of inspection time.	37.0

SUBJECT	SCOPE	HOURS
USAF Aircraft	General aircraft descriptions, jumpmaster procedures, and aircraft preparations. Critical elements of airborne operations are the aircraft and drop altitude. These aircraft are service tested and approved for troop drops. Minimum jump altitude and considerations that apply to basic airborne jumps, tactical training jumps, and combat jumps are discussed.	1.0
Prejump Training	Discussion of the five points of performance and methods of activating the reserve, towed parachutist procedures, collisions, entanglements, and the three types of emergency landings.	1.0
A-Series Containers	Discussion of the characteristics, capabilities, and methods of packing, rigging, and inspecting the A-series containers, and how to attach and inspect the cargo parachute.	2.0
Duties and Responsibilities of the DZSO/DZSTL	Discussion of the prerequisites to perform duties as the drop zone support team leader (DZSTL), drop zone safety officer (DZSO), assistant drop zone safety officer (ADZSO). Discussion of tactical DZ assessment and composition of the team for different airdrop scenarios. Scoring procedures using MAC Form 168 are included.	1.0
Jump Commands, Door Procedures, and Door Bundle Ejection Procedures	Demonstration and practical exercise of the proper sequence of jump commands and time warnings with proper hand-and-arm signals; the door procedures used by a jumpmaster; and door bundle ejection procedures using aircraft mock-ups. Remaining time is spent for practical exercise in the mock-ups (or actual aircraft).	4.0

SUBJECT	SCOPE	HOURS
CARP Drop Zones	Additional discussion includes drop zone surveys (USAF Form 3823, DZ Survey, formerly MAC Form 339). Discussion of the methods of marking drop zones for computed air release point (CARP). Discussion of drop zone selection factors. Practical exercise on determining the mean effective wind and use of authorized anemometers. Familiarization with drop zone marking requirements for day and night airborne operations. Discussion of ADEPT options (1, 2, and mass exits).	2.0
Nomenclature Examination	Written nomenclature examination.	0.5
Review and Critique	Brief review of the previous week's instruction to ensure all questions and tasks are clear.	1.0
Nomenclature Examination Retest	Personnel who failed the initial test are retested.	0.5
Written Examination	Written comprehensive examination which covers all instruction that has been presented.	1.0
Jumpmaster Personnel Inspection Pre-Examination	Each student inspects three parachutists—two with no combat equipment and one combat-equipped parachutist (5-minute time limit).	4.0
Jumpmaster Personnel Inspection Examination	Each student inspects three parachutists—two with no combat equipment and one combat-equipped parachutist (5-minute time limit).	3.0
Jumpmaster Personnel Inspection Retest	Personnel who failed the initial test are retested.	2.0
Prejump Training Examination	Oral presentation by each student to determine his ability to effectively conduct prejump training.	6.0

SUBJECT	SCOPE	HOURS
Jumpmaster Briefing	A briefing on all airborne operations and grading procedures.	.5
Aircraft Inspection Class	Practical exercise on aircraft inspection procedures using the C-130 or C-141.	1.0
Written Examination Retest	Personnel who failed the initial test are retested.	1.0
Day Practical Work in Aircraft Examination	Graded practical exercise jumpmastering personnel with attached equipment. Half the students are graded while the other half act as their jumpers. A second lift is required to test the second half of the class.	10.0
Maintenance of Air Items	Inspection, maintenance, and turn-in of all air items used during the airborne operations.	1.0
Out-processing	Administrative out-processing the class.	2.0
Graduation		1.0

APPENDIX C
JUMPMASTER REFRESHER COURSE

The Jumpmaster Refresher Course is designed to update qualified jumpmasters who are not in a current status IAW Chapter 7. It ensures that standardization of course content is maintained. The unit designated to conduct this course ensures that equipment normally used is available for this training. The course will be instructed and documented by a current and qualified JM. All JMs should attend a refresher course on a yearly basis, no matter the level of experience or currency. This will ensure all personnel are familiar with current changes, procedures, and standardization. Any unit having special requirements, such as nonstandard aircraft or special items of equipment, may add periods of instruction to the course as needed. Individual service components may modify these requirements depending on training aids and equipment availability.

SUBJECT	SCOPE	HOURS
Individual Equipment	Characteristics and nomenclature of individual equipment containers. Demonstration of the correct rigging, attaching, and lowering procedures for approved and tested equipment in use by the training unit.	1.0
Jumpmaster Personnel Inspection	Demonstration of the correct method for inspecting a parachutist with attached combat equipment. Practical exercise uses two-man buddy teams, where the individual is required to conduct a personnel inspection and find/report major and minor rigging deficiencies that have been placed in the parachute assembly and attached equipment. Personnel changed often to ensure all receive the same amount of inspection time.	2.0
Pre-Jump Training	Discussion of the five points of performance and methods of activating the reserve, towed parachutist procedures, collisions, entanglements, and the three types of emergency landings.	0.5

SUBJECT	SCOPE	HOURS
Drop Zone Support Teams, DZ Procedures, and DZ Formulas	Discussion and familiarization of the prerequisites to perform duties as the drop zone support team leader (DZSTL), drop zone safety officer (DZSO), assistant drop zone safety officer (ADZSO); and composition of the team for different airdrop scenarios. Scoring procedures using MAC Form 168 are included.	1.0
Jump Commands, Rigging Procedures, Spotting Procedures, Door Procedures, and Door Bundle Ejection Procedures, CRCC	Discussion and practical exercise (PE) of the proper sequence of jump commands and time warnings with appropriate hand-and-arm signals, door procedures used by the jumpmaster, and door bundle ejection procedures. A demonstration is given using aircraft mock-ups. The remaining time is spent for practical exercise in the mock-ups (or actual aircraft).	1.0
CARP Drop Zones	Additional discussion of methods of marking drop zones for computed air release point (CARP). Students are familiarized with drop zone marking requirements for day and night airborne operations.	1.0

APPENDIX D

PLANNING CONSIDERATIONS FOR WATER, WIRE, AND TREE EMERGENCY LANDINGS

An airborne airdrop is an inherently high-risk operation. When water, wire, or tree obstacles are on or close to the intended drop zone, the jumpers face an even higher risk of injury. This appendix assists the commander in conducting a DZ risk assessment analysis. This appendix also provides the commander, DZSO/DZSTL, and JM with operational and logistical planning measures to lower the risks jumpers may encounter during airborne operations. Individual service components may modify these requirements depending on training aids and equipment availability. Individual service components will follow their own service risk assessment procedures and regulations.

D-1. WATER OBSTACLES

A water obstacle is any body of water (for example, a lake, pond, river, stream, or canal) that has a depth of 4 feet or more, is 40 feet wide or wider, and is located within 1,000 meters of any edge of the surveyed or tactically assessed DZ.

a. **Risk Assessment Analysis for a Drop Zone with Water Obstacles**. When making a training parachute jump DZ risk assessment, the commander should consider the proximity of the water obstacle to the DZ, the depth of the water obstacle, and the width of the water obstacle. Additionally, the following factors may enter into the water obstacle risk assessment: the condition of the water obstacle bottom, the current of a free-flowing water obstacle, water temperature, the number of obstacles, the equipment available to reduce the risk level, jumper experience levels, jump time (day or night and percent of illumination), and whether or not the selected DZ is critical to mission success. The following risk categories are assigned to DZs with water obstacles:

(1) ***High Risk.*** A high-risk condition exists if a water obstacle is within 1,000 meters of any edge of the DZ, water depth is 4 feet or more, and water is 40 feet wide or wider. If a high-risk condition exists, it will be necessary to use a boat detail and have approved life preservers on the jumpers. If the water is 4 feet deep or more, but not over 40 feet wide, a boat detail is not required. However, approved life preservers are still required for the jumpers.

(2) ***Medium Risk.*** A medium-risk condition exists if a water obstacle is more than 1,000 meters but less than 1,500 meters from any edge of the DZ, water depth is 4 feet or more, and water is 40 feet wide or wider.

(3) ***Low Risk.*** A low-risk condition exists if a water obstacle is more than 1,500 meters from any edge of the DZ, water depth is 4 feet or more, and water is 40 feet wide or wider.

b. **Planning Considerations**. The commander, DZSO/DZSTL, and JM perform the following actions to reduce the risks associated with water obstacles:

(1) ***Commander.***

(a) Ensure a risk assessment analysis has been conducted to determine the unintentional water landing risk level for jumpers. If a high risk exists, select (if possible) an alternate DZ that allows mission conduct at a lower risk level.

(b) Ensure a follow-on assessment has been made to determine whether the jumpers' risk level has changed.

(c) Ensure that key leaders, jumpmasters, and jumpers have been informed of the water obstacle risks and the risk level (high, medium, or low).

(d) Ensure USAF Form 3823, DZ Survey (formerly MAC Form 339), is current and available. Ensure the DZSO/DZSTL and JM have read it and completely understand the unintentional water landing risk level and the safety measures that are to be used.

(e) If a boat detail is used, ensure the DZSO/DZSTL and the unit providing the detail have properly conducted initial or refresher training. Ensure the OIC/NCOIC of the boat detail knows where and when the detail begins its duties and how to contact the DZSO/DZSTL.

(f) Ensure that approved life preservers are coordinated for if used.

(g) Ensure the DZSO/DZSTL, JM/AJM, safeties, and jumpers are informed of all water obstacle risks and that the DZSO/DZSTL and JM complete their duties.

(2) ***DZSO/DZSTL.***

(a) Determine if a follow-on assessment of the DZ has been conducted to confirm the current status.

(b) If the risk assessment indicates high risk and a boat detail is necessary, ensure the OIC/NCOIC is fully briefed on the plan. Ensure all boat detail personnel have been trained and have all necessary equipment available to conduct the mission.

(c) Read all applicable regulations, FMs, and SOPs. Ensure copies are present throughout mission.

(3) ***Jumpmaster.***

(a) If approved life preservers are to be used, ensure they have been inspected within the last 180 days and are serviceable, and that all jumpers have been trained on life preserver wear, fit, and use (to include manual inflation).

(b) Ensure all personnel have received prejump training within 24 hours prior to drop time, with special emphasis on unintentional water landings.

c. **Water Obstacle Coverage**. Each water obstacle may require a different type of coverage. The following is an example composition of a boat detail. Equipment should be altered to best accomplish the mission.

(1) OIC/NCOIC (qualified as a boat operator) and assistant boat operator.

(2) Qualified boat operators - 1 primary and 1 assistant for each boat.

(3) Recovery personnel - 2 for each boat (one may be lifeguard qualified and combat lifesaver certified). All boat detail personnel should be strong swimmers.

(4) Each recovery boat team may need the following equipment:

- Boat (Zodiac RB-10 or solid-bodied boat of comparable size) with operable outboard motor.
- Enough fuel/oil to complete the mission.
- Life vest/floatation device for each boat detail member and additional floatation devices for jumpers.
- Life ring with attached rope—1.
- Radio with spare battery—1.
- Shepherd's crook—1.
- Grappling hook—1.
- Long backboard to facilitate CPR—1.

- Aid bag with resuscitation equipment—1.
- Rope, 120 feet long—1.

(5) For night recovery operations, each recovery boat must have operational night vision devices with spare batteries - 2 for each boat.

d. **Optional Training**. The following optional training is suggested:

(1) Suspended harness training on second through fifth points of performance may be given. Step-by-step training on the procedures jumpers will take for an unintentional water landing may be conducted.

(2) An optional dunk tank training device may be constructed to allow the lowering of jumpers (wearing parachute harness and B-5 or B-7 life preserver) into the water. This training familiarizes the jumpers with the proper emergency water landing procedures.

D-2. WIRE OBSTACLES

A wire obstacle is a wire or set of wires (regardless of height or type) located within 1,000 meters of any edge of the surveyed or tactically assessed DZ. The types of wire obstacles that could pose a risk to jumpers are power, telephone, or cable television wires. Wire fence can be regarded as an obstacle if it will pose a hazard to jumpers. Power line capacity (voltage or amperage) is **not** a factor when determining the risk to jumpers who may come in contact with a wire obstacle on or near the DZ.

CAUTION

Regardless of voltage or current-carrying capacity, if a power line is located within 1,000 meters of any edge of the surveyed or tactically assessed DZ, the power should be cut off before using the DZ, if possible.

a. **Risk Assessment Analysis for a Drop Zone with Wire Obstacles**. When making a training parachute jump DZ risk assessment, the commander should consider the proximity of the wire obstacle to the DZ, the height of the wire obstacle, the number of obstacles, the equipment available to reduce the risk level, jumper experience levels, drop time (day or night and percent of illumination), and whether or not the selected DZ is critical to mission success. The following risk categories are assigned to DZs with wire obstacles:

(1) ***High Risk.*** A high-risk condition exists if a wire obstacle is within 1,000 meters of any edge of the DZ. If a high-risk condition exists, it may be necessary to have a recovery detail at the DZ.

(2) ***Medium Risk.*** A medium-risk condition exists if a wire obstacle is more than 1,000 meters but less than 1,500 meters from any edge of the DZ.

(3) ***Low Risk.*** A low-risk condition exists if a wire obstacle is more than 1,500 meters from any edge of the DZ.

b. **Planning Considerations**. The commander, DZSO/DZSTL, and JM perform the following actions to reduce the risks associated with wire obstacles:

(1) ***Commander.***

(a) Ensure a risk assessment analysis has been conducted to determine the unintentional wire landing risk level for jumpers. If a high risk exists, select (if possible) an alternate DZ which allows mission conduct at a lower risk level.

(b) Ensure a follow-on assessment has been made to determine whether the jumpers' risk level has changed.

(c) Ensure that key leaders, jumpmasters, and jumpers have been informed of the wire obstacle risks and the risk level (high, medium, or low).

(d) Ensure USAF Form 3823, DZ Survey (formerly MAC Form 339), is current and available. Ensure the DZSO/DZSTL and JM have read it and completely understand the wire landing risk level and the safety measures to be used.

(e) If a recovery detail is used, ensure the DZSO/DZSTL and the unit providing the detail have properly conducted initial or refresher training. Ensure the OIC/NCOIC of the recovery detail knows where and when the detail begins its duties and how to contact the DZSO/DZSTL.

(f) Ensure the DZSO/DZSTL, JM/AJM, and jumpers have been informed of all wire obstacle risks. Ensure the DZSO/DZSTL and JM complete their duties.

(2) ***DZSO/DZSTL.***

(a) Determine if a follow-on assessment of the DZ has been conducted to confirm the current status.

(b) If the risk assessment indicates high risk and a recovery detail is utilized, ensure that the OIC/NCOIC is fully briefed on the plan. Ensure all recovery personnel have been trained and have all necessary equipment available to them to conduct the mission.

(c) Ensure that coordination with the power company has been made to cut off the power not later than 1 hour prior to drop time, if possible.

(d) Read all applicable regulations, FMs, and SOPs. Ensure copies are present throughout mission.

(3) ***Jumpmaster.***

(a) Ensure all personnel have been briefed on the wire obstacles.

(b) Ensure all personnel have received prejump training within 24 hours prior to drop time with special emphasis on unintentional wire landings.

c. **Wire Obstacle Coverage**. Each wire obstacle may require different types of coverage. The following is an example composition of a recovery detail. Equipment should be altered to best accomplish the mission.

(1) OIC/NCOIC and assistant.

(2) Enough personnel to recover jumpers who may become entangled in the wire obstacles.

(3) A recovery team may need the following equipment:

- Radios with spare batteries - 2 (1 for OIC/NCOIC and 1 for recovery team).
- Grappling hook—1.
- Tree-climbing kit—1.
- Long backboard to facilitate CPR—1.
- Aid bag with resuscitation equipment—1.
- Ropes, 120 feet long—2.
- Wood poles 15 feet long—2.
- **Wood** extension ladder, 20 feet long—1.

- Snap links—4.

(4) For night recovery operations, the following equipment should be added:

- Night vision devices (with spare batteries) for each team.
- Operational flashlights (with spare batteries) for each team.

d. **Optional Training**. Optional training includes suspended harness training on the second through fifth points of performance. Step-by-step training on the procedures jumpers take for unintentional wire landings may be conducted.

D-3. TREE OBSTACLES

A tree obstacle is any tree or group of trees that are on, around, or within 1,000 meters of any edge of the drop zone.

a. **Risk Assessment Analysis for a Drop Zone with Tree Obstacles**. When making a training parachute jump DZ risk assessment, the commander should consider the proximity of the tree obstacles to the DZ, the number of obstacles, the equipment available to reduce the risk level, jumper experience levels, drop time (day or night and percent of illumination), and whether or not the selected DZ is critical to mission success. The following risk categories are assigned to DZs with tree obstacles:

(1) ***High Risk.*** A high-risk condition exists if a tree or group of trees are within 1,000 meters of the DZ or are on any edge of the DZ and have a height of 35 feet or more. If a high-risk condition exists, it may be necessary to have a recovery detail present at the DZ.

(2) ***Medium Risk.*** A medium-risk condition exists if a tree or group of trees are on or within 1,000 meters of any edge of the DZ and have a height of 20 to 35 feet.

(3) ***Low Risk.*** A low-risk condition exists if a tree obstacle having a height of less than 20 feet is on or within 1,000 meters of any edge of the drop zone.

b. **Planning Considerations**. The commander, DZSO/DZSTL, and JM perform the following actions to reduce the risks associated with tree obstacles.

(1) ***Commander.***

(a) Ensure a risk assessment analysis has been established to determine the tree landing risk level for jumpers. If a high risk exists, select (if possible) an alternate DZ that allows mission conduct at a lower risk level.

(b) Ensure a follow-on assessment has been made to determine whether the jumpers' risk level has changed.

(c) Ensure that key leaders, jumpmasters, and jumpers have been informed of the tree obstacle risks and the risk level (high, medium, or low).

(d) Ensure USAF Form 3823, DZ Survey (formerly MAC Form 339), is current and available. Ensure the DZSO/DZSTL and JM have read it and completely understand the tree landing risk level and the recovery measures to be used.

(e) If a recovery detail is used, ensure DZSO/DZSTL and the unit providing the detail have properly conducted initial or refresher training. Ensure the OIC/NCOIC of the recovery detail knows where and when the detail begins its duties and how to contact the DZSO/DZSTL.

(f) Ensure that the DZSO/DZSTL, JM/AJM, and jumpers have been informed of all tree obstacle risks and that the DZSO/DZSTL and JM complete their duties.

(2) ***DZSO/DZSTL.***

(a) Determine if a follow-on assessment of the DZ has been conducted to confirm the current status.

(b) If the risk assessment indicates high risk and a recovery detail is used, ensure that the OIC/NCOIC is fully briefed on the plan. Ensure all recovery personnel have been trained and have all necessary equipment available to conduct the mission.

(c) Read all applicable regulations, FMs, and SOPs; ensure copies are present throughout mission.

(3) ***Jumpmaster.***

(a) Ensure all personnel have been briefed on the tree obstacles.

(b) Ensure all personnel have received prejump training within 24 hours prior to drop time, with special emphasis on unintentional tree landings.

c. **Tree Obstacle Coverage**. Each tree obstacle may require a different type of coverage. The following is an example composition of a recovery detail. Equipment should be altered to best accomplish the mission.

(1) OIC/NCOIC and assistant.

(2) Enough personnel to recover jumpers who may become entangled in the tree obstacles.

(3) A recovery team may need the following equipment:

- Radios with spare batteries - 2 (1 for OIC/NCOIC and 1 for recovery team).
- Grappling hook—1.
- Tree-climbing kit—1.
- Long backboard to facilitate CPR—1.
- Aid bag with resuscitation equipment—1.
- Ropes, 120 feet long—2.
- Wood poles 15 feet long—2.
- **Wood** extension ladder 20 feet long—1.
- Snap links—4.

(4) For night recovery operations, the following equipment should be added:

- Night vision devices (with spare batteries) for each team.
- Operational flashlights (with spare batteries) for each team.

d. **Optional Training**. Optional training includes suspended harness training on the second through fifth points of performance. Step-by-step training on the procedures jumpers will take for unintentional tree landings may be conducted.

D-4. DROP ZONE RISK ASSESSMENT DECISION MATRIX AND LEADER'S CHECKLISTS

Following are Table D-1, drop zone risk assessment decision matrix, and Figures D-1 through D-3 (pages D-7 through D-8), leaders' checklists.

	HIGH RISK	MEDIUM RISK	LOW RISK
WATER	Within 1,000 meters of any edge of DZ, more than 4 feet deep, and more than 40 feet wide.	More than 1,000 meters but less than 1,500 meters from any edge of DZ, more than 4 feet deep, and more than 40 feet wide.	More than 1,500 meters from any edge of DZ, more than 4 feet deep, and more than 40 feet wide.
WIRE	Within 1,000 meters of any edge of DZ.	More than 1,000 meters but less than 1,500 meters from any edge of DZ.	More than 1,500 meters from any edge of DZ.
TREE OR TREES	Within 1,000 meters of any edge of DZ and 35 feet tall or taller.	Within 1,000 meters of any edge of DZ.	Within 1,000 meters of any edge of DZ, but less than 20 feet tall.
OBSTACLE TRAINING AND EQUIPMENT	Required.	Recommended, but at commander's discretion.	Not required.

Table D-1. Drop zone risk assessment decision matrix.

COMMANDER

_____ Risk assessment and follow-on assessment have been conducted.

_____ Key leaders, jumpmasters, and jumpers have been informed of water obstacle risks and the risk level (high, medium, or low).

_____ USAF Form 3823, DZ Survey (formerly MAC Form 339), is current and available; the DZSO/DZSTL and JM have read it and completely understand the unintentional water landing risk level and the safety measures that are to be used.

_____ The DZSO/DZSTL and OIC/NCOIC of the boat detail have been briefed and understand their mission.

_____ B-5/B-7 life preservers are coordinated for if used.

_____ The DZSO/DZSTL, JM/AJM, safeties, and jumpers have been informed of all water obstacle risks.

_____ The DZSTL/DZSO and JM have completed their duties.

DZSO/DZSTL

_____ Risk assessment analysis has been conducted.

_____ If a boat detail is used, the OIC/NCOIC of the boat detail is fully briefed on the plan.

_____ All personnel have been trained and have all necessary equipment available to conduct the mission.

_____ All applicable regulations, FMs, and SOPs have been read.

_____ The boat detail maintains communications throughout the mission. Communications are established 1 hour prior to drop time and checked 15 minutes prior to drop time.

JUMPMASTER

_____ All personnel have been briefed on the water obstacles.

_____ B-5/B-7 life preservers have been inspected within the last 180 days and are serviceable.

_____ All jumpers have been trained on wear, fit, and use (to include manual inflation) of life preservers.

_____ All personnel receive prejump training within 24 hours of drop time, with special emphasis on unintentional water landings.

Figure D-1. Leader's checklist for possible water landings.

COMMANDER

_____ Risk assessment and follow-on assessment analysis have been conducted.

_____ Key leaders, jumpmasters, and jumpers are informed of wire obstacle risks and the risk level (high, medium, or low).

_____ USAF Form 3823, DZ Survey (formerly MAC Form 339), is current and available. The DZSO/DZSTL and JM have read it and completely understand the wire landing risk level and the safety measures that are to be used.

_____ The DZSO/DZSTL and the OIC/NCOIC of the recovery detail are briefed and understand their mission.

_____ The DZSO/DZSTL, JM/AJM, safeties, and jumpers are informed of all wire obstacle risks.

_____ DZSTL/DZSO/JM complete their duties.

DZSO/DZSTL

_____ Risk assessment analysis has been conducted.

_____ If a recovery detail is used, the OIC/NCOIC of the recovery detail is fully briefed on the plan.

_____ All personnel have been trained and have all necessary equipment available to conduct the mission.

_____ All applicable regulations, FMs, and SOPs have been read.

_____ The recovery detail maintains communications throughout the mission. Communications are established 1 hour prior to drop time and checked 15 minutes prior to drop time.

JUMPMASTER

_____ All personnel have been briefed on the wire obstacles.

_____ All personnel have received prejump training within 24 hours of drop time with special emphasis on wire landings.

Figure D-2. Leader's checklist for possible wire landings.

COMMANDER

_____ Risk assessment and follow-on assessment analysis have been conducted.

_____ Key leaders, jumpmasters, and jumpers are informed of tree obstacle risks and the risk level (high, medium, or low).

_____ USAF Form 3823, DZ Survey (formerly MAC Form 339), is current and available. The DZSO/DZSTL and JM have read it and completely understand the tree landing risk level and the safety measures that are to be used.

_____ The DZSO/DZSTL and the OIC/NCOIC of the recovery detail are briefed and understand their mission.

_____ The DZSO/DZSTL, JM/AJM, safeties, and jumpers are informed of all tree obstacle risks.

_____ The DZSTL/DZSO and JM have completed their duties.

DZSO/DZSTL

_____ Risk assessment analysis has been conducted.

_____ If a recovery detail is used, the recovery detail OIC/NCOIC is fully briefed on the plan.

_____ All personnel have been trained and have all necessary equipment available to conduct the mission.

_____ All applicable regulations, FMs, and SOPs have been read.

_____ The recovery detail maintains communications throughout the mission. Communications are established 1 hour prior to drop time and checked 15 minutes prior to drop time.

JUMPMASTER

_____ All personnel have been briefed on the wire obstacles.

_____ All personnel receive prejump training within 24 hours of drop time, with special emphasis on unintentional tree landings.

Figure D-3. Leader's checklist for possible tree landings.

GLOSSARY

ADEPT	alternate door exit procedure for training
ADP	airdrop personnel
ADZSO	assistant drop zone safety officer
AF	Air Force
AFCCT	Air Force combat control team
AFJMAN	Air Force joint manual
AFR	Air Force regulation
AGL	above ground level
AJM	assistant jumpmaster
AIRPAC	all-purpose weapons and equipment container system
ALICE	all-purpose, lightweight, individual carrying equipment
AMC	Air Mobility Command
AMCR	Air Mobility Command regulation
ANGLICO	air and naval gunfire liaison company
APFT	Army Physical Fitness Test
AR	Army regulation
ARS	automatic release system
ASOP	airborne standing operating procedure
ASP	ammunition supply point
ATC	air traffic control
AWADS	adverse weather aerial delivery system
AZAR	assault zone availability report
BCU	battery coolant unit
CARP	computed air release point
CCP	circulation control point
CCT	combat control team
CDS	container delivery system
CFM	cubic feet per minute
CO	commissioned officer
CO2	carbon dioxide (gas)
COMALF	commander of the airlift forces
CPR	cardiopulmonary resuscitation
CWIE	container, weapon, individual equipment
DA	Department of the Army
DACO	departure airfield control officer
DC	District of Columbia
DMJP	Dragon missile jump pack
DOD	Department of Defense
DTG	date-time group
DZ	drop zone
DZSO	drop zone safety officer

DZST	drop zone support team
DZSTL	drop zone support team leader
ETLBV	enhanced tactical load-bearing vest
FHT	field handling trainer
FLA	front line ambulance
FM	field manual
FPLIF	field pack, large, internal frame
FS	fuselage station
GLO	ground liaison officer
GMRS	ground marking release system
GTA	ground-to-air
GUC	ground unit commander
HAHO	high-altitude, high-opening
HALO	high-altitude, low-opening
HD	heavy drop
HE	heavy equipment; high explosive
HEAT	high-explosive antitank
HPT	hook-pile tape
HQ	headquarters
hrs	hours
HSPR	harness, single-point release
IAS	indicated airspeed
IAW	in accordance with
ICS	intercommunication system
IFF	identification, friend or foe
IMC	instrument meteorological conditions
intercomm	intercommunication (radio)
JA/ATT	Joint Airborne/Air Transportability Training
JACRS	code letters used on CARP drop zones
JM	jumpmaster
JMPI	jumpmaster personnel inspection
JPI	joint preflight inspection
JSJR	jumpmaster spotted, jumpmaster released
JTF	joint task force
KT	knot
LAPES	low-altitude parachute extraction system
LAW	light antitank weapon
lb	pound

LCE	load-carrying equipment
LDA	lateral drift apparatus (training apparatus)
MAC	Military Airlift Command
MACOM	major Army command
MCWP	Marine Corps Warfighting Publication
METT-T	mission, enemy, terrain, troops and time available
MEW	mean effective wind
MFF	military free fall
MG	machine gun
MIRPS	modified improved reserve parachute system
MO	malfunction officer
MOD	modified
mph	miles per hour
MSL	mean sea level
MTT	mobile training team
MTOE	modified table of organization and equipment
NAVAIDS	navigational aids
NCO	noncommissioned officer
NET	not earlier than
NLT	not later than
No.	number
NOPO	nonporous
NVIS	night vision imaging systems
OD	olive drab
OIC	officer in charge
OPCON	operational control (the authority delegated to direct those personnel needed to accomplish a specific mission)
PASGT	personnel armor system, ground troops
PE	personnel
PI	point of impact
PIBAL	pilot balloon
PIE/R2	parachutist's individual equipment rapid release
PIL	parachutist's impact liner
PIR	Parachute Infantry Regiment
PLD	personnel lowering device
PLF	parachute landing fall
port	left side of aircraft (direction of flight)
PP	parachute personnel
PRB	parachute recovery bag
RAAWS	ranger antiarmor/antipersonnel weapon system
RAM	raised angle marker

reps	repetitions
RMP	reprogrammable microprocessor
RP	release point
SAW	squad automatic weapon
SCUBA	self-contained underwater breathing apparatus
SFQC	Special Forces Qualification Course
SH	suspended harness (training apparatus); student handbook
SKE	station-keeping equipment
SLT	swing landing trainer (training apparatus)
SMJP	Stinger missile jump pack
SOF	special operations force
SOLL	special operations low level
SOWT	special operations weather team
SOI	signal operation instructions
SOP	standing operating procedure
SSN	social security number
STABO	a system for extracting personnel by helicopter (combined first letters of the surnames of the five tactical airdrop personnel who designed the system)
starboard	right side of aircraft (direction of flight)
station time	time the aircraft is to be loaded and prepared for flight (normally, 35 minutes prior to takeoff)
STT	special tactics team
TACSAT	tactical satellite
TALCE	tanker/airlift control element
TAP	tactical airdrop personnel
TC	training circular
TDA	table of distribution and allowances
TM	technical manual
TO	technical order (USAF)
TOE	table of organization and equipment
TOT	time on target
TP	training practice
TTB	tactical training bundle
US	United States
USA	United States Army
USAF	United States Air Force
USAIS	United States Army Infantry School
USASOC	United States Army Special Operations Command
USMC	United States Marine Corps

VIRS	verbally initiated release system
VMC	visual meteorological conditions
VS	visual signal
WDZ	water drop zone
WDZSO	water drop zone safety officer
WDZSTL	water drop zone support team leader
WO	warrant officer
WSVC	wind streamer vector count

REFERENCES

Documents Needed

None required.

Readings Recommended

These readings contain relevant supplemental information.

AF Form 3823	Drop Zone Survey. February 1994.
AFI Reg 13-217	Assault Zone Procedures, 1 June 1999.
AR 59-4/AFJI 13-310/ OPNAVINST 4630.24C/ MCO 13480.1B	Joint Airdrop Inspection Records, Malfunction Investigations and Activity Reporting. 1 May 1998.
AR 350-1	Army Training. 1 August 1981, with Change 1, 1 August 1983.
AR 350-2	Opposing Force Program. 15 June 1983.
DA Pam 351-4	U.S. Army Formal Schools Catalog. 31 October 1995.
DA Form 1306	Statement of Jump and Loading Manifest. May 1963.
DD Form 1574	Serviceable Tag - Materiel. October 1966.
DD Form 1748-2	Joint Airdrop Malfunction Report (Personnel-Cargo). November 1997.
FM 10-500-2/TO 13C7-1-5	Airdrop of Supplies and Equipment: Rigging Airdrop Platforms. 1 November 1990, with Change 1, 14 December 1992, and Change 2, 23 December 1996.
FM 10-500-3/TO 13C7-1-11	Airdrop of Supplies and Equipment: Rigging Containers. 8 December 1992, with Change 1, 26 September 1996.
FM 10-550/TO 13C7-22-71	Airdrop of Supplies and Equipment: Rigging Stinger Weapon System and Missiles. 29 May 1984, with Changes 1-3, 14 February 1989 - 28 June 1996.

FM 10-542	Airdrop of Supplies and Equipment: Rigging Loads for Special Operations. 07 October 1987.
FM 21-20	Physical Fitness Training. 30 September 1992, with Change 1, 1 October 1998.
FM 3-21.32(FM 57-38)	Pathfinder Operations. 1 October 2002.
FM 31-19/MCWP 3-15.6 AFI 11-411(I)/ NAVSEA SS400-AG-MMO-010	Military Free-Fall Parachuting Tactics, Techniques, and Procedures. 1 October 1999.
FM 31-71	Northern Operations. 21 June 1971.
FM 100-27/AFM 2-50	US Army/US Air Force Doctrine for Joint Airborne and Tactical Airlift Operations. 31 January 1985, with Change 1, 29 March 1985.
MAC Form 168	Airdrop/Airland Extraction Zone Control Log. 1 December 1992.
MAC Form 339	Drop Zone Survey. February 1989 (obsolete but still in use; replaced by AF Form 3823).
TC 31-24	Special Forces Air Operations. 9 September 1988.
TC 31-25	Special Forces Waterborne Operations. 3 October 1988.
TM 1-1520-237-10	Operator's Manual for UH-60A, UH-60L Helicopters, and EH-60A Helicopters. 31 October 1996, with Changes 1-6, 30 June 1997 - 3 April 2000.
TM 9-1425-429-12	Operator's and Organizational Maintenance Manual for Stinger Guided Missile System Consisting of Weapon Round Basic, Round Post, Weapon Round RMP, Trainer Handling Guided Missile Launcher M60, Interrogator Set AN/PPX-3A/3B, Interrogator Set Programmer AN/GSX-1/1A. 21 April 1992, with Change 1, 30 October 1992.
TM 10-1670-201-23/ TO 13C-1-41/ NAVAIR 13-1-17	Organizational and Direct Support Maintenance Manual for General Maintenance of Parachutes and Other Airdrop Equipment. 30 October 1973, with Changes 1-7, 7 March 1974 - 17 May 1990.

TM 10-1670-251-12&P	Operator and Unit Maintenance Manual for Personnel/Cargo Lowering Device 500-lb Capacity. 11 June 1991, with Change 1, 30 June 1994.
TM 10-1670-262-12&P	Operator and Unit Maintenance Manual Including Repair Parts and Special Tools List Personnel Insertion/Extractions Systems for STABO, Fast Rope Insertion/Extraction System, and Anchoring Device. 25 September 1992, with Change 1, 24 November 1993, and Change 2, 30 June 1994.
TM 10-1670-298-20&P	Unit Maintenance Manual Including Repair Parts and Special Tools List for Container Delivery System A-7A Cargo Sling; A-21 Aerial Deliver Cargo Bag; A-22 Aerial Delivery Cargo Bag; A-23 Aerial Delivery Cargo Bag; Capsule, Cargo, CTU-2/A; Strap Connector, 60 Inches Long; Strap Connector, 120 Inches Long. 15 September 1995.
TM 38-250/ AFJMAN 24-204/ NAVSUP PUB 505/ MCO P4030.29G/ DLAI 4145.3	Preparing Hazardous Materials for Military Air Shipments 1 March 1997.
US Army Special Operations Command Regulation 350-2	Training, Airborne Operations Manual. 1 June 1995.

INTERNET WEB SITES

U.S. Army Publishing Agency
http://www.usapa.army.mil

Army Doctrine and Training Digital Library
http://www.adtdl.army.mil

INDEX

Table of Contents

Part I. Scouting

PART I. SCOUTING

CHAPTER 1. FUNDAMENTALS OF SCOUTING

Scouting involves observing terrain and/or the enemy, and accurately reporting those observations. Scouting requires proficiency in the use of weapons, cover and concealment, route selection, and skill in unobserved day or night movement.

1001. PURPOSE

When an infantry unit is not actively fighting the enemy, it should be actively searching for the enemy, which is the primary purpose of scouting. The unit attempts to keep the enemy off balance while making preparations for further attacks. Physically locating and keeping the enemy off balance are normally accomplished by small units ranging from a two-person scouting party to a squad-size patrol.

Infantrymen are sent out as scouts or as members of a patrol because the commander needs information about the enemy, terrain, and the location of friendly troops. The lives of the entire unit may depend upon the success or failure of a scout or patrol and the accuracy and timeliness of the report. The success of the scout or patrol will depend upon their training, preparation by the commander, and understanding their mission and the commander's requirements.

To wage combat successfully, a commander must have accurate, detailed, and timely information about the enemy, the terrain, and adjacent friendly units. Well-trained scouts and capably led patrols are among the most effective means the commander has for acquiring the information necessary to plan tactical actions and make decisions in execution.

1002. REQUIRED SCOUTING SKILLS

To be effective, a scout must be able to—

- Recognize terrain features.
- Read a map and determine direction.
- Practice and implement the principles of cover and concealment.
- Fully utilize movement and route selection.
- Know the enemy (estimate enemy unit composition and strength).
- Observe and report information accurately.
- Select routes and move through numerous types of terrain.

CHAPTER 2. TERRAIN, MAPS, AND DIRECTION

A scout must understand map symbols, identify elevations from contour interval lines, scale distance on a map, relate natural and man-made features shown on the map to the actual features on the ground, plot a course from one point to another, and locate his current position. To relate a map to the actual terrain and its features, a scout must be able to orient it to the ground using a compass, two points, a watch, and the sun or the stars.

2001. TERRAIN FEATURES

Since the infantry works and fights on the ground, terrain ashore information that scouts gather and report on is of great importance. Hills, valleys, woods, and streams are the forms and growths commonly referred to as natural land features. Artificial or man-made features include houses, bridges, and railroads. Figure 2-1 shows some important terrain features.

2002. THE LENSATIC COMPASS

The best method of finding direction, during both day and night, is with a compass. The lensatic and M2 are the two types of compasses issued to infantry units. Both work on the same principle.

The standard compass for general use in the Marine Corps is the pivot-mounted lensatic compass, so called because azimuths are read through a magnifying lens in the eyepiece. Figure 2-2, on page 2-2, shows the lensatic compass and its nomenclature. The meter graphic scale on the side of the compass is 1:50,000, which is the most used scale in military mapping. The graphic scale is useful in the field as a straightedge, an aid in orienting the map, and a means of reading map grid coordinates. The plastic dial is graduated in both degrees and mils. Numbers on the dial are printed in black. There are luminous markings on the bezel, floating dial, and on both ends of the sighting wire, plus a 3-degree bezel serration and clicking device that permit reading azimuths at night. The compass is carried in a nylon case that may be attached to the cartridge belt.

Magnetic compasses are affected by the presence of iron or magnetic fields. Consequently, the scout should not be within the influence of local magnetic

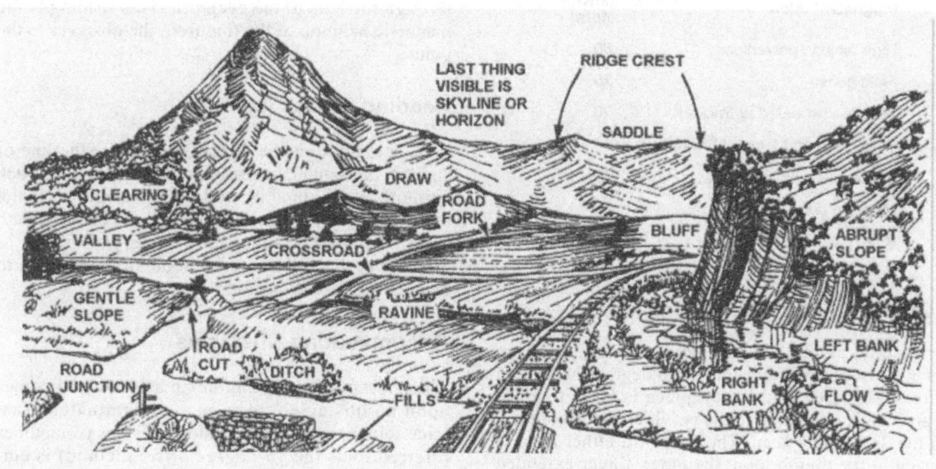

Figure 2-1. Natural and Artificial Terrain Features.

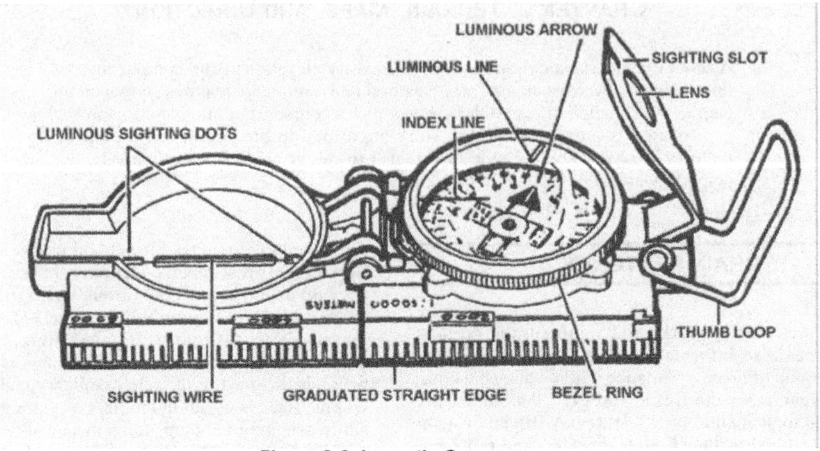

Figure 2-2. Lensatic Compass.

attraction while using a compass to determine direction. The rifle, pistol, and other metal objects must be laid aside when reading the compass. The minimum distances the scout should be from visible masses of iron and electrical fields of magnetism for the compass to provide accurate readings are provided below.

Magnetic Fields	Minimum Distances (meters)
High tension powerlines	60
Field guns	20
Vehicles (wheeled or tracked)	20
Telephone and telegraph wires	10
Barbed wire	10
Machine gun	3
Rifle, pistol	1

Reading an Azimuth

To read an azimuth to any point, the cover of the compass is raised to an angle of 90 degrees in relation to the index face, and the eyepiece is lifted to a 45-degree angle in relation to the bezel, or so the numbers on the dial can be seen. The thumb of either hand is placed in the thumb loop, the index finger extended along the side of the compass case, and the remainder of the hand closed. The closed hand and wrist are grasped with the other hand. The elbows are drawn in close to the body, forming a firm foundation for the compass. The eye is placed to the lens on the eyepiece. The compass is pointed at the object or point to which the azimuth is to be read. A sighting is taken through the sighting slot in the eyepiece, and the point is lined up with the sighting wire in the cover. The compass is held until the dial steadies; then the reading is taken through the lens of the eyepiece. This reading is the magnetic azimuth of the line from the observer to the point.

Reading a Back Azimuth

A back azimuth is the direction opposite the line of sight. If the azimuth is less than 180 degrees, the back azimuth is obtained by adding 180 degrees. If the azimuth is greater than 180 degrees, the back azimuth is obtained by subtracting 180 degrees. Back azimuths are used to determine a return route or to resection to determine a current position.

Circumventing Obstacles

When a scout is traveling on an azimuth and comes upon an obstacle—such as a contaminated area, minefield or swamp—the following steps (sometimes referred to as the *90-degree offset* method) is employed to go around or circumvent the obstacle and resume movement along the original azimuth (see fig. 2-3). The steps are as follows:

1. Move up to the obstacle and make a full 90-degree turn to the right (or left).

2. Walk beyond the obstacle, keeping track of the distance in paces or meters.

3. Stand at the end of the obstacle, face in the original direction of march, and follow that azimuth until the obstacle has been passed.

4. Make a 90-degree turn to the left (or right) and move the distance previously measured to return to the original line of march.

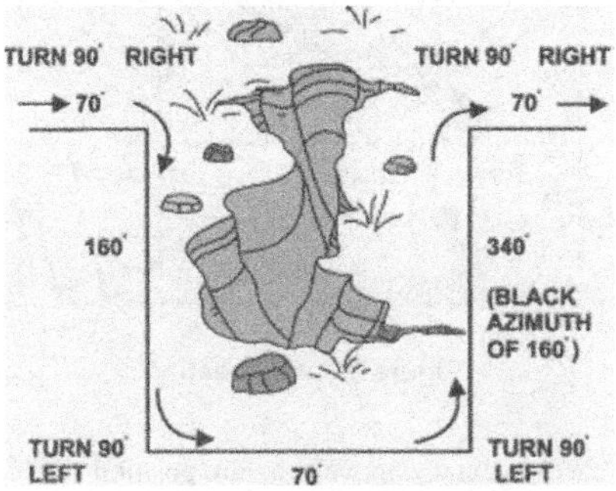

Figure 2-3. Circumventing an Obstacle.

Following an Azimuth During the Day

The eye is placed to the lens in the eyepiece and the compass moved until the desired azimuth reading is visible beneath the fixed index. Without moving the compass, the vision is shifted from the lens through the sighting slot in the eyepiece, and a sighting is taken out beyond the sighting wire in the cover. A prominent terrain feature on this line of sight is selected, the compass closed, and the landmark approached. When the scout reaches the landmark, the procedure is repeated.

Following an Azimuth at Night

It is necessary to prepare and set the compass before departing on a night movement because at night only the luminous parts of the compass can be seen. To prepare the compass for night use, the luminous parts must be fully charged by sunlight or artificial light, such as a flashlight. To set a compass—

1. Move the compass so that the desired azimuth on the dial is directly under the index line on the lower glass.

2. Rotate the upper movable glass so that the luminous line is directly above the north arrow of the dial.

3. Set the compass for marching at night on the specified azimuth.

—OR—

1. Face the general direction of movement.

2. Line up the north arrow and the luminous line on the bezel with the luminous sighting dots.

3. Hold the compass still with one hand and grip the knurled bezel ring with the other hand.

4. Turn the bezel ring the prescribed number of clicks in the proper direction, remembering that each click equals 3 degrees. For example, to set an azimuth of 21 degrees, the bezel ring would be turned seven clicks to the left.

5. Turn the whole compass until the north needle lines up with the luminous line. The compass is then set on the desired azimuth. The azimuth is the line formed by the two luminous sighting dots on the inside of the cover.

To march on a preset azimuth during night movement, open the compass and move it so the north arrow is directly below the luminous line (see fig. 2-4). Move

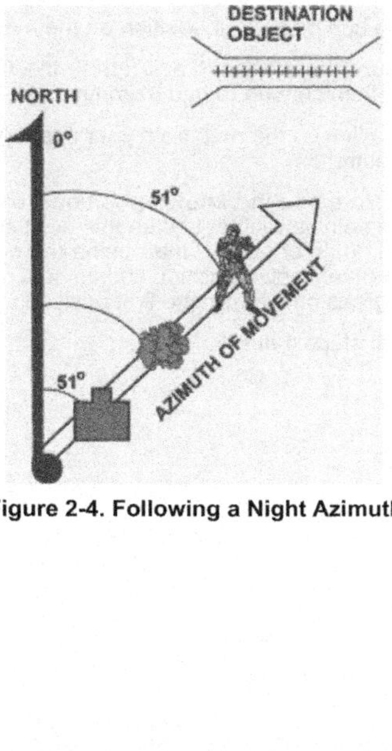

Figure 2-4. Following a Night Azimuth.

in the direction of the line formed by the two luminous sighting dots. It is necessary to refer to the compass more frequently at night than during the day. If stars are visible, find a prominent star along the azimuth of movement to use as a reference point. When the view of the sky is restricted by overcast conditions or vegetation, send a scout forward along the azimuth of movement to the limit of visibility. This scout is guided along the azimuth of movement by a stationary navigator. When the scout reaches the limit of visibility, the navigator moves to the scout's location. This process is repeated until the destination is reached.

A more rapid method for reaching the scout's destination is to equip the navigator with a compass. The navigator can set the compass as explained earlier and the scout proceeds providing security 180 degrees to the front on the specified azimuth, receiving right and left corrections from the navigator while both are on the move. The point scout must stay within visual range of the navigator. If available, a strip of white or luminous tape on the back of the point scout's helmet will assist.

Intersection

Intersection is the location of an unknown point by successively occupying at least two, preferably three known positions and sightings on the unknown point. It is used to locate features not depicted on the map or not readily identifiable. To determine an intersection, perform the following steps (see fig. 2-5):

1. Orient the map using the compass.

2. Locate and mark your position on the map.

3. Measure the magnetic azimuth to the unknown position; then convert to grid azimuth.

4. Draw a line on the map from your position on this grid azimuth.

5. Move to a second known position from which unknown point is visible. Locate this position on the map and again orient the map using the compass. The second unknown position should be a minimum of 30 degrees offset from the first position.

6. Repeat steps 4 and 5.

Figure 2-5. Intersection.

To check accuracy, move to a third position and repeat steps 1 thorough 4. Where the lines cross is the location of the unknown position. Using three lines, a triangle is sometimes formed—called the *triangle of error*—instead of an intersection. If the triangle is large, recheck your work to find the error. Do not assume that the position is at the center of the triangle.

Resection

Resection is the location of the user's unknown position by sighting on two or three known features that are identifiable on the map. To determine a resection, perform the following steps (see fig. 2-6):

1. Orient the map using the compass.

2. Locate two or three known positions on the ground and mark them on the map.

3. Measure the magnetic azimuth to a known position then convert to grid azimuth.

4. Change the grid azimuth to a back azimuth and draw a line on the map from the known position back toward the unknown position.

5. Repeat step 3 and step 4 to determine a second known position.

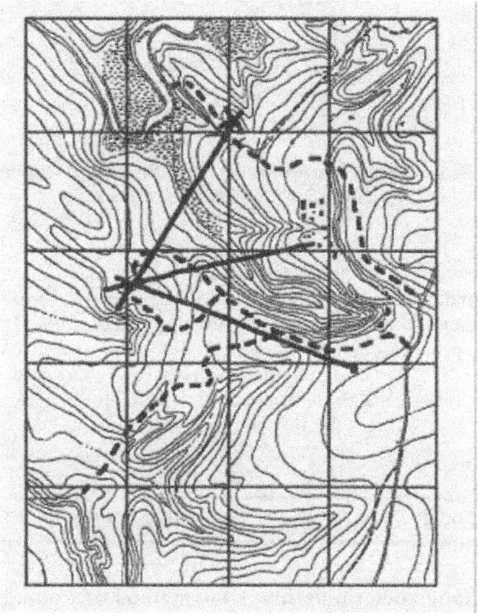

Figure 2-6. Resection.

To check accuracy, repeat the steps above for a third known position. The intersection of the lines is your location. Using three lines, a triangle of error may be formed. If the triangle is large, recheck.

2003. STARS, SUN, AND OTHER FEATURES

In rare cases when a scout is without a compass, the following examples are alternate means to determine direction. When using constellations to determine direction, identify your location's Temperate Zone. A Temperate Zone is the area between the tropics and the polar circles.

At night, the stars provide an excellent means of maintaining a line of march. In the North Temperate Zone (north of the equator), the Big Dipper constellation is one key to determining direction of true north. It is made up of seven fairly bright stars in the shape of a dipper with a long curved handle (see fig. 2-7). The two stars that form the side of the cup farthest from the handle, used as pointers, are situated in the direction of a bright star that is about five times the distance between the two stars of the dipper cup. This bright star is the North Star and is directly over the North Pole. The pointers always designate the North Star, which is the direction of true north.

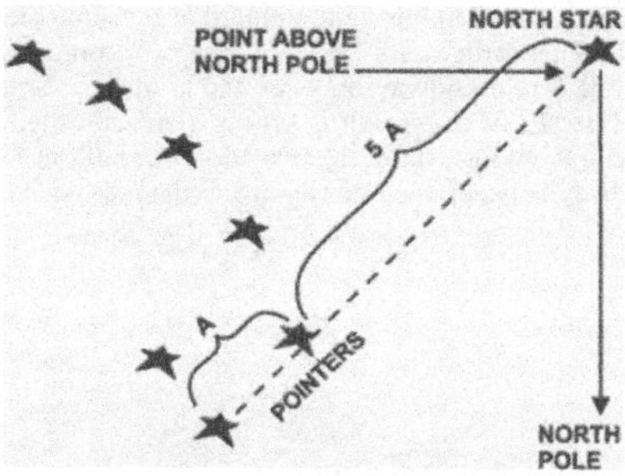

Figure 2-7. Locating the North Pole.

In the Southern Hemisphere, true south is determined in relation to the Southern Cross, a constellation composed of five stars. Two bright pointer stars in the vicinity of the Southern Cross serve as locators to help locate true south (see fig. 2-8). The outer four stars are

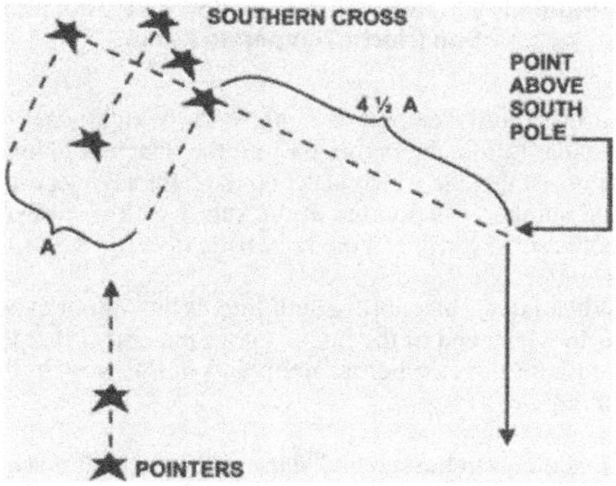

Figure 2-8. Locating the South Pole.

fairly bright and form a cross. This cross is imagined as the frame of a kite. A straight tail, four and one half times as long as the length of the kite itself, is put on the kite using finger widths for a measuring stick. The end of this tail will be close to a position directly over the South Pole. Usually, it will not be possible to see a star in the immediate vicinity, because there is no bright star visible directly above the South Pole.

During daylight hours, a watch and the sun can be used to determine direction within 8 degrees. In the North Temperate Zone, the watch is held horizontally, face up, and the hour hand pointed at the sun (see fig. 2-9). The north-south line and the direction of south can be found midway between the hour hand and the number 12, if the watch is set on standard time. If in daylight savings time, the direction of south is found midway between the hour hand and the number 1.

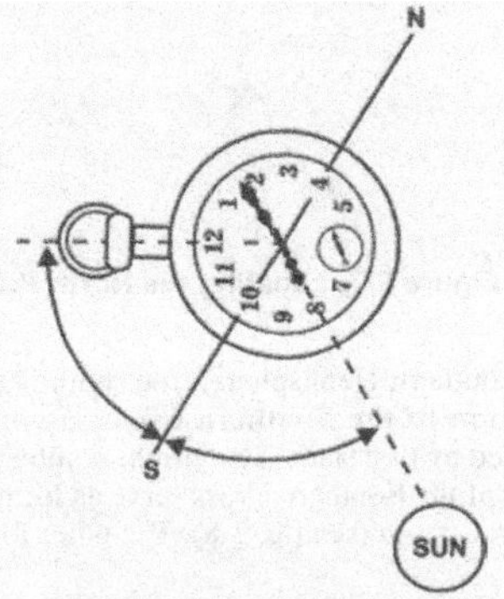

Figure 2-9. Determining Direction by Watch and Sun (North Temperate Zone).

In the South Temperate Zone, if the watch is set on standard time, the number 12 on the watch is pointed at the sun; if the watch is set on daylight savings time, the number 1 is pointed at the sun. North is midway between 12 (or 1) and the hour hand (see fig. 2-10).

When laying in a north-south line, if any doubt exists as to which end of the line is north, remember that the sun is in the east before noon and in the west in the afternoon.

In addition to the sun and stars, other methods a scout without a compass can use to determine direction include determining prevailing wind direction and

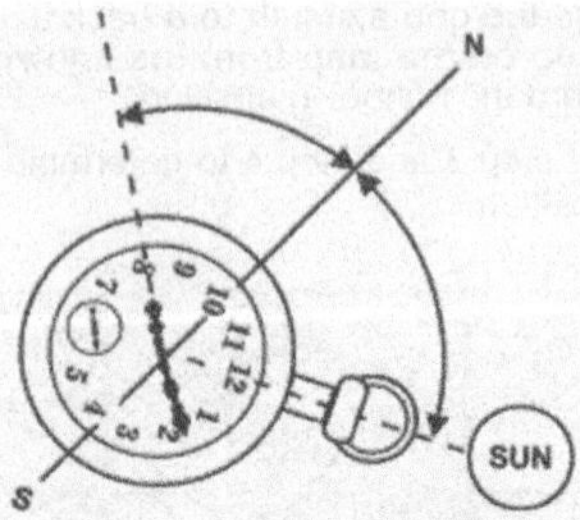

Figure 2-10. Determining Direction by Watch and Sun (South Temperate Zone).

using a mountain for orientation. By previous study of maps and photographs, a scout can keep informed of location and direction by using a distinctive edge of woods, a deep ravine or the direction of a stream's flow. A scout should constantly evaluate and memorize both the immediate terrain and general area for prominent features and landmarks.

2004. RANGE DETERMINATION

Range determination is the method of finding the distance between an observer and an enemy target or an object. By accurate range determination, the members of a given unit can set their sights correctly and place effective fire on enemy targets. The degree of accuracy is dependent on several factors, such as terrain relief, time available, and experience of the observer.

Mental Estimation

A mental distance estimate is made using a known unit of measure. Distance is estimated to the nearest 100 meters by determining the number of known units of measure between the observer's position and a target. For example, a football field, which is 100 yards, can be used as a known unit of measure for determining the distance between an observer's position and a target. For longer distances, progressive estimation may be necessary. To do this, the observer determines the number of units of measure to an intermediate point and doubles the value. The observer should consider the effects in table 2-1 in estimating distances.

Table 2-1. Effects to Consider in Mental Estimation of Distances.

Objects Appear Nearer	Objects Appear More Distant
In bright light.	In poor light or in fog.
In clear air at high altitude.	Only a small part of the object can be seen.
The background is in contrast with the color of the object.	The background is similar in color to that of the object.
The observer is looking down from a height.	The observer is looking over a depression, most of which is visible.
The observer is looking over a depression, most of which is hidden.	The observer is kneeling or sitting, especially on a hot day, when the ground is moist.
The observer is looking down a straight feature such as a road.	
The observer is looking over water, snow, or a uniform surface such as a cultivated field or desert.	

Estimating in Good Visibility

When visibility is good, distances can be estimated by using the appearance of tree trunks, branches, and foliage (as seen by the naked eye) in comparison with map data. Table 2-2 is a guide for wooded terrain. Table 2-3 is a guide for urban environments.

Estimating From a Terrain Study

The Marine should always use terrain/map analysis to assist in estimating distances. When the Marine is looking in a specific direction, the estimation of distance can be enhanced by studying the terrain and comparing it with the map. Particular emphasis should be given to color contrasts of terrain features seen along the observer-target line (OTL). For example, the distance across successive ridge lines or depressions in the distance may be identifiable by only slight changes of color to the eye. Different colors of grass might reveal a hidden terrain feature such as a stream.

Table 2-2. Estimating Distance in Wooded Terrain.

Distance in Meters	Tree Description
1,000	Trunk and main branches are visible. Foliage appears in cluster-like shape. Daylight may be seen through the foliage.
2,000	Trunk visible, main branches distinguishable, foliage appears as smooth surface. Outline of foliage of separate trees distinguishable.
3,000	Lower half of trunk visible. Branches blend with foliage. Foliage blends with adjoining trees.
4,000	Trunk and branches blend with foliage and appears as a continuous cluster, smooth in appearance. Movement of foliage due to wind cannot be deleted.
5,000 and beyond	Whole area covered by trees and appears smooth and dark.

Table 2-3. Estimating Distance in Urban Terrain.

Distance in Meters	Object Identified by the Unaided Eye
1,000	Lone tree trunk
1,500	Individuals and horsemen
3,000	Chimneys on rooftops
4,000	Windows in houses
4,000–5,000	Individual houses in populated area
8,000–9,000	Villages and individual houses
15,000–18,000	Large houses, towers, and steeples

Chapter 3. Enemy Activity

A commander often acts on information furnished by scouts. Therefore, scouts must aim at absolute accuracy in reporting enemy activity. This chapter discusses estimating enemy strengths, interpreting signs and tracks, and knowing the enemy.

3001. Estimating Enemy Strengths

If troops cannot be counted, their strength may be estimated by: noting the length of time it takes various types of moving columns to pass given point, the area required of a unit in camp or bivouac, or the front on which they are deployed. When the ground is dry, infantry on the march raise a low, thick cloud of dust, and motor vehicles or mechanized units raise a thick, rapidly moving cloud. Additionally, through practice, a scout may gain information as to the strength and composition of enemy forces by listening to noises and observing lights, fires, and smoke. A scout gains valuable experience in estimating enemy strengths by observing friendly forces in camp, on the march, and deployed. The knowledge scouts gain during field exercises of the appearance and tactical dispositions of squads, platoons, companies, and larger units will be of great assistance in estimating the strength and composition of enemy units observed under various conditions.

3002. Interpreting Signs and Tracks

In addition to estimates made through direct observation, a scout may often be able to estimate size, composition, direction, rate of movement, condition, discipline, state of training, and morale of enemy forces through signs and tracks left behind.

Signs

The examination of vacated enemy positions provides valuable information. The size of a bivouac or defense area ordinarily indicates the number of enemy occupants. Clothing, ration containers, dumps, etc., further indicate the quantity of the departed enemy force. The condition of the bivouac area and amount of material abandoned give an indication of the enemy morale, training, and discipline. A well-policed area indicates good discipline. Rubbish, ration and smoking residue, and nonessential personal items of equipment adrift indicate a lower state of morale, training, and discipline. Stores and material left behind in good condition may indicate a hasty movement or withdrawal. Burned or destroyed materials indicate a deliberate, orderly withdrawal or movement. Letters, insignia, and other articles may reveal the identity of the enemy unit.

In the case of a moving enemy, the distance between periodic halts indicates the rate of march if enemy habits relative to marches and halts are known. Condition of the halt areas indicates the state of morale, training, and discipline.

The physical condition of enemy dead and wounded and their personal equipment and weapons are reported. The general condition and state of maintenance of destroyed or abandoned vehicles should also be reported.

Tracks

A track is a mark left on the ground by the passage of a person or object. Examination of tracks reveals information about the enemy.

Troops

A few tracks overlapping each other on both sides of a road or trail may indicate a patrol in staggered formation. A large number of tracks indicates troops in column formation. A large column will wear a dry road smooth and flat. In damp terrain, a freshly made track will have sharp edges; ordinarily, signs of moisture will disappear in about 15 minutes. A runner's toes are dug into the ground; a walker's footprint is fairly even.

Vehicles

The type of track indicates whether the vehicle is wheeled or tracked. A scout acquires the necessary experience to make the proper determination by observing vehicle tracks during training.

The direction of travel can be determined by the way tracks pass across ruts, by impressions on the edges of holes in the ground, how water is splashed from puddles, or by the way grass, twigs, and branches are bent; for example—

- A vehicle (wheeled or tracked) entering a rut pushes dirt into the rut and leaves an indentation on the exit side of the rut.
- A wheel going over holes in the ground leaves a deeper impression on the edge toward the direction of travel.
- The side of a puddle with the greater splash indicates the direction of travel of the vehicle.
- When traveling cross-country, the direction in which grass is bent and/or twigs, branches, and bushes are broken indicates direction of travel.

A general rate of speed can be estimated by the amount of water or mud splattered. A fast-moving vehicle will throw larger amounts of water or mud a greater distance to the front and sides than a slow-moving vehicle, and it leaves a deeper impression on the exit edges of holes. The faster the travel, the deeper the impression.

3003. KNOWING THE ENEMY

A scout should learn as much as possible about enemy psychology, habits, organization, and tactics. The more knowledge gained about the enemy, particularly the enemy's normal security measures, the better the scout's chances are to observe and obtain accurate information with minimum risk to the mission's success. Scouts gain much of this knowledge through experience, but they also gain a great deal of their preliminary information, particularly that pertaining to enemy organization and tactics, during training and may be updated by unit commanders and intelligence officers.

CHAPTER 4. DAYLIGHT SCOUTING

A scout must be able to operate in all types of terrain and under all conditions of visibility. He must be thoroughly familiar with the principles for using cover and concealment, camouflage, individual movement, and route selection, both to and from the objective.

4001. COVER AND CONCEALMENT

Cover is protection from the fire of hostile weapons. Concealment is protection from observation or surveillance from hostile air and ground observation, but not from hostile fire. Both cover and concealment are divided into two main categories: natural and artificial. Natural cover includes small hills, ditches, rocks or vegetation. Fighting holes, bunkers, and brick walls are examples of artificial cover. Some features, such as buildings, provide both cover and concealment. In deciding whether to seek cover or concealment, a scout must make the best choice to complete the mission (see fig. 4-1).

Figure 4-1. Correct Use of Cover.

Concealment Principles

Concealment principles are as follows:

- Remain motionless while observing. Anything in motion attracts the eye.
- Use all available concealment.
- Observe from the prone position (it offers a low silhouette and makes detection by the enemy difficult.
- Expose nothing that reflects light.
- Blend with the background because contrasting colors are noticeable.
- Remain in the shade because moving shadows attract attention.
- Distort or change the regular outline of objects. Most military objects have distinctive shapes that make obvious shadows and silhouettes.
- Avoid the skyline. Figures on the skyline can be seen from great distances and are easily identified by their outlines.

Concealment Techniques

Concealment techniques are as follows:

- When observing, the scout looks around an object's side (unless it is transparent) and prepares to fire, if necessary, around the side of or, if possible, through an object.
- Looking or firing over an object can make the scout an easily visible target for the enemy. If the scout must fire over the top of concealment or cover, the outline of the head or helmet should be broken or distorted.
- Upon the approach of an airplane, the scout takes a prone position, turns face-down, and remains motionless. If surprised by an airplane, the scout remains in place and does not look up.

- The scout covers exposed body parts such as the face, back of the neck, and hands with grease paint, mud or other materials to reduce sun reflection.
- Camouflage for equipment can be improvised from garnishing or sandbags to prevent sun reflection.
- In snowy terrain, white overgarments are worn.
- The helmet cover outline should be distorted.

4002. CAMOUFLAGE

Camouflage is the use of concealment and disguise to minimize the possibility of detection and/or identification of troops, material, equipment, and installations. The purpose of camouflage is to provide concealment of military objects from enemy observation. Camouflage is also used to conceal an object by making it look like something else. A scout's mission usually requires individual and equipment camouflage. If natural camouflage is not adequate, the position is camouflaged. In using camouflage, remember that objects are identified by their form (outline), shadow, texture, and color. The principal purpose of camouflage in the field is to prevent direct observation and recognition.

Individual Camouflage

Successful individual camouflage involves the ability to recognize and take advantage of all forms of natural and artificial concealment available (vegetation, soil, debris, etc.) and knowledge of the proper use of artificial camouflage materials.

Aids to Individual Camouflage

A scout must recognize the terrain's dominant color and pattern and must change the appearance of clothing and equipment accordingly in order to blend and not contrast with the terrain (see fig. 4-2).

The helmet is camouflaged by breaking up its shape, smooth surface, and shadow. Use of a helmet cover works best. In the absence of a helmet cover, mud can be irregularly blotched on the helmet to disguise its form and dull the surface. A helmet cover may be improvised from irregularly colored cloth or burlap to blend with the background. Foliage can be draped to prevent the visor of the helmet from casting a dark shadow across the face. Foliage should not stick up like plumes because any head movement will give away the position.

A small, thin bush in the shadow of a large bush makes a good observation point. Lone trees, rocks, fence corners, and outstanding landmarks are easily picked up by the enemy as obvious observation posts.

If camouflage clothing is not available, other available clothing can be attached in irregular splotches of appropriate colors.

Exposed skin reflects light and attracts the enemy's attention. Even very dark skin will reflect light because of its natural oil. The buddy system is recom-mended when applying camouflage. Standard Marine Corps issue camouflage face paint sticks are two toned:

- Loam and light green for light-skinned troops, in all but snow regions.
- Sand and light green for dark-skinned troops.
- Loam and white for troops in snow-covered terrain.

Shiny areas (forehead, cheekbones, nose, and chin) are painted with a dark color. Shadow areas (around the eyes, under the nose, and under the chin) are painted

Figure 4-2. Avoid Contrasting Backgrounds.

with a light color. Skin that is exposed on the back of the neck and hands is painted with a two-color combination in an irregular pattern (see fig. 4-3).

When standard issue face paint sticks are not available, burnt cork, charcoal or lamp black can be used to tone down exposed areas of skin.

Mud is used only in an emergency because it changes color as it dries and may peel off, leaving the skin exposed. Since mud may contain harmful bacteria, mud should be washed off as soon as possible.

Any equipment that reflects light should be covered with a nonreflective material that aids in the concealment of the weapon (for example, black electrical tape or mud). The straight line of the rifle or other infantry weapons may be very conspicuous to an enemy observer. The barrel and hand guard should be wrapped with strips of contrasting colored cloth or tape to break the regular outline. Mud or dirt dulls the reflecting surface of the stock, barrel, and bayonet where coloring has been worn. Lamp black may also be used on metal parts. The function of the weapon must not impaired.

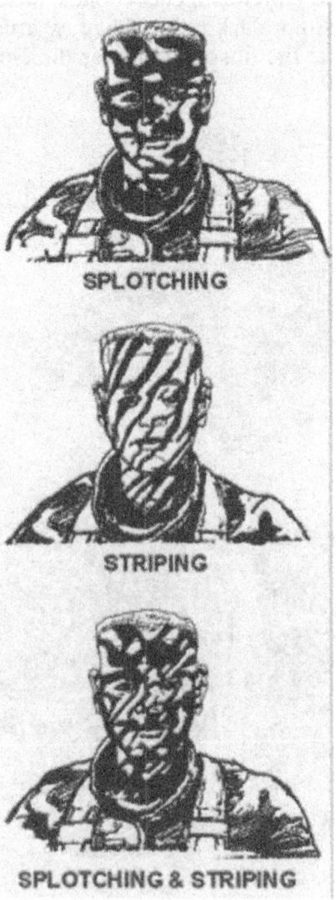

Figure 4-3. Face Camouflage.

If time, material, and surroundings permit, a ghillie suit should be constructed. (Refer to MCWP 3-15.3, *Scout Sniping*.)

Aids to Camouflage a Position

To successfully camouflage a position, the scout must remember to—

- Camouflage the position as soon as it is occupied.
- Avoid using too much material for camouflage. Even though natural materials are used, too much may make the object and its shadow stand out from its surroundings, thus attracting the attention of a hostile observer.
- Inspect completed camouflage work from the enemy's point of view to check effectiveness.

Continuous Camouflage

Camouflage around and on the scout's position must be maintained in a fresh condition as wilted and dead foliage can give the position away. If the mission dictates that the position should be occupied for longer periods, wilted foliage should be replaced during periods of reduced visibility.

4003. INDIVIDUAL MOVEMENT

Principles

The principles of individual movement are as follows:

- Scouts move from one concealed position to another. When not changing positions, they remain motionless.
- The scout's head is lifted slowly but steadily, without abrupt movements, to search for a new position.
- Scouts select the next stopping place before moving and ensures it is not contained by the enemy.
- Scouts change position on the run: springs up, runs with the body bent low, zigzags, quickly drops to the ground slightly to the right or left of the objective, then rolls or crawls to the desired position.

(Remember the phrase, "I'm up—he sees me—I'm down.")

Rushing

When starting from the prone position—

- Raise the head slowly and steadily and select a new position.
- Lower the head slowly, draw arms inward, cock right leg forward, and prepare to rush.
- Use one movement to raise the body by straightening both arms.
- Spring to your feet, stepping off with the left foot.
- Bend forward as low as possible when running. Never advance directly to the next position; always zigzag.

When hitting the deck—

- Stop.
- Plant both feet in place.
- Drop quickly to the knees and slide the hand to the heel of the rifle.
- Fall forward, breaking your fall with the butt of the rifle. (To confuse the enemy, roll over after hitting the deck and roll into firing position with feet, knees, and stomach flat on the ground.)
- Keep head down if you do not intend to fire.

When rolling over—

- Hit the deck and assume the prone position.
- Bring the rifle in close to the body, placing the rifle butt in the crotch.
- Roll over swiftly to confuse any enemy observers as to final intended location. Never reappear at the same place you went down.

Low Crawl

The low crawl is used when—

- Cover and concealment are scarce.
- The enemy has good observation over the area in which the scout is moving.
- Speed is not essential.

To perform the low crawl, keep the body as flat as possible against the ground. Grasp the rifle sling at the upper sling swivel. Let the balance of the rifle rest on the forearm and let the butt of the rifle drag on the ground. Keep the muzzle off the ground.

To start forward, push arms forward and pull right leg forward. To move forward, pull with arms and push with right leg. Change the pushing leg frequently to avoid fatigue.

High Crawl

The high crawl is used when—

- Cover and/or concealment are available.
- Poor visibility reduces enemy observation
- Greater speed of movement is required.

To perform the high crawl, keep body off the ground. Rest weight on forearms and lower legs. Cradle rifle in arms, keeping the muzzle off the ground. Keep knees well behind the buttocks to stay low.

Move forward, alternately advancing right forearm and left knee; then left forearm and right knee.

Movement Aids

Aids to movement include—

- Carrying only necessities. Additional weight causes premature fatigue and impedes free movement.
- Not disturbing birds or animals whose flight would betray your presence. If birds or animals are alerted, remain motionless under cover for a few minutes, as the enemy's attention may also be attracted.
- Moving during an incident that diverts attention, such as an airplane flight, a distant disturbance or sudden bursts of fire.
- Fog, smoke, or even light haze offer concealment for movement; however, the enemy may have thermoimagery and night vision devices. Therefore, darkness and smoke cannot be used as easily.
- Following a stream or road by staying as far away from them as possible while still keeping them in sight. Keep close to the dune line when moving along a beach.
- When moving through tall grass or similar growth, move when the wind blows, changing direction frequently. A straight route will be more readily noticed.
- Whenever possible, avoid areas of soft ground so as not to leave tracks.

- When crossing a road or water obstacle, choose crossing sites where the enemy's observation is restricted (an area in shadows or near a bend) and cross rapidly.

4004. ROUTE SELECTION

Prior to Movement

A scout and the immediate commander conduct a map reconnaissance before starting on a mission. This assists them in selecting the route according to available cover and concealment and any indicated enemy activity.

Prior to and during the course of the mission, move to an observation point to visually reconnoiter the terrain for movement and select the tentative route. It may be necessary to make wide detours around open spaces or those containing enemy patrols or other enemy activity.

Carefully study the country to be traversed and pay close attention to the general features, streams, ridges to be crossed, and their relation to the general direction to be taken (see fig. 4-4).

Make notes of terrain features and landmarks along the proposed route and rely on notes for guidance (see fig. 4-5 on page 4-6). Additionally, determine the compass direction and readings for each change of direction at the start. Finally, learn the location of unit boundaries and observation/listening posts as well as general location of other friendly or scouting parties. Be sure to avoid man-made and natural obstacles as they will slow progress and overall success of the mission. If possible, use the local populous as a source of intelligence. When returning to friendly lines, avoid using the same route.

En Route

En route, the actual advance will be a series of movements from one observation point to the next. The distance and route will depend on cover and terrain. Assess the cover, terrain, and any enemy or civilian activity to determine whether or not to modify the approach or return routes. Unless the mission requires it, avoid danger areas (for example, houses, villages, potential assembly or bivouac areas, roads, and streams) that may give away your position by being

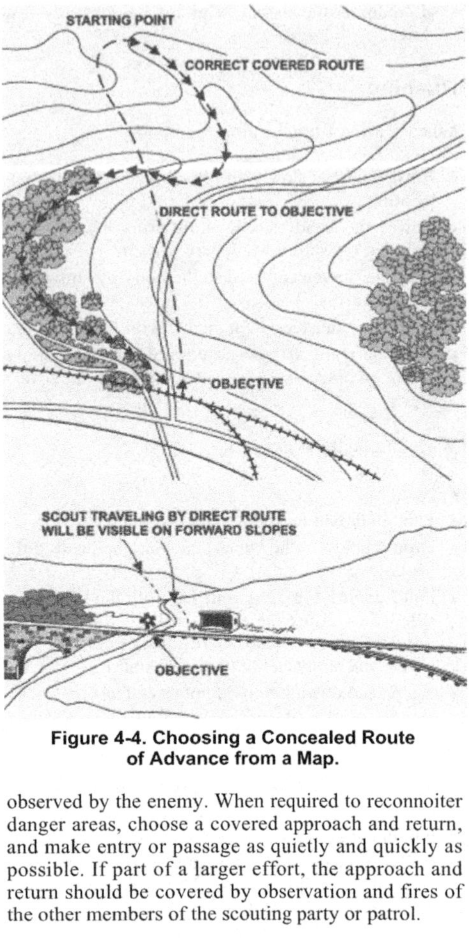

Figure 4-4. Choosing a Concealed Route of Advance from a Map.

observed by the enemy. When required to reconnoiter danger areas, choose a covered approach and return, and make entry or passage as quietly and quickly as possible. If part of a larger effort, the approach and return should be covered by observation and fires of the other members of the scouting party or patrol.

Stream Crossings

When the crossing does not appear to be held by the enemy, advance upon it rapidly. If there are two or more scouts, one crosses while the other(s) provide protection. Note the length, width, depth, and approaches to a crossing. Observe the condition of the road or trail that crosses the stream, and report on the suitability of the crossing for use by tracked and wheeled vehicles. If the crossing is under observation by enemy, seek another crossing site or dash across to avoid detection.

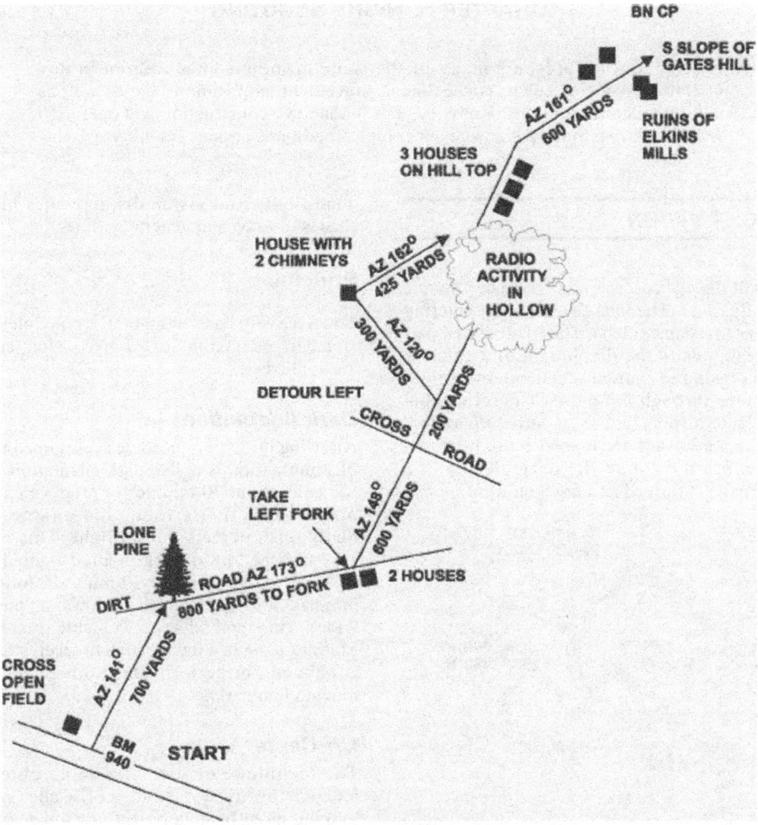

Figure 4-5. Proposed Route Sketch.

CHAPTER 5. NIGHT SCOUTING

Night scouting presents many of the same problems encountered in day operations-such as cover, concealment, movement, and camouflage-as well as additional considerations. Knowledge of human eye construction and operation will enable maximum advantage under night conditions or poor visibility.

5001. NIGHT VISION

Certain parts of the eye correspond to parts of a simple camera (see fig. 5-1). The lens focuses light entering the eye similar to a camera lens. The iris (colored part of eye) corresponds to the diaphragm of a camera, opening and closing to regulate the amount of light entering the eye through the pupil. The retina corresponds to camera film. Light rays strike the retina, form an image, and cause an impression to be transmitted to the brain through the optic nerve. In a camera, the image is formed and fixed on film.

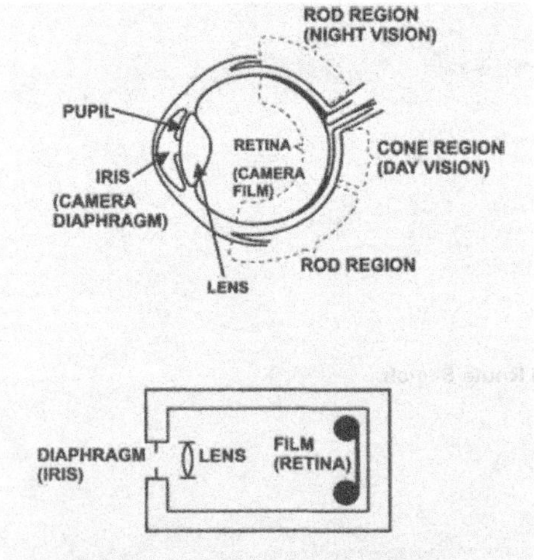

Figure 5-1. The Eye is Like a Camera.

The retina is composed of cone cells and rod cells, so-called because of their shapes. Cone cells distinguish color, shape, and sharp contrast. Because they are activated by light conditions, they are blind during periods of low illumination. Rod cells produce a chemical substance called visual purple that makes them active in darkness, low illumination or night conditions. Rod vision distinguishes black, white, shades of gray, and general outlines.

Principles

To effectively "see" at night, the principles of night vision dark adaptation, off-center vision, and scanning are applied.

Dark Adaptation

Allowing the eyes to become accustomed to low levels of illumination is called dark adaptation. It takes the rod cells about 30 minutes to produce enough visual purple to activate them and enable the eye to distinguish objects in dim light. This may also be accomplished by staying in a red-lighted area, or by wearing red goggles for 20 minutes, followed by 10 minutes in darkness (which allows the pupils to open wide). This method saves valuable time by allowing Marines to be in a lighted area to receive orders, check equipment, or perform some other function before moving into darkness.

Off-Center Vision

The technique of focusing on an object without looking directly at it is called off-center vision. When looking directly at an object, the image is formed on the cone region, which is not sensitive at night (see fig. 5-2 on page 5-2). When looking slightly to the left, right, above or below an object, the image is formed on the area of the retina containing rod cells, which are sensitive in darkness. The most sensitive area varies in individuals, but is usually found by looking 6 to 10 degrees away from an object; in effect, out of the corner of the eye (see fig. 5-3 on page 5-2).

Scanning

Off-center vision used to observe an area or an object is called scanning. When using rod vision, the visual purple in the rod cells bleaches or blacks out in 4 to 10 seconds and the object observed disappears. As the visual purple in the rod cells in one area bleaches out, the eyes must slightly shift to use fresh rod cells. Eyes

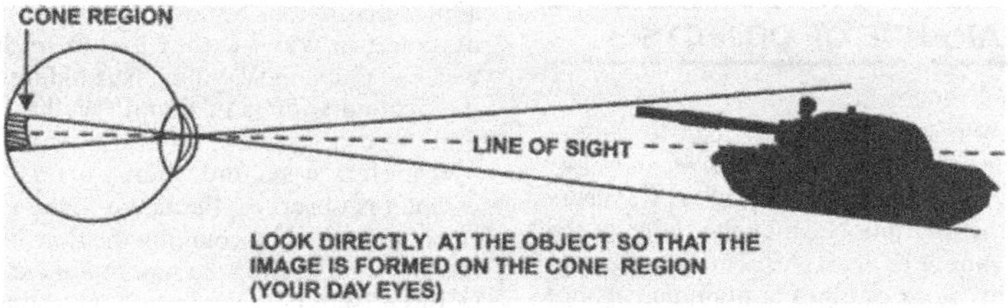

Figure 5-2. Day Vision.

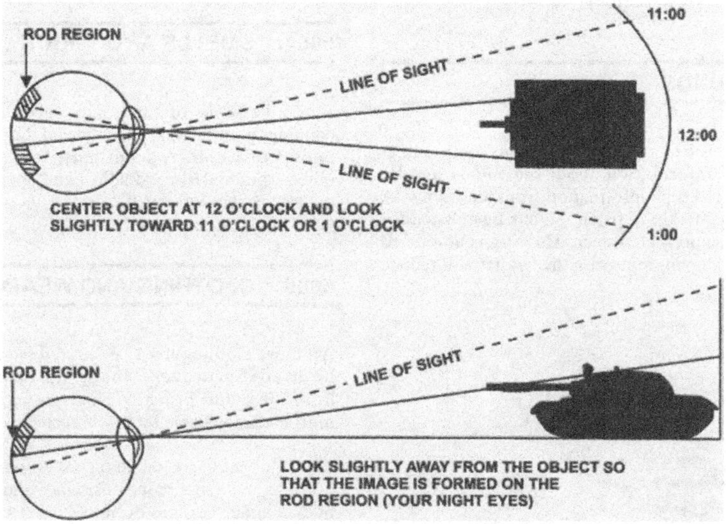

Figure 5-3. Night Vision.

should be moved in short, abrupt, irregular movements over and around the target (see fig. 5-4).

Preserving Night Vision

Night vision is quickly destroyed if bright light is allowed to enter the eye. When entering a lighted area or when observing in a temporarily lighted area (illumination, flares), one eye should be closed and covered to preserve its night vision. When the light goes off, fades or the lighted area is exited, the night vision retained by the protected eye enables it to see until the other eye adapts to the darkness. Red light helps preserve night vision, but like white light, it can be observed at long distances.

Factors that decrease night visual acuity include fatigue, lack of oxygen, long exposure to sunlight, alcohol, nicotine (within the past 48 hours), and age. When night vision has been attained, straining will not improve effectiveness; however, practice identifying objects at night will improve perception.

5002. APPEARANCE OF OBJECTS

Darkness not only makes objects difficult to see but also changes their appearance, distorts size, and blots out details. A tree visible against the night sky appears smaller than in the daytime because the twigs at the end of branches cannot be seen. A scout must train to identify objects by block outlines at night and cannot rely on details visible in daylight. Binoculars enlarge objects or parts of objects otherwise too small to be seen and help identify objects already spotted. Night observation devices increase night visibility and should be used whenever possible.

5003. SOUNDS

At night, sounds become very important. By listening, a scout gains information about the enemy and by exercising care, keeps information from the enemy. A scout stops frequently to listen. Scouts must listen for long periods in perfect silence. Hearing is amplified with the mouth open; removing the helmet will reduce sound distortion. Sounds are transmitted a greater distance in wet weather and at night than in dry weather and in the daytime. By holding the ear close to the ground sounds of people walking and vehicles moving can be heard. Sound travels approximately 370 meters a second. When a flash from a fired weapon is observed, the range to the weapons can be easily estimated by counting the time interval between the flash and hearing the report. For example, counting to three (one thousand one, one thousand two, one thousand three), indicates the distance is 1,110 meters. The cadence is determined by actual practice at known ranges.

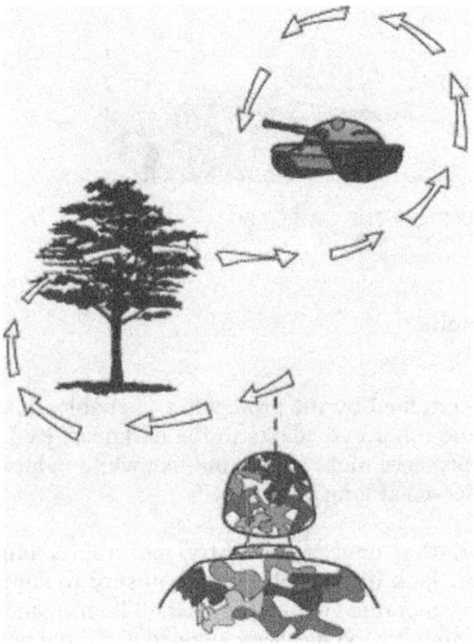

Figure 5-4. Scanning

5004. SMELLS AND TOUCH

A scout's sense of smell can warn of enemy fires, cooking, motor parks, gasoline and diesel engines, and bodies of water. A scout must feel and recognize objects in the dark, and adjust and operate equipment quietly by sense of touch.

5005. CLOTHING AND WEAPONS

All loose clothing must be secured (string or tape can be used) to prevent snagging on barbed wire, brambles, and brush. Helmet covers are worn to muffle sounds made by low branches.

The belt buckle should be turned around to the side in order to move in a prone position without scraping the buckle against stones or hard surfaces. Identification tags can be taped together to prevent rattling. Hands, face, and neck can be blackened so skin does not reflect light or appear as white spots in the darkness. (Refer to para. 4002.)

Scouts are normally armed with rifles. Rifle slings should be taped to prevent rattling. All weapons parts should be checked for glare elimination measures.

5006. CONCEALMENT

Although total darkness provides concealment, scouts must observe the same principles of concealment during moonlight conditions as in the daytime. Scouts should assume enemy employment of night observation devices and observe the principles of night movement such that presence will not be disclosed by noise when close to the enemy.

5007. AIDS TO NIGHT SCOUTING

Aids to night scouting include the following:

- Carry out scouting missions close to or within hostile positions on dark or rainy nights.
- Stifle a threatening sneeze by pressing fingers upward against the nostrils.
- Stifle a threatening cough by applying slight pressure with the finger on the Adam's apple.
- Stop a ringing sound that interferes with hearing by yawning.
- Speak softly rather than whispering if voice communication is necessary.
- Move boldly and rapidly when taking advantage of any sound-such as shelling, rustling wind or distant firing-to push forward if firing is taking place.
- Avoid shell craters and depressions in damp weather conditions if the enemy has employed chemical munitions.
- Move the eyes constantly; concentrating on one object too long will strain them.
- Take notice of the enemy's use of flares. When the enemy employs flares, few enemy patrols are apt to be out; when flares are not employed, the enemy's patrols are likely to be numerous.
- Drop to a prone position upon hearing a flare being fired and before it illuminates. Remain motionless while it is burning. If open or moving when a flare bursts in the air, freeze or drop quickly in the split second after the flare illuminates while the enemy is blinded. You are an easy target for the enemy if the flare bursts in the air or on the ground behind you. Never look at a flare. If you activate a trip flare, drop to the ground and crawl away from the illuminated area.
- Consider all patrols or individuals encountered as hostile until proven friendly. If encountering someone, crouch low, silhouetting the approaching person against the sky. At the same time, make yourself an indistinct target in case the person encountered is an enemy.
- Return fire only to avoid capture if fired on when close to enemy positions.

5008. AIDS TO NIGHT MOVEMENT

- Aids to night movement include the following:
- Move silently.
- Advance in stealthy legs. Each leg should follow some terrain feature that serves as a guide. When there are no terrain features to serve as guides, move in a straight or nearly straight line from one defined point to another, or maintain direction by using a compass.
- Avoid running, except in an emergency.
- Take advantage of sounds that may distract the enemy.
- Fall silently without making an outcry.

Walking

When walking at night—

- Place the heel down first. Balance the weight of the body on the rear foot until a secure spot is found.
- Lift the forward foot high to clear any stiff grass, brush, or other obstruction.
- Continue to balance body weight on the rear foot, lower the forward foot gently, toe first, to explore the ground for objects that might make noise. Step over fallen logs and branches, not on them.
- Lower the heel of the forward foot slowly; gradually transferring body weight to that foot.

Creeping

The low crawl and high crawl are not suitable at night when very near the enemy because an easily heard shuffling noise results. Creeping is the recommended method of movement:

- Creep at night on the hands and knees.
- Use your hands to feel for twigs, leaves or other substances that might make a noise. Clear a spot to place your knee. Keeping your hand at that spot,

bring your knee forward until it meets your hand. Then place your knee on the ground and repeat the action with the other hand and knee.

- Lay the rifle on the ground at your side and clear an area for it. Lift the rifle up and move it forward. Movement is slow and tedious, since it must be done silently.

Hitting the Deck at Night (Right-Handed Shooter)

To safely hit the deck at night from the standing position—

- Advance your left leg, place the butt of the rifle in your right armpit with the hand remaining on the pistol grip, and grasp it with the right hand at the balance.
- Quietly drop down on the right knee and left hand.

- Move the left leg carefully to the rear, and then move the right leg to the rear.

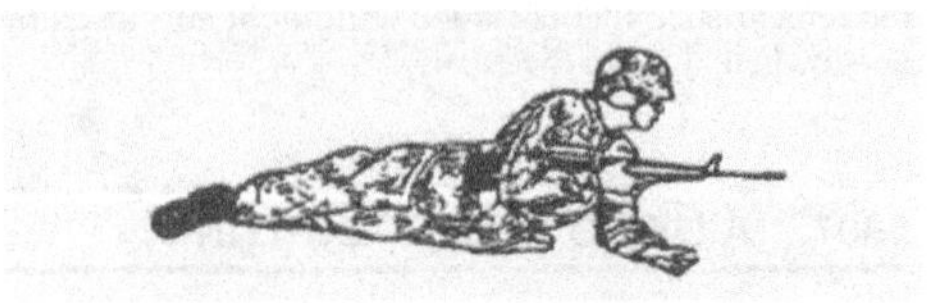

- Lie flat on the ground, or take up a firing position if necessary.

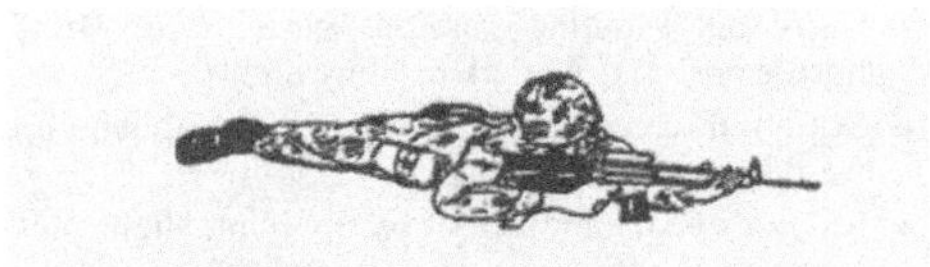

Wire Obstacles

A mission often requires a scout to pass through and work behind enemy positions. To accomplish this, the scout must be able to quietly pass through enemy wire obstacles and cross trenches. Cutting a gap in wire is time-consuming. If possible, walk over the low bands of enemy wire and crawl under the high bands (see fig. 5-5). Avoid movement along wire barriers, as enemy

Figure 5-5. Crossing Wire Silently at Night.

covering fires are generally planned parallel to them to take advantage of canalization and enfilade fire.

To step over low wire at night, crouch low to view the strands against the sky. Grasp the top strand with one hand; with the other hand, reach forward and feel for a clear spot for foot placement without stepping on other strands or any object apt to make a noise. Raise the body up, still grasping the top strand of wire. To avoid catching the foot in another strand, lift the foot up and over, passing it close to the hand grasping the wire.

If a high wire obstacle is encountered at night and wire cutters are not available, pass under the wire with your back on the ground. Grasp the lowest strands in your hands and hold them clear of the body while you slide under them.

When cutting wire and working solo, cut a wire near a post (see fig. 5-6), then dispose of all but one loose end. Grasp the wire close to a post and cut between your hand and the post, muffling the sound and keeping the loose wire in your grasp. When cutting wire in tandem, one firmly holds the wire with the hands positioned close to the cutters, in order to muffle the sound and prevent the loose ends from flying back, while the other one cuts. In both instances, the loose ends of the wire are bent back to form a passage.

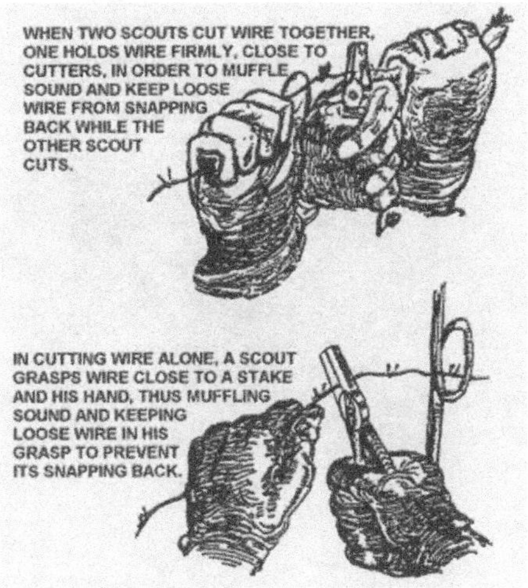

Figure 5-6. Cutting Wire Silently at Night.

Wrap a sandbag around the wire cutters and wire to deaden the sound.

Do not cut a complete gap in the wire; cut only the bottom wire(s). Leave the top wire(s) intact to lessen the chance of discovery by the enemy.

Crossing Trenches

Before approaching a trench, wait outside the trench for awhile and listen. Do not enter or cross a trench near its junction with a communication trench. Crawl silently up to the edge of the trench and look into it. Remove all loose dirt and rocks from the edge. If it is a narrow trench, spring up and jump across, sinking quietly to the ground on the other side and remaining there a moment to listen before proceeding. If the trench is wide, climb silently and slowly down into it and out the other side, using the revetment for support (see fig. 5-7). Do not enter enemy trenches unless it is absolutely necessary in order to accomplish the mission. Ordinarily, work is better accomplished from outside the trench. Sentries usually pay more attention to sounds in front of them; therefore, if it is necessary to enter a trench, cross it first at the place where enemy observation is restricted, then approach from the rear.

5009. LOCATING AND PLOTTING THE ENEMY AT NIGHT

For night work, a scout must understand the use of a lensatic compass. (Refer to para. 2002.) Using the lensatic compass, the scout can guide the platoon into position, locate adjoining elements of the command, keep direction when on patrol, determine the location of gaps in the enemy wire and the position of enemy out guards.

Locating Gaps in Enemy Wire

When searching for gaps in enemy wire, at least two lensatic compasses are needed: one to register the gap in the wire and the other for navigation. When a gap in enemy wire is located, lie outside the gap, keeping a distance of 10 meters from the barbed wire. Sight with the lensatic compass on a prominent point on the skyline in line with the gap. In selecting the prominent point in the skyline, pick one that appears on the map (i.e., hill mass, house, road junction). If the only prominent point available is one not identifiable on a map (i.e., a tree, destroyed vehicle, enemy position),

Figure 5-7. Crossing Trenches Silently at Night.

use it. The exact location of this point can be fixed the next day by visual reconnaissance of the area from an observation point. When the needle rests—

- Clamp it in place by lowering the eyepiece to the closed position.
- Rotate the movable bezel ring until the luminous line is directly over the north end of the needle. The azimuth of the gap from the prominent point is now registered.

The compass is carried back without further adjustment. The azimuth setting can be recorded later on a map (see fig. 5-8).

Locating Enemy Out Guards

At night, locate enemy guards by their sounds and failure to maintain light discipline. As sounds of the enemy are heard, and/or observations of the enemy made, shoot an azimuth with one compass. When the needle comes to rest, clamp the compass by lowering the eyepiece to the closed position.

Figure 5-8. Locating a Gap in Enemy Wire at Night.

Note the time and nature of each sound, the estimated distance, and which compass was used to fix the location. Plot this data on a map then wait until the debrief to turn in the notes and compasses. See figure 5-9.

Figure 5-9. Locating Enemy Outguards at Night.

50010. ROUTES OF MOVEMENT

Prior to beginning a night missions, a scout studies the ground in detail from an observation point, air photos, and a map during daylight. The route of advance should be below the skyline. Avoid becoming a silhouette (see fig. 5-10).

Unless the moon is bright, avoid, if possible, passing through woods, ditches, ravines, and brush, because noises of movement may lead to discovery. If the enemy is known to have night observation device capability, avoiding these kinds of terrain may not be possible. To avoid enemy ambushes, return by a different route from the advance and change routes on successive nights.

Figure 5-10. Difference Between Correct Routes by Day or Night.

CHAPTER 6. OBSERVING AND REPORTING

When sent out on a mission, a scout's duties are to observe and report. the first five chapters covered scout protection measures and movement in enemy territory. This chapter discusses performing the assigned mission and reporting the mission after its completion.

6001. OBSERVATION POSTS

Positions

The following guidelines apply to observation posts. See figure 6-1.

- When selecting observation posts, scouts should chose the least prominent position.
- Scouts may occupy one or more observation posts.
- An observation post should not be manned for more than 24 hours.
- A selected observation post should be observed for 10 to 15 minutes to ensure it is not occupied.
- Scouts move to the chosen observation post by a concealed route.
- If the post is located on a hill, crawl to a position where the skyline is broken.
- If a tree is used, the position should have a background so as not to be silhouetted against the sky while climbing or observing.
- When leaving the observation post, a different route from that of the approach should be used.
- If a radio is used, its antenna should be located to provide clear communication to the controlling commander but masked from enemy observation and direction-finding equipment. Upon departure, scouts should remove the antennae from the observation post so as not to give away the position.

Observing

Using all senses available, be particularly alert for movement, objects, sounds, and smells inappropriate to the surroundings. While observing, avoid all unnecessary movement. If observing from a building, keep back from doors and windows.

In daylight, look first at the ground nearest you. Begin observing close to your post and search a narrow strip 50 meters or less deep, going from right to left parallel to your front; then search from left to right a second and similar strip farther away but overlapping the first. Continue to observe until the entire field of view has been searched (see fig. 6-2 on page 6-2).

At night, use a night observation device. If one is not available, search the horizon with short, jerky movements, and short pauses. Look a little to one side of an object and then to the other. Lower the head close to the ground to view the object more clearly. Use low-powered field glasses to increase sight range.

Figure 6-1. Method of Approaching an Observation Post.

Figure 6-2. Method of Searching Ground.

6002. REPORTING

It is imperative that the scout accurately and completely report who, where, when, and what was observed to the proper authority upon mission completion. In most cases, a scout will not be equipped with a radio.

Verbal Reports

Verbal reports should be made when writing is impractical, when the information is not complicated or when the enemy is likely to intercept a messenger carrying a written message.

Written Messages

Written messages, preferred to verbal reports, are recorded in message book blanks issued for that purpose (see fig. 6-3) and delivered to a higher authority as soon as possible. The NATO spot report (SPOTREP) should be reviewed as the written message is a NATO format.

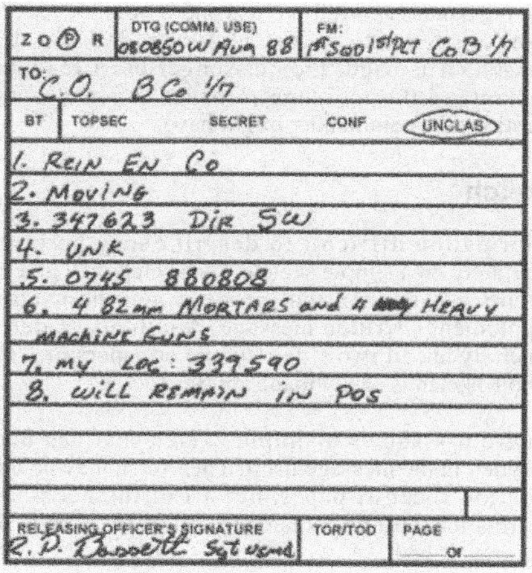

Z O (P) R	DTG (COMM. USE) 080850W Aug 88	FM: 1st Sqd 1st Plt Co B 1/7
TO: C.O. B Co 1/7		
BT TOPSEC SECRET CONF (UNCLAS)		
1. Rein En Co		
2. Moving		
3. 347623 Dir SW		
4. Unk		
5. 0745 880808		
6. 4 82mm Mortars and 4 Heavy		
Machine Guns		
7. My Loc: 339590		
8. Will Remain in Pos		
RELEASING OFFICER'S SIGNATURE R. P. Bassett Sgt USMC	TOR/TOD	PAGE ___ OF ___

Figure 6-3. Sample Field Message.

The message body is brief, accurate, and clear; facts and opinions are distinguished. If secondhand information is reported, its source is included. Reports include all information of value, first about the enemy, and then about the originator. Information about the enemy should cover—

- Size and/or strength.
- Actions or activity.
- Location and direction of movement.
- Unit identification. (The designation of the enemy unit may be derived from unit markings, uniforms worn or through prisoner interrogation.)
- Time of observation.
- Equipment and weapons.

Messages are printed in block letters. Individual items of information are numbered and separated into paragraphs. If doubt exists as to message receipt by the commander, a summary of its contents is included in the next message. Information about the originator or writer should cover—

- Location at the time of enemy observation (reference to an important terrain feature, by map coordinates, by the back azimuth from each of two definitely located points, or the back azimuth and distance from one known point).
- Intentions. (Remain in position? Continue on the mission? Take other action?)

The message is carefully reread and if possible read by another person to ensure understandability. If a messenger is used, the messenger must read and understand the message in order to answer any questions the commander might have.

Sketch

Information difficult to describe may be given accurately on a simple sketch. The sketch may give all the necessary information or it may be used to supplement a written message. A military sketch is generally one of two types: simple or panoramic. The simple sketch is easily made and read.

Figure 6-4 shows a simple sketch that has been included in the message itself. The sketch may be on a separate sheet of paper, but all of the necessary information must be contained in either the sketch, the message or both.

The panoramic sketch is a picture of the terrain's elevation in perspective, as seen from one point of observation. Although a panoramic sketch is not difficult to create, skill and training are necessary to enhance usefulness. Figure 6-5 on page 6-4 illustrates how to make a panoramic sketch.

Overlay

The same information sent back to higher headquarters on the sketch may be sent on an overlay, if the sender and the person to whom the message is to be sent have copies of the same map. Figure 6-6 on page 6-6 illustrates a simple overlay. The overlay is drawn on transparent paper as follows:

- Orient the map and place it on a hard, flat surface.
- Place the transparent paper over the part of the map of the object or information to be transmitted and hold the paper in this position.
- Orient the overlay to the map by tracing in the intersecting grid lines at two opposite corners of the overlay. Write the correct number designation on the overlay. The cross made by the intersection is called a *tick mark* and enables the receiver to locate the exact area on the map covered by the overlay.
- Sketch the object seen or the information to be transmitted on the tracing paper (the sheet on top of the map) in the exact location it would appear on the map (the sheet underneath the tracing paper).

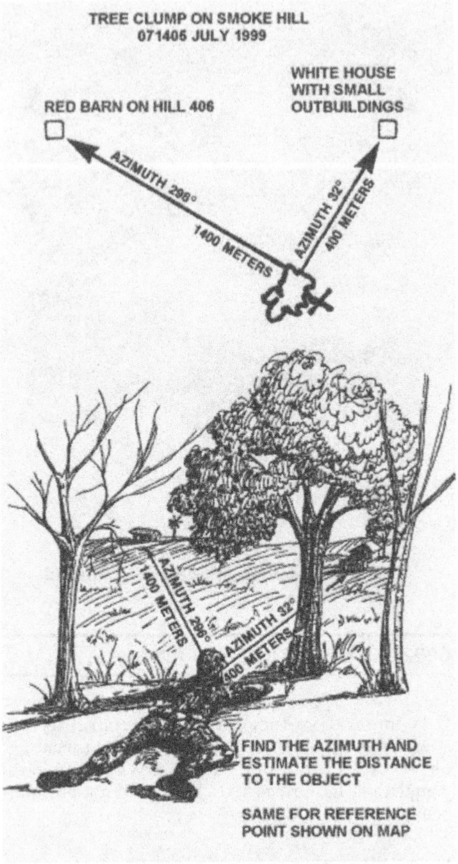

Figure 6-4. Making a Simple Sketch.

Explanatory notes are annotated in the overlay's margin, arrows point to the objectives.

- Indicate with an "X" and an appropriate explanation the position from which the observer saw the object or obtained information.
- Include title and scale of the map from which the overlay was made, date and hour the information was obtained, and signature of the observer on the overlay in the lower right-hand corner.

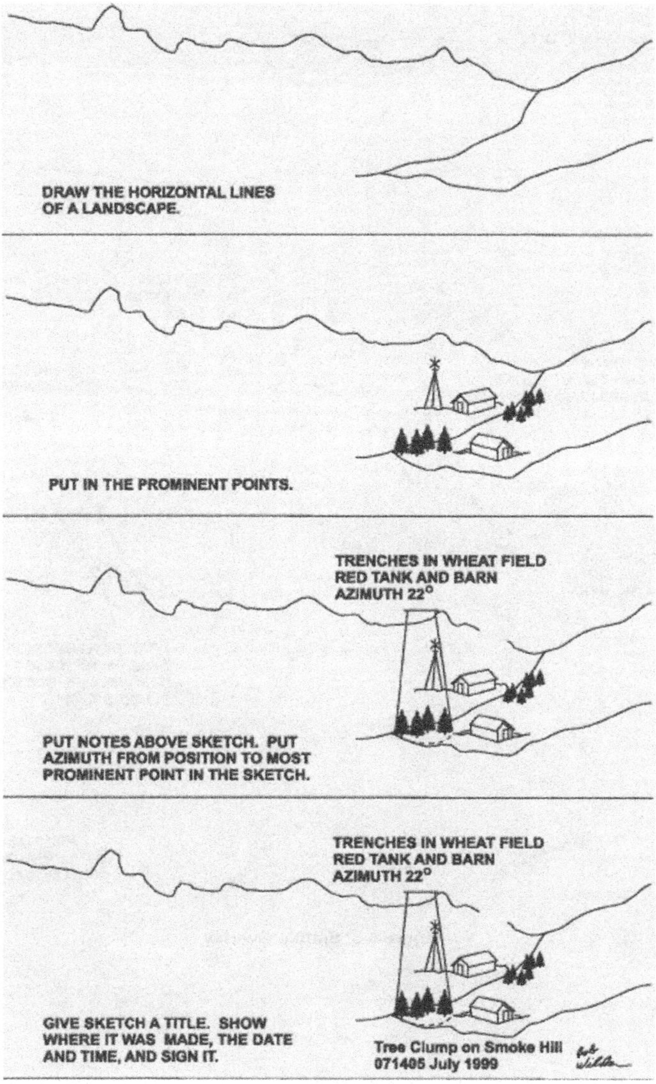

Figure 6-5. Panoramic Sketch.

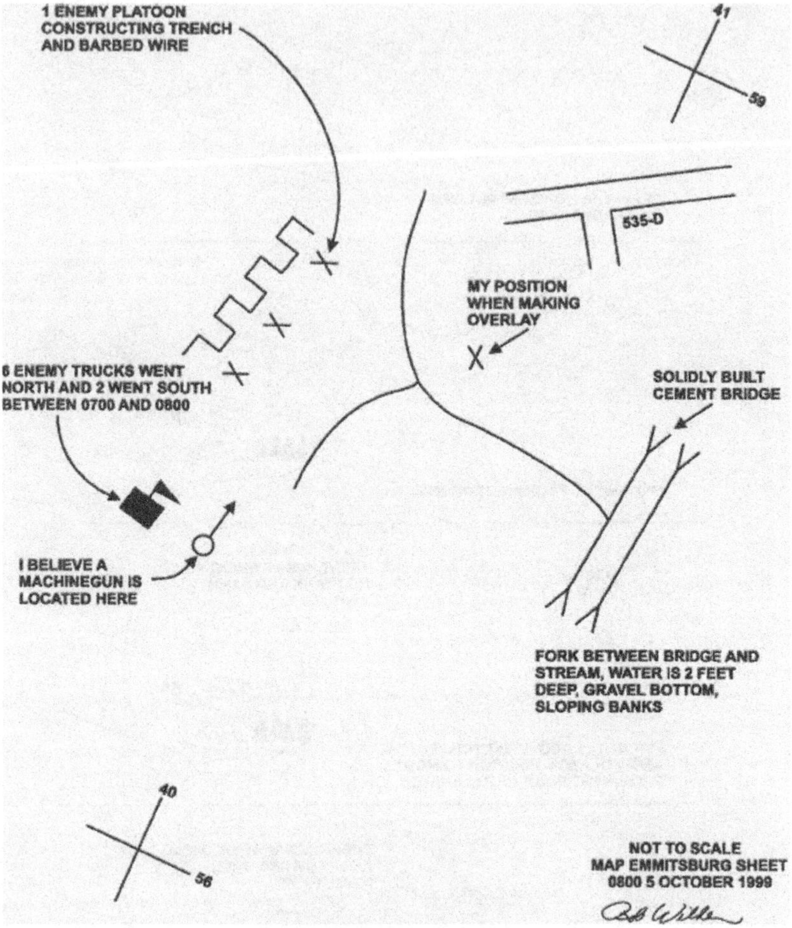

Figure 6-6. Simple Overlay.

CHAPTER 7. SCOUTING FIRE TEAMS

Upon completion of individual training, a Marine adapts this training to the functioning of the unit. The basic tactical unit for scouting is the fire team. Within the fire team, scouts normally work in pairs to furnish security and gather information. Information gathered by the individual scouts is passed to the fire team leader.

7001. POSITIONING

When a rifle platoon in the approach march is not preceded by adjacent units (left, right, front, and rear), it employs its own scouting elements. The scouting element for a platoon is usually one fire team; however, an entire squad may be used.

A fire team used as a scouting element is called a scouting fire team and is controlled by the platoon commander, assisted by the squad leader. A squad leader whose squad is providing the scouting fire teams normally marches near the platoon commander to assist in the control of the scouting fire teams.

A scouting fire team moves aggressively to cover the front of the advancing platoon and to locate the enemy's position(s). It generally moves in a wedge or skirmisher's formation. Normally, a scouting fire team is deployed on a frontage of 50 to 75 meters (10 to 17 meters between each individual scout). The entire squad may be employed to cover a wider frontage. The platoon commander coordinates the movement of the scouting fire team(s) so as to protect the main body of the platoon from enemy fire from points within 400 to 600 meters away, or in close terrain from points within the limits of enemy observation.

Scouting fire team(s) should have enough firepower to overcome resistance from small enemy advanced posts and patrols; the intent is to make enemy riflemen and machine gunners open fire and disclose their positions. Without scouts in advance, the platoon may move into areas where enemy fire may prevent further advance or maneuver and inflict heavy casualties (see fig. 7-1).

Scouting fire teams are covered by the platoon or, when the platoon is masked, the fire team leader ensures individual scouts are maneuvered and coordinated so that the fire team covers its own advance. The fire team leader constantly watches for signals from the platoon commander and remains in visual contact at all times.

The distance between the scouting fire team and the platoon is terrain-dependent. The scouting fire team should not be beyond visibility of the platoon. In open terrain, the platoon commander usually directs the scouting fire team to move by bounds along a succession of locations designated by the platoon commander as intermediate objectives.

Individual scouts should advance as stealthily as possible, while remaining consistent with their mission of reconnaissance to the front, taking advantage of cover without delaying the advance. An occasional glimpse of scouts constantly advancing over a wide front can make the enemy uneasy. It is this activity, and not the target the scouts offer, that may cause the enemy to open fire and disclose its location.

When fired upon, scouts must drop to cover and return fire only when necessary to complete the mission. If

Figure 7-1. Position of Scouts Preceding an Attacking Platoon.

necessary, one scout reports back to the fire team leader who informs the platoon commander.

7002. LOCATING ENEMY POSITIONS

The key terrain of defensive positions are those points that afford extended observation over the ground where the attack must advance. The enemy will place machine guns and infantry to defend critical points.

Members of a scouting fire team preceding an attacking platoon identify the probable enemy infantry and machine gun positions (see fig. 7-2). They use concealment and cover to conduct their advance in order to discover the exact location of enemy positions.

7003. ACTION WITH AN ATTACKING PLATOON

The scouting fire team reconnoiters to the front of the advancing platoon. As soon as the scouting fire team leader indicates the area is secure, the platoon advances and the scouting fire team moves forward. Squads within the platoon advance by bounds; at least one squad is positioned to support the other(s) by fire. Successive positions along the line of advance are selected and designated by the platoon commander as intermediate objectives, and reconnoitered by the scouting fire team before occupation. By conducting proper reconnaissance, surprise by the enemy or movement in the wrong direction may be prevented.

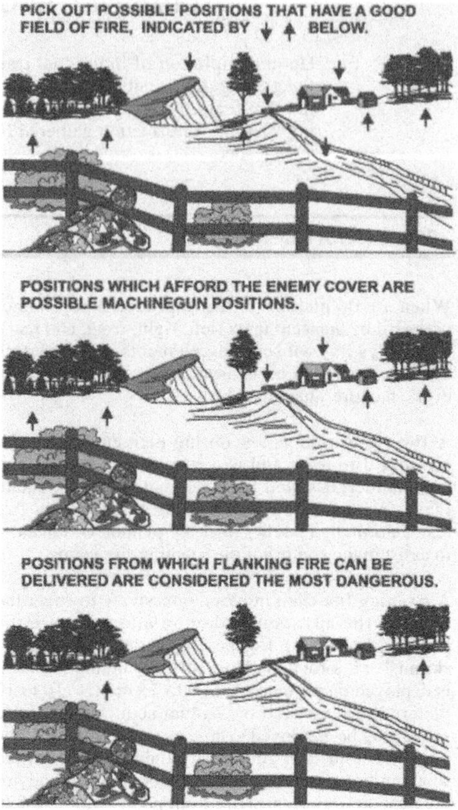

Figure 7-2. Assessing Probable Enemy Machine Gun Positions.

Movement

The distance between the scouting fire team and the front of the main body of the platoon is dependent on the mission, enemy, terrain and weather, troops and support available, and time available (METT-T). In close terrain, such as dense woods, the scouting fire team's movements closely resemble those used for night operations. In approaching houses, woods, and villages, one scout of each pair covers the other while the latter reconnoiters (see fig. 7-3).

A scouting fire team moves forward aggressively to cover the front of the advancing platoon, usually adopting either skirmisher's or wedge formation in order to be prepared to go into action immediately and to cover a wide frontage of 50 to 75 meters.

As a scouting fire team advances in open terrain, it is supported when possible by elements of the platoon; in close terrain, by mutual support within each fire team. Mutual support within the fire team is accomplished by the fire team leader and the automatic rifleman forming a team that supports by fire the advance of the rifleman and assistant automatic rifleman until they reach a location designated by the fire team leader. The rifleman and assistant automatic rifleman then support by fire the movement of the team leader and automatic rifleman. These successive points to which the sub elements of the fire team move are normally designated as fire team intermediate objectives by the fire team leader. The fire team leader sets as many fire team intermediate objectives as necessary to maintain mutual support within the team. This process is repeated until the team can be covered by other elements of the rifle platoon.

A scouting fire team takes advantage of available cover and concealment without delaying its advance. The orders of the platoon commander govern the distance at which it precedes the platoon. The terrain and the probable position of the enemy affect the scouting fire team's distance in front of the platoon. It may be as much as 400 to 600 meters in advance of the platoon. In open terrain, the platoon commander usually directs that the scouting fire team move by bounds to a succession of intermediate objectives. In close terrain or conditions of limited visibility, the scouting fire team is normally ordered to precede the platoon at the limit of visibility, maintaining visual contact with the platoon commander.

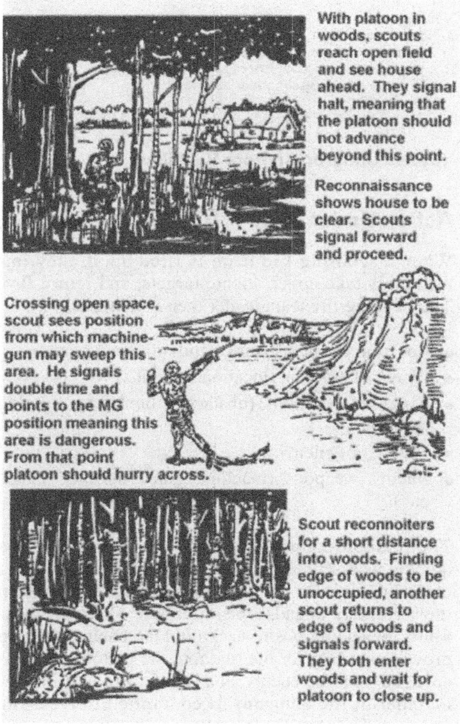

Figure 7-3. Conduct of Scouts Preceding a Platoon.

Action in Woods

When a scouting fire team is directed to advance over open ground to the edge of a woodline, two members of the team, preferably the rifleman and assistant automatic rifleman, reconnoiter inside the woodline while the remainder of the fire team covers them. It is not recommended they separate until finished with their reconnaissance of the far side of the danger area. Both members staying together can cover the same area using a zigzag reconnaissance and they are better equipped to overcome any opposition.

In heavy underbrush and/or poor visibility, the rifleman and assistant automatic rifleman proceed into the woodline together for 50 to 60 meters. The two then separate, searching out either flank to the first high ground or limits of observation, probably 50 to 100 meters (see fig. 7-4 on page 7-4). After the initial search and out posting the limit of advance, the remaining scout signals the fire team forward.

In light underbrush and/or good visibility, the assistant automatic rifleman remains at the edge of the woodline while the rifleman searches the woodline. (see fig. 7-5 on page 7-5). The rifleman searches the woodline in a zigzag pattern, reports back to the assistant automatic rifleman, then moves to an outpost position at the limit of advance. In turn, the fire team leader signals the platoon commander that it is safe for the platoon to move forward.

The fire team leader then moves the remainder of the fire team into the woods, joining up with the forward scout manning the outpost. The scouting fire team occupies and holds a line 50 to 75 meters within the woods and observes toward the direction of movement until the platoon closes up. The scouting fire team leader awaits further word from the platoon commander before moving the team further into the woods.

When directed, the scouting fire team leader moves the team forward until they reach the far edge of the woods. The team is held at the edge of the woods and the fire team leader notifies the platoon commander of the situation. The platoon commander moves the platoon to a position where it can cover the scouting

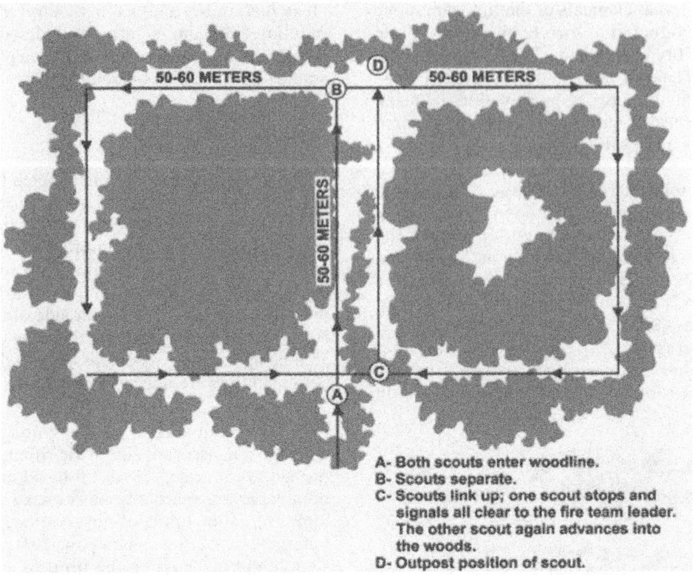

Figure 7-4. Searching Edge of a Woodline (Dense Underbrush/Poor Visibility).

fire team as it exits the woods and directs the team leader to move out and continue the scouting mission.

A scouting fire team passing through woods ahead of its unit maintains a distance allowing visual and oral communications. If an obstacle is encountered, reconnaissance to its front and flanks must be carried out. When advancing along a road or path, scouts precede the platoon to provide necessary protection and to prevent surprise fire on the platoon. When crossing a road or path, they reconnoiter well to the flanks before signaling "all clear" to the platoon.

The scouting fire team will not exit the woods until the arrival of the platoon commander, who will then be given an opportunity to alter the disposition or direction of march. The point where the platoon exits the woods is considered a danger area where the platoon is vulnerable to enemy fire. The scouting fire team is sent ahead to reconnoiter the danger area, as well as the next area to be occupied by the platoon. They signal back whether conditions require a halt, an advance or a quick rush across the open area. The scouting fire team leader must be continually on the lookout for signals from the rear.

Action Under Fire

When a scouting fire team is fired upon, they immediately take cover, locate targets, and return fire. The scouting fire team leader then determines—

- Location of enemy (range and reference points).
- Extent of position (location of flanks).
- Types of positions (obstacles, bunkders, fighting holes, etc.).
- Number of enemy.
- Enemy weapons (machine guns, mortars, tanks, etc.).

The platoon commander assesses the situation as quickly as possible based on the limited information obtained. Usually the platoon commander brings up the remaining squads, sets up a base of fire, and assaults the enemy position. Should the enemy position prove too strong for the platoon, the platoon remains engaged with the enemy as a base of fire until the remainder of the company is committed to clear the enemy resistance.

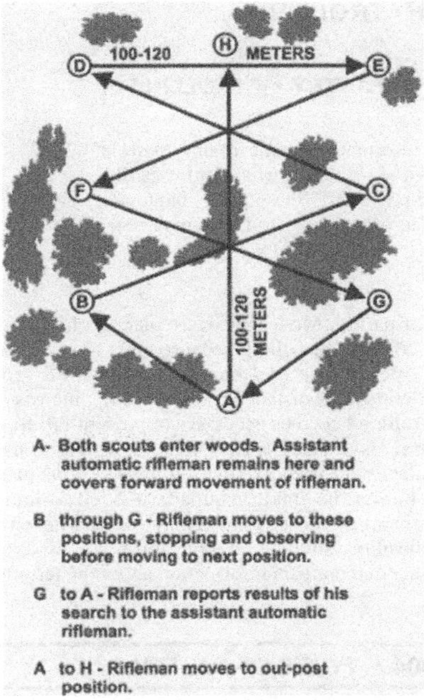

Figure 7-5. Searching Edge of a Woodline (Light Underbrush/Good Visibility).

7004. ACTION WITH AN ENVELOPING UNIT

When a platoon is given the mission to envelop an enemy position, a scouting fire team is employed for protection and reconnaissance in the same way as when the platoon is advancing in the approach march.

PART 2. INFANTRY PATROLLING

CHAPTER 8. FUNDAMENTALS OF INFANTRY PATROLLING

This chapter begins Part 2 and provides basic information about infantry patrols; specifically, their purpose, types, and missions. Infantry patrol training is also addressed and keys to successful patrolling are presented. Subsequent chapters of this part cover patrol organization, preparation, movement, and reconnaissance actions.

8001. DEFINITIONS

A patrol is a detachment of ground, sea or air forces sent out for the purpose of gathering information or carrying out a destructive, harassing, mopping-up or security mission (Joint Publication [JP] 1-02). The mission to conduct a patrol may be given to a fire team, squad, platoon or company.

8002. RELATION OF PATROLLING TO SCOUTING

Each patrol member must be knowledgeable in the principles of scouting and maintain membership of a larger team. To develop the teamwork skills required among the members of a patrol, additional training beyond the basic principles is necessary to become a well-trained scout. A patrol member must respond quickly to the decisions and orders of the patrol leader. There must be complete confidence among all members of the patrol and the confidence that they, as a team, will be successful in their mission.

8003. PURPOSE

A commander must have current information about the enemy and the terrain in order to employ the unit effectively. Patrols are an important means of gaining this information and are used to destroy enemy installations, capture enemy personnel, perform security missions or prevent the enemy from gaining information. Modern warfare places a high premium on effective patrolling because units have larger areas of operations and can be threatened from all directions. As distances between units increase, more patrolling becomes necessary to prevent infiltration by guerrillas or small enemy units, as well as to maintain contact with friendly adjacent units. Active patrolling by numerous small groups is needed to locate the enemy and gather information on the enemy's disposition, strength, morale, and weapons, as well as gather and confirm information about the terrain.

8004. TYPES OF PATROLS

Classification as to Mission

Reconnaissance Patrol

Reconnaissance patrols gather information about the enemy, terrain or resources. Relying on stealth rather than combat strength, they gather this information and fight only when necessary to complete the mission or to defend themselves. The distance covered by reconnaissance patrols varies based on the terrain and mission. The squad is ideally suited for reconnaissance patrol missions because of its relative small size and its experience of working together.

Combat Patrol

A combat patrol is a fighting patrol assigned missions that require engagement with the enemy in combat. Larger and more heavily armed than reconnaissance patrols, combat patrols have a mission to capture enemy documents, provide security, and capture or destroy enemy equipment and installations. Such action is ordinarily followed by a return to friendly positions. Regardless of the mission, the patrol reports any information concerning the enemy and terrain acquired during the accomplishment of the assigned mission. There are four types of combat patrols: raid, contact, ambush, and security (normally conducted by a Marine rifle platoon). A rifle platoon reinforced with crew-served weapons is normally considered the minimum size for contact, economy of force or ambush patrols. In some situations, such as the capture of a small enemy outpost, a rifle platoon could conduct a raid. However, a raid is a complex mission and, due to the organization of a raid force (command, reconnaissance, assault, support, security, and reserve elements), a rifle company is normally the smallest force assigned to a raid.

Classification as to Means of Movement

Foot Patrols

Movement by foot is the most common means; however, there are inherent disadvantages. Foot patrols travel slowly and carry limited quantity and types of equipment and supplies. Range and area coverage is relatively restricted. Foot patrols also have apparent advantages in that they have fewer restrictions as to terrain that can be covered; are more difficult for the enemy to detect; provide thorough coverage within limits of range; and are generally not inhibited by weather.

Motorized Patrols

Where terrain and road networks permit, a motorized patrol overcomes the inherent disadvantages of the foot patrol. Mechanized forces require patrolling units that can keep pace with them. However, motorized patrols are restricted to certain types of terrain, and tend to bypass areas that may be advantageous to and occupied by enemy infantry.

Waterborne Patrols

Waterborne patrols move over seas, lakes, rivers and streams, canals, and other inland waterways. The water is either used as a medium of entry to an objective area or is the actual patrol route. Waterborne patrols are limited by the location of water routes in the terrain and tend to bypass areas that may be advantageous to and occupied by the enemy.

Helicopterborne Patrols

Where terrain is extremely difficult or the enemy situation precludes the use of vehicle or motorized patrols, helicopterborne patrols are a method or means to conduct a patrol.

8005. TRAINING

Training is essential to successful patrolling. Premature and unordered actions by members of the patrol destroy coordination and control. Leaders are trained to issue their orders calmly to inspire confidence and discipline, and to avoid misunderstanding. Patrol members must work together and fight as a team. Training should develop the following skills:

- Expertise in handling individual and special weapons, and familiarity with enemy weapons that may be captured.
- Recognize camouflaged personnel, equipment, and defensive positions; ability to pick up fleeting targets fire the rifle from any firing position.
- Understand fire discipline and, after weapon firing, immediately change position (see fig. 8-1).
- Quick and accurate observation skills, and the ability to recall and transmit clearly and briefly, both orally and in writing.

Figure 8-1. Changing Position After Firing.

- Recognize and quickly respond to improvised signals, visually or by sound.
- Ability to swim with weapon and equipment.
- Use issued or improvised camouflage suits and garnish helmet in order to blend with the surroundings. Smudge face, hands, and any bright surfaces of weapons and equipment with some substance, such as mud or charcoal, to prevent the reflection of light.
- Silence self, equipment, and weapon.
- Use antimalarial and water purification tablets.
- Acclimation to temperature extremes.
- Develop a sense of direction and learn how to follow a course by compass, stars, sun, flow of streams, prominent terrain features, and by observing other natural phenomenon. Learn to determine the distance traveled from a known point and to keep a record of azimuths and the distance traveled on each azimuth (dead reckoning).
- Call for and adjust indirect fire assets.
- Familiarization with all communications assets and the use of field expedient antennae.

8006. KEYS TO SUCCESSFUL PATROLLING

Regardless of the category or means of conducting a patrol, the keys to successful patrolling are—

- Detailed planning. Every portion of the patrol must be planned, all possible contingencies considered.
- Productive, realistic rehearsals. Each phase of the patrol is rehearsed, beginning with actions in the objective area. Similar terrain and environmental conditions are used when conducting rehearsals.
- Thorough reconnaissance. Ideally, the patrol leader will physically conduct a reconnaissance of the route and objective. Photographs and/or maps will be used to supplement the reconnaissance.
- Positive control. The patrol leader must maintain positive control, this includes supervision during patrol preparations.
- All-around security. Security must be maintained at all times, particularly near the end of the patrol where there is a natural tendency to relax.

CHAPTER 9. PATROL ORGANIZATION

Organizing a patrol is a two-step process: the general organization of the entire patrol and the task organization of various patrol elements. Normally, the nature of patrolling does not permit long preparation periods and rehearsals to specifically build a unit for each mission. Accordingly, the patrol leader must combine unit integrity considerations with proven concepts of patrol organization.

9001. GENERAL ORGANIZATION

The patrol leader establishes a patrol headquarters and elements to accomplish the mission.

The headquarters is composed of the patrol leader and the personnel who provide support for the entire patrol, such as a forward observer, corpsman, and radio operator.

The major subdivisions of reconnaissance and combat patrols are elements. The existing infantry structure (squads and fire teams) is reinforced as required. For example, a reinforced platoon tasked to conduct a combat patrol that will raid an enemy outpost could be organized as follows:

- Platoon headquarters (command element):
  - Patrol leader (platoon leader).
  - Assistant patrol leader (platoon sergeant).
  - Navigator.
  - Radio operator (company tactical net).
  - Radio operator (patrol tactical net).
  - Corpsman.
- The first squad (security element) provides security en route to the objective area (the point, flank security, and rear security) and at the objective area (the flanks and the objective rally point).
- The second squad (support element) supports by fire for the attack, covering fire for the withdrawal, and supporting fires to cover the crossing of danger areas.
- The third squad (assault element) provides the assault force to attack and seize the objective; searchers to clear the objective; pacers, compass man, navigator, and the assistant patrol leader en route and back from the objective area.

Any attachments a patrol may have (i.e., demolition team, scout snipers, and machine gun squad) will be added to the element that supports its function. For example, the demolition personnel should go with the unit conducting the attack, and scout snipers and machine gunners should stay with the support squad.

9002. TASK ORGANIZATION

The preceding paragraphs described the elements necessary for a patrol to accomplish its mission. These elements reflect the internal functions or tasks required for the patrol to succeed. Depending on the METT-T, there are various methods of grouping these elements together. Task-organization is the further subdivision of patrol elements into teams that are required to perform essential tasks. In creating teams, unit integrity of infantry units should be maintained.

The patrol is organized so each individual, team, and element is assigned a specific task, but capable and prepared to perform other tasks. This may not be possible for certain specialist tasks requiring a trained technician.

Chapter 10. Patrol Preparation

For a patrol to succeed, all members must be well trained, briefed, and rehearsed. The patrol leader must have a complete understanding of the mission and a thorough understanding of the enemy and friendly situations. The patrol leader should make a complete reconnaissance of the terrain to be covered (either visual or map), and must issue an order to the patrol, supervise preparations, and conduct rehearsals.

10001. MISSION

The mission assigned to a patrol must be clear and oriented toward one objective with a specific task and purpose. More than one primary objective or indefinite missions invites confusion, casualties, and failure.

10002. FACTORS INFLUENCING PATROL SIZE

The circumstances under which patrolling by infantry units is conducted make it necessary for combat patrols to be able to fight offensively, security patrols to defend themselves, and reconnaissance patrols to move quickly and only fight if necessary to break contact or defend themselves. The size of a patrol depends on METT-T.

Generally, a patrol should contain the least number of members needed to accomplish the mission. Combat missions ordinarily require larger patrols than reconnaissance missions.

10003. COMMANDER'S DUTIES

Determining Patrol Requirements

The need for conducting patrols derives from the commander's stated mission (issued by higher level commander) and other specified and implied tasks (secondary or supporting efforts necessary to accomplish the stated mission). This requires analyzing the unit's mission and determining the necessary reconnaissance and/or combat tasks that must be performed. By considering the mission, along with time available, the commander develops the overall concept of operations to include the patrol plan and the specific tasks for each patrol.

Assigning Units

When assigning patrol missions, maintaining the integrity of the existing unit while considering the skills and experience of the unit and its leader are critical factors to the infantry commander. To provide operational depth and equitable apportionment of hazardous assignments, the commander ensures that each of the subordinate leaders and units develops the skills and experience necessary to conduct successful patrols. Assignment of patrol units must consider the commander's concept of operations as a whole and the plans for subsequent employment of assigned forces after completion of the assigned patrolling mission.

Providing Adequate Time

The commander must allow the patrol sufficient preparation time by completing the mission analysis, estimate, and preparation of orders. The commander should use warning orders to alert subordinates to possible requirements and afford them the opportunity for concurrent planning.

Providing the Patrol Leader Information

The following information should be provided:

- A simple, straightforward explanation of the mission, particularly for night patrols.
- General routes (defined by checkpoints) or exact routes (defined by avenues of approach or other terrain features) to follow.
- Enemy composition, disposition, and strength.
- Location and activities of friendly troops.
- Outposts or other security elements through which the patrol is to pass.
- Terrain conditions.
- Missions and routes of other patrols.

- Time patrol is to depart and return.
- Method of reporting information while on patrol (radio, messenger), place where messengers are to be sent, and place where the patrol leader is to report upon completion of the patrol.
- The challenge and password to be used for exiting and reentering friendly lines.
- Special instructions such as locations to be avoided and essential elements of information higher headquarters is seeking.
- Report barrier/obstacle plan of friendly units if known; support available from friendly unit (e.g., medical evacuation [MEDEVAC], react force).
- Fire support available.

Providing Special Skills and Equipment

There will be situations when the unit assigned to conduct a patrol does not have the necessary technical skills or equipment organic to it to successfully accomplish the patrol mission. In these cases, additions (attachments) are made to the unit, such as—

- Machine gun and/or shoulder-launched, multipurpose assault weapon (SMAW) team and/or squad.
- Forward observers (mortar/artillery).
- Radio operator(s).
- Combat engineers.
- Tracked vehicle crewman to assess the ability to traverse the terrain.
- Corpsman.
- Other personnel (snipers, translators) as required.

Providing Miscellaneous Support

The commander must ensure that the patrol leader is provided with the food, water, ammunition, radios and batteries, maps, special clothing, and any other items required by the unit (includin attachments) needed for the mission. Post-patrol support such as debriefings by intelligence personnel must also be planned.

Reviewing the Patrol Leader's Plan and Preparations

Once the patrol leader receives the mission, conducts visual and/or map reconnaissance, and develops the plan, the commander confirms the patrol leader's understanding of the mission and plan for accomplishing it. This discussion between the patrol leader and commander ensures that the patrol leader understands the commander's intent and is properly prepared to provide the patrol leader an opportunity to ask for clarification or additional support, if required.

Debriefing the Patrol

Upon return of the patrol, the commander receives the patrol report at a debriefing attended by the patrol leader and all patrol members. The debriefing should be conducted as soon as possible following the patrol's return, while information is still fresh in the minds of the patrol members. A patrol report, based on the information collected during the debrief, is generated and forwarded to the next higher com-mander.

10004. PATROL LEADER DUTIES

The patrol leader organizes and prepares the patrol by using the six troop-leading steps to make the best use of resources available. These steps are to—

1. Begin planning.
2. Arrange for reconnaissance and coordination.
3. Make reconnaissance.
4. Complete the plan.
5. Issue the order.
6. Supervise.

Begin Planning

The patrol leader begins by evaluating all factors affecting the mission. He looks for possible courses of action that lead to a decision, and then transforms this decision into an order. The first step includes making an initial assessment and decision on using available time, issuing a warning order and initial preparatory tasks, and initiating his estimate.

The initial planning effort assesses the time, assistance, and information available, and plans the proper use of each. Time allowances include reconnaissance; completion of the estimate and order; troop preparation; and such briefings, rehearsals, and inspections as required before beginning the patrol.

The patrol leader reviews the mission and the attachments and/or support available and decides what preparatory efforts must begin immediately. Proper use of subordinates to manage these initial tasks during this period reduces preparation time and frees the patrol leader for proper planning and reconnaissance. A warning order is issued using a modified five-paragraph order format (situation, mission, execution, administration and logistics, and command and signal [SMEAC]) as a checklist. (Refer to app. A for specific information that a patrol warning order should contain within each checklist item.

Once the initial preparations are set in motion, the patrol leader begins estimating by analyzing the mission; considering the friendly, enemy, and operating environments; considering each course of action available against what the enemy might do; comparing the courses of action in terms of mission accomplishment, capabilities, and probable casualties; and choosing one that becomes the basis for concept and order. While the patrol leader should use and organize notes, estimates must be done quickly and accurately, particularly for immediate situations. To organize thinking, the patrol leader uses METT-T, which consists of:

- Mission—the mission assigned to the patrol and how it relates to the mission of the commander who is sending the patrol.
- Enemy—what is known or suspected of enemy presence and capabilities, habits and characteristics, and fighting techniques.
- Terrain and weather—including ground, vegetation, drainage, weather, and visibility.
- Troops and support available—friendly situation and support available.
- Time available—the constraints and impact of time on preparation and mission accomplishment.

The estimate begins with mission analysis, which is the most important part of the entire planning process. Here, the patrol leader considers the specified tasks of the mission assigned and identifies other significant actions (specified and implied tasks) that must be undertaken to accomplish the stated mission. These tasks are arranged in sequence of accomplishment. The tasks and sequence create the framework for developing courses of action for the patrol concept of operations.

Arrange for Reconnaissance and Coordination

The patrol leader arranges a personal reconnaissance to observe as far forward as possible and also coordinates with the appropriate commanders for the patrol's "passage of lines" (see paragraph 11001 for complete definition) and supporting fires. The patrol leader also coordinates with other patrol leaders who may be operating in the same or adjacent areas and requests that the commander assigning him the patrol mission coordinate the patrol action with adjacent commanders, local security, and night defensive fires, as appropriate. The patrol leader may delegate any or all of these arrangements to the assistant patrol leader if the patrol leader requires the time for planning.

Make Reconnaissance and Complete the Estimate

The patrol leader uses personal reconnaissance to answer questions that arise from the map reconnaissance and METT-T evaluation. Specific points include passage points, lanes through obstacles, locations of friendly listening posts and observation posts, possible approach and return routes, enemy positions (if any), and intermediate observation points on the way to the objective.

In selecting approach and return routes, the patrol leader chooses routes that best use concealment and avoid opposition and obstacles. To lessen the chances of ambush by the enemy, the return trip is planned along a different route. In addition to personal reconnaissance and review of the map and aerial photographs, the advice of other patrol leaders who already are familiar with the terrain and the objective area should be considered.

After compiling information about the situation and possible time constraint, the patrol leader completes an estimate. The first step is developing courses of action, each of which will provide for movement to the objective area, mission accomplishment, and the return, based on the tasks and their sequencing identified in mission analysis. While the eventual concept of operations is presented in order of occurrence, the patrol leader must develop the courses of action by either backward or forward planning. In situations where the objective is well defined and there is sufficient information to plan the action for mission accomplishment (reconnaissance or combat), the patrol leader begins the scheme for accomplishing the mission at the objective and then, planning backwards, considers the options for getting there and back.

The following sections on movement to and return from the objective area, reconnaissance missions, and combat missions discuss methods and options available to the patrol leader in developing the courses of action. The principal variables between courses of action will be who, where, and how in the following:

- Patrol task organization.
- Routes to the objective area.
- Observation point(s) (reconnaissance patrol), ambush site, form of maneuver, type of ambush (combat patrol), fire support plan.
- Return routes.

The patrol leader then mentally considers the progress of each course of action (a map or simple sketch is a useful aid) against expected and unexpected enemy action. By comparing the options against each other and prospective enemy opposition, the patrol leader chooses the course of action that has the best chance of success. Includes in this mental preview process is the time to determine the patrol plan for unexpected contingencies (enemy attack/counterattack, casualty handling). These contingency actions, together with the selected course of action, become the patrol concept of operations.

Once the patrol leader determines the scheme of maneuver, the fire support required to accomplish the mission is addressed including the fire power organic to the patrol and what additional indirect fire support will have to be provided by other units. When planning for indirect fire support, the patrol leader considers the following questions:

- Will artillery, mortar, or close air support be required at the objective area (combat patrols)?
- What artillery and mortar targets exist along the routes to and from the objective area that can be employed by the patrol if it encounters the enemy during movement (reconnaissance patrols and combat patrols)?
- What additional fire support will be required to cover the patrol's movement from the objective area back to the friendly area once the enemy is aware of the patrol's actions at the objective area (combat patrols)?

The effect that casualties have upon the patrol depends upon many factors. Generally, more casualties can be expected in a combat patrol than in a reconnaissance patrol. A patrol may continue on to the objective carrying its casualties, send them back with a detail of Marines, abort the mission and return the entire patrol with the casualties, or call their parent unit for assistance.

Some factors that determine what action the patrol leader takes are: patrol's mission; unit's standing operating procedure for handling wounded; number of casualties and nature of their injuries; availability of aid, helicopters or other means of casualty evacuation. Helicopter evacuation should only be used for the most serious casualties. For infantry units conducting patrols in proximity to the enemy, helicopter evacuation of casualties may compromise the patrol's mission and force the patrol to return to friendly positions before the mission is completed.

The patrol leader determines the requirement for nuclear, biological, and chemical (NBC) defense equipment. Gas masks should always be carried due to the availability of riot control agents (RCAs) to the enemy. If chemical or biological agents have been employed in the area that the patrol must pass through, protective garments will have to be worn by patrol members for part of or the entire patrol. Wearing extra clothing and carrying extra equipment affects the speed of the patrol's movement. A contingency plan for post patrol decontamination must be developed.

Complete the Plan

At this point, the patrol leader has completed the basic thinking necessary for accomplishing the assigned mission. The patrol leader prepares the patrol order to spell out the details, assign tasks to subordinates, and explain the entire endeavor for ease of understanding by the other members of the patrol.

Prepare the Order

The patrol leader's order contains more detailed information than discussed in the warning order. Orders follow the prescribed five-paragraph order format but contain greater detail. The patrol order is a modified 5-paragraph order; the major modifications are to paragraphs 3a and 3c. The format for the order is contained in appendix B.

Issue the Order

The patrol leader asks for a status report on the initial preparatory tasks assigned to subordinate leaders and specialists when the warning order was issued. When the patrol leader has completed planning and initial preparations have progressed to the point where the patrol order may be issued, the members of the patrol are assembled. Roll call is taken to ensure all patrol members are present, then the prepared order is issued.

This will be the only opportunity for the patrol leader to issue detailed instructions. The mission, in particular, must be unmistakably clear so that once the patrol is committed, all subordinate leaders can act with unity of purpose.

Whenever possible, the patrol leader should have a Marine, such as the navigator, build a terrain model using dirt, sand, twigs, etc., explaining the concept of operations for movement to the objective area, actions at the objective area, and the return. Terrain models provide patrol members with a clear and simple layout of the area of operations and key terrain.

Supervise

Inspections and rehearsals, vital to proper preparation, are conducted even though the patrol leader and patrol members are well experienced in patrolling. Inspections determine the patrol's state of physical and mental readiness.

The patrol leader inspects before rehearsals to ensure completeness and correctness of uniform and equipment. The following areas are checked:

- Camouflage.
- Identification tags, Geneva Convention cards.
- Prescribed equipment, weapons, and ammunition are available and serviceable.
- Tape and other items are used to "silence" equipment (prevent noise produced during movement).
- Items that could provide information to the enemy (e.g., letters and papers) remain behind.
- Unnecessary equipment and excess weight remain behind.

The patrol leader questions each patrol member to ensure the following is known:

- The mission, planned routes (primary and alternate), and the fire support plan of the patrol.
- The individual's role: what to do and when to do it.
- What others are to do and how their actions impact.
- Challenges and passwords, codes, reporting times, radio call signs, frequencies, and any other pertinent details.

There is usually a period of time between final rehearsal and departure. The patrol leader reinspects just before departure to ensure all equipment is still in working order and the unit is ready to embark on the mission.

Rehearsals ensure the operational proficiency of the patrol. Plans are checked and needed changes are made. The patrol leader verifies the suitability of equipment. It is through rehearsals that patrol members become thoroughly familiar with the actions to take during the patrol.

If the patrol will operate at night, both day and night rehearsals are conducted. Terrain similar to that over which the patrol will operate is used. All actions are rehearsed. If time is limited, the most critical phases are rehearsed. Action at the objective is the most critical phase of the patrol and is always rehearsed.

An effective method is to talk the patrol through each phase, describing the actions and having each member perform individual duties. When satisfied, the patrol leader walks the patrol through all phases of the patrol using only the signals and commands to be used during the actual conduct of the patrol. Rehearsals continue until the patrol is thoroughly familiar with the plan. The rehearsal is also used to test the soundness of the patrol order and patrol organization.

After the rehearsal, the patrol leader makes final adjustments to the plan and patrol organization based on what was learned during the rehearsal and from other sources, such as the S-2 and adjacent patrols. When this is completed, the patrol leader issues final instructions to subordinate leaders noting any changes made in the patrol organization or plan. While the subordinate leaders are briefing the remainder of the patrol members, the patrol leader reports to the commander stating that the patrol is ready to begin the mission. The patrol leader also coordinates the location and time that the patrol can test fire all weapons prior to departure.

CHAPTER 11. MOVEMENT TO AND RETURN FROM THE OBJECTIVE AREA

This chapter provides guidance to patrol leaders for movement to and return from the objective area. Action in the objective area depends on whether the patrol is assigned a reconnaissance mission, combat mission or security mission. Chapters 12 and 13 provide detailed guidance on these types of missions.

11001. PASSAGE OF LINES

A passage of lines is an operation in which a force moves forward or rearward through another force's combat positions with the intention of moving into or out of contact with the enemy. (JP 1-02)

During the initial preparation for the patrol, the patrol leader selects a patrol assembly area and reconnoiters the area of passage designated by the commander. In coordination with the unit commander responsible for the area of passage, the patrol leader identifies gaps or lanes in minefields and wire obstacles and locates local security elements through which the patrol will pass. The patrol leader also checks the route from the patrol assembly area to the passage point or contact point where the patrol will depart friendly lines. If possible, both the route to the passage point and the route through the frontlines should be concealed from the enemy's view.

The patrol leader also reconnoiters the area for return passage of lines and coordinates with the unit commander responsible for the area of passage for passage points and lanes as necessary. The patrol leader observes these points from the direction that the patrol will use upon return to friendly lines, if possible, to aid in recognition upon return. The patrol leader provides the forward unit with information about the size of the patrol, general route, and expected time of return. The manner of challenge and recognition of the returning patrol should be coordinated in detail.

Upon return to friendly forward local security squads and/or frontlines, the patrol leader leaves the patrol in a covered position and moves forward with a radio operator and at least one Marine for security to make contact with the friendly unit in the manner agreed. After contact is made and recognized, the patrol leader rejoins the patrol and takes them to the passage point, personally checking in each member.

11002. ORGANIZATION FOR MOVEMENT

The patrol's task organization establishes the elements and teams needed to accomplish the mission in the objective area and to and from the objective. The patrol leader determines the formation(s) in which the patrol moves to the objective area.

Formations

The proper use of patrol formations is critical to the patrol's success. The squad and fire team formations described in FMFM 6-5 (proposed MCWP 3-11.2), *Marine Rifle Squad*, also apply to infantry patrolling formations. Because the movement of the patrol must be concealed from the enemy, the patrol normally moves through terrain that provides concealment. Control of the patrol in this type of terrain is difficult; thus the column formation, which is easily controlled, is normally used. However, as various types of terrain are encountered, the patrol leader uses the same considerations in determining the appropriate formation used in other infantry operations.

The standard squad and fire team formations are adaptable to a patrol. The patrol may change formations en route to match the situation and terrain. The patrol leader may have to sacrifice some control for better dispersion or give up some speed for greater stealth and more security. Other considerations include—

- Visibility, weather, terrain, and vegetation will influence dispersion and control of individuals and units. These factors may also affect the enemy; if visibility is good for the patrol, it is also good for the enemy. Two pieces of luminous tape worn on the back of the collar will aid in control and movement on dark nights. The collar is turned down when near the enemy. The tape can also be worn on the back of the cap, but should be covered or removed when near the enemy.

- Preserving the integrity of fire units (fire teams and, if attached, machine-gun teams) is of primary importance. If team members are detached from a machine-gun team, the unit integrity is lost and effectiveness is reduced. The patrol leader must also position fire units so as not to mask their fires.
- All-around defense of the patrol must not be sacrificed under any circumstances. The conventional squad and platoon formations provide adequate firepower in any direction required. When attachments are made to the patrol, the attachments are positioned within the formation to enhance the firepower of the patrol. If a fire unit, such as a machine-gun team or squad, is attached, it is incorporated into the all-around defense of the patrol by modifying the conventional formation (see fig. 11-1). It is permissible to employ machine-gun teams individually during movement. However, assault rocket squads and teams should be employed primarily as rifle fire units for movement. Firing the SMAW or AT-4 from within a patrol formation can present a serious back-blast danger to patrol members and should only be fired on the patrol leader's direction.
- Time allotted for mission accomplishment is also a major consideration. In selecting the formations, the patrol leader must consider the speed of the movement required to meet the time constraints (if any) imposed on the patrol. If required to meet a time schedule, a formation that permits rapid movement should be used. Speed, however, must never be permitted to force the patrol leader to make rash tactical decisions.

Exercise of Control

The patrol leader is positioned for best control over the patrol. The assistant patrol leader moves at or near the rear of the patrol. Other subordinate leaders move with their elements. All patrol members assist by staying alert and passing on signals and orders. A signal to halt may be given by any patrol member, but the signal to resume movement is given only by the patrol leader.

Arm-and-hand signals are the primary means of communication within a patrol and should be used exclusively when near the enemy. All members must know the standard infantry signals (refer to FMFM 6-5), as well as any special signals required, and be alert to receive and pass them to other members.

The patrol leader should speak just loudly enough to be heard. At night, or when close to the enemy, the patrol leader halts the patrol and has subordinate leaders come forward. They speak in a low voice and then pass the information to their subordinates by moving from member to member.

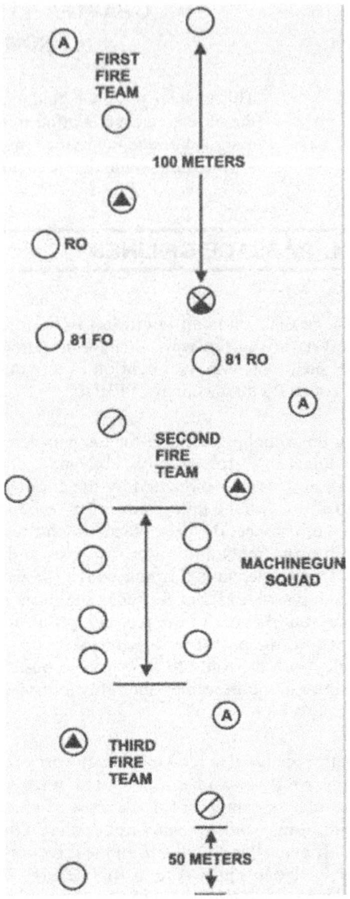

Figure 11-1. Squad Patrol with Attached Mortar Forward Observer Team and Machine Gun Squad.

Radios provide a means of positive control within a large patrol; however, radios should be used only when arm-and-hand signals or face-to-face contact between the patrol leader and subordinate leaders is impractical. When close to the enemy, words are spoken into the microphone with hands cupped over it in a low voice.

Other sound signals may be used if the patrol leader is sure they serve the purpose intended. Planned sound signals are rehearsed before starting on the patrol. Sound signals used must be simple, natural sounds that are few in number and easily understood. Bird and animal calls are seldom satisfactory.

Infrared equipment, such as the sniper scope and infrared filters for the flashlight, may be used as means of sending and receiving signals and maintaining control at night.

Luminous tape may be used to assist in control at night. Small strips on the back of the cap or collar of patrol members aid in keeping visual contact with the front member. However, the luminous tape must be covered when near the enemy.

An important aspect of control is accounting of personnel, especially after crossing danger areas, enemy contacts, halts, and exiting and re-entering friendly lines. The patrol leader may arrange for the last member to send up the count automatically after crossing danger areas, enemy contact, and halts. In large patrols or when moving in a formation other than a column, subordinate leaders check members and report the count to the patrol leader by the quickest method appropriate to the circumstances.

Navigation

One or more members are assigned as navigators for the patrol. Their function is to assist the patrol leader in maintaining direction by the use of the compass.

The patrol leader assigns at least two members as pacers to keep track of the distance from point to point. The average of their count is used to approximate the distance traveled. Pacers are separated so they do not influence each other's count. Pacers must know how to convert their own paces to meters.

The route is divided into increments, with each increment starting at a recognizable point on the ground. The pacers begin their counts from zero at the beginning of each leg. This makes the pace count easier to keep and checks for accuracy.

The pace count is sent forward when the patrol leader turns to the member behind and in a low voice says, "send up the pace" or uses the arm-and-hand signal of tapping his boot. This signal is passed to both pacers, who in turn send up the pace count in meters; for example, "two-hundred" or "one-seven-five."

Patrol members must understand that the counts of both pacers are sent forward. The patrol leader must know the counts of both men in order to check them.

Security

The patrol is dispersed consistent with control, visibility, cover, and concealment. Scouts are employed to the front, flanks, and rear of the patrol to provide security. For the patrol members traveling in the main body of the patrol, areas of responsibility are assigned to the front, flanks, and rear. Scouts are the eyes and ears of the patrol leader. They move when and where directed by the patrol leader and maintain contact with the patrol leader at all times, except when momentarily obscured by vegetation or other terrain features.

Front

Small patrols (squad size) may employ from one scout up to a fire team as the point, depending on the enemy situation, terrain, and route being followed by the patrol. Normally, squad-size patrols will use two scouts as the point. The point is responsible for investigating the route of advance immediately to the front of the patrol.

The point moves as far ahead of the patrol as visibility and terrain permit. When visibility is good, the point may precede the main body by as much as 100 meters. The point travels right and left ahead of the patrol, searching the area over which the patrol will pass.

The point maintains direction by knowledge of the general route to be followed and visual contact with the patrol leader. The patrol leader or the navigator ensures that the point is proceeding correctly.

The point, which stays far enough ahead of the patrol to provide security, is not a trail breaker for the patrol. If the point loses contact with the patrol, the point waits for the main body to catch up or moves rearward if contact is not quickly regained.

One of the navigators may be positioned with the point. One or more members works as the point while the other is the navigator.

Flank

Flank security for a patrol of squad size or less may be provided by using one or two members on either flank. If two scouts are assigned to a flank, one is positioned to observe the patrol leader and the other works farther out from the patrol. The scout who must observe the patrol leader remains within a maximum distance of 100 meters. The scout farther out remains in sight of the inside scout but normally does not move more than 20 or 25 meters away and remains prepared to relieve flank security regularly. Moving through dense woods or jungle may render the use of flank security impractical because of reduced visibility. In such cases, it moves with the patrol itself, but maintains observation to its assigned flanks.

Rear

A small patrol normally has only one rifleman assigned as rear security. An interval between the member assigned as rear security and the last Marine of the patrol is maintained at the limit of visibility, up to 50 meters. This member maintains rear security for the patrol by constantly observing to the rear.

Halting

Speed of movement is slower at night than in day patrols and reduces the danger of a Marine becoming separated from the patrol. The patrol occasionally halts to observe and listen for enemy activity; this is called a security halt. Upon signal, when reaching a danger area and periodically throughout movement en route, every member freezes in place, remains quiet, observes, and listens. It may be necessary to call a security halt just after departing friendly areas and just before entering friendly areas.

The patrol may halt briefly to send a message, eat, rest, check direction, or make a reconnaissance. The area selected should provide adequate concealment and cover, as well as favor the defense. All-round security is established and the patrol leader ensures all members move out when the patrol resumes movement. (For extended halts, see paragraph 11005.)

Infiltration

The disposition of enemy forces may sometimes prevent a patrol from entering the enemy occupied area as a unit; however, pairs of scouts or fire teams may slip through without being discovered. (Refer to FMFM 6-5 for infiltration techniques and procedures).

11003. CONTROL MEASURES FOR MOVEMENT

Checkpoints

A checkpoint is a predetermined point on the surface of the Earth used as a means of controlling movement, a registration target for fire adjustment or reference for location (JP 1-02). Checkpoints are a means of control between the parent unit and the patrol. These locations are decided upon and coordinated before the patrol leaves, so that both the patrol members and parent unit will know the patrol's location when it reports in. The parent unit can follow the progress of the patrol without transmitting coordinates over the radio that the enemy could monitor.

Rally Points

A rally point is an easily identifiable point on the ground where units can reassemble and reorganize if they become dispersed (MCRP 5-12C). It should provide cover and concealment, be defensible for at least a short time, and be easily recognized and known to all patrol members. All rally points are considered and identified as tentative rally points until they are occupied, found to be suitable, and designated as rally points by the patrol leader.

Initial Rally Point

The initial rally point is within friendly lines where the patrol can rally if it becomes separated before departing the friendly area or before reaching the first en route rally point. It may be the assembly area where the patrol waits while the patrol leader contacts the last friendly position through which the patrol will pass. The location of the initial rally point must be coordinated with the forward unit commander.

En route Rally Points

En route rally points are between the initial rally point and the objective rally point; and from the objective rally point back to the point where the patrol reenters friendly lines. They are determined as the patrol passes through a likely area that is suitable for a rally point.

Objective Rally Point

The objective rally point is located nearest the objective where the patrol makes final preparations prior to approaching the objective. It also serves as a location where the patrol reassembles after completing actions on the objective. The objective rally point must be suitable to accommodate those activities accomplished prior to actions on the objective. This position must provide the patrol concealment from enemy observation and, if possible, cover from enemy fires. It may be located short of, to a flank or beyond the objective. It should be out of sight, sound, and small-arms range of the objective area. The patrol leader's reconnaissance of the objective is made from this position; it is the release point from which patrol elements and teams move to the objective to accomplish the mission.

Rally Point Selection

The patrol leader selects likely locations for tentative rally points during reconnaissance or map study. A tentative rally point must be confirmed and announced after examination proves suitability. A tentative initial rally point and a tentative objective rally point are always selected and identified in the patrol order. If necessary, the patrol leader selects additional rally points en route as suitable locations are reached. When the patrol reaches a danger area that cannot be bypassed, such as an open meadow or stream, the patrol leader selects a rally point on both the near and far side. If good locations are not available, the patrol leader designates the rally points in relation to the danger area. The patrol leader will say, for example, "50 meters this side of the trail" or "50 meters beyond the stream."

Rally Point Use

The initial rally point and en route rally points are designated to enable the patrol to reassemble if it is unavoidably separated or dispersed. Identifying features are pointed out. The patrol leader ensures that the information is passed to all patrol members. When crossing a danger area, a near side rally point and a far side rendezvous point are designated. These rally points should only be used when all other methods of retaining control of the patrol have failed. The success of the patrol may be jeopardized if it is dispersed and cannot rally expeditiously.

If the patrol has left the friendly area and becomes dispersed, patrol members return to the last designated rally point (the initial or an en route rally point) unless the patrol leader gives other instructions.

As previously noted, the patrol leader selects two rally points at the near and far sides of danger areas that cannot be bypassed. If the patrol becomes separated or dispersed at a danger area, and there has been no enemy contact, the patrol should reassemble at the rally point on the far side of it. If, however, the patrol is separated or dispersed at a danger area as a result of enemy contact, members who have already crossed the danger area assemble and reorganize at the rally point on the far side; members who have not crossed assemble and reorganize at the rally point on the near side. In the absence of the patrol leader and unless directed otherwise, the senior member at the rally point on the near side takes charge, attempts to move the rallied patrol members to the rally point on the far side, and rejoins the remainder of the patrol.

Rally Point Actions

The patrol leader plans the actions to be taken at rally points and instructs the patrol accordingly in the patrol order. Planned actions at the initial rally point and en route rally points must provide for the continuation of the patrol as long as there is a reasonable chance to accomplish the mission. Plans for actions at rally points should provide for—

- Recognition signals for assembly at rally points.
- Minimum number of members and maximum amount of waiting time required before the senior member at the rally point moves the rallied patrol members onward toward the objective or returns to friendly lines.
- Instructions for patrol members who find themselves alone at a rally point.

11004. PRECAUTIONS AT DANGER AREAS

A danger area is any place where the patrol is vulnerable to enemy observation or fire (open areas, roads, trails, and obstacles such as barbed wire, minefields, rivers and streams, and lakes). Any known or suspected enemy position the patrol must pass is also a danger area. The patrol leader plans for crossing each danger area and includes these plans in the order.

The patrol reconnoiters the near side of a danger area first, then the patrol leader sends scouts to reconnoiter the far side. Once the scouts report that the far side is clear of the enemy, the remainder of the patrol crosses the danger area. As each individual or group crosses the danger area, they are covered by those remaining and by those who have successfully crossed. Enemy obstacles are avoided since they are usually covered by fire.

In crossing a river, the near bank is reconnoitered first; then the patrol is positioned to cover the far bank. Scouts are sent across to the far bank. After the far bank has been reconnoitered and the scouts report that it is clear of the enemy, the patrol crosses as rapidly as possible. This may be done individually or in pairs. If crossing the river requires swimming, the patrol uses improvised rafts to float equipment, weapons, and ammunition across. (Refer to MCRP 3-02C, *Water Survival Handbook.)*

A road or trail is crossed at or near a bend or where the road is narrow. Observation is restricted and, if the enemy is present, the patrol is exposed as short a time as possible. The near side is reconnoitered first, then scouts are sent across to reconnoiter the far side. This includes reconnoitering the tentative rally point on the far side. Once the scouts report "all clear," the remainder of the patrol crosses rapidly and quietly.

If the patrol must pass close to an enemy position, it takes advantage of battlefield noises to cover the sounds of movement. If supporting fires are available, the patrol leader can call for them to divert the enemy's attention as the patrol passes.

11005. HIDE

When a patrol is required to halt for an extended period in an area not protected by friendly troops, the patrol moves into a location which, by the nature of the surrounding terrain, provides passive security from enemy detection. Such an assembly area is termed a *hide*. To establishment a hide—

- Cease all movement during daylight hours to avoid detection.
- Hide the patrol for an extended period while the patrol leader conducts a detailed reconnaissance of the objective area.
- Rest and reorganize after extended movement.
- Reorganize after a patrol has infiltrated the enemy area in small groups.

The patrol leader's plan must include tentative hide locations when the patrol's mission dictates an extended halt within enemy areas. These tentative locations must be confirmed by actual ground reconnaissance prior to occupation by the patrol. The plan for a hide includes both passive and active security measures.

Passive security measures are—

- Avoid built-up areas.
- Select an area remote from all human habitation.
- Avoid known or suspected enemy positions.
- Avoid ridge lines, topographic crests, valleys, lakes, and streams.
- Avoid roads and trails.
- Avoid open woods and clearings.
- Select areas offering dense vegetation, preferably bushes and trees that spread out close to the ground.

Active security measures—

- Establish security covering all likely avenues of approach into the site.
- Establish communications (wire, radio, signal, runner) with posted security to provide early warning of enemy approach.
- Select an alternate area for occupation if the original hide is compromised or found unsuitable.
- Plan for withdrawal in the event of discovery.
- Establish an alert plan with a certain percent of the personnel awake at all times.
- Organize the elements of the patrol so necessary activities can take place with a minimum amount of movement.

The size of the area physically occupied by a patrol in a hide and the number of security posts required are governed by the terrain, quantity and quality of cover and concealment, and size of the patrol.

If the situation permits, a hide can also be used as the final preparation position and/or objective rally point.

11006. IMMEDIATE ACTIONS UPON ENEMY CONTACT

A patrol may make contact with the enemy at any time. Contact may be through observation, a meeting engagement or ambush. Contact may be visual, in which the patrol sights the enemy but is not itself detected. When this is the case, the patrol leader can decide whether to make or avoid physical contact, basing his decision on the patrol's assigned mission and capability to successfully engage the enemy unit.

When a patrol's assigned mission prohibits physical contact (except that necessary to accomplish the mission), its actions are defensive in nature. Physical contact, if unavoidable, is broken as quickly as possible and the patrol, if still capable, continues its mission.

When a patrol's assigned mission permits or requires it to seek or exploit opportunities for contact (as in the case of a combat patrol), its actions are offensive in nature, immediate, and positive.

In patrolling, contacts (visual or physical) are often unexpected at very close ranges, and short in duration. Effective enemy fire often provides leaders little or no time to fully evaluate situations and issue orders. In these situations, immediate action provides a means for swiftly initiating positive offensive or defensive action, as appropriate.

Two types of physical contact with the enemy are meeting engagement and ambush. Meeting engagement is a combat action that occurs when a moving force, incompletely deployed for battle, engages an enemy at an unexpected time and place. It is an accidental meeting where neither the enemy nor the patrol expect contact and are not specifically prepared to deal with it. An ambush is a surprise attack from a concealed position.

Immediate Actions

Immediate actions are designed to provide swift and positive small unit reaction to visual or physical contact with the enemy. They are simple courses of action in which all Marines are well trained. Minimal signals or commands are required and they are developed as needed for the combat situation. The signals can, in many cases, be initiated by any member of the unit. It is not feasible to attempt to design an immediate action drill to cover every possible situation. It is better to know the immediate action drill for each of a limited number of situations that may occur during a patrol. Arm-and-hand signals associated with immediate actions—such as FREEZE, ENEMY IN SIGHT, and HASTY AMBUSH RIGHT or LEFT—are contained in FMFM 6-5 (proposed MCWP 3-11.2).

Immediate Halt

When the patrol detects the enemy but is not itself detected, the situation requires the immediate, in-place halt of the patrol. The first member visually detecting the enemy gives the silent signal for FREEZE. Every member halts in place, weapon at the ready, and remains absolutely motionless and quiet until further signals or orders are given.

Air Observation and/or Attack

These actions are designed to reduce the danger of detection from aircraft and casualties from air attack.

When an enemy or unidentified aircraft that may detect the patrol is heard or observed, the appropriate immediate action drill is FREEZE. The first member hearing or sighting an aircraft that may be a threat signals FREEZE. Every member halts in place until the patrol leader identifies the aircraft and gives further signals or orders. Members of the patrol must not look up at the aircraft as sunlight can reflect off their faces even when camouflaged.

When an aircraft detects a patrol and makes a low level attack, the immediate action drill air attack is used. The first member sighting an attacking aircraft shouts, "AIRCRAFT," followed by the direction of the incoming attack: FRONT, LEFT, REAR or RIGHT. The patrol moves quickly into line formation, well spread out, at right angles to the aircraft's direction of travel. As each member comes on line, the member hits the ground, using available cover, then positions the body perpendicular to the aircraft's direction of travel, to present the shallowest target possible (see fig. 11-2 on page 11-8). Between attacks (if the aircraft returns or if more than one aircraft attacks), patrol members seek better cover. Attacking aircraft are fired upon only on command of the patrol leader.

Meeting Engagement

Hasty Ambush. This immediate action is used to avoid contact and to prepare to initiate an unplanned ambush on the enemy. It may often be a subsequent action after the command freeze. When the signal HASTY AMBUSH is given (by the point member, patrol leader or another authorized patrol member), the entire patrol moves quickly to the right or left of the line of movement, as indicated by the signal, and takes up the best available concealed firing positions (see fig. 11-3 on page 11-9). The patrol leader initiates the ambush by opening fire and shouting, "FIRE"; thus ensuring initiation of the ambush if the weapon misfires. If the patrol is detected before this, the first member aware of detection initiates the ambush by firing and shouting. The patrol leader may decide not to initiate the ambush in order to avoid contact unless the patrol is detected. When used as an offensive

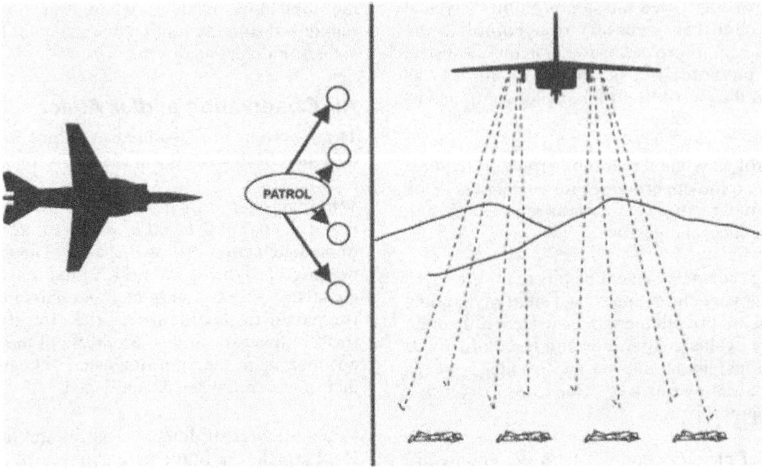

Figure 11-2. Immediate Action, Air Attack.

measure, the enemy is allowed to advance until he is in the most vulnerable position before the ambush is initiated. An alternate means for initiating the ambush is to designate an individual (for example, point or last member) to open fire when a certain portion of the enemy unit reaches or passes that member.

Immediate Assault. This immediate action drill is used defensively to make and quickly break undesired but unavoidable contact (including ambush) and offensively to decisively engage the enemy (including ambush). When used in a meeting engagement, members nearest the enemy open fire and shout, "CONTACT," followed by the direction of the incoming attack: FRONT, LEFT, REAR or RIGHT. The patrol moves swiftly into line formation and assaults (see fig. 11-4).

Defensive Measures

When used defensively, the assault is stopped if the enemy withdraws and contact is broken quickly. If the enemy stands fast, the assault is carried through the enemy positions and movement is continued until contact is broken.

Offensive Measures

When used offensively, the enemy is decisively engaged. Escapees are pursued and destroyed until orders to break contact are given by the patrol leader. If the patrol is fired upon from beyond 50 meters, the patrol must break contact as quickly as possible and continue the mission. If it engages the enemy any longer than necessary to break contact, it may put the mission in jeopardy.

Fire and Maneuver

Fire and maneuver is one means to break contact. One portion of the patrol returns the enemy fire while another portion moves by bounds away from the enemy. Each portion of the patrol covers the other by fire until contact is broken by all.

Clock System

The clock system is another means to break contact. Twelve o'clock is the direction of movement of the patrol. The patrol leader shouts a direction and a distance. For example: "TEN O'CLOCK-TWO HUNDRED," means the patrol should move in the direction of ten o'clock for 200 meters. Patrol members keep their same relative positions as they move so the original formation is not disrupted. Subordinate leaders must be alert to ensure that the members of their elements and teams receive the correct order and move as directed.

Counter Ambush

When a patrol is ambushed, the immediate action drill used depends on whether the ambush is a near ambush

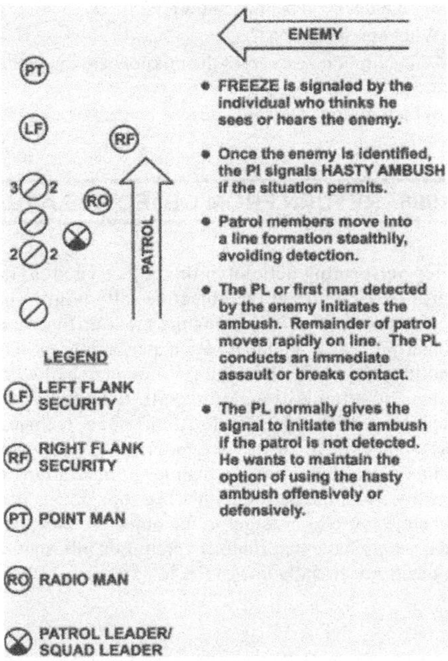

Figure 11-3. Immediate Action, Hasty Ambush.

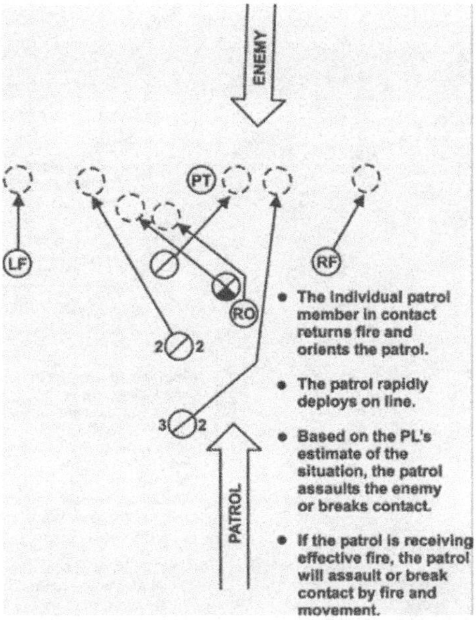

Figure 11-4. Immediate Action, Immediate Assault.

(the enemy is within 50 meters of the patrol) or a far ambush (the enemy is beyond 50 meters of the patrol). Fifty meters is considered the limit from which the ambushed patrol can effectively launch an assault against the enemy.

Near Ambush. In a near ambush, the killing zone is under very heavy, highly concentrated, close range fires. There is little time or space for members to maneu-ver or seek cover. The longer they remain in the killing zone, the greater the chance they will become casualties. Therefore, if members in the kil-ling zone are attacked by a near ambush, they immediately assault without order or signal directly into the ambush position, occupy it, and continue the assault or break contact, as directed. This action moves them out of the killing zone, prevents other elements of the ambush from firing on them without firing on their own members, and provides positions from which other actions may be taken (see fig. 11-5 on page 11-10). Members not in the killing zone maneuver against the enemy as directed. The assault continues until all patrol members are outside of the killing zone.

Far Ambush. In a far ambush, the killing zone is also under very heavy, highly concentrated fires, but from a greater range. This greater range provides members in the killing zone maneuver space and some opportunity to seek cover at a lesser risk of becoming a casualty. If attacked by a far ambush, members in the killing zone, without order or signal, immediately return fire, take the best available positions, and continue firing until directed otherwise. Members not in the killing zone maneuver against the ambush force, as directed (see fig. 11-6 on page 11-10). The assault is continued against the enemy or until the order to break contact is given.

In each situation, the success of the counter ambush employed depends on the members being well trained in quickly recognizing the distance from which an ambush is initiated and well rehearsed in the proper reaction.

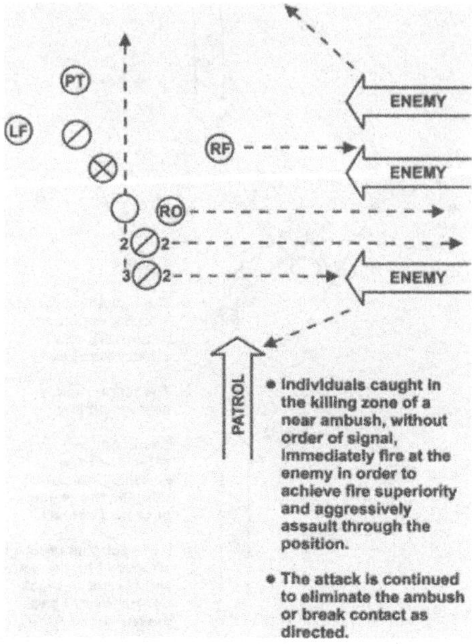

Figure 11-5. Near Ambush.

11007. PATROL LEADER'S ACTION IN A DEVELOPING SITUATION

While good patrolling depends on good planning, the patrol leader's plan must be flexible. Every combat situation develops differently than expected, and the patrol leader must be ready to quickly adapt to the situation as it develops. One of the most difficult tasks in battle is to recognize the correct moment for making a decision. Generally, it is more difficult to determine the moment for making a decision than it is to formulate the decision itself. When the situation demands, decisions must be made promptly without waiting for more information. In a developing situation, the patrol leader should use the following questions as a guide for battlefield decisionmaking:

- How has the situation changed?
- How does the change affect mission accomplishment and the immediate superior's mission?
- Must a decision be made now?
- What are the options?
- Which option best serves the mission and the unit as a whole?
- Which option offers the greatest chance of success?

11008. RETURN FROM OBJECTIVE AREA

After performing actions in the objective area, the patrol reassembles at the objective rally point. This phase of the patrol is perhaps the most difficult and dangerous. Patrol members are experiencing fatigue, emotional letdown, and wounds; they may be low on water and ammunition. Above all, the enemy was likely alerted if not in pursuit. At this point, the patrol leader must move the patrol rapidly but carefully and maintain patrol security at a high level. In returning to friendly lines, the patrol neither uses nor travels near the same route used to get to the objective area. The enemy may have that route covered. For information on reentry of friendly lines, refer to paragraph 11001.

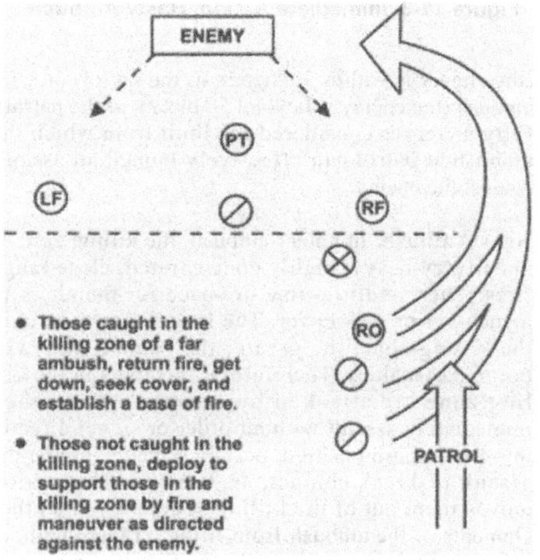

Figure 11-6. Far Ambush.

Chapter 12. Reconnaissance Patrols

The commander needs accurate, timely information about the enemy and the terrain the enemy controls to assist in making tactical decisions. Reconnaissance patrols are one of the most reliable means for obtaining this information. These patrols engage the enemy only when necessary to accomplish their mission or for protection. They generally avoid combat, accomplish the mission by stealth, and do not maintain contact with the unit that sent them out. A reconnaissance patrol is capable of carrying the search for information into the area occupied by enemy forces—usually beyond the range of vision of friendly ground observation posts—and examining objects and events at close range.

12001. General Missions

General missions for reconnaissance patrols include gathering information about location and characteristics of friendly or hostile positions and installations, terrain (routes, stream crossings), and obstacles.

12002. Specific Missions

Locate the Enemy

Reconnaissance patrols try to determine the—

- Location of enemy forces, installations, and equipment.
- Identification of enemy units and equipment.
- Strength of enemy forces.
- Disposition of enemy forces.
- Movement of enemy personnel and equipment.
- New or special types of weapons.
- Presence of mechanized units.
- Unusual enemy activity.
- Presence of NBC equipment.

Reconnaissance of Enemy Wire Obstacles

A patrol with the mission of investigating enemy-emplaced wire obstacles employs a formation providing for all-around security and takes precautions against being observed by the enemy. The patrol leader and one Marine inspect each gap and establish its location by means of compass bearings to prominent objects in the rear of enemy or friendly positions or through the use of global positioning satellite receivers.

Reconnaissance of Contaminated Areas

A patrol with the mission of investigating a contaminated area reconnoiters and marks the area's boundaries. Patrol members wear protective clothing and field protective masks. The patrol leader reports the extent of the area, the ttype of agent used, the terrain and vegetation and the method of marking the contaminated area. A sketch of the contaminated area, prepared by the patrol leader, should be included in the patrol report.

Reconnaissance of Enemy Minefields

Patrols assigned to reconnoiter enemy minefields are often composed of specially trained personnel. The type of mine is identified and the possibility of bypassing the area determined.

Reconnaissance of Terrain

The reconnaissance patrol is particularly suited for gathering information about the terrain within their area of operations. (Refer to MCRP 2-15.3B, *Reconnaissance Reports Guide*.) Commanders must know the location and condition of bridges, streams, and roads in order to make proper plans for the movement of troops. The following information should be obtained by a reconnaissance patrol assigned such missions:

- Bridges:
  - Maximum load capacity.
  - Material used in construction (wood, stone, concrete or steel).

- Material used for abutments (dirt, wood, masonry, concrete).
- Type and number of supporting members of the bridge.
- Condition and dimensions of bridge flooring.
- Presence or absence of mines and demolition charges on the bridge and at entrances/exits.
- Terrain crossed (ravine, stream, road).
- Primary use (rail, auto, foot).
- Location and type of an alternate crossing if bridge is unusable.

- Streams and fords:
  - Width and depth of stream.
  - Composition of the bottom (mud, sand, gravel, rock).
  - Speed of current in miles per hour.
  - Composition and gradient of banks.
  - Surrounding terrain and vegetation cover.
- Roads:
  - Width.
  - Composition of surface.
  - Condition.
  - Road blocks.
  - Gradient/degrees of slopes.
  - Curves (location and length).
  - Pitch and culvert locations.
  - Primary use and frequency of use.

12003. TYPES OF RECONNAISSANCE

Route

Route reconnaissance is a directed effort to obtain detailed information of a specified route and all terrain from which the enemy could influence movement along that route. (Refer to MCRP 5-12C.) It is also described as a form of reconnaissance focused along a specific line of communications—such as a road, railway, or waterway—to provide new or updated information on route conditions and activities along the route. (Refer to MCRP 5-12A, *Operational Terms and Graphics*.)

A route reconnaissance normally precedes the movement of forces. It provides detailed information about a specific route and the surrounding terrain that could be used to influence movement along that route. Considerations include trafficablility, danger areas, critical points, vehicle weight and size limitations, and locations for friendly obstacle emplacement.

The objective area is normally defined by a line of departure, a route, and a limit of advance.

Area

An area reconnaissance is a directed effort to obtain detailed information concerning the terrain or enemy activity within a prescribed area such as a town, ridge line woods or other features critical to operations. (Refer to MCRP 5-12C.) An area reconnaissance could also be made of a single point, such as a bridge or installation. (Refer to MCRP 5-12A.)

Emphasis is placed on reaching the area without being detected. Enemy situations encountered en route are developed only enough to allow the reconnoitering unit to report and bypass.

Zone

A zone reconnaissance is a directed effort to obtain detailed information concerning all routes, obstacles (to include chemical or radiological contamination), terrain, and enemy forces within a zone defined by boundaries. A zone reconnaissance is normally assigned when the enemy situation is vague or when information concerning cross-country trafficability is desired. (Refer to MCRP 5-12C.)

The commander specifies specific routes or areas of interest within the zone. The zone to be reconnoitered usually is defined by a line of departure, lateral boundaries, and a limit of advance.

12004. TASK ORGANIZATION

A reconnaissance patrol is organized around the current structure of the Marine rifle squad with one or more of the squad's fire teams assigned as the reconnaissance element to reconnoiter or maintain surveillance over the objective. At least one fire team acts as a security element whose functions are to—

- Secure the objective rally point (see para. 11003).
- Give early warning of enemy approach.
- Protect the reconnaissance unit.

Reinforcing the squad is based on METT-T. If two squads are required, then the mission should be assigned to a platoon.

A company may send a reconnaissance patrol to specifically reconnoiter a given area (a riverbed or a bridge), or to maintain surveillance on a particular streambed for the next four nights. Reconnaissance patrols perform three basic functions en route to and from the objective:

- Provide control.
- Provide security while moving and in the objective area.
- Conduct reconnaissance or surveillance.

Depending on the size of the reconnaissance patrol, an element must be tasked to reconnoiter the area or zone. In the case of a surveillance mission, two teams are recommended to share the same task (so one could lay up and rest, while the other lays low and hides to maintain strict silence and no movement while surveillance is being maintained on the objective). Whatever the case, each unit that has a surveillance mission or reconnaissance mission must be thoroughly briefed as to what essential elements of information are to be collected as well as their location in the objective area.

The patrol should be organized with one or two fire teams to actually conduct the reconnaissance mission and the remaining fire team to provide security. Each fire team should be prepared to assume either mission. A small area reconnaissance patrol needs only one fire team for the assigned mission. A patrol with a wider area reconnaissance mission should use two fire teams to physically conduct the mission and one fire team for cover and/or security.

The security element for a reconnaissance patrol should be organized to cover the likely avenue of approach into the objective area, to protect the units conducting the reconnaissance, and to cover the objective rally point.

The variations of the special organization for reconnaissance patrols illustrate the patrol leader's flexibility in organizing the patrol to meet mission requirements. The patrol leader is not limited to the variations reflected herein but can choose any special organization as long as that task organization accommodates the requirement for command and control, reconnaissance, and security.

12005. SIZE OF RECONNAISSANCE PATROLS

A reconnaissance patrol should be kept to the minimum number of personnel required to accomplish the mission. A mission requiring a patrol to remain away from its unit for a considerable period of time, or one requiring a patrol to send back information by messenger, increases the size of the patrol. Reconnaissance patrols seldom exceed a squad in strength. Unit integrity should be preserved whenever possible. Intelligence personnel, interpreters, and other specialists, such as radio operators or engineers, are assigned to a patrol if the particular mission requires.

12006. RECONNAISSANCE EQUIPMENT

Patrol members are armed and equipped as necessary for accomplishing the mission. The automatic rifle in each fire team provides a degree of sustained firepower in case of enemy contact. The patrol should have at least two: pairs of binoculars, pairs of wire cutters, maps, compasses, and watches. Night observation equipment may be used. Pencils and small notebooks are carried so notes and sketches can be made. A message book with message blanks and overlay paper is mandatory.

12007. RECONNAISSANCE PATROL ACTIONS AT THE OBJECTIVE AREA

Route

The patrol leader halts and conceals the patrol near the objective area in the final preparation position. The patrol leader then conducts leader's reconnaissance to confirm the plan for positioning the security teams and employing units assigned to the reconnaissance mission. The patrol leader returns to the patrol and positions the security to provide early warning of enemy approach and secure the objective rally point. The reconnaissance unit(s) then reconnoiters the objective area (route). The reconnaissance unit may move to several positions, along or adjacent to the specific route, in order to conduct a thorough reconnaissance. After completing the reconnaissance, each reconnaissance team moves to the objective rally point and reports to the patrol leader. The patrol then returns to friendly lines and the leader makes a full report.

Area

The patrol leader halts and conceals the patrol near the objective area in the final preparation position. The patrol leader then conducts leader's reconnaissance to pinpoint the objective and confirm the plan for positioning the security teams and employing units assigned the reconnaissance mission. The patrol leader returns to the patrol and positions security to provide early warning of enemy approach and secure the objective rally point. The reconnaissance unit(s) then reconnoiters the objective area. The reconnaissance unit may move to several positions, perhaps making a circle around the objective area, in order to conduct a thorough reconnaissance. When the reconnaissance is completed, the patrol leader assembles the patrol and tells members what has been observed and heard. Other patrol members contribute observations. The patrol then returns to friendly lines and the patrol leader makes a full report.

Zone

The patrol leader halts the patrol at the final preparation position, confirms the plan, and conducts leader's reconnaissance. The patrol leader positions the security team and sends out the reconnaissance team. When the entire patrol is used to reconnoiter the zone, it provides its own security. After completing the reconnaissance, each reconnaissance team moves to the objective rally point and reports to the patrol leader. The patrol then returns to friendly lines.

Chapter 13. Combat Patrols

Combat patrols are assigned missions that usually include engaging the enemy. They are fighting patrols. Every combat patrol has a secondary mission: gaining information about the enemy and terrain. Combat patrols are employed in both offensive and defensive combat operations and they assist the parent unit in accomplishing its mission by inflicting damage on the enemy; establishing and/or maintaining contact with friendly and enemy forces; denying the enemy access to key terrain; and probing enemy positions to determine the nature and extent of enemy presence.

13001. Task Organization

A combat patrol is organized around the current structure of the Marine rifle platoon. A combat patrol leader should use the unit's normal organization (fire team, squad, and platoon) in assigning functions, patrol missions, and chain of command. Combat patrols must be able to perform the following four basic functions en route to and from the objective:

- Provide control.
- Provide security.
- Provide support by fire.
- Attack or assault the objective.

A rifle platoon could task-organize as a combat patrol as follows:

- Platoon headquarters (patrol headquarters).
- First squad (security).
- Second squad (support).
- Third squad (assault).

Every combat patrol must—

- Provide a control mechanism in the form of a headquarters.
- Designate a unit (a fire team or squad) to provide security while moving en route to the objective or while at the objective. At the objective area, this unit isolates the objective area, secures the objective rally point, and covers the withdrawal.
- Designate a unit to act as support. This unit provides the base of fire in the attack or covers withdrawals or advances.
- Designate a unit(s) to conduct the attack or assault. This unit(s) engages the enemy at the objective area by fire and maneuver or movement. It also operates immediately in the objective area (searching, demolition, prisoners of war, etc.). Paragraph 9001 outlines the general organization of combat patrols. As in the case with reconnaissance patrols, the task organization of a combat patrol depends on the specific mission assigned. If any special requirements are generated because of the specific mission, the patrol is task-organized to fit the needs of the mission.

13002. Equipment

Combat patrols are armed and equipped as necessary for accomplishing the mission. In addition to binoculars, wire cutters, compasses, and other equipment generally common to all patrols, it usually carries a high proportion of automatic weapons and grenades. Communications with higher headquarters is important as success of the mission may depend on being able to call for supporting fires. Also, internal radio communications with the units and teams may be useful. However, the patrol must not be so overburdened with equipment as to impede movement or mission accomplishment.

13003. Raid Patrols

A raid is a surprise attack on an enemy force or installation with the attacking force withdrawing after accomplishing its mission. Raids destroy or capture enemy personnel or equipment, destroy installations, or free friendly personnel who have been captured by the enemy. Patrolling techniques are used in planning and when moveing to and from the objective. (Refer to MCWP 3-41.2, *Raids*.) Surprise, firepower, and violence of action are the keys to a successful raid. Patrols achieve surprise by attacking—

- When the enemy is least prepared (e.g., during periods of poor visibility such as darkness, rain, fog, or snow).
- From an unexpected direction. (This might be accomplished by approaching through a swamp or other seemingly impassable terrain.)
- With concentration of firepower at critical points within the objective.

Planning

A successful raid requires detailed planning. The leader of a combat patrol engaged in raiding must anticipate probable situations and decide upon definite courses of action to meet them. Rehearsals are imperative.

A raid patrol conducts such missions as destroying an enemy outpost or seizing prisoners from an observation post or lightly defended position.

While preparing for the mission, the patrol leader requests fire support required for the accomplishment of the mission. If practical, artillery and mortars should be employed to isolate the objective to prevent movement of enemy reinforcements into the area.

Execution

The leader's plan must be detailed and complete. All of the considerations outlined in chapter 11 must be covered. Patrol formations must provide for ease of control and all-around security while moving to and from the objective area and provide for rapid and coordinated deployment of the various units once the objective area is reached. The leader's plan usually includes the encirclement of the hostile position—either physically or by fire—in order to isolate it during the assault.

The final simultaneous assault against the objective develops when enemy defensive fires at the objective are suppressed by either friendly fire superiority or surprise. The assault is covered by the fire of the unit assigned the function of support by fire.

Grenades, SMAWs, and demolitions are most effective for clearing bunkers.

Security units are posted to isolate the objective. The patrol leader signals them when the withdrawal begins. As a minimum, security is on each flank and to the rear (at the objective rally point).

Actions at the Objective

The patrol leader halts the patrol near the objective at the final preparation position. Security is established and the leader's reconnaissance is made with appropriate subordinate leaders. When the leaders return to the patrol, they confirm previous plans or announce any changes. Movements are arranged so all units reach their positions simultaneously. This improves the patrol's capability for decisive action, if prematurely detected by the enemy.

The teams of the security element move to positions to secure the objective rally point, give early warning of enemy approach, block avenues of approach into—and prevent enemy escape from—the objective area. As the assault element moves into position, the security element informs the patrol leader of all enemy activity, firing only if detected or on the patrol leader's order. Once the assault element commences action, the security element prevents enemy entry into or escape from the objective area. The security element covers the withdrawal of the assault element (and support element, if employed) to the objective rally point, withdrawing only upon order or prearranged signal.

As the assault element approaches the objective, it deploys early enough to permit immediate assault if detected by the enemy. Each team uses stealth while moving into proper position. On command, or if one or more of the assault element is detected and fired upon by the enemy, the support element opens fire to neutralize the objective, then ceases or shifts fire according to prearranged plans and signals. As supporting fires cease or shift, the assault element assaults the objective. Demolition, search, and other teams are protected by the assault element while they work. On order, the assault element withdraws to the objective rally point.

If a support element is employed, its leader deploys teams to provide fire support for the assault element. Each member of the support element must know the scheme of maneuver to be used by the assault element, specific targets or areas to be neutralized by fire, and the signals that will be employed to commence, shift, and cease fires. The support element withdraws on order of the patrol leader. At the objective rally point, the patrol leader quickly reorganizes the patrol and begins the return movement to friendly lines.

13004. CONTACT PATROLS

General

Contact patrols establish and/or maintain contact to the front, flanks or rear by—

- Establishing contact with an enemy force when the definite location of the force is unknown.
- Maintaining contact with enemy forces through direct and/or indirect fires, or observation.
- Avoiding decisive engagement with the enemy.

Task Organization and Equipment

Task organization and equipment depend on the known enemy situation and anticipated enemy contact. A patrol sent out to establish contact with an enemy force is organized, armed, and equipped to overcome resistance of light screening forces in order to gain contact with the main enemy force. It is not organized and equipped to engage the main enemy forces in combat. Communication is paramount; radios must be reliable over the entire distance covered.

Actions at the Objective

The patrol leader selects a series of objectives. Once an objective is reached, the patrol leader initiates a planned set of actions in order to establish and maintain contact with the enemy. These plans and actions are guided by the missions to establish or maintain contact—not to engage in decisive combat. Contact with the enemy is maintained for the purposes of surveillance, applying pressure, and preventing seizure of the initiative. If the contact patrol becomes decisively engaged with the enemy, many of the tasks originally assigned to the patrol cannot be accomplished, since the enemy has seized the initiative and friendly forces are not forced to react.

13005. AMBUSH PATROLS

General

An ambush is a surprise attack from a concealed position upon a moving or temporarily halted target. It is one of the oldest and most effective types of military actions. Ambush patrols conduct ambushes of enemy patrols, resupply columns, and convoys. The intent of an ambush is to place the enemy in a dilemma where staying in the kill zone or attempting to move out of it prove equally lethal. The ambush may include an assault to close with and decisively engage the enemy, or the attack may be by fire only.

Purpose of Ambushes

Ambushes are executed for the general purpose of reducing the enemy's over-all combat effectiveness and for the specific purpose of destroying its units. The cumulative effect of many small ambushes on enemy units lowers enemy troop morale and harasses the enemy force as a whole.

Destruction is the primary purpose of an ambush because loss of men killed or captured, and loss of equipment and supplies destroyed or captured, reduces the overall combat effectiveness of the enemy.

Harassment, though less apparent than physical damage, is a secondary purpose of ambushes. Frequent ambushes force the enemy to divert men from other missions to guard convoys, troop movements, and carrying parties. When enemy patrols fail to accomplish their mission because they are ambushed, the enemy is deprived of valuable information. A series of successful ambushes causes the enemy to be less aggressive and more defensive minded. His men become apprehensive, overly cautious, reluctant to go on patrols, seek to avoid night operations, are more subject to confusion and panic if ambushed, and in general, decline in effectiveness.

Classification of Ambushes

A *deliberate ambush* is one in which prior information about the enemy permits detailed planning before the patrol departs for the ambush site. Information needed to plan a deliberate ambush includes the size, composition, and organization of the force to be ambushed; how the force operates; and the time it will pass certain points or areas. A deliberate ambush may be planned for such targets as—

- Any force if sufficient prior information is known.
- Enemy patrols that establish patterns by frequent use of the same routes or habitually depart and re-enter their own areas at the same point.
- Logistic columns.
- Troop movements.

An *ambush of opportunity* is conducted when available information does not permit detailed planning before the patrol departs. This is the type of ambush that an infantry unit normally conducts. An ambush of opportunity should not be confused with a hasty ambush. An ambush of opportunity is a planned ambush; a hasty ambush is an immediate action. In planning for an ambush of opportunity, the patrol must be prepared to execute any of several courses of action based on the types of targets that may be ambushed and must rehearse prior to departure. The course of action taken is determined when the opportunity for ambush arises.

The patrol leader may be directed to reconnoiter an area for a suitable ambush site, set up at the site selected, and execute an ambush against the first profitable target that appears.

The patrol may depart just after dark, move to a specific point, observe until a designated time, ambush the first profitable target after that time, and return before daylight.

A *hasty ambush* is an immediate action where the patrol makes visual contact with an enemy force and has time to establish an ambush without being detected. The actions for a hasty ambush must be well rehearsed and accomplished through the use of hand and arm signals given from the patrol leader.

Types of Ambushes

There are two types of ambushes: point and area. The *point ambush* is one where forces are deployed to attack along a single killing zone. The *area ambush* is one where forces are deployed as multiple related point ambushes.

A point ambush, whether independent or part of an area ambush, is positioned along the enemy's expected route of approach. Formation of the forces conducting the ambush is important because, to a great extent, the formation determines whether a point ambush is able to deliver the heavy volume of highly concentrated fire necessary to isolate, trap, and destroy the enemy.

The ambush formation to be used is determined by careful consideration of possible formations and the advantages and disadvantages of each in relation to—

- Terrain.
- Visibility.
- Forces.
- Weapons and equipment.
- Ease or difficulty of control.
- Target to be attacked.
- Combat situation.

For a detailed discussion of ambush formation, see appendix D.

Ambush Operation Terms

The *ambush site* is the location where an ambush is established.

The *killing zone* is that portion of an ambush site where fires are concentrated to trap, isolate, and destroy the target. On little-traveled roads, an obstacle placed in a defile, in the woods, on a bridge or on a steep upgrade can be used effectively to force vehicles to halt, and thus render the occupants vulnerable to attack. Antitank mines may be emplaced and the occupants of the wrecked vehicle killed or captured while still dazed by the explosion.

A *near ambush* is a point ambush where the attacking force is located within reasonable assaulting distance of the killing zone (50 meters is a guide). A near ambush is most often conducted in close terrain, such as a jungle or heavy woods.

A *far ambush* is a point ambush where the attack force is located beyond reasonable assaulting distance of the killing zone (beyond 50 meters is a guide). A far ambush may be more appropriate in open terrain offering good fields of fire or when the target will be attacked by fire only.

Factors for a Successful Ambush Patrol

There are many factors that give the ambush its best chance of success. The ideal situation would be to position the ambush on favorable terrain and have detailed planning completed beforehand.

The patrol leader of an ambush looks for the most favorable terrain in which the enemy is canalized between two obstacles with limited opportunity to attack or escape. Suitable areas include defiles, small clearings, bends in trails, and steep grades. Dense undergrowth adjacent to the ambush site permits observation from concealed positions. The ambush patrol should have maximum cover and concealment, not only for the firing positions, but for the routes of withdrawal. The enemy should be in an area offering as little protection from fire as possible. Favorable fields of fire include stretches of road, trail or open ground of at least 100 meters for machine guns and 15 meters for rifle fire and grenades. The ambush site can be improved by constructing obstacles—such as felled trees, wire, land mines, or booby traps—to impede the enemy.

Planning

A deliberate ambush or an ambush of opportunity requires thorough planning.

A deliberate ambush plan is based on extensive knowledge of the enemy and terrain, and is planned and rehearsed in great detail. A physical reconnaissance of the ambush site is made during the preparation phase and information gained is incorporated into the plan. All likely immediate actions of the enemy when ambushed are examined. Planned counteractions are developed and rehearsed.

In planning an ambush of opportunity, any available information on the enemy and terrain is used. A tentative plan for the ambush that incorporates all anticipated actions is developed and rehearsed. However, the bulk of planning is done concurrently during the patrol leader's reconnaissance of the prospective ambush site. In a rapidly developing situation, hasty ambush immediate action is employed.

The route and ambush site considerations apply to both deliberate ambushes and ambushes of opportunity. A primary route that allows the patrol to enter the ambush site from the rear is planned. Entering the prospective killing zone is avoided. If the killing zone must be entered to place mines or explosives, care is taken to remove any tracks and signs that might alert the enemy and compromise the ambush. If mines or explosives are to be placed on the far side of the ambush site, or if the appearance of the site from the enemy's viewpoint is to be checked, a wide detour is made around the killing zone. Care is taken to remove any tracks that might reveal the ambush. An alternate route from the ambush site to the objective rally point, as in other patrols, is planned.

Maps and aerial photographs are used to carefully analyze the terrain. When possible, an on-the-ground reconnaissance of the ambush site is made prior to occupation. Obvious ambush sites are avoided as the element of surprise is even more difficult to achieve in these areas. An ambush site must provide for—

- Favorable fields of fire.
- Occupation and preparation of concealed positions.
- Canalization of the target into the killing zone. (An ideal killing zone restricts the enemy on all sides, confining him to an area where he can be quickly and completely destroyed. Natural obstacles, such as cliffs, streams, embankments, or steep grades, are used whenever possible to force vehicles to slow down. Man-made obstacles, such as barbed wire, mines, and craters in the roads, are used to supplement natural obstacles.)
- Covered routes of withdrawal that enable the ambush force to break contact.
- Avoidance of enemy pursuit by fire.

Occupation of Ambush Site

The surrounding area is searched for enemy patrols prior to occupation of the ambush site. Ambush formations are used to physically deploy the patrol in a manner to inflict maximum destruction upon the enemy and to provide maximum security to the patrol. Ambush formations are contained in appendix D.

Positions

The patrol is moved into the ambush site from the objective rally point. Security is positioned first to prevent surprise while the ambush is being established. Automatic weapons are then positioned so each can fire along the entire killing zone. If this is not possible, overlapping sectors of fire are provided to cover the entire killing zone. The patrol leader then selects his position, located so he can tell when to initiate the ambush. Riflemen and grenadiers are positioned and sectors of fire are assigned to cover any dead space left by the automatic weapons. The patrol leader sets a time by which positions are to be prepared. Patrol members clear fields of fire and prepare positions in that order, with attention to camouflage for both.

Suitable Objective Rally Point

An easily located objective rally point is selected and made known to all patrol members. The objective rally point is located far enough from the ambush site so that it will not be overrun if the enemy assaults the ambush. Routes of withdrawal to the objective rally point are reconnoitered. Situation permitting, each man walks the route he is to use and picks out checkpoints. When the ambush is to be executed at night, each man must be able to follow his route in the dark. After the ambush has been executed, and the search of the killing zone completed, the patrol is withdrawn quickly but quietly, on signal, to the objective rally point where it reorganizes for the return march. If the ambush was not successful and the patrol is pursued, withdrawal may be by bounds. The last group may arm mines, previously placed along the withdrawal route, to further delay pursuit.

Local Security

Security must be maintained. Security elements do not usually participate in the initial attack, but protect the rear and flanks, and cover the withdrawal.

Patience

The Marines of the ambush force must control themselves so that the ambush is not compromised. Patience and self-discipline are exercised by remaining still and quiet while waiting for the target to appear, particularly if the patrol occupies the ambush site well ahead of the arrival of the enemy. Patience is necessary so as not to alert the enemy to the presence of the ambush.

Surprise

Surprise must be achieved, or the attack is not an ambush. If complete surprise cannot be achieved, it must be so nearly complete that the target is not aware of the ambush until too late for effective reaction. Surprise is achieved by careful planning, preparation, and execution so that targets are attacked when, where, and in a way for which they are least prepared.

Coordinated Fires

Properly timed and delivered fires contribute heavily to the achievement of surprise, as well as to destruction of the enemy. The lifting or shifting of fires must be equally precise; otherwise, the assault is delayed and the enemy has an opportunity to recover and react. All weapons, mines, and demolitions are positioned and all fires, including those of available artillery and mortars, are coordinated to achieve the following results:

- Isolation of the killing zone to prevent the enemy's escape or reinforcement.
- Surprise delivery of a large volume of highly concentrated fires into the killing zone.

Control

Close control of the patrol is maintained during movement to, occupation of, and withdrawal from the ambush site. This is best achieved through rehearsals and establishment and maintenance of good communications. When the enemy approaches, the temptation to open fire before the signal is given is resisted. The patrol leader must effectively control all elements of the ambush force. Control is most critical at the time the enemy approaches the killing zone. Control measures must provide for—

- Early warning of enemy approach.
- Fire control. Withhold fire until the enemy has moved into the killing zone, then open fire at the proper time.
- Initiation of appropriate action, if the ambush is prematurely detected. Individual patrol members must be prepared to react if detected by the enemy prior to the initiation of the ambush.
- Timely and orderly withdrawal of the ambush force from the ambush site and movement to the objective rally point.

It is important to remember that an ambush patrol should have four distinct signals: one to open fire (with an alternate signal to open fire to be used at the same time as the primary); a signal to cease fire or shift fire; a signal to assault or search the killing zone; and a signal to withdraw. The signal to open fire should meet two criteria: first, it should be the firing of a weapon that will kill the enemy; secondly, it should be a weapon reliable in any weather condition. A good primary signal is a Claymore mine, and an alternate signal would be a closed bolt weapon (M16A2). Open bolt weapons (M240G, M249) should not be relied upon to initiate an ambush.

Execution of an Ambush

The manner in which the patrol executes an ambush depends primarily on whether the ambush's purpose is harassment or destruction. To a lesser degree, the execution of the ambush is determined by whether the ambush is deliberate or an ambush of opportunity.

When the primary purpose is harassment, the patrol seals off the area with security teams to prevent enemy reinforcement and escape. Maximum damage is inflicted with demolitions and automatic weapons fire. The patrol delivers a very heavy volume of fire for a short time and withdraws quickly and quietly. The patrol avoids being seen by the enemy.

When the primary purpose of the patrol is destruction, the area is sealed off with security units. Maximum damage is inflicted with demolitions, antitank weapons, and automatic weapons fire from the support team or element. When these fires cease or shift, an assault is launched into the killing zone with heavy fire and violence to complete destruction. The assault unit provides security, while designated teams search and/or capture personnel and destroy vehicles and equipment. On the patrol leader's command, or by prearranged signal, all units withdraw to the objective rally point and move out quickly.

When the patrol's primary purpose is to obtain supplies or capture equipment, security units seal off the area. Demolitions and weapons are used to disable vehicles. The assault unit must use care to ensure its fire does not damage the desired supplies or equipment. Designated teams secure the desired items; other teams then destroy enemy vehicles and equipment.

The most successful ambush is one where the attacker is deployed and concealed in such a way that the enemy will unknowingly be surrounded by fire. The usual method is for the attackers to deploy themselves along a trail or route the enemy will travel. The enemy is permitted to pass by the center of the attacker's force so that the attack can be made from the front, flank, and/or rear. One or two men are posted well forward and to the rear along the route to prevent any enemy from escaping. All fires should be delivered simultaneously on a prearranged signal.

An effective method of luring the enemy is for an ambushing patrol to cut communication or electrical wire. The patrol then deploys and ambushes the line repair crew when it arrives. Since the line crew may be protected by riflemen, the attackers must be careful to engage the entire party.

Vehicles and foot personnel moving on well-established transportation routes can sometimes be captured by altering or moving directional signs so as to divert the enemy into an area where he can be more readily attacked. The attack can best be accomplished at an obstacle, such as a stream or gully, that forces the enemy to stop or slow down.

After the enemy has been ambushed and destroyed, the unit quickly withdraws over a prearranged route to the objective rally point. Speed is very important, since the noise of the ambush could alert other nearby enemy units.

13006. SECURITY PATROLS

General

Security patrols are assigned missions that may or may not require them to engage the enemy. They are used in proximity to defensive positions, on the flanks of advancing units or in rear areas. Purposes of security patrols are to detect infiltration by the enemy, destroy infiltrators, and protect against surprise and ambush.

In any situation where there is a threat of attack, such as a rear area threatened by guerrillas or a facility that is under threat of a terrorist attack, all Marines, not just the infantry, must know how to conduct a security patrol.

In just the offensive operations, infantry units provide security patrols to screen their flanks, areas, and routes. Whereas, in defensive operations, security patrols are used to prevent the enemy from infiltrating an area, detect and destroy infiltrators, and prevent surprise attacks. In rear areas, particularly when there is guerrilla or terrorist threat, the requirement to conduct security patrols increases for all Marine air-ground task force (MAGTF) units ashore, particularly aviation and combat service support units.

Task Organization and Equipment

Generally, a Marine rifle squad or similar sized organization is considered ideal for security patrols. Communications are important to higher headquarters so that they receive information from the patrol; and communications are important to the patrol to request fire support, etc. The radio the patrol carries must have the range necessary for higher headquarters to be able to receive transmissions from anywhere along the patrol route, and the patrol must have a secondary means of reporting (i.e., flare signals upon contact).

Patrol Procedures

All of the procedures presented in previous sections are to be used in security patrols.

Patrol Planning

Security patrol planning includes—

- Rehearsing prior to departing friendly lines.
- Maintaining communications.
- Support by organic weapons.
- Reinforced if necessary.
- Using varied routes and never establishing a routine pattern.
- Staying within proximity of friendly units.

Patrol Techniques

Within rear areas, an irregular pattern of patrol is established and changed daily. Outside of friendly lines it would be prudent to establish a definite preplanned route for the patrol, of which all adjacent units know the route. The parent unit commander establishes frequent checkpoints for control. If checkpoints are designated, the patrol leader treats them as individual objectives to be searched and cleared.

The patrol has a definite plan as to what to do if contact with enemy is made, how to break contact, how to defend itself, and how to call for supporting fires. It is imperative that patrol members know what to do if they become split or separated; i.e., location of rally points and how to be recovered.

13007. URBAN PATROLS

General

As national strategy continues to focus on regional vice global conflicts, the Marine Corps will continue to conduct urban operations in various operational environments. Cities and towns are often the center of economic and political power and are therefore extremely vulnerable to urban insurgent activities and violence. The lessons learned from recent operations in Somalia, as well as experiences gained by British forces in Northern Ireland, provide the foundations of urban patrolling.

Enemy Forces

Enemy forces in urbanized areas range from organized military forces to low intensity engagements with insurgents, such as terrorists or local gangs. For Marine Corps doctrine and tactics, techniques, and procedures (TTP) for dealing with the higher intensity threat in urbanized areas, refer to MCWP 3-35.3, *Military Operations on Urbanized Terrain.*

Common Threat Tactics

Cities provide cover and concealment for both friendly forces and enemy forces. However, enemy forces generally find active support only in certain areas of the town or city. The urban-based insurgent or terrorist usually lives in a friendly community or in one where the people are too frightened to withhold support or inform anyone about the situation. The insurgent or terrorist normally maintains close contact with leaders and others friendly to the cause. The enemy will often have efficient communication and intelligence systems, sometimes involving women and children to provide cover for its activities.

The urban insurgent normally cannot, like his rural counterpart, establish bases and recruit large military units. He is generally an individual or a member of a relatively small group. He relies on the cover afforded by the city's populace and terrorizing them to coerce loyalty or support.

Urbanized areas tend to give the insurgent and/or terrorist many opportunities to initiate action and gain advantage. The normal presence of large numbers of people in cities provides the foe an opportunity to mass crowds quickly and manipulate demonstrations easily. The presence of women and children during mass demonstrations may restrict the courses of action available to friendly forces. Major incidents stemming from overreaction or excessive use of force by friendly forces may provide the insurgent with propaganda material. Publicity is easily gained in an urbanized area because major incidents can't be completely concealed from the local population. Insurgent successes can be exploited to discredit the ability of host nation police, friendly forces, and the civil government, and gain recognition for the insurgents' cause.

The urban insurgent or terrorist can usually be expected to operate more boldly than his rural counterpart. This is reflected in the enemy's tactics. A single sniper or bomber may be the norm in the urbanized area, whereas the rural threat is generally the more conventional ambush. In urbanized areas, explosive devices can be easily emplaced and used effectively against large groups or select individuals. Enemy forces may be expected to employ the following tactics in urbanized areas:

- Using local communications, such as radio and newspapers, for propaganda purposes.
- Disrupting industry and public services through strikes and sabotage.
- Generating widespread disturbances designed to stretch the resources of the security force.
- Creating incidents or massing crowds in order to lure the patrol or reaction force into a trap.
- Provoking security forces in the hope that they may react improperly, therefore discrediting the security force by means of propaganda.
- Sniping at roadblocks, outposts, sentries, and patrols.
- Attacking friendly bases with rockets and mortars.
- Planting explosive devices, either against specific targets or indiscriminately, to cause confusion and destruction, and lower public morale and confidence.
- Using ambush patrols.
- Firing on friendly helicopters.

Principles of Urban Patrolling

Patrolling in an urban environment often presents conditions considerably different and often more complex than those encountered in rural and less inhabited areas. While the principles of patrolling are still relevant in an urban situation, the nature of urban patrolling has led to the development of six specific urban patrolling principles. They are—

1. **Depth.** The restrictive, canalizing nature of urbanized terrain usually limits a patrol's ability to disperse laterally. To prevent the patrol from bunching up, patrols normally maintain dispersion along the length of a patrol formation.

2. **Mutual Support.** The positioning of units in-depth within the patrol enables one unit to cover another unit's movement and facilitates immediate action during various situations. Aircraft, vehicles, and snipers also provide good mutual support.

3. **Deception and Pattern Avoidance.** Deception and pattern avoidance are normally a planning consideration of the headquarters directing the entire patrolling plan. By varying patrol routes, durations, and departure times, hostile actions commonly used against urban patrols—such as enemy ambushes, and roadblocks—are more difficult to plan and may be preempted.

4. **Intra-patrol Communication.** Elements of an urban patrol must have the means to communicate with each other. Ideally, each element will possess a radio enabling it to remain in continuous communication that facilitates rapid response and reporting to higher headquarters, reaction force coordination, and coordination of actions with other patrols or fire support agencies.

5. **Establishment of a Reaction Force.** The nature of urbanized terrain (its compartmentalization) makes urban patrols more vulnerable to a wider range of hostile actions. The requirement for immediate, coordinated reinforcement of a patrol is best satisfied by employment of an established reaction force. The reaction force requires superior mobility (relative to the enemy's) and fire support to be effective for this task.

6. **Three-Dimensional Threat.** Patrolling in an urban environment requires constant attention to its three-dimensional aspect; hostile actions can originate from rooftops, streets, subsurface levels or combinations of all levels at once.

Classification of Urban Patrols

Mission

The vast majority of urban patrols are overt in nature, with their presence readily apparent to the local populace. Most urban patrols are combat vice reconnaissance patrols. The vast majority of urban patrols are security patrols. However, units may be assigned secondary tasks of reconnoitering specific or general areas along the patrol routes.

Raids normally involve a swift penetration of an objective to secure information, confuse the enemy or destroy installations. Raids include a planned withdrawal upon completion of the assigned mission.

Movement

Means of movement are as follows:

- Dismounted: movement on foot.
- Vehicular: movement by motorized, mechanized or armored vehicle.
- Helicopterborne: movement by helicopter, however, helicopterborne patrols will usually involve dismounted or vehicular movement after patrol insertion.
- Combination: movement using a combination of methods.

Task-Organization

Units task-organize to meet the specific requirements of the mission and situation. The Marine rifle squad is ideally suited for urban patrolling and can easily integrate attached specialists required by specific missions. Specialists who may accompany urban patrols include—

- Interrogator-translator team (ITT) and counterintelligence team (CIT) Marines.
- Explosive ordnance disposal (EOD) personnel.
- Members of host nation or allied military forces.
- Interpreters.
- Local community leaders.
- Local law enforcement officers.
- Public affairs personnel escorting media representatives.

Dismounted Patrol Organization

Patrols should maintain unit integrity (fire team and/or squad) in organizing elements for an urban patrol. Similar to reconnaissance patrols, the vulnerability of urban patrols necessitates that all elements must provide for their own security in addition to the combat aspect of the mission. Combined assault and security teams are an effective method to organize for all-around security.

Urban Patrol Base Operations

Urban patrols may operate from an established patrol base that may be located within the unit's assigned area of operations or an area designated for a patrol. The patrol base should be located in a building used exclusively for this purpose. Patrol bases may also be located within a larger site that houses other agencies, such as a higher headquarters' command echelon.

If located within a larger site, the patrol base will be included in the overall facility security plan. If isolated, the patrol base must consider the following security factors:

- External security:
  - Barrier plan.
  - Sentry posts.
  - Local security patrols.
- Internal security:
  - Covered positions for all Marines.
  - Contingency plan for hostile actions against patrol base.

The Reaction Force

The high probability of an urban patrol becoming involved in a hostile or volatile incident requires the establishment of a dedicated reaction force for rapid reinforcement, support or extraction of the patrol. Ideally, the reaction force is—

- Large enough and task-organized in a manner that it can meet and quickly defeat the expected threat. Reaction forces are normally tiered with a lead element (normally one-third of the unit's size) and a main body force (the other remaining two-thirds of the force). Reaction force response times routinely are determined in advance by higher headquarters.
- Ready to respond immediately.
- Motorized or mechanized and supported by close air support and other fire support.
- Familiar with the area of operations.
- Briefed on the patrol's plans and monitors the status of patrols in progress.
- Task-organized to be multimission capable.
- Able to communicate with the higher headquarters, fire support assets, patrol base, and the patrol.
- Controlled by higher headquarters, once employed.

Patrol Preparation

Planning

Higher headquarters will—

- Designate the area for patrol.
- Provide intelligence briefs and updates.
- Ensure liaison with allied forces and the civilian populace.
- Provide special equipment and personnel required for the mission (scout snipers, public affairs officer, interpreters, etc.)
- Provide urban maps, photos, terrain models as required.
- Consider deception and pattern avoidance when issuing mission.
- Prescribe rules of engagement (ROE).

Intelligence Brief

An intelligence brief is conducted by the S-2 officer or representative prior to a patrol conducting its mission. The brief adresses the situation relevant to the specific patrol (e.g., routes, areas, updated enemy situations).

Coordination

Higher headquarters will effect liaison with adjacent and allied forces, as well as civil authorities and other agencies, having a possible effect on the patrol. The patrolling unit generally follows the same procedures as those used during patrol planning and execution in a jungle or forest environment.

The Urban Patrol Order or Warning Order

The Urban Patrol or Warning Orders use the same format and considerations as noted in this publication for patrolling. They rely heavily on a detailed terrain model, photographs, and subterranean construction to ensure complete understanding of the plan.

Rehearsals

The limited size of the patrol base usually precludes the need for full-scale rehearsals. Immediate action drills, such as crossing danger areas, are rehearsed in as much detail as possible, despite the limited available space.

Inspections

Initial and final inspections are conducted in the same manner as other patrols. Attached personnel must be fully integrated into the patrol and familiar with the plan and unit standing operating procedures (SOPs).

Conducting an Urban Patrol

Movement

Individual and unit movement considerations are generally the same as those for other patrols. However, urban environments require consideration of additional factors. Because of these factors, an urban patrol leader should—

- Ensure that each movement within a patrol takes place under the observation or cover of another individual or element of the patrol.
- Know where cover can be taken in the event of a hostile incident or action.
- Be prepared for contact with civilians, especially children, during the patrol and be aware that they may intentionally attempt to distract patrol members.
- Expect the presence of vehicles (both moving and stationary) along the patrol route.
- Expect members of the patrol to be approached by dogs and what action to take if threatened.

Patrol Formations

Squad-sized Patrols. The need for immediate fire power outweighs the dangers of becoming canalized. In contrast to other types of patrols, the headquarters element of an urban patrol will normally locate at the lead of the patrol column. This allows the patrol leader greater flexibility through control of two combined assault and security (A&S) teams. The leaders of these elements tactically stagger their members on each side of the road (see fig. 13-1 on page 13-12).

A&S teams follow in trace of the headquarters element and maintain unit integrity on separate sides of the street. One unit will remain slightly to the rear to create a staggered interval between Marines on either side of the street. This allows A&S teams to take lateral routes in support of headquarters element without having to cross a street to do so.

Platoon-sized Patrols. Squads will generally travel abreast of each other, moving along parallel routes. The interval between squad-sized units and/or teams is situation-dependent, but is usually between 100 and 150 meters (roughly two city blocks; this often prevents visual contact between the units). The intent is to create less of a target to an aggressor, yet still allow the patrol to quickly react to an incident. Individuals within units or teams will move in a staggered column as in a squad-sized patrol. See figure 13-2, which is located on page 13-13).

Night

Night patrols will generally be at least squad-sized and will generally use the same formation as that for day patrols. At night, it may be necessary to close distances between individuals or elements to maintain control. Consideration should be given to the use of night vision devices and thermal weapons sights.

Navigation, Control, and Security Measures

The patrol leader is ultimately responsible for the navigation. The headquarters element normally functions as the base unit during movement. The designated navigator is normally assigned from within the headquarters element. City maps are often inaccurate or outdated; however, when used with aerial photographs and other navigational aids, they can be effective for urban navigation.

Checkpoints and phase lines should be related to major streets (alleys, buildings, bridges) for easy identification. Arrival at checkpoints and crossing of phase

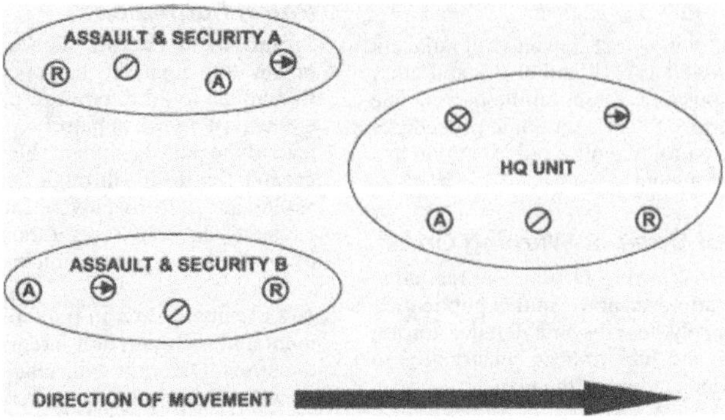

Figure 13-1. Squad-sized Dismounted Urban Patrol Formation.

lines should be relayed to higher headquarters using established brevity or codewords. A detailed patrol overlay is issued to both the reaction force and higher headquarters to keep them advised of routes and control measures used. A terrain model should be maintained at higher headquarters to aid in supervision and control of the patrol.

The use of camouflage should be limited to avoid frightening and confusing the local populace. The often overt nature of urban patrols may negate the need for camouflage. Patrols are sometimes deployed to show force presence and usually move on the urban street in plain view.

During daylight, patrols will routinely vary their rate of movement ranging from short halts to brief periods of double-timing. The British term for this urban patrolling technique is *hard-targeting*, meaning it makes the patrol harder for an enemy to target. Altering the rate of movement is intended to frustrate the enemy's ability to coordinate an attack or ambush against a targeted patrol.

Patrols should use short security halts, with Marines taking up mutually supporting firing positions. Marines must always work in pairs, ensuring mutual support. The last Marine in the element will provide rear security, but stays in his buddy's sight.

Individual Tasks

Individuals may be assigned collateral tasks performed throughout the patrol that may increase the patrol's situational awareness. Individual tasks may include—

- Vehicle spotter: looks for suspicious or known insurgent vehicles.
- Personnel spotter: observes and attempts to recognize previously identified enemy in crowds.
- Talker: attempts to gain information from casual conversations with the local populace. (Talkers are usually subordinate leaders or Marines with foreign language skills.)
- Searcher: conducts physical searches of vehicles and personnel while other patrol members provide cover and security.
- Marksman: engages point targets when the tactical situation does not permit massed or high volume fire. Other patrol members provide security to cover the marksman's engagement.

Departure of Friendly Lines

Urban patrols must vary their departure times to prevent being ambushed while exiting the patrol base.

Individual elements will usually depart exits at staggered times and at different movement rates, especially where sentries cannot provide cover. (This technique is used when the environment adjacent to the patrol base is dangerous.)

Once an element has exited friendly lines, a short halt is conducted in a predetermined, covered initial rally point 50 to 100 meters from the base. This ensures all elements are in position before the patrol continues.

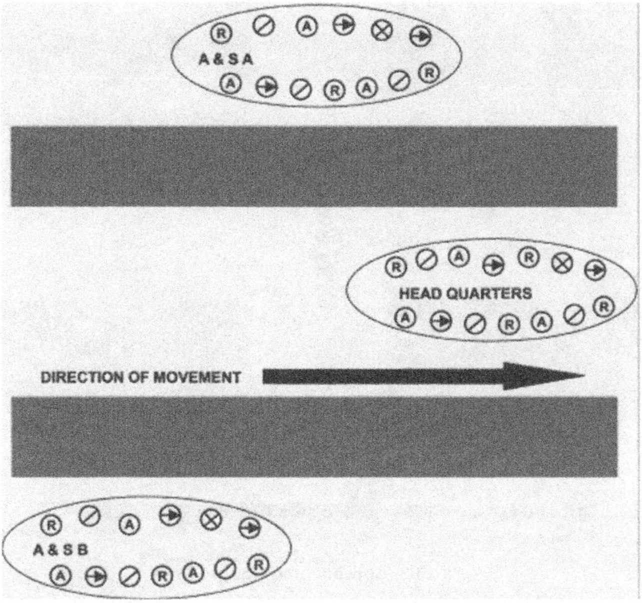

Figure 13-2. Platoon-sized Dismounted Urban Patrol Formations.

Exit points, routes from the base, departure techniques, and locations of IRPs should be varied constantly to avoid setting a pattern. This pattern avoidance may also include using vehicles to insert patrols away from the urban patrol base and employing empty vehicles as part of a deception plan.

Danger Areas

Urban patrols may encounter hundreds of danger areas during a single patrol. The three-dimensional threat requires keen situation awareness by every patrol member. Many danger areas can be dealt with simply by avoidance, while others require an adjustment of patrol formation, movement rate, etc. In the urban environment, places to be treated as danger areas are points that pose a major threat to the patrol, such as local political and religious headquarters, weapons containment areas, roads and routes that canalize movement and direct fire, and any area with a history of repeated contact.

Near and far side rally points are designated and briefed during the issuance of the patrol order. Squad-sized formations may use the A&S teams to provide flank security for the headquarters element and for each other. The headquarters element identifies the danger area and takes up a position on the near side of intersection. Individuals provide all-around security (see fig. 13-3 on page 13-14).

Two Marines (one from each A&S team) are designated to move through the headquarters element and establish respective firing positions on the near side of the danger area covering the patrol's near side flanks. They are followed by a second pair (again, one Marine from each A&S team) that moves across to the far side of the danger area and establishes respective firing positions covering the patrol's far side flanks (see fig. 13-4 on page 13-14).

Once near and far side flank security is established, the headquarters element moves across to the far side of danger area (see fig. 13-5 on page 13-15).

The remaining A&S team members then cross the danger area and join the headquarters element on the far side (see fig. 13-6 on page 13-15).

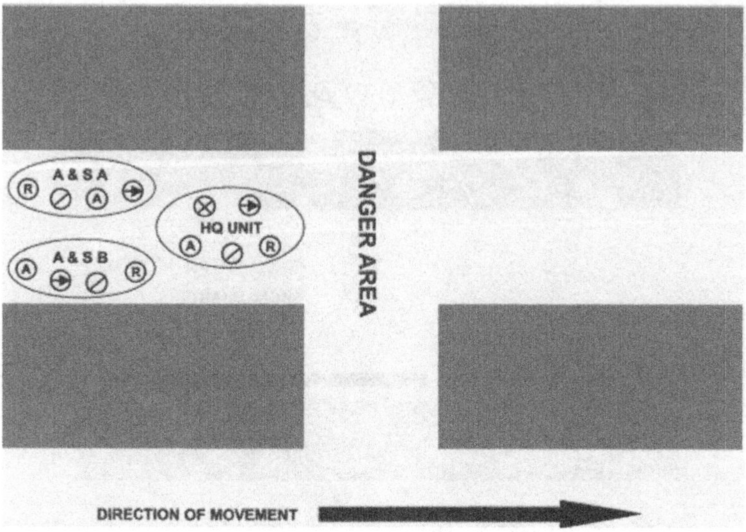

Figure 13-3. Approaching Danger Area.

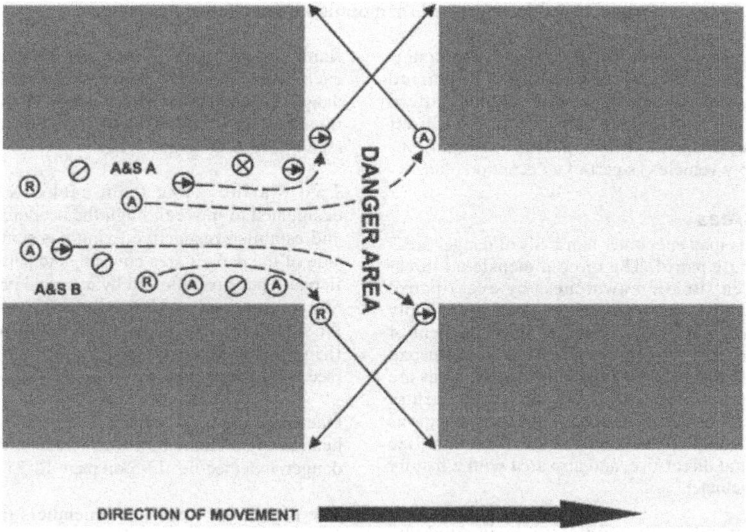

Figure 13-4. Securing Flanks.

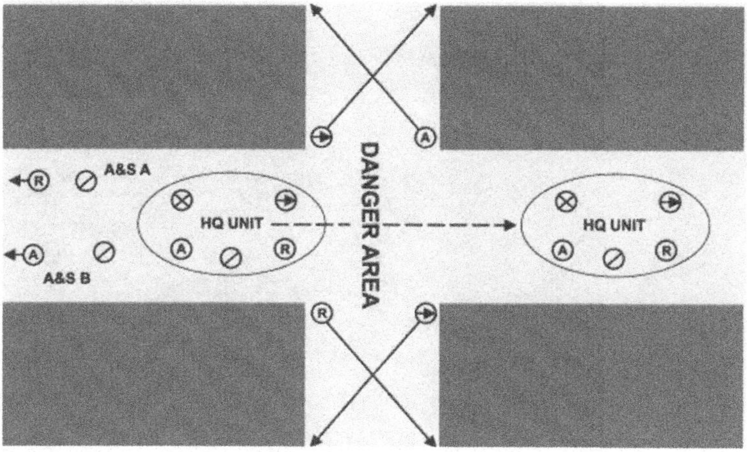

Figure 13-5. HQ Element Moves Through.

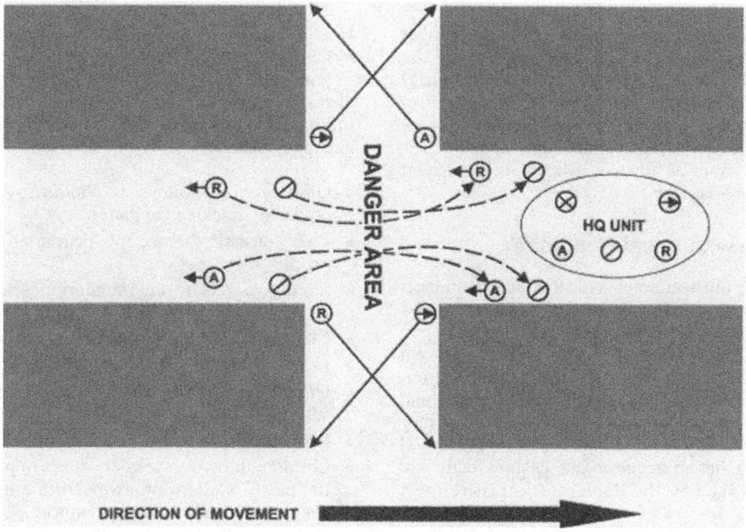

Figure 13-6. A&S Elements Move Through.

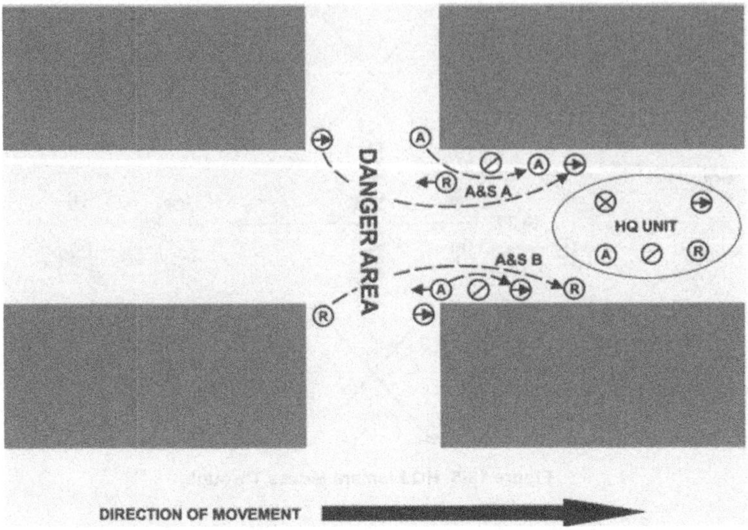

Figure 13-7. Patrol Resumes Movement Along Intended Route.

Once the trail A&S team members cross the danger area, they take up rear security and cover the movements of the near and far side flank security teams as they return to their positions in the patrol formation. The near side security team should collapse back first, followed by their far side counterparts. The patrol then resumes its advance away from the danger area (see fig. 13-7).

Interaction with Local Populace

Urban patrols must interact with the local populace. Patrols are at first a novelty to the civilians but can quickly become an unwelcome intrusion. The movement of the patrol must be fast enough to prevent the enemy from massing their fires upon it, but deliberate enough to ensure adequate security and mutual support. Patrol members must realize that they are usually the only Marines the local populace will encounter and that an inappropriate gesture, comment or act could lead to the deterioration of rapport between U.S. forces and the general population. Marines must remember that the vast majority of the individuals with whom they come in contact will be noncombatants attempting to survive in trying political, economic, and social situations.

Hostile incidents often seem to occur spontaneously, but there are usually indications that can alert Marines to imminent danger. The most obvious are the sudden alteration of normal routines, patterns, and attitudes of the local populace or other unusual activity. Some examples include—

- Observers on rooftops, in windows, etc., who are obviously tracking the patrol.
- The unusual absence of pedestrian traffic and people on porches.
- Stores, markets or street vendors closed suddenly or without explanation.
- Changes in civilian attitude toward patrol members.
- Unknown individuals or vehicles in the patrol area.
- Unfamiliar vehicles parked in the patrol area (possible car bomb).
- Roadblocks.
- Children throwing rocks at patrols to possibly draw the patrol's attention away from a more serious danger, such as a deliberate ambush.
- Vehicles riding unusually low due to overloading (possibly ferrying people, weapons, explosives).
- Agitators trying to provoke an incident with patrol members.

- Absence of the usual stray dogs (dogs are adept at sensing danger and avoiding it).
- Anti-American graffiti suddenly appearing in the patrol area.
- Pictures of enemy leaders and martyrs posted in the patrol area.
- Civilian workers failing to appear at U.S. or friendly bases.
- Normal deliveries and pick-ups conducted late or early without reason.
- Sudden change of civilian sentiment in newspaper articles, radio broadcasts or other media.
- Women and children leaving to live elsewhere.

Immediate Actions Upon Enemy Contact

Reaction to Sniping

Snipings are often executed from a single firing point, but coordinated snipings delievered from multiple points are not uncommon.

Patrol element or team leaders should constantly try to identify likely firing points and anticipate their own reactions to a possible shooting. Normally, snipers in an urban environment have a detailed withdrawal plan. Once a patrol comes into contact with a sniper, the patrol leader must immediately assess the situation and maneuver his patrol accordingly. The patrol's mission, location, size, ROE, and location of the threat often determine whether the patrol will attempt to neutralize the targeted sniper. If the patrol leader decides to kill or capture the sniper, he uses planned and rehearsed immediate actions to maneuver and counter the sniper's assault. The goal is to kill the sniper or cut off his escape and capture him. There are three immediate reactions to neutralizing a sniper: initial contact, immediate follow-up, and subsequent follow-up.

Initial Contact. The initial contact is made when the sniper fires the first shot. The patrol must react immediately and positively to get behind the firing position in order to kill or capture the gunman. The period of contact ends when the gunman is killed or captured, or the patrol element or team leader on the scene ends it. The following technique is the same for both squad- and platoon-sized patrols:

- The element or team in contact attempts to identify the firing position and maneuvers designated marksman into position to return well-aimed and controlled fire. Other members of the patrol take up positions to cover the marksman's engagement.
- The patrol element or team leader in contact sends initial contact report to the patrol leader, who notifies higher headquarters.
- The element or team leader in contact determines appropriate cut-off positions and relays them to flanking elements or teams.
- The patrol element or team leader in contact continues to observe the firing point, but does not enter it due to the possibility of booby traps. Flank elements or teams set up along likely escape routes.
- The incident ends when either the sniper ceases fire or is neutralized.

Immediate Follow-Up. Regardless of the fate of the gunman, isolation of the firing point is necessary to prevent reinforcement and preserve forensic evidence (scent, spent casings, etc.). If not under fire, members of the patrol element or team cordon off the area surrounding the firing point. Flanking elements or teams maintain their positions and prevent civilians from entering the area. The patrol leader moves to link-up with the element or team in contact (if not his own), and makes an estimate of the situation. The patrol leader sends a SPOTREP to higher headquarters.

Subsequent Follow-Up. The aim of the subsequent follow-up is to use follow-on forces to clear the building of remaining resistance or to obtain evidence that can be used to capture the gunman. The patrol leader establishes a position where he can brief arriving units (reaction force commander, S-2 representative, EOD personnel, etc.). Once the arriving units have been briefed, recommendations are made to higher headquarters via radio. No one is allowed into the cordon without the patrol leader's approval.

Reaction to Becoming Decisively Engaged

If a patrol becomes decisively engaged from numerous firing positions, the following immediate action should be taken:

- All patrol members move to available cover and return accurate fire on identified firing points.
- The patrol leader assesses the situation and makes a decision to either request the reaction force or break contact.
- If the reaction force is requested, the patrol will maintain its position until the reaction force arrives. The patrol should use fire and maneuver to gain better tactical positioning and support the arrival of the reaction force. When the reaction force arrives,

its commander may decide to either clear occupied buildings or cover the patrol during its extract.

Reaction to Bomb Threat or Discovery

The use of command-detonated explosive devices is a common ambush tactic employed by a terrorist or insurgent in an urban environment. The appropriate response to a reported threat or an actual discovery is generally involves four steps (known as the four Cs)—

1. The patrol leader **CONFIRMS** the presence of the suspicious item.

2. Without touching or moving anything suspicious, patrol elements or teams **CLEAR** the immediate danger area to a minimum of 100 meters. The area is cleared from the suspected device outward, inform civilians as to the reason for evacuation.

3. A&S teams establish a **CORDON** to secure the cleared area. Avenues of approach are cordoned off to keep people out and to protect EOD or engineer personnel clearing the device. The assistant patrol leader acts as the cordon commander and informs the patrol leader when the cordon is secure. An effective cordon technique is to tape off the area with engineer tape, creating both a physical and psychological boundary.

4. **CONTROL** of the area is maintained throughout the bomb clearing operation by the patrol leader. The patrol leader sends a report to a superior concerning details of the device (if known) and the area affected. The patrol leader coordinates with arriving personnel (EOD, engineers, etc.). The patrol leader maintains communication with the assistant patrol leader and keeps the Marines informed of the progress of the clearing operation.

Reaction to a Bomb Detonation

Bombs may be used by an insurgent as a means of initiating an ambush on mounted or dismounted patrols, in which case the actions for decisive engagement apply. Immediate action in response to an isolated explosion is similar to that used in reaction to a sniping and breaks down into the same three phases:

Initial Contact. The patrol leader attempts to identify the likely initiation point and sends an initial contact report to higher headquarters. If the bomb was command-detonated, the patrol leader sends his A&S teams deep to cut off the bombers' escape routes. Any casualties are moved a minimum of 100 meters from explosion and out of the line of sight to it.

Immediate Follow-Up. The A&S teams may need to maneuver to positions behind the likely initiation point to cut off escape. Once in position, personnel checks are conducted and any suspects are detained. The patrol leader coordinates requests for required support (MEDEVAC, reaction force, etc.)

Subsequent Follow-Up. Due to the possibility of secondary detonations, the four Cs (confirm, clear, cordon, and control) can be conducted as in reaction to a bomb discovery or bomb threat.

Civil Disturbances

Urban patrols must prepare to react to spontaneous aggression by the local populace. In many cases, civil disturbances are organized by the enemy to draw dismounted patrols into a targeted area, or to distract them from enemy activity occurring elsewhere. Civil disturbances are generally divided into two categories: minor aggressive actions, and full-scale rioting.

Minor aggressive actions are activities characterized by rock-throwing or use of devices such as Molotov cocktails and may either be directed at the patrol or take place between different ethnic factions of the population. Minor aggressive actions are normally spontaneous in nature and may have minimal or limited objectives for the insurgents.

Full-scale rioting events are usually in response to another major event or incident that may enflame the populace. Full-scale riots are well-planned and orchestrated, with clear objectives or targets in mind. At times, patrols will need to attempt to maintain control of a civil disturbance situation; however, dismounted and mobile small unit patrols should generally avoid potential flashpoints. Procedures to handle civil disturbances are as follows:

- The patrol leader reports the incident to headquarters and attempts to diffuse the crowd by talking to crowd leaders.
- If the patrol leader determines the size of the disturbance is too large for the force to handle, the patrol should move away from the disturbance to a safer, more remote covered area and occupy positions to observe and report the situation to higher headquarters. To prevent the patrol from being pursued by the crowd, the patrol should move quickly and change direction, often at road junctions, to gain distance from it.

Patrol members should maintain dispersion to create a more difficult target. They should face the crowd at all times to see and avoid any projectiles thrown. Individual self-discipline must be maintained throughout the disturbance. Marines charging into the crowd or throwing objects back at the crowd will only worsen the situation. If pursued or trapped, the patrol leader may consider using riot control agents (combat support, pepper gas, etc.) to disperse or slow a crowd's movement. The use of riot control agents must be authorized under the established ROE. If the patrol leader believes the situation is deteriorating beyond the patrol's ability to control it, the patrol leader should request the reaction force, which may be better equipped to handle a large riot or mob.

Break Contact

As with patrols in rural areas, the patrol leader may be forced to break contact as a result of decisive engagement with the enemy. On the basis of his estimate of the situation, the patrol leader will normally break contact in one of the following ways:

- As a patrol, with elements providing cover for movement as defined by clock direction and distance.
- As individual units/teams taking separate routes out of the area, then linking up at a designated rally point a safe distance away from the engagement.

As in any contact with enemy forces, smoke may be employed to screen movement. Fire support agencies can be utilized to suppress targets; riot control agents can be employed to disrupt enemy movement.

Reentry of Friendly Lines

The reentry of a dismounted patrol into an urban patrol base is no different from that of a patrol conducted in a rural area. The same planning considerations and control methods apply.

Missions Related to Urban Patrolling

House Calls

House calls missions are usually part of a coordinated effort to collect information within the area of operations. They involve obtaining up-to-date information on particular houses and occupants. When possible, local police should accompany patrols to do the actual talking to the occupants. If this is not possible, a technique that may be used by an urban patrol is—

- A&S elements move to provide cover around the target house.
- The headquarters element provides security just outside the house.
- The patrol leader and one Marine for security enter the house, if invited. If not invited, they talk to the occupants on the doorstep. Patience and tact are required in requesting information. An interpreter should be present when language differences exist.

Vehicle Checkpoints

Urban terrorists or insurgents commonly use vehicles to transport personnel, weapons, explosives, and equipment. Civilian vehicles are often used for these purposes, creating the requirement to check as many vehicles as possible. While permanent, fortified checkpoints may be conducted along approaches into an urbanized area, dismounted patrols can be employed to establish hasty vehicle checkpoints to stop vehicles and to keep the enemy off-guard. The two common types of vehicle checkpoints patrols establish are hasty and deliberate.

Hasty checkpoints are deployed anywhere based upon the decision of the patrol leader. Patrols must not set patterns through the frequent use of the same sites.

Deliberate checkpoints are tasked by higher headquarters to achieve a specific purpose. Time and locations are carefully considered to avoid setting patterns.

The general layout for a squad-sized, two-way dismounted checkpoint is depicted in figure 13-8 on page 13-20. The technique is as follows:

- The A&S teams are positioned stealthily in blocking positions on both sides of the road.
- Both the patrol leader and assistant patrol leader act as "talkers" for each direction of traffic (with local police or interrogator or translator Marines acting as interpreters) while a two-man team from the headquarters element physically searches the vehicles.
- Obstacles or parked vehicles may be employed to create a staggered roadblock in center of the checkpoint to slow approaching vehicles.
- The checkpoint location should be sited so that approaching vehicles cannot see it until they have passed a security team, and they have no escape route then available.
- Signs announcing the checkpoint should be displayed a safe distance from the search area for safety to both drivers and Marines.
- Normally higher headquarters will issue criteria that determines which vehicles are searched, but random checks of cars should normally be made as

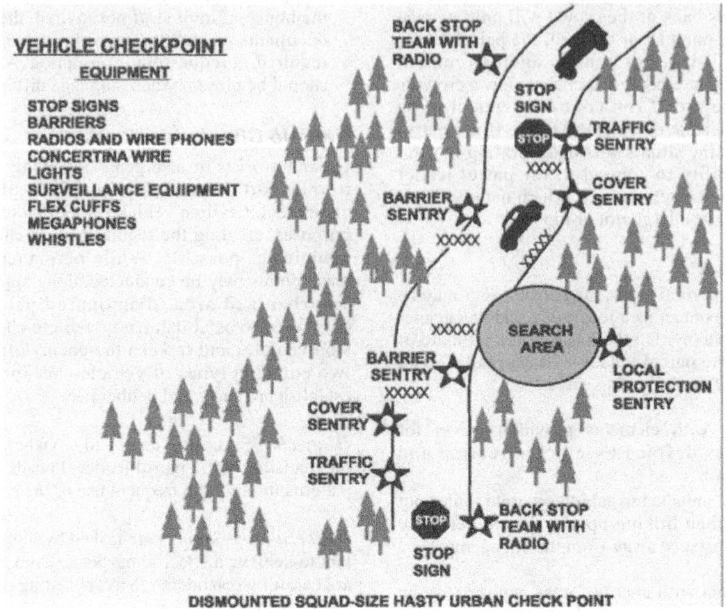

Figure 13-8. Dismounted Squad-sized Hasty Urban Vehicle Checkpoint.

well. While the vehicle is being searched the driver should accompany the searcher around the vehicle.

- Vehicle occupants should be made to exit the vehicle and then searched. Whenever possible, women should be searched by female police or female Marines.
- All patrol members should conduct themselves with courtesy and politeness. If nothing is found, an apology for the inconvenience is recommended.
- A hasty vehicle checkpoint should not be conducted any longer than 30 minutes for security reasons.
- The ROE should dictate whether or not action should be taken against vehicles that fail to stop at the checkpoint. Failure of a vehicle to stop does not automatically give authorization to fire.

Observation Posts

Urban observation posts are established to provide extended security, not only for patrol bases but also for patrols operating within the observation post's sector of observation. Observation posts can be established in conjunction with sniper operations and for providing overwatch for patrols operating within their sector of observation. Observation posts are nor-mally positioned on dominating terrain or in buildings outside the patrol base itself.

Insertion to the observation posts and conduct of observation may be either overt or clandestine in nature. Overt observation posts usually will be hardened positions to increase security. A patrol provides cover while the observation post is being inserted. A clandestine observation post relies on stealth of insertion and occupation for protection. It is normally positioned in abandoned buildings to cover sectors of observation that overt observation posts cannot. Because of their nature, clandestine observation posts are difficult to successfully establish and should not be manned for an extended period of time. Orders establishing observation posts (and patrols) must address the method of extraction as well as actions upon compromise/attack.

Cordon and Search

The cordon and search mission involves isolating a predesignated area by cordoning it off and systematically searching for enemy personnel, weapons, supplies, explosives or communications equipment. While large-scale cordon and search operations are planned and rehearsed in advance and normally entail extensive coordination with local law enforcement agencies, a squad-sized urban patrol may often conduct a cordon and search of a point target—searching one house or building identified by intelligence as a possible weapons cache.

The basic principle of a search of a populated area is to conduct it with limited inconvenience to the population. The populace may be inconvenienced to the point where they will discourage urban guerillas or insurgents from remaining in the area, but not to the point that they will assist the enemy as a result of the search.

Upon receiving intelligence that warrants the searching of a building or a specific tasking from higher headquarters, A&S elements of the patrol move to establish an inner cordon around the target building to seal it off, with the primary intent of preventing movement out of the targeted building.

On order, the designated reaction force deploys to establish an outer cordon, oriented outward some distance from the inner cordon and covering routes leading into the area in order to prevent outside interference/reinforcement. The reaction force maintains a reserve to reinforce either cordon or react to unfolding events (civil disturbance in response to the operation).

Once the cordons have been established, the patrol leader, with the assistance of local police or interpreters, informs the local populace that a building is about to be searched, that a house curfew is in effect (if permitted by higher headquarters), and that all occupants should remain indoors. Occupants of the target house are instructed to gather at a central location to stay out of the way of the search party.

The headquarters element, having linked up with any required assistance (explosive ordnance disposal (EOD), ITT, etc.) now acts as the search party and accompanies local police. A female searcher should be included in the party, if necessary.

Occupants are searched and screened first for possible enemy personnel. Apprehended persons are evacuated as soon as possible.

The head of the household should accompany the search party throughout the operation to be able to counter incriminating evidence and possible accusations of theft and looting against Marines. If possible, a prominent member of the local community should act as a witness.

Buildings are best searched from top to bottom. Ideally, the search is conducted with the assistance of combat engineers using mine detectors to locate hidden arms and ammunition.

If the targeted building is empty or the occupant refuses entry, it may be necessary to forcefully enter the premises to conduct a search of the dwelling. If an unoccupied house containing property is searched, arrangements should be made with the local community to secure it until its occupants return. Unnecessary force and damage to property should be avoided during the search.

Motorized Urban Patrols

The advantages of a motorized urban patrols is their ability to capitalize on the speed, mobility, and protection offered by various vehicles. They may be motorized, mechanized or armored vehicles or a combination. Generally, motorized urban patrols possess greater combat power than dismounted patrols and can cover larger areas faster than dismounted patrols.

The disadvantages of motorized urban patrols is that they are restricted to roads and are vulnerable to ambush by the enemy. They are also restricted in their ability to interact with the local populace.

Motorized patrols are generally organized in the same manner as dismounted patrols (see fig. 13-9). Unit integrity is maintained when assigning personnel to specific vehicles.

The urban patrolling principles apply to motorized patrols in much the same manner as dismounted patrols. Mutual support and depth are achieved by maintaining constant observation between vehicles and coordinating support with any dismounted patrols in the area. All-around security is achieved through the use of constant observation as well as the vehicle's mobility and firepower. Positive communications between units or teams are maintained through vehicle radios. Patrol routes and speeds are varied to promote deception or pattern avoidance.

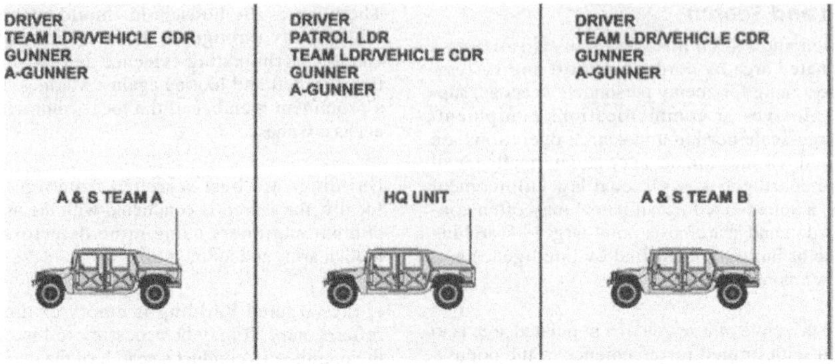

Figure 13-9. Organization of a Squad-sized Motorized Patrol.

The canalizing nature of streets and alleys force vehicular patrols to use a traveling overwatch movement technique to reduce vulnerability to ambushes. All vehicles travel at a moderate rate of speed with the lead vehicle stopping only to investigate potential danger areas. If vehicles must stop in danger areas, designated crew members' will dismount to provide security. The gunner will remain at the ready and in the turret while the driver remains in the driver's seat with the vehicle running.

Vehicles should move at a high rate of speed only when responding to an incident. At all other times, vehicle speed should be between 15 to 20 mph to allow for quick reaction and good observation. Distances between vehicles should be approximately 50 meters (one half to one city block) or such that visual contact and mutual support are ensured. Particular care is taken at major road junctions and other danger areas to ensure individual vehicles do not become isolated.

Vehicles with doors removed generally enhance observation and overall security, yet expose Marines to thrown objects, theft and concealment.

CHAPTER 14. INFORMATION AND REPORTS

It is necessary that patrol leaders and all patrol members be trained in observing and reporting their observations accurately. The leader of a patrol should have all members of the patrol immediately signal or report any information obtained. These reports should not be restricted to information about the enemy, but should also include information about the terrain, such as newly discovered roads, trails, swamps, and streams. The leader includes all information in the report to the officer dispatching the patrol. (Refer to MCRP 2-15.3, *Reconnaissance Reports Guide,* for detailed report formats.)

14001. REPORTING

The officer dispatching the patrol instructs the patrol leader on whether and when messages are to be sent back during the patrol and what communication means to use. Messages may be oral or written. They must be accurate, clear, and complete. Every message should answer the question what, where, and when. For a detailed discussion on reporting, see paragraph 6002.

Verbal Messages

A patrol leader sending a verbal message should make it simple, brief, and avoid using numbers and names. The messenger should accurately repeat the message back to the patrol leader before leaving.

Written Messages

In preparing written messages, the patrol leader must distinguish between fact and opinion. Information about the enemy should include: strength; armament and equipment; actions; location and direction of movement; unit destination, if known; time enemy was observed; and the patrol's location when the observation was made. Use of an overlay or sketch may often simplify the message.

Messengers

A messenger team is given exact instructions as to where to deliver the message and the route to take. Any information obtained along the route should be reported at the time the message is delivered. Messengers must be given all practical assistance. If in danger of capture, the messenger immediately destroys the message.

Radio and Signals

If the patrol is provided with a radio, a definite radio schedule for checking in must be arranged before departure of the patrol. The patrol leader takes every precaution to ensure that codes and copies of messages are not captured by the enemy. If a close reconnaissance of enemy lines is required, the radio should be left in a concealed location at a safe distance from the enemy. Once a report is sent by radio, the patrol should immediately leave the area to avoid the possibility of detection by enemy locating devices. Pyrotechnics (flares, colored smoke, grenades) and air-ground panels may also be used by patrols for reporting information by a prearranged signal.

SALUTE Report

Information must be reported as quickly, accurately, and as completely as possible. An established method to remember how and what to report about the enemy is to use the acronym SALUTE:

Size
Activity
Location
Unit
Time
Equipment

An example of such a report is: "Seven enemy soldiers, unit unknown, traveling SW, crossed road junction on BLACK RIDGE at 211300 August carrying one machine gun and one rocket launcher."

14002. CAPTURED ITEMS

Every patrol should make a practice of searching enemy casualties, prisoners, and installations first for booby traps, then for equipment, papers, maps, messages, orders, diaries, and codes. Search techniques should be practiced often in order to reduce the time exposed to potentially dangerous situations. Items found are collected by the patrol leader and turned in with the patrol report. The items found are marked as to time and place of capture. When possible, captured items should be linked to a specific prisoner who possessed the items or to the place where the items were found. When this is done, the enemy prisoner of war tag and item tag are marked accordingly. The patrol leader must impress upon the members of the patrol the importance of turning in all documents and equipment. Furthermore, the patrol leader must ensure all information gathered by the patrol is quickly disseminated.

14003. PRISONERS

A patrol normally does not capture prisoners unless required by the mission. If prisoners are taken, the "5S and T" rule applies. This memory aid stands for: search, segregate, silence, speed, safeguard, and tag.

Search. Prisoners are body-searched thoroughly for weapons and documents as soon as they have been captured. This search must include the helmet, body armor, and gas mask. These items are left with the prisoner for protection until the patrol is completed. Weapons, equipment, and documents are tagged and immediately sent to the patrol leader.

Segregate. Prisoners are segregated into isolated groups: officers, noncommissioned officers (NCOs), privates, deserters, and civilians. By segregating prisoners, it makes it more difficult for leaders to organize escapes and issue orders to subordinates.

Silence. Silence is essential. Do not allow prisoners to talk to each other.

Speed. Speed is required in getting prisoners to the commander who dispatched the patrol. Timely information secured from prisoners is essential.

Safeguard. Prisoners are safeguarded as they are moved. They are restrained, but not abused. If the patrol will soon reach friendly positions, prisoners are not given cigarettes, food or water until they have been questioned by interrogators. If the patrol will not return to a friendly position for a lengthy time, the prisoners are given food, water and medical aid.

Tag. Equipment and personal effects recovered from prisoners are tagged so that they may later be matched back to specific individuals.

14004. PATROL REPORT

Every patrol leader makes a report when the patrol returns. Unless otherwise directed, the report is made to the person ordering the patrol. If the situation permits, the report is written and supported by overlays and/or sketches. The patrol leader's report should be a complete account of everything of military importance observed or encountered by the patrol while on the assigned mission. It should include the following information:

- Size and composition of patrol.
- Tasks and purpose (mission).
- Time of departure.
- Time of return.
- Routes, out and back (show by sketch, azimuth, trace on map).
- Terrain (general description to include any man-made or natural obstacles and critical terrain features which, if occupied by either enemy or friendly forces, would allow them to control the surrounding area).
- Enemy (size, activity, location, unit, time, equipment).
- Any map corrections (show on map).
- Miscellaneous information not covered elsewhere in report.
- Results of enemy encounters.
- Condition of patrol, including disposition of any dead or wounded.
- Conclusion and recommendations.

14005. PATROL CRITIQUE

After the patrol has rested and eaten, the patrol leader should hold a critique. Constructive criticism is made. It is an excellent time to prepare for future patrols by going over lessons learned as a result of the patrol.

Appendix A. Patrol Warning Order

The warning order is issued as soon as practical with all available information included to assist patrol members in preparation.

1. Situation. Friendly and enemy situation information necessary for initial preparation.

2. Mission. Statement of what the patrol is to accomplish, and the purpose for accomplishing it. When, how, and where will be discussed in the patrol order.

3. Execution

a. Task Organization. General patrol organization and assignment of responsibilities if known; otherwise, promulgate in the patrol order.

b. Tasks. Alert subordinate leaders to patrol tasks (e.g., stream crossing, helicopter rappelling, demolitions) requiring preparation prior to departure.

c. Coordinating Instructions

(1) Time schedule and location for individual preparation, rest, briefings, in-spection, rehearsal, and departure.

(2) Time, place, uniform, and equipment for receiving the patrol order.

(3) Tasks for subordinate leaders to direct and supervise the initial preparation which may include drawing ammunition, rations, and special equipment; conducting immediate action drills or other necessary individual or unit training; meeting and briefing attachment personnel; reconnoitering the area for passage of lines; and coordinating with the necessary unit leaders.

(4) Preliminary guidance to specialists and key individuals regarding their roles and organization within the patrol.

4. Administration and Logistics

a. Individual uniform, equipment, weapons, and prescribed load of rations, water, and ammunition.

b. Crew-served weapons (if required) and guidance regarding distribution of weapons and ammunition during movement.

c. Special equipment requirements (wire cutters, demolitions, radios, flashlights, infrared equipment, mines, binoculars) and their distribution during movement.

d. Restricted or prohibited items.

5. Command and Signal

a. Designation of assistant patrol leader and his role in preparation.

b. Designation of navigators and radio operators (if required).

c. Brief outline of patrol leader's schedule for preparation, and where he can be reached.

Appendix B. Patrol Order

A patrol order follows a warning order. Any of the following subjects that have been addressed in the warning order may be omitted with the exception of the mission statement. A patrol order is more detailed than a 5-paragraph squad order in that a great deal of attention is given to individual duties.

The patrol order also provides orientation information, both enemy and friendly, that impacts the patrol and enemy forces. This information can include, but is not limited to, weather, terrain, visibility, NBC considerations, local population situation, terrain model and/or map orientation, and behavior.

1. Situation

   a. Enemy Forces

      (1) Composition, disposition, and strength are based on size, activity, location, unit, time, and equipment (SALUTE).

      (2) Capabilities and limitations to defend, reinforce, attack, withdraw, and delay (DRAW-D).

      (3) Enemy's most probable course of action.

   b. Friendly Forces

      (1) Mission of next higher unit (task and commander's intent).

      (2) Adjacent unit missions (task and intent). Also identify left, front, right, and rear.

   c. Attachments and Detachments. (date and time effective).

2. Mission. The mission is the task to be accomplished, and its purpose (who, what, where, when, and why). For patrols, specify if the mission or time has priority.

3. Execution

   a. Commander's Intent and Concept of Operations

      (1) Commander's intent.

      (2) The concept of operations tells the where, how, and who and lays out the patrol leader's general scheme of maneuver and fire support plan. It outlines the following:

(a) Task organization of the patrol.

(b) Movement to the objective area, to include navigation method.

(c) Actions in the objective area.

(d) The return movement, to include navigation method.

(e) Use of supporting forces (including illumination, if required).

b. Subordinate Element Missions. Subordinate element missions (task and purpose) are assigned to elements, teams, and individuals, as required.

c. Coordinating Instructions. This paragraph contains instructions common to two or more elements, coordinating details, and control measures applicable to the patrol as a whole. At a minimum, it includes—

(1) Time of assembly in the assembly area.

(2) Time of inspections and rehearsals (if not already conducted).

(3) Time of departure and estimated time of return.

(4) Location of departure and reentry of friendly lines and the actions associated with departure and reentry.

(5) Details on the primary and alternate routes to and from the objective area.

(6) Details on formations and order of movement.

(7) Rally points and actions at rally points.

(8) Final preparation position and actions at this position.

(9) Objective rally point and actions at this point.

(10) Actions at danger areas.

(11) Actions in the event of enemy contact.

(12) Details on actions in the objective area not covered elsewhere.

(13) Estimated time of patrol debriefing upon return.

4. Administration and Logistics

a. Changes/additions to uniform, equipment, and prescribed loads from that given in the warning order.

b. Instructions for handling wounded and prisoners.

5. Command and Signal

a. Command Relationships. Identify key leaders and chain of command.

b. Signal. Challenge and password, arm and hand signals, special signals, and radio frequencies and call signs.

APPENDIX C. PATROL EVALUATION CHECKLIST

This appendix provides a comprehensive checklist of critical patrolling steps, techniques, and procedures to aid unit leaders to critique patrol performance during training. Unit leaders may use the list as they observe the performance of a patrol to provide the patrol leader with a detailed analysis of the performance.

Warning Order	Yes	No
Did the PL ensure all patrol members were present before issuing the warning order?		
Did the PL issue a brief statement on the enemy situation?		
Did the PL issue a brief statement on the friendly situation?		
Did the PL state the mission in a clear, positive manner?		
Did the PL list all members of the patrol including attachments?		
Was the chain of command for the patrol covered fully?		
Were all members of the patrol asigned positions and duties in particular squads and teams?		
Were all the necessary individual duties assigned?		
Did the PL follow established principles in organizing the patrol into squads and teams?		
Was each patrol member assigned a particular weapon to carry on patrol?		
Did the PL list all special equipment needed to accomplish the mission?		
Was required special equipment assigned to the proper element to carry?		
Did the PL select uniform and equiment common to all based on METT-T?		
Coordination with Adjacent Units		
Did the PL coordinate with other patrols operating to the right and left?		
Was the route out and back coordinated?		
Was the time of departure and return cocordinated?		
Were call signs and frequencies coordinated?		
Was a signal for the FPF coordinated so as not to approach friendly lines during this time?		
Coordination With Front Line Units		
Did the coordinator pass the size of the patrol?		
Was the time of departure and return coordinated?		
Did the coordinator give a general area of oeprations for the patrol?		
Did the coordinator ask for information on known or suspected enemy positions and/or obstacles?		
Did the coordinator ask about information on the latest enemy activity?		
Did the coordinator ask for detailed information on friendly fire support avalable and the unit's barrier plan?		
Was the location of the IRP established and coordinated?		
Did the coordinator ask the forward unit to monitor their patrol frequency?		
Was the current challenge and password confirmed?		
Did the coordinator request that all information coordinated be passed on to any relieving unti?		
Patrol Order		
General		
Did the PL check to ensure all patrol members were present before issuing the patrol order?		

	Yes	No
Did the PL issue the patrol order in a forceful, confident manner?		
Did the PL make maximum use of available visual aids in issuing the patrol order (i.e., terrain model, sand table, map board, chalkboard)?		
Did the PL issue the patrol order in correct sequence?		
Did the PL issue the entire patrol order without allowing interruptions by patrol members?		
Did the PL adequately answer all questions asked by patrol members		
Did paragraph 1a (Enemy Situation) include—		
A weather forecast for the period of operation?		
A description of the terrain over which the patrol was to operate?		
Identification or description of enemy units known to bein the area of operations?		
Known locations of enemy units?		
Recent activity of enemy units?		
Stength of enemy units in the operating area?		
Did paragraph 1b (Friendly Situation) include—		
Mission of the next higher unit?		
Locations, missions, and planned actions of units on right and left?		
Fire support available to support the patrol (as per the coordination)?		
Missions and routes of other patrols operating in the immediate area?		
Did paragraph 1c (Commander's Intent) include—		
All views and ides of what the commander wants to be accomplished?		
The final result desired by the commander?		
Did paragraph 1c (Attachments and Detachments) include—		
All attachments to patrol and effective time of attachment?		
All detachments from patrol and effective time of detachment?		
Did paragraph 2 (Mission) include, at a minimum—		
Who was to conduct the patrol?		
What the patrol was tasked to do (i.e., conduct a point recon patrol)?		
Where the action was to take place?		
Why the action was to take place (i.e., the purpose)?		
Did paragraph 3 a (Commander's Intent and Concept of Operations) include—		
A complete concept of operations?		
A detailed description of the mission of elements?		
A detailed description of the mission of teams?		
A detailed description of the duties of specific individuals (i.e., navigator, compass man, pace m an, corpsman, APL)?		
Time of departure and time of return?		

	Yes	No
Type of formations and order of movement to be used?		
Description of the route and alternate routes(s) to include azimuths and distances between checkpoints as a minimum?		
Techniques to be used in the departure from friendly areas?		
Techniques to be used for the reentry into friendly areas?		
Location of the IRP and a tentative ORP, to include grid coordinates and recognizable terrain features?		
Method for designating and passing rally points?		
Actions to be taken at rally points if their use became necessary?		
Actions to be taken in the event of enemy contact?		
Actions to be taken at danger areas?		
Complete, detailed description of actions to be taken at the objective?		
Times and locations for rehearsals and the order of priority for rehearsals?		
Times and lcoations for inspections and the methods of conduct (to include uniform and equipment to be worn and/or carried)?		
Where, when, and by whom the debriefing is to be conducted?		
Did paragraph 4 (Administration and Logistics) include—		
Rations to be carried or reference to the warning order if there is no change?		
Arms and ammunition to be carried or reference to the warning order if there is no change?		
Uniform and equipment to be worn and carried or reference to the warning order if there is no change?		
Method for handling wounded or dead?		
Method for handling prisoners?		
Did paragraph 5a (Signal) include—		
Complete description of all signals, code words, frequencies, and call signs to be used within the patrol?		
Were signals adequate?		
Radio call signs to be used?		
Primary and alternate frequencies to be used?		
Call signs and frequencies to be used with other units in the area of oeprations (if applicable)?		
Required reports to be transmitted to higher headquarters, along with occasions for reporting?		
Code words and brevity codes to be used between the patrol and higher headquarters?		
Current challenge and password to be used in friendly held areas?		
Current challenge and password to be used forward of FEBA?		
Did paragraph 5b (Command) include—		
Chain of command or reference to the warning order if there is not change?		
Location of the PL during all stages of operation?		

	Yes	No
Location of the APL during all stages of operation?		
Inspection Rehearsal		
General		
Did the P L adequately use the allotted inspection time?		
Was the inspection conducted in an orderly manner?		
During the inspection, did the PL check the following:		
Completeness and correctness of uniform?		
Completeness of all equipment necessary to accomplish the mission?		
Operational condition of equipment?		
Did the PL question members to ensure they knew—		
The mission of the patrol?		
The concept of operations?		
Their individual duties and responsibilities?		
Chain of command and succession?		
Duties and responsibilities of key personnel?		
Did the PL utilize the rehearsal area to conform as much as possible to area of operations?		
Were the following major actions rehearsed:		
Actions at the objective?		
Actions at danger areas?		
Actions on enemy contacts?		
Departure and reentry of friendly lines?		
Did the PL critique each action after it had been rehearsed?		
Were interpatrol communications and control measures checked at rehearsal?		
Did the PL maintain control of the patrol during the rehearsal?		
Movement		
Did the PL use proper formations for movement?		
Did the PL adequately control the point team?		
Did the PL check the compass men?		
Was the PL aware of his position at all times?		
Was light and noise discipline enforced?		
Was the rate of movement appropriate for the mission?		
Did the PL make full use of pace?		
Did the PL make full use of count?		
Was proper security maintained during movement?		
Did the PL select and pass proper rallying points?		
Did the PL ensure that all members knew location of rallying points?		

	Yes	No
Did the PL properly use arm and hand signals?		
Did the PL recognize and halt the patrol a safe distance from a danger area?		
If the PL conducted a map check, did he conduct it properly (light discipline, security, etc.)		
Did designated individuals reconnoiter the far side of a danger area properly?		
Were support teams emplaced properly?		
Did the PL use a proper formation in crossing the danger area?		
Were reports made on enemy contact and at checkpoints?		
Actions on Enemy Contact		
Did the PL halt patrol a safe distance from a tentative ORP?		
Did the PL issue an adequate frag order before departing to look for a tentative ORP?		
Did the PL take appropriate personnel to look for the ORP?		
Did the PL select a suitable ORP?		
Did the PL secure the site adequately?		
Did the patrol occupy the ORP as stated in the patrol order or subsequent frag order?		
Did the patrol occupy the ORP in an orderly manner?		
Did the PL maintain control of the patrol during occupation of the ORP?		
Did the PL issue a satisfactory contingency plan to the APL before departing on a leader's recon?		
Did the PL take appropriate personnel on a leader's recon?		
Did the PL maintain or ensure that the patrol leader had communications with the patrol?		
Did the patrol avoid being detected by the enemy during the leader's recon?		
If contact was made, did the PL take appropriate action?		
Was security maintained during the leader's recon?		
Did the PL ensure the objective was kept under surveillance?		
Were appropriate orders given to surveillance teams?		
If a leader recon proved a tentative ORP unsuitable, did the PL move the ORP?		
Was a satisfactory leader's recon conducted?		
Was the ORP move conducted in a satisfactory and orderly manner?		
On return to the ORP, did the PL issue appropriate frag order and allow enough time for dissemination?		
Actions at Objective		
Did the PL issue a frag order for action at the objective?		
Was security emplaced prior to the teams departing from the ORP?		
Did the PL employ the terrain at the objective to the best advantage?		
Was stealth maintained while moving into the objective?		

	Yes	No
Did the Pl exercise positive control of squads, teams, and individuals at the objective?		
Were actions at the objective in accordance with the details outlined in the patrol order?		
Did the PL make use of supporting arms at the objective?		
Was the action at the objective successful?		
Was withdrawal from the objective accomplished quickly and orderly?		
Did units withdraw according to the details reflected in the patrol order?		
Was reorganization at the ORP completed in an expeditious and orderly manner?		
Did the patrol withdraw from the ORP quickly and quietly?		

Appendix D. Ambush Formations

Formations discussed here are identified by names which correspond to the general pattern formed on the ground by deployment of the assault element.

Line Formation

The assault element is deployed generally parallel to the route of movement of the enemy. This positions the assault element parallel to the long axis of the killing zone and subjects the target to heavy flanking fire. The size of the force which can be trapped in the killing zone is limited by the area the assault element can effectively cover. The enemy is trapped in the killing zone by natural obstacles, mines, demolitions, and direct fires. (See fig. D-1.)

A disadvantage of the line formation is the chance that lateral dispersion of the target may be too great for effective coverage. The line formation is appropriate in close terrain that restricts enemy maneuver, and in open terrain where one flank is protected by natural obstacles or can be protected by mines and demolitions. Similar obstacles can be placed between the attack force and the killing zone to provide protection from possible enemy counter attack. When an ambush is deployed in this manner, access lanes are left so that the enemy can be assaulted. (See fig. D-2.)

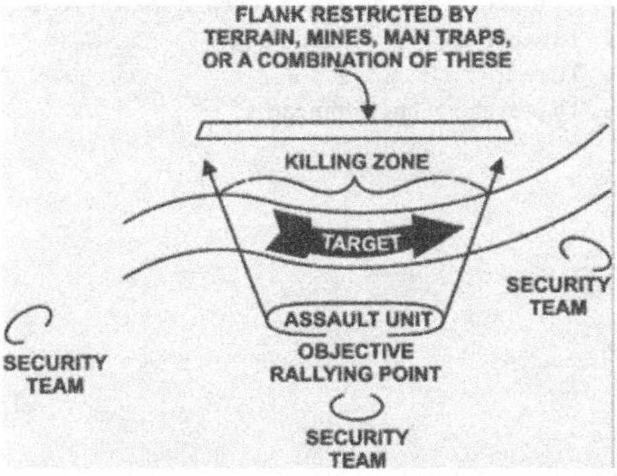

Figure D-1. Line Formation (Harassing or Destruction Ambush).

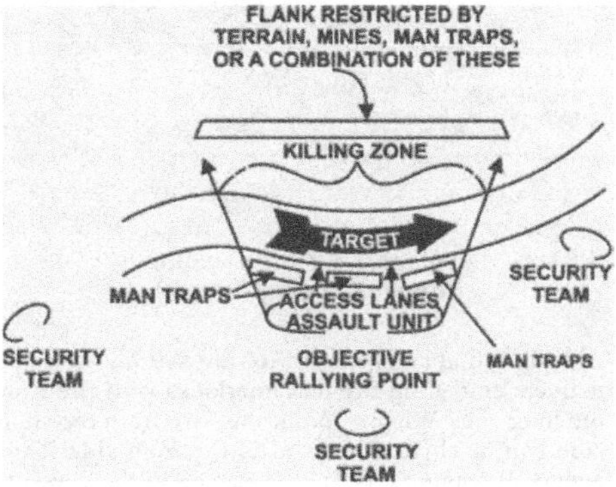

Figure D-2. Line Formation (Access Lanes for Assault of Target).

The main advantage of the line formation is its relative ease of control under all conditions of visibility.

L Formation

The "L" shaped formation is a variation of the line formation. This formation is very flexible because it can be established on a straight stretch of a trail or road (see fig. D-3) or at a sharp bend in a trail or a road (See fig. D-4 on page D-2).The long side of the assault element is parallel to the killing zone and delivers flanking fire. The short side of the attack force is at the

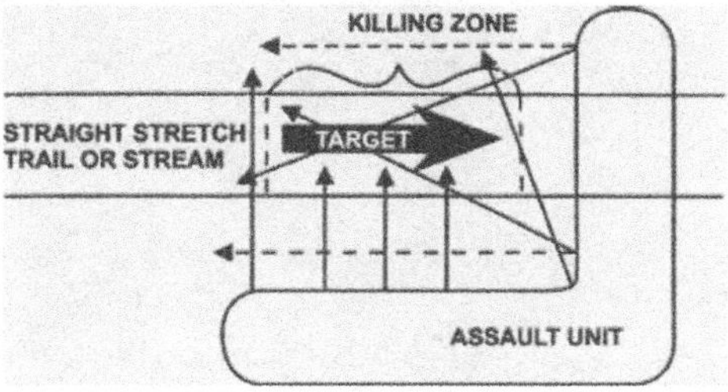

Figure D-3. "L" Formation (Destruction Ambush).

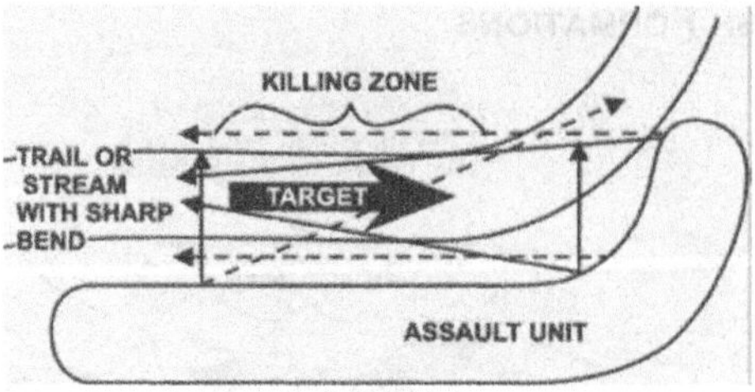

Figure D-4. "L" Formation (Bend of Trail or Stream).

end of, and at right angles to, the killing zone and delivers enfilading fire that interlocks with fire from the long side. When appropriate, fire from the short side can be shifted to parallel the long side if the enemy attempts to assault or escape in the opposite direction. In addition, the short side prevents escape and reinforcement. (See fig. D-5.)

"Z" Formation

The "Z" shaped formation is another variation of the line formation.

The assault element is deployed as in the "L" formation, but with an additional side so that the formation resembles a "Z". (See fig. D-6).

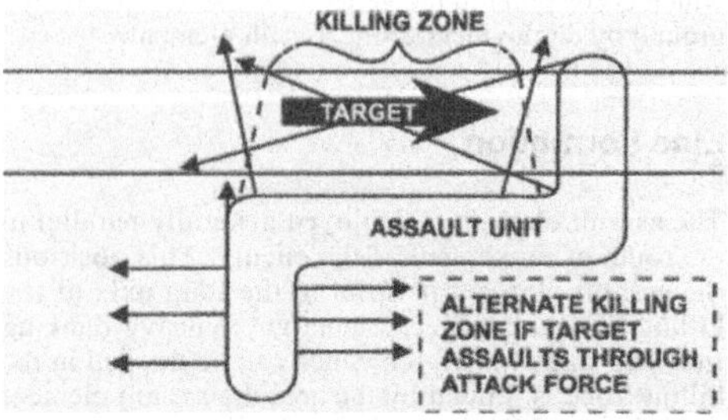

Figure D-6. "Z" Formation.

The additional wing may serve any of the following purposes:

- To engage an enemy force attempting to relieve or reinforce the enemy unit engaged in the kill zone.
- To seal the end of the killing zone.
- To restrict a flank.
- To prevent an envelopment.

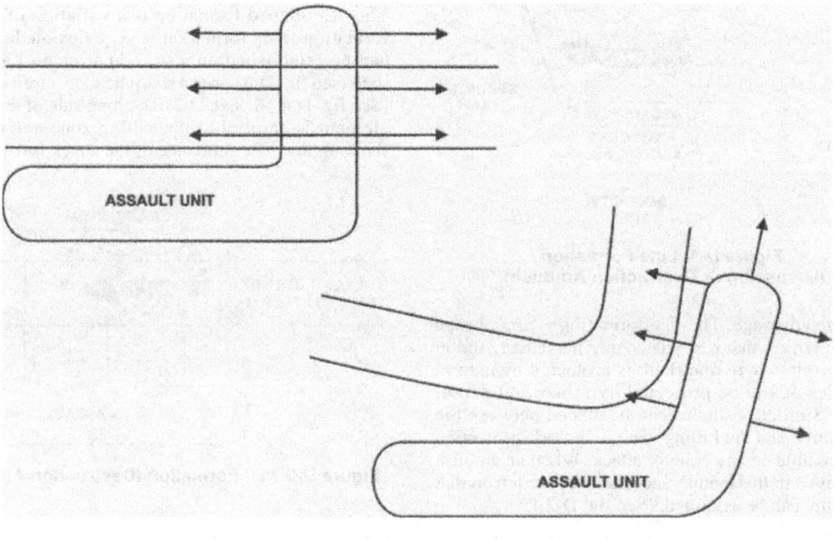

Figure D-5. "L" Formation.

"T" Formation

In the "T" shaped formation, the assault element is deployed across (perpendicular to) the enemy's route of movement so that its position forms the crossing of a "T" at the top. (See fig. D-7.)

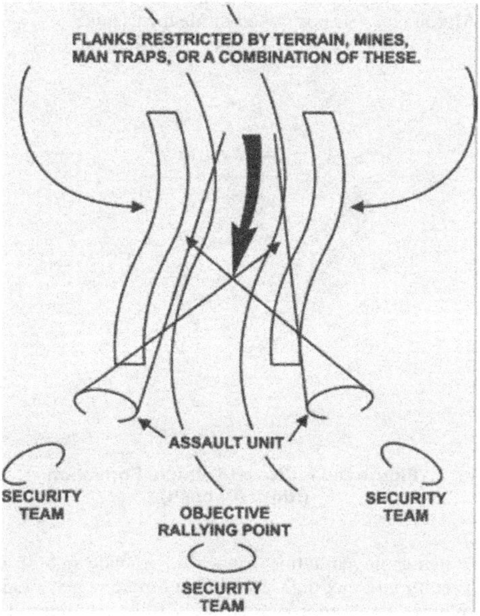

Figure D-7. "T" Formation.

This formation can be used day or night to establish an ambush to interdict movement through open areas that are hard to seal off.

A small force can use the "T" formation to harass, slow, and disorganize a larger force. When the lead elements of the enemy are engaged, they will normally attempt to maneuver right or left to close with the ambush. Mines and other obstacles placed to the flanks of the killing zones slow the enemy's movements and permit the ambush force to deliver heavy fire and withdraw without becoming decisively engaged.

The "T" formation can be used to interdict small groups attempting night movement across open areas. For example, the assault element is deployed along an avenue of approach with every second man facing the opposite direction. The attack of the enemy approaching from either direction requires only that every second man may shift to the opposite side of the formation. Each man fires only to his front and only when the enemy is at a very close range. Attack is by fire only and each man keeps the enemy under fire as long as it remains to his front.

If the enemy attempts to escape in either direction along the killing zone, each Marine takes the enemy under fire as the enemy comes into the Marine's sector of fire. The "T" formation is very effective at halting infiltration. But it has one chief disadvantage; there is a possibility that the ambush will engage a superior force at night while spread out. (See fig. D-8.)

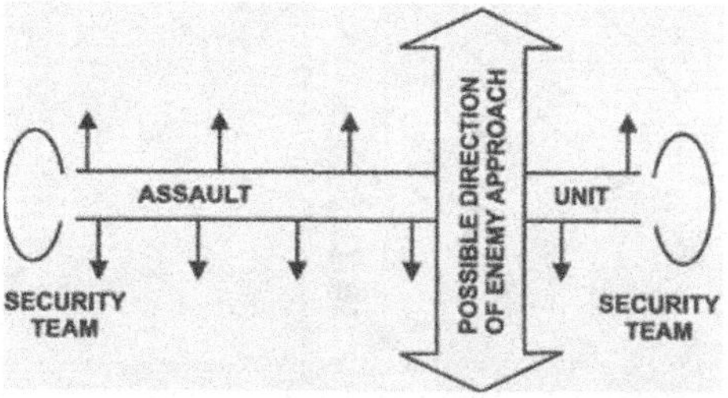

Figure D-8. "T" Formation (Target Approaching from Either Direction).

The "V" Formation

The "V" shaped formation is deployed along both sides of the enemy's route of movement so that it forms a "V"; care is taken to ensure that neither group (within the "V") fires into the other.

This formation subjects the enemy to both enfilading and interlocking fire. The "V" formation is best suited for fairly open terrain but can also be used in close terrain. When established in close terrain, the legs of the "V" close in as the lead element of the enemy force approaches the apex of the "V", and opens fire at a close range.

Here, even more than in open terrain, all movement and fire must be carefully coordinated and controlled to ensure that the fire of one wing does not endanger the other wing. The wider separation of forces makes this formation difficult to control, and there are few sites that favor its use. Its main advantage is that it is difficult for the enemy to detect the ambush until it is well into the killing zone. (See figs. D-9 and D-10.)

Triangle Formation

The triangle is a variation of the "V" formation and can be varied in three ways:

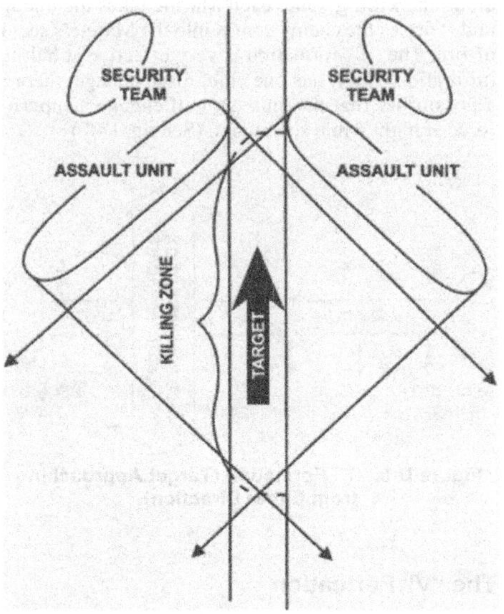

Figure D-9. "V" Formation (Open Terrain).

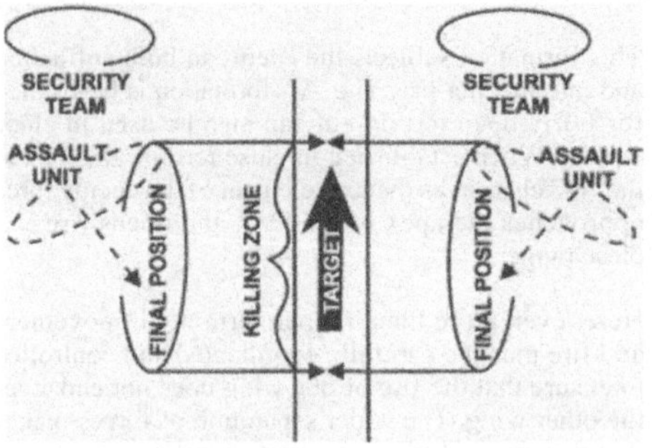

Figure D-10. "V" Formation (Close Terrain).

Closed Triangle Formation

(See fig. D-11.) The assault element is deployed in three teams, positioned so that they form a triangle (or closed "V"). An automatic weapon is placed at each point of the triangle and positioned so that it can be shifted quickly to interlock with either of the others. Men are positioned so that their sectors of fire overlap. Mortars may be positioned inside the triangle.

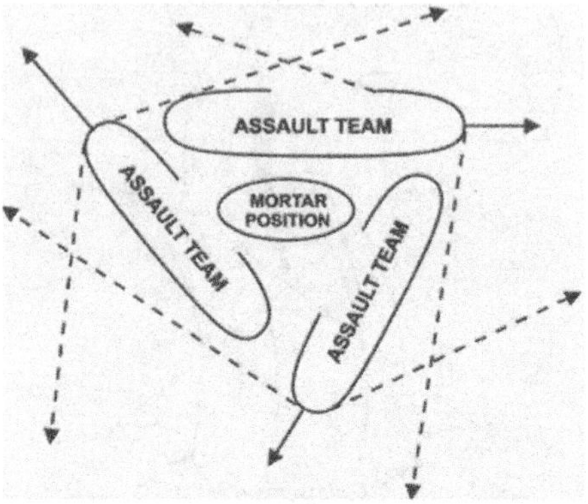

Figure D-11. Closed Triangle Formation (Night Ambush).

When deployed in this manner, the triangle ambush is used to interdict night movement through open areas. When enemy approach is likely to be from any direction, this formation provides all-around security, and security forces are deployed only when they can be positioned so that if detected by an approaching enemy, they will not compromise the ambush. Attack is by fire only, and the enemy is allowed to approach within close range before fire is initiated.

The advantages of the closed triangle formation are ease of control, all-around security, and the enemy can be brought under the fire of at least two automatic weapons, regardless of the direction they approach.

Disadvantages are that it requires a force of platoon size or larger to reduce the danger of being overrun by an unexpectedly large force and that one or more sides of the triangle may come under enfilade fire. The lack of dispersion, particularly at the points, increases the danger from enemy mortar fire.

Open Triangle Harassing Formation

This variation of the triangle ambush enables a small force to harass, slow, and inflict heavy casualties upon a large force without itself being decisively engaged. The assault element is deployed in three teams, positioned so that each team becomes a corner of a triangle containing the killing zone. (See fig. D-12.)

When the enemy enters the killing zone, the team to the enemy's front opens fire on the leading element. When the enemy counterattacks, the group withdraws and the team on the enemy's flank opens fire. When this team is attacked, the team to the opposite flank opens fire. This process is repeated until the enemy is pulled apart. Each team reoccupies its position, if possible, and continues to inflict the maximum damage possible without becoming decisively engaged.

Open Triangle Destruction Formation

The assault element is again deployed in three teams, positioned so that each team is a point of the triangle, 200 to 300 meters apart. The killing zone is the area within the triangle. The enemy is allowed to enter the killing zone; the nearest team attacks by fire. As the enemy attempts to maneuver or withdraw, the other teams open fire. One or more teams, as directed, assault or maneuver to envelop or destroy the enemy. (See fig. D-13 on page D-6.)

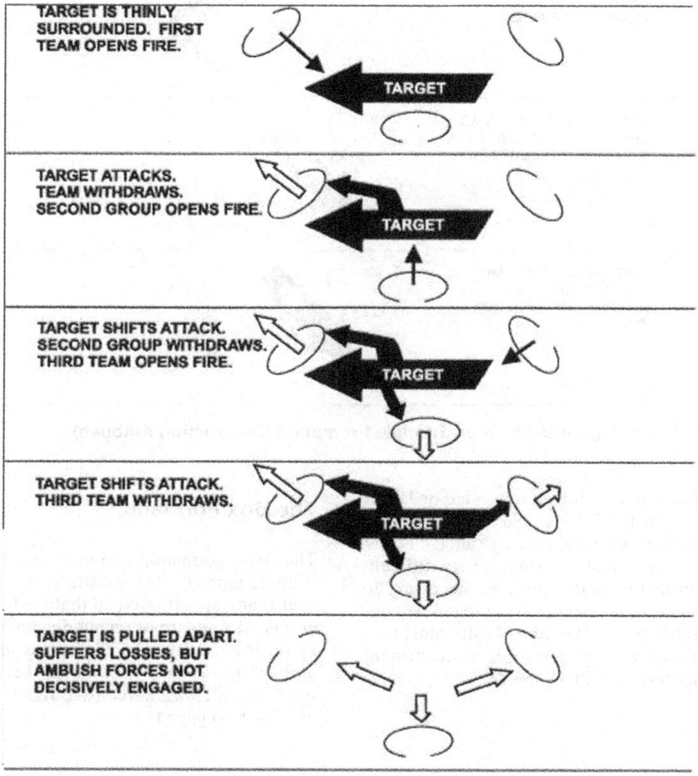

Figure D-12. Open Triangle Formation.

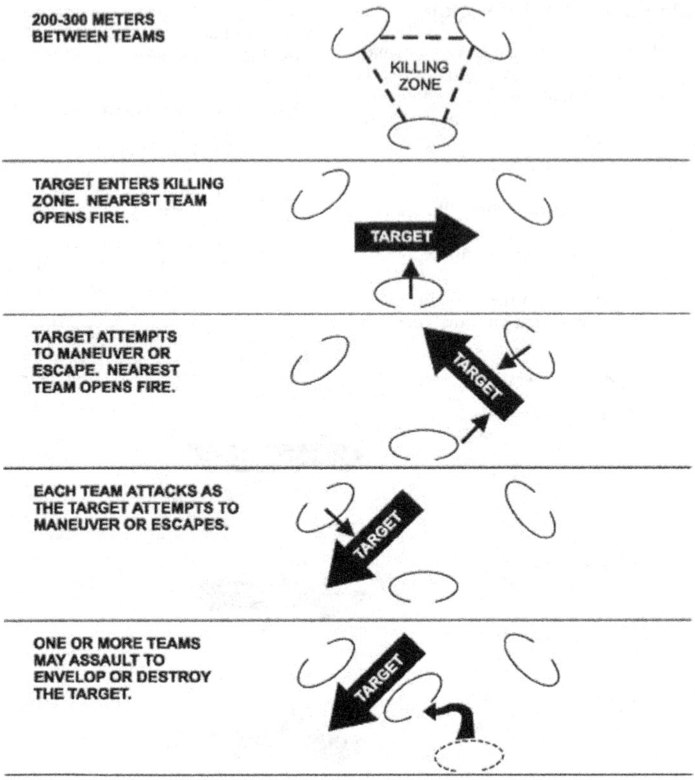

Figure D-13. Open Triangle Formation (Destruction Ambush).

This formation is suitable for platoon-size or larger ambush forces. A smaller force would be in too great a danger of being overrun. Another disadvantage is that control, in assaulting or maneuvering, is very difficult. Very close coordination and control are necessary to ensure that assaulting or maneuvering teams are not fired on by another team. The ambush site must be a fairly level open area that provides concealment around its border for the ambush force.

The Box Formation

The "box" formation is similar in purpose to the open triangle ambush. The assault element is deployed in four teams, positioned so that each team becomes a corner of a square or rectangle containing the killing zone. It can be used for a harassing or destruction ambush in the same manner as the variations of the open triangle formation. (See figs D-14 on page D-7 and D-15 on page D-8.)

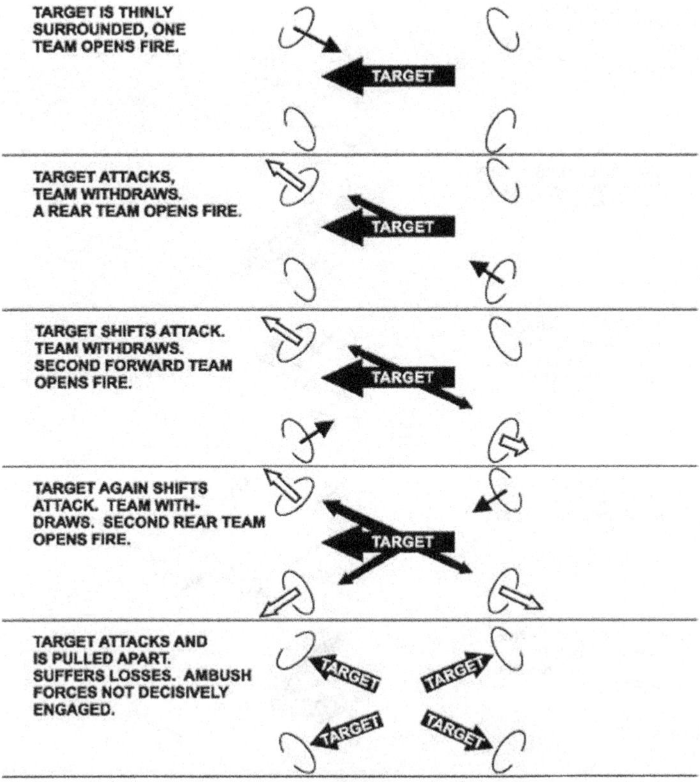

Figure D-14. Box Formation (Harrassing Ambush).

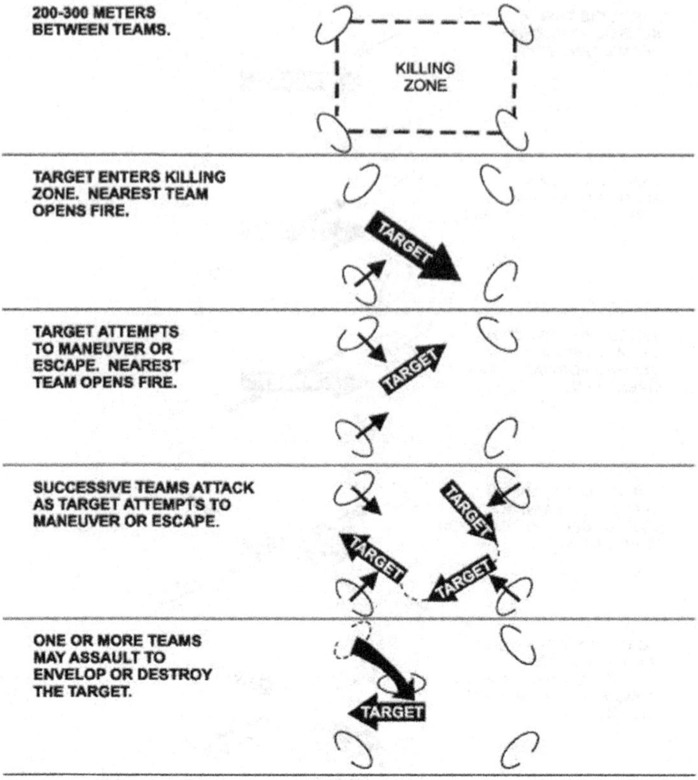

Figure D-15. Box Formation (Destruction Ambush).

Appendix E. Acronyms

A&S . . . . . . assault and security

CIT . . . . . . counterintelligence team

EOD . . . . . . explosive ordnance disposal

ITT . . . . . . interrogator-translator team

MAGTF . . . . . . Marine air-ground task force
MEDEVAC . . . . . . medical evacuation
METT-T . . . . . . mission, enemy, terrain and weather, troops and support available

NBC . . . . . . nuclear, biological, chemical
NCO . . . . . . noncommissioned officer

OTL . . . . . . overserver-target line

RCA . . . . . . riot control agents
ROE . . . . . . rules of engagement

SMAW . . . . . . shoulder-launched multipurpose assault weapon
SMEAC . . . . . . situation, mission, execution, administration and logistics, and command and signal
SOP . . . . . . standing operating procedures
SPOTREP . . . . . . spot report

TTP . . . . . . tactics, techniques, and procedures

Appendix F. References

Joint Publication (JP)

1-02 DOD Dictionary of Military and Associated Terms

Fleet Marine Force Manual (FMFM)

6-5 Marine Rifle Squad (proposed MCWP 3-11.2)

Marine Corps Warfighting Publications (MCWP)

3-15.3 Scout Sniping (under development)

3-35.3 Military Operations on Urbanized Terrain (MOUT)

3-41.2 Raids (under development)

Marine Corps Reference Publications (MCRP)

2-15.3A Reconnaissance Patrol Leader's Planning Handbook (under development)

2-15.3B Reconnaissance Reports Guide

3-02C Water Survival Handbook (under development, currently exists as Fleet Marine Force Manual 0-13)

5-12A Operational Terms and Graphics

5-12C Marine Corps Supplement to the DOD Dictionary of Military and Associated Terms

Made in the USA
Monee, IL
05 November 2026

67820238R00295